Whether therefore ye eat, or drink, or whatsoever ye do, do all to the glory of God.

(First epistle of Paul the Apostle to the Corinthians, chapter 10:31)

Foreword

Computer simulation of electronic circuits has become an indispensable tool in verifying circuit design. Practically all of the designed ICs are simulated with programs like SPICE (its derivatives or equivalents) in the early design stages and after the complete layout design. Hence, we can safely assume that they are the most widely used and trusted CAD tools. These programs are being used by circuit and even system designers in semiconductor/system houses all over the world. Most of the universities have established VLSI circuit design programs in which the students design ICs and verify their performance. However, only a small fraction of circuit simulator users fully understands the foundations on which these programs are developed. Increasing levels of understanding is of crucial importance since these simulators are still not robust enough and the knowledgeable users have a definite advantage. Thus, the topic and scope of this book, *Circuit Simulation: Methods and Algorithms* are so timely and important.

The amazing feature of this book is that it deals with several different levels of abstraction which should appeal to a very broad spectrum of readers. This book can be extremely useful to the circuit simulator users to enhance their understanding of the tools and guide them through many examples. The book can also be an excellent text on the foundations of circuit simulation useful for advanced undergraduate level and graduate courses as well. Finally, the most sophisticated users may take advantage of the advanced examples of application of circuit simulation techniques.

Another unique feature of this book is the automatic license for the OPTIMA-3 simulator for everybody buying this book. OPTIMA-3 is a very comprehensive simulation system with capabilities exceeding by far the capabilities of SPICE. However, the readers are not pushed into using this simulator as the application examples are presented simultaneously for PSPICE and OPTIMA-3. This feature makes the book a very complete and useful educational vehicle which should meet the needs of many educators and students in the electrical engineering field.

Professor Andrzej J. Strojwas
Carnegie Mellon University
Pittsburgh, PA

Preface

Contemporary design and manufacturing ...

... of large and complicated integrated circuits requires extensive computer support. This gives rise to such dynamically developing areas as Computer-Aided Design and Computer-Aided Manufacturing (CAD/CAM), bridging the gap between microelectronics and computer science. The CAD software available on present computers includes, at the very least, such design tools as: a layout editor, circuit extraction from that layout, extraction of device parameters, logic and circuit simulators, libraries of device models, and standard subcircuits. All programs are supported by a graphical user-interface. Such a basic toolset enables engineers to design Application Specific Integrated Circuits (ASICs). Similar, though more powerful tools are used by companies manufacturing VLSI circuits. Simulation programs, such as the ubiquitous SPICE, form an important part of these CAD packages.

By simulation we mean different forms of circuit analysis, able to predict circuit responses and to display them in a user-friendly form. The first purpose of any such simulation is verification of circuit performance. Simulators also are used in conjunction with optimization programs to achieve designs which optimize certain aspects of circuit quality; this is done by minimizing deterministic and/or statistical objective-functions. Another role is in Monte Carlo estimations of quality statistics for mass-production of integrated circuits. There, they are accompanied by simulators for the fabrication process (such as FABRICS), that can predict the spreads of device parameters while the circuit simulation predicts the effects of these. We can see therefore, that circuit simulators are vital tools in the commercial design and manufacturing of integrated circuits.

Simulation requires efficient device modeling to resolve the conflict between the complexity of any full description of solid-state physics and the need to minimize computational effort, particularly when modeling large circuits. Such simulation programs as HSPICE employ complicated models of electronic devices, suitable only for very small circuits, while very large circuits can only be simulated using simple models. At each level, a compromise has to be struck between accuracy and efficiency.

Over recent decades, much research effort has gone into developing computerized circuit theory, numerical analysis, computer science and, not least, computing technology. As a result, circuit simulation has now become fast, accurate and user-friendly. Nonetheless, in most areas of technology the possibilities for creating complex circuits still exceed the bounds of simulation and it is necessary to consider the need for partitioning circuits and concurrent simulation of their parts.

The role of circuit simulation is not limited to design and manufacture, or even research. It is a very useful tool in education and, indeed, has been used as a partial substitute for practical laboratory work. By experimenting with SPICE, students can become acquainted with electronic circuits without becoming distracted by measurement problems. The evaluation version of PSPICE is free and is often adequate for small student problems; it also introduces them to one version of an industry-standard simulator.

The great importance of circuit simulation means that it must be taught effectively in University Electronics Courses. This book "Circuit Simulation Methods and Algorithms" aims to help with that.

This book is ...

... dedicated to the electrical simulation of coupled circuits. Electromagnetic field problems and the simulation of microwave circuits are beyond its scope. The book offers a complete presentation of circuit simulation methods and algorithms, from basic principles to advanced methods for large-scale circuits. As a book on circuit simulation it entirely neglects the issues peculiar to logic simulation, such as gate-level, switch-level and mixed-mode simulators. The whole course is illustrated with many examples, which are usually mixed in with the theoretical treatment. The study of both theory and examples can be accompanied by computer experiments. In that sense the book is quite unusual and attractive.

The experimental work in this book is based on two simulators: PSPICE (evaluation version, which is assumed to be readily available to the reader and OPTIMA-3. This book incorporates two 3.5" diskettes. On Diskette A we have a variety of sample tasks for PSPICE and for OPTIMA. On Diskette B we have the general-purpose circuit simulator OPTIMA-3 itself. This simulator is useful for a variety of educational purposes, and will hopefully be of interest to our readers. OPTIMA-3 is almost compatible with SPICE. In those places where it is not, it considerably extends SPICE possibilities in line with a flexible idea of compact behavioral modelling on the analog level. It incorporates slightly improved algorithms, a bank of behavioral models for devices, and, which should be stressed, a variety of education-oriented options. This simulator makes possible a number of experiments which cannot be conducted in PSPICE.

The book works on several levels of abstraction, in teaching circuit simulation. Each level is addressed to a different group of readers.

Simulator user level. Section 1.1, TUTORIAL.DOC and other data given on the diskettes can be viewed as addressed first to those of our readers who are only users of simulators. They will find in Section 1.1 an explanation of basic concepts helping them to understand how SPICE operates. To become more familiar with its language, they can study the TUTORIAL.DOC (its ASCII file is given on Diskette A) and then experiment with the examples in directory \SPICE of this diskette. In the TUTORIAL.DOC readers will find out that the use of the simulator OPTIMA-3 is almost as simple as PSPICE and that the differences between them are only due to the extended possibilities of OPTIMA-3. Then they are invited to experiment with this program using installing it from Diskette B and running some examples given in the directory \OPTIMA of Diskette A.

Basic level. Our book covers problems ranging from the foundations of circuit simulation (where even some concepts of circuit theory are explained) up to advanced methods for large-scale circuits. Therefore, the topics can be suitable for undergraduate, graduate and, in places, even for postgraduate courses. The basic level consists of studying only the theoretical treatment and those examples provided. To check his/her understanding the reader can after each section review the topics by means of problems and questions listed at the end. However, on this basic level, those places where we refer to computer simulations and information given in diskette files can be ignored.

Thorough level. Problems elaborated on the basic level can be studied more thoroughly by simultaneous application of PSPICE and OPTIMA-3. In our book we often refer to sample tasks. On finding such a reference we can proceed to a computer, edit one of task-files S*i**.CIR for PSPICE or O*i**.CIR for OPTIMA (*i* is a chapter number) and read the explanations contained in its comments. Then we may carry out experiments guided by these explanations and run the appropriate simulator. Finally, we turn to a graphical postprocessor and/or to text results. In some cases the task-file is accompanied by a file *.TXT incorporating further information connected with the experiment. This practice enables us to study circuit simulation much more thoroughly.

This book covers a very wide range of topics.

Chapter 1 is dedicated to foundations of computerized circuit theory and basic circuit simulation concepts. It introduces a formalism based on canonical circuit equations and then a circuit variables transformation.

Chapter 2 presents numerical methods of linear circuit analysis, including the formulation of equations and principles of a.c. analysis. Problems of symbolic analysis are not considered.

Chapter 3 is dedicated to numerical methods for solution of linear algebraic equations. This area of numerical linear algebra is especially important in circuit simulation and has therefore been covered in a separate chapter. Its presentation is not formal, but rather focused on circuit simulation aspects. Sparse matrix techniques are also included.

Chapter 4 covers methods and algorithms for d.c. analysis. The treatment starts from a basic presentation of numerical methods for nonlinear algebraic equations and then goes on to the most efficient and common modified Newton-Raphson algorithms.

Chapter 5 lays the foundations for numerical integration of ordinary differential equations, and then discusses their practical applications to time-domain circuit analysis. Problems of charge conservation, local accuracy, stability and accuracy control are presented thoroughly.

Chapter 6 is dedicated to methods of periodic steady-state analysis. The two most efficient methods are presented: the Newton-Raphson approach and a new secant algorithm, which is also implemented in OPTIMA-3.

Chapter 7 gives extensive coverage of methods of sensitivity analysis. These include the theory and implementation of sensitivity analysis for static and dynamic circuits. Finally, the large-change sensitivity analysis of linear circuits is described.

Chapter 8 is dedicated to the direct methods of large-scale circuit analysis exploiting decomposition and latency. First, we consider linear circuits solution and then proceed to algorithms for nonlinear circuits.

Chapter 9 contains a presentation of concurrent relaxation-based circuit simulation methods for large-scale circuits. The following areas are discussed: timing-simulation, iterated timing-simulation, waveform-relaxation simulation, block-relaxation simulation, and event-driven simulation. The treatment in this chapter is not limited to a general overview, but enables us to really understand the newest approaches to large-scale circuit simulation.

OPTIMA-3 license ...

... is automatic for everybody buying this book. At the same time we hope that the restrictions on copying this program are well understood.

Anyone having any problems or suggestions and comments is encouraged to contact the author by e-mail: jogr@ipe.pw.edu.pl.

Acknowledged gratefully are ...

... many persons who gave me outstanding assistance and so enabled me to carry out this work.

I wish to thank Prof. Andrzej Filipkowski from the Electronics Department, Warsaw University of Technology, Warsaw, Poland for his friendship and constant moral support during this work. Gratitude is also due to Prof. Kel Fidler of the Department of Electronics, University of York, Heslington, York, U.K. and to Prof. Andrzej Strojwas from Electrical and Computer Engineering at the Carnegie-Mellon University, Pittsburgh, PA, USA for his friendly reception of this work.

I am especially grateful for the friendship of Dr. William Fawcett from the Department of Electronics, University of York, Heslington, York, U.K. who has offered me his help in improving the English of this book and to Dr Jerzy Rutkowski from the Silesian University of Technology, Gliwice, Poland who also helped in ensuring that my English attained a satisfactory level.

I also wish to express my gratitude to those of my colleagues from the Institute of Electronics Fundamentals, Department of Electronics, Warsaw University of Technology, who supported the work on OPTIMA-3:

First of all to Mr. Dariusz Bukat, without whom there would be no OPTIMA-3 at all, for his friendship, diligence, and cooperation during several years spent improving this simulator.

Moreover, to other colleagues: to Dr Leszek Opalski, who was one of contributors of the ideas developed in this program; to Dr Marian Bukowski, contributor of a new procedure for periodic steady-state analysis; to Dr. Wojciech Bandurski from Poznań University of Technology, Poznań, Poland, who contributed a new model of a lossy-nonuniform transmission line. I must also thank Mr. Ryszard Rogocki, creator of Node-Viewer, Mr. Dariusz Andziak, Mr. Artur Iterman, and Mr. Jorge Cardenas who, as students, worked on parts of this program.

It is also essential to thank my Wife and Son for their splendid patience and moral help during those years of my work when this book, as well as the enclosed software, were taking shape.

5 December 1993 Jan Ogrodzki
Warsaw, Poland

Table of Contents

Chapter 1
Introductory Topics in Circuit Simulation

Chapter 2
Numerical Analysis of Linear Circuits

Chapter 3
Numerical Solution of Linear Algebraic Equations

Chapter 4
D.c. Analysis of Nonlinear Circuits

Chapter 5
Time-domain Analysis of Nonlinear Circuits

Chapter 6
Periodic Steady-state Time-domain Analysis

Chapter 7
Sensitivity Analysis

Chapter 8
Decomposition-based Methods for Large-scale Circuits

Chapter 9
Relaxation-based Simulation Methods

List of Main Acronyms

BDF	Backward Differentiation Formula
CCE	Canonical Circuit Equations
CCCS	Current-Controlled Current Source
CCVS	Current-Controlled Voltage Source
CD	Current-Defined
DF	Differentiation Formula
GS	Gauss–Seidel
GSN	Gauss–Seidel–Newton
IEDF	Implicit Euler Differentiation Formula
KCL	Kirchhoff's Current Law
KVL	Kirchhoff's Voltage Law
LTE	Local Truncation Error
MNE	Modified Nodal Equations
NE	Nodal Equations
NR	Newton-Raphson
OADE	Ordinary Algebraic-Differential Equation
ODE	Ordinary Differential Equation
TDF	Trapezoidal Differentiation Formula
VCCS	Voltage-Controlled Current Source
VCVS	Voltage-Controlled Voltage Source
VD	Voltage-Defined

Chapter 1

Introductory Topics in Circuit Simulation

1.1 Main concepts

1.1.1 Systems, models, circuits, equations and responses

Real electrical **systems** consist of electrical devices (e.g. resistors, transistors, integrated circuits) fixed to certain boards (or even made on them) and electrically assembled. If a mathematical characterization of these devices is provided, it is possible to set up a formal description of the whole system and so predict its behavior. The formal characterization of a device by means of a set of ideal elements and mathematical expressions is called a **device model**. Once we have models of all devices involved, we can combine them and then use a computer to solve the whole system for signals that are of interest. This procedure is called **circuit analysis** or **circuit simulation**. Proper modeling is a crucial element in simulation, on which the ability to satisfactorily predict physical reality depends, yet practical simulation must involve a compromise between computational complexity and inherent inaccuracy.

In this section the path from a physical system to prediction of its electrical response is outlined via an extremely simple example shown in Figure 1.1(a). The modeling stage relies on the assignment of certain models to the various devices involved. At this point we have to choose which physical phenomena should be incorporated in the model, and which are to be neglected. In the system shown in Figure 1.1(a) we assume the following models: resistance R_1 for the resistor, voltage source $V_G(t) = 1(t)A\sin(\omega t)$ switched on at the instant $t=0$ for the a.c. generator GEN, constant current source I for the current supplier CS and, finally, an assembly of nonlinear conductance $i_D(v_D) = i_s[\exp(v_D/V_T) - 1]$ and nonlinear capacitance $C_D(v_D) = C_{D0}\exp(v_D/V_T)$ for the diode D. Thus we have carried out the modeling procedure that results in the circuit in Figure 1.1(b).

The circuit is a mathematical object subject to computer analysis. Before passing to analysis itself, we have to answer three important questions. The first two are whether the accuracy of the models is sufficient for our specific engineering problem, and the whether parameters of the models might be easily identified by an appropriate measuring and

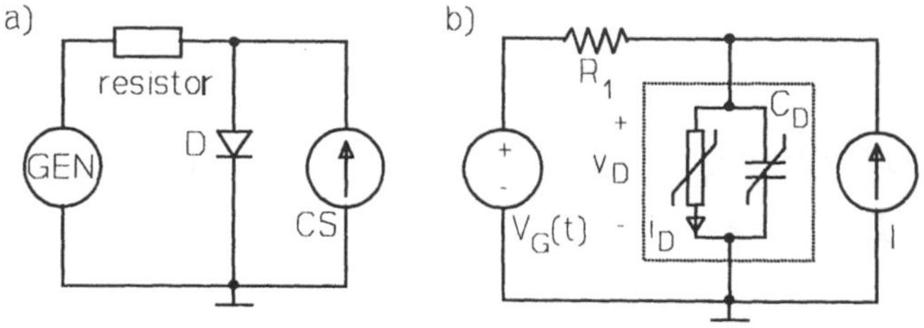

Figure 1.1 (a) A simple system and (b) its circuit model

fitting procedure. The third question is whether, due to the complexity of the model, the whole circuit is not too large to be solved by the computer available. To our surprise, these problems are neither marginal, nor obvious. In practice, engineers make use of proprietary models as, for example, those implemented in the ubiquitous circuit simulator SPICE [1.1-1.4], and often employ measured sets of parameters provided by companies as libraries. They should know, however, that even common device models are sometimes not suitable in certain ranges of signals, and the extraction of parameters by taking measurements in those ranges may involve considerable inaccuracy. Discussion of these important topics is beyond the scope of our book.

Coming back to our circuit in Figure 1.1(b) we are able to set up its mathematical characterization. Though practical circuits are quite complicated, even such a small example exhibits what is essential in circuit analysis - a formal description. From the formulae introduced for circuit elements, after the voltage source has been transformed into a current counterpart and the sum of currents in the node has been equated to zero, we obtain the circuit **equation**

$$R_1 C_D(v_D) \frac{\mathrm{d}v_D}{\mathrm{d}t} + R_1 i_D(v_D) + v_D = V_G(t) + R_1 I \qquad (1.1)$$

which enables us to predict the circuit behavior in time. Signals, measured in a physical system or derived from a mathematical characterization of the circuit (i.e. from a system model) have been called **responses** of the system or circuit respectively. Usually it is necessary to make use of numerical solution methods to get responses from the system equations. Only in a few cases, where a circuit is quite simple, is the symbolic derivation of responses possible. General-purpose methods, implemented in a variety of circuit simulation programs, are extensively discussed in this book. Algorithms implementing these methods usually generate circuit responses in the form of numerical tables and/or plots.

Simulation techniques can be used as black boxes by providing input files for general-purpose simulators, e.g. for SPICE. For instance, our circuit in Figure 1.1(b) might be drawn up in SPICE input language as follows [1.1-1.4]

```
S1-1a - THE CIRCUIT FROM FIGURE 1.1 - SPICE ANALYSIS
* excitation VG = 1V*sin(2*pi*100kHz*t)
VG 1 0 SIN(0,1,100K,0,0)
* resistor
R1 1 2 1K
* diode - let parameters of its model in Figure 1.1(b) be:
*    Is  = 1e-15A,   Cdo = 6.37e-22F;
* we constrain the SPICE model of diode to be equivalent;
* thus let temperature be equal to a nominal 27 C;
* since a zero bias diffusion capacitance is equal to
*    Cdo = Is*TT/(2*pi*Ut),
* hence, the corresponding transit time TT = 100ns;
* junction capacitance CJO is set to 0;
* series resistance RS is also set to 0;
D1 2 0 MDIO
.MODEL MDIO D (IS=1E-15, RS=0, N=1, TT=100N, CJO=0)
.TEMP 27
* bias excitation I=0.5mA
I 0 2 0.5M
* time-domain transient analysis from 0 to 50 us
* with a display increment 0.1 ns
.OPTION TRTOL 0.01
.TRAN 0.1N 50U
* graphical postprocessing: we observe V(2)
.PROBE V(2)
.END
```

The voltage response V(2) at the node number 2 $V(2)=v_D$, appears to be a considerably distorted periodic signal shown in Figure 1.2. We can obtain it by running the above listed file S1-1a.cir on PSPICE and by observing the binary output file with the graphical postprocessor PROBE. The file S1-1a.CIR is included on Diskette A. From these results, we notice that at the instant 0, when the excitation $V_G(t)$ is switched on, our waveform (called the transient (TR) response) begins with a certain value, derived from a response to the direct current excitation I (the so-called direct current (d.c.) response). From this simple experiment it is clear that circuit responses contain distinct components: one resulting from d.c. excitations, and another related to time-varying sources. As is seen from the results S1-1a.out, PSPICE first accomplishes d.c. analysis and then proceeds to TR. It is essential for users to have deep insight into a variety of analyses and different types of responses. Before we pass to more advanced topics we address therefore the basic concepts.

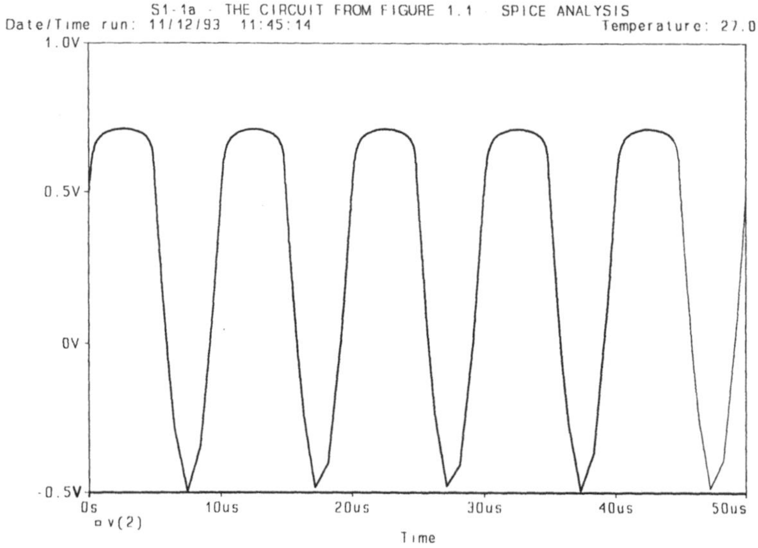

Figure 1.2 The voltage V(2) obtained from SPICE simulation of the circuit S1-1a.cir

1.1.2 Direct current, transient and alternating current responses

In this section we deal with a variety of circuit responses, that are derived not only from the types of excitation provided but also from the amount of energy stored in capacitors and inductors. Let us recall the circuit in Figure 1.1(b). Before the instant 0 the voltage source $V_G(t)$ is switched off, that is its value is zero, and so stands for a short-circuit. The circuit is driven by the constant current I which theoretically started at minus infinity. Just as this excitation charges up the capacitance to a fixed voltage, its current reaches zero, and hence the capacitance becomes an open circuit. Thus as a counterpart of the circuit in Figure 1.1(b) we obtain a d.c. equivalent resistive circuit shown in Figure 1.3(a) and called the nonlinear substitute d.c. circuit. In operation it relies on the direct current I splitting into two parts and flowing through the conductance $i_D(v_D)$ and the resistance R_1. As a result, a certain voltage, say v_{D0}, is set up in the circuit. This is the **direct current (d.c.) response**.

The d.c. response may be obtained in black-box manner, that is, by means of PSPICE. Turning to the example S1-1a.cir, we see that the d.c analysis is carried out automatically before calculating the TR response. We have obtained $v_{D0} = V(2) = 0.4998\,\text{V}$, and the current flowing through VG is equal to 4.998E-4 A. The user may also constrain PSPICE to work

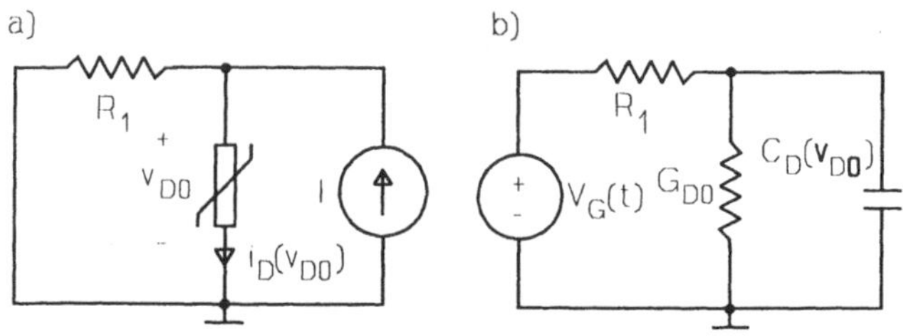

Figure 1.3 (a) Substitute nonlinear d.c. and (b) linear a.c. circuit corresponding to the example of Figure 1.1(b)

out d.c. responses by introducing the .OP command. Most nonlinear systems are designed primarily to yield a satisfactory d.c. response. Hence d.c. analysis is essential in engineering practice.

The essence of d.c. analysis is to have a mathematical characterization of the substitute d.c. circuit, as shown for instance in Figure 1.3(a). By simple inspection we can derive its nonlinear algebraic equation

$$R_1 i_D(v_{D0}) + v_{D0} = R_1 I \tag{1.2}$$

which may be solved for the d.c. response v_{D0}, while $i_D(v_{D0})$ is known from Section 1.1.1.

A substitute d.c. circuit, like that of Figure 1.3(a), is, in general, a result of open-circuiting capacitors, short-circuiting inductors and setting excitations to their d.c. components (often treated as their values before and at the instant 0). A mathematical description of this substitution circuit is a set of algebraic, usually nonlinear, equations. Comparison of Equations (1.1) and (1.2) shows that d.c. equations arise when time derivatives, occurring in full circuit characterization, are all set to zero, and functions of time are replaced by their d.c. components. Thus two rules have been extracted: the former for circuits and the latter for equations. In fact both exhibit the same principle.

From d.c. responses we proceed to time variable signals, that are established in circuits while excited with sources varying in time. This so-called transient (TR) response depends mostly on the time-profile of the excitation but the initial state of the circuit at the instant when a non-d.c. component of the excitation starts is also of great importance. In most cases, the TR response begins with the d.c. component of the response, i.e. the so-called **bias point**, as it was in the simulation task S1-1a.cir. The TR response is unique for the given circuit if its initial state is additionally defined. As will be shown in this book, implementation of TR analysis relies on the numerical solution of ordinary differential equations (ODE), such as Equation (1.1). Even from a theoretical point of view such equations have unique solutions only if initial conditions for unknown functions (and corresponding derivatives) at the instant 0 are provided. Initial conditions derived from the bias point are said to be dependent on circuit components since the bias is a function of these components.

In TR analysis fixed initial conditions, hence independent of any circuit response, may also be considered. In PSPICE, "manual" initial voltages across capacitances and currents through inductances are available. We may consider, for instance, a quite simple RC circuit with an instantaneous voltage excitation jump from 2 to 10 volts given in the file S1-2.cir. It enables us to make two experiments. The first one includes initial bias point analysis to provide the automatic initial condition $V(2) = 1\ \text{V}$. After a slight change in the input file S1-2.cir we have the case of "manual" initial conditions. The initial voltage condition of 2 V across the capacitance is defined by setting IC=2 in the capacitor line. Then the keyword UIC (Use Initial Conditions) is added to the command .TRAN to cause the analyzer to start from the initial condition defined.

At this point, recalling simulation results of the example from Figure 1.1(b), we see that the TR response is nearly periodical. A periodical steady state is achieved after a number of periods. It is important to notice that the values of steady state signals before and after the period are the same. Thus to get the steady state from analysis over one period, proper initial conditions should be imposed. This observation is essential for developing TR analysis dedicated to periodical steady states.

Results of S1-1a.cir show that the steady state is quite distorted and so very far from the sinusoidal character of both the excitation and the expected steady state. In the file S1-1b.cir the reader is encouraged to experiment with a gradual reduction of the amplitude of excitation $V_G(t)$. As the amplitude decreases, the output becomes less and less distorted. For a sufficiently small amplitude the response in fact becomes sinusoidal, and hence the circuit operates for alternating current (a.c.) signals as a linear circuit.

As is well known from basic circuit theory, a linear circuit subjected to sinusoidal excitations reaches a steady state with sinusoidal signals of certain amplitudes, phase shifted with respect to the excitations. These are known as the **alternating current (a.c.) responses**. In nonlinear circuits with a.c. (i.e. sinusoidal) sources of a single frequency the same effect arises when amplitudes of excitations are small enough to obtain negligible nonlinearities of elements in the vicinity of the bias point. In such a case we have the so-called **small-signal a.c. response**. Calculation of a.c.

responses in both linear and nonlinear, hence small-signal, cases is called **a.c. (or small-signal a.c.) analysis.**

In small-signal a.c. analysis a useful signal (called an incremental component) is a difference between an instant value of a TR response and its d.c. component (a bias point). Recalling the circuit shown in Figure 1.1(b) we observe that in the sinusoidal steady state the TR response $v_D(t)$ appears against a background of the bias v_{D0}. The smaller the amplitude A, the smaller is the incremental response $\hat{v}_D(t) = v_D(t) - v_{D0}$ and hence the incremental slope of the conductance $(i_D(t) - i_{D0})/\hat{v}_D(t)$ lies closer to the differential slope $G_{D0} = \mathrm{d}i_D/\mathrm{d}v_D$. In small-signal a.c. analysis infinitely small increments, and hence the differential case, are taken. Thus the nonlinear conductance $i_D(v_D)$ operates as the linear conductance G_{D0}, while the nonlinear capacitance $C_D(v_D)$ is represented by the linear capacitance $C_D(v_{D0})$. Moreover the d.c. excitations are deleted. Therefore the circuit of Figure 1.1(b) for small-signal a.c. response becomes equivalent to the substitute circuit given in Figure 1.3(b). Hence, in general, small-signal a.c. analysis requires solution of an a.c. substitute circuit including components obtained from the original circuit by means of the linearization.

We can use PSPICE to carry out small-signal a.c. analysis without going deeply into the implemented methods. The provided file S1-1c.cir includes the command .AC producing a.c. analysis of the circuit of Figure 1.1(b). Alternating current analysis consists in examining linear properties of a circuit and so the absolute amplitudes and phases of excitations are of secondary importance. Description of an a.c. excitation may often be reduced to: "AC 1" written in a source line to set the amplitude to unity, and the phase to zero. Results of a.c. analysis are usually presented as amplitudes and/or phases of the output signals plotted against frequency. To view results of the example run PSPICE and then handle the file S1-1c.dat by means of the graphical postprocessor PROBE.

Coming back to the concept of the a.c. substitute circuit, as for example the circuit in Figure 1.3(b), the substitute circuit is a counterpart of a mathematical characterization of the steady state subjected to a small sinusoidal excitation. Let us recall the circuit in Figure 1.1(b) and related Equation (1.1). Even an intuitive consideration shows that to obtain an equation to be satisfied by the infinitely small incremental response we have to subtract Equation (1.2) (standing for the bias) from (1.1) and then to substitute the true increment of i_D for the differential increment $G_{D0}(v_D - v_{D0})$. The latter corresponds to the assumption of infinitely small increments and we call it the linearization dedicated to a.c. analysis. After the subtraction we have the formula

$$R_1 C_D(v_D)\frac{\mathrm{d}v_D}{\mathrm{d}t} + R_1[i_D(v_D) - i_D(v_{D0})] + (v_D - v_{D0}) = V_G(t). \tag{1.3}$$

The linearization of i_D and the introduction of hats to mark the small increments yields

$$R_1 C_D(v_{D0})\frac{\mathrm{d}\hat{v}_D}{\mathrm{d}t} + [R_1 G_{D0} + 1]\hat{v}_D = V_G(t). \tag{1.4}$$

Notice that the derivative of v_D in Equation (1.3) is equal to the derivative of its incremental component $\hat{v}_D$, as both differ in a d.c. constant only.

From basic circuit theory we know that the solution of such equations can be found by the so-called symbolic method, where unknown sinusoidal signals are represented in the complex plane by phasors, whose lengths correspond to amplitudes while angular coordinates stand for phases. In phasor algebra the sinusoidal signal $v_D(t)$ transforms into an unknown complex number $\bar{v}_D$. Its derivative is the complex number $\mathrm{j}\omega\bar{v}_D$, while the excitation $A\sin(\omega t)$ is represented simply by the number A. From this algebra the following complex equation arises

$$[\mathrm{j}\omega R_1 C_D(v_{D0}) + R_1 G_{D0} + 1]\,\bar{v}_D = A \tag{1.5}$$

where the phasor $\bar{v}_D$ stands simultaneously for the amplitude and phase of the a.c. response $\hat{v}_D(t)$. Hence we obtain them in the following form

$$|v_D| = \frac{A}{\sqrt{[1 + R_1 G_{D0}]^2 + [\omega R_1 C_D(v_{D0})]^2}}$$
$$\operatorname{Arg} v_D = -\arctan\frac{\omega R_1 C_D(v_{D0})}{1 + R_1 G_{D0}}. \tag{1.6}$$

The response (1.6) represents, in a symbolic manner, the same result as has been produced by the PSPICE task S1-1c.cir. Certainly PSPICE does not pursue our mathematical development but sets up general complex circuit equations, such as (1.5), and solves them using general methods.

The aim of this section is to provide an insight into introductory concepts in circuit simulation. We want to understand basically how a simulation, as implemented in PSPICE, deals with circuits. To test your understanding, try to answer the following questions:

- What do we mean by the physical system, model and circuit?

- How can we explain the concept of the circuit response and the essence of the d.c., a.c. and TR response components?
- What sorts of circuit equations and correspondent substitute circuits are subject to d.c., TR and a.c. responses respectively?

1.2 Principles of circuit characterization

Circuits are basically composed of branches, i.e. two-poles characterized by a pair consisting of a current and a related voltage. Moreover, branches are connected to certain circuit nodes. Thus electrical phenomena, other than at high frequencies where the size of the system is considerably smaller than the wavelength of any signals, obey Kirchhoff's Laws. These lead to a mathematical characterization of circuits which will be treated later on in this book.

1.2.1 Circuit branches

Electrical circuits are composed of **branches**, that is, of elements stretched between two terminals, one of which is defined as positive and the other as negative. The branch is described by a pair of so-called branch variables: the voltage v established between the positive and negative terminals and the current i, which by convention, flows from the positive terminal to the negative one, as shown in Figure 1.4(a). This conventional opposition of the directions of the current and voltage results in an interpretation of the sign of the branch power $p = vi$. A positive sign stands for absorption of the power by the branch (for further storage or

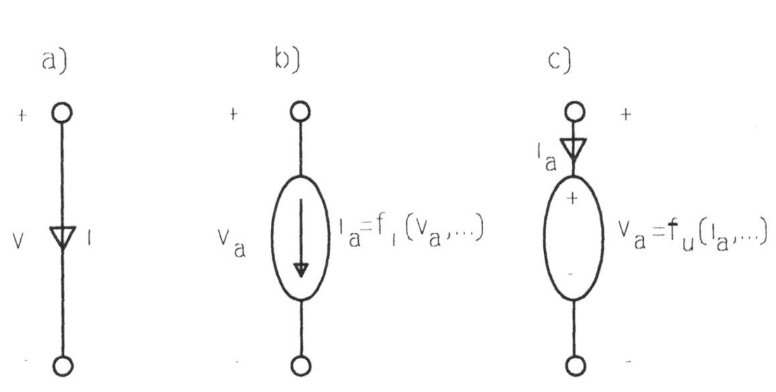

Figure 1.4 (a) Convention of notation for voltages and currents of branches, (b) a current-defined and (c) a voltage-defined branch

dissipation outside as e.g. heat), while a negative sign denotes that the power is supplied to the circuit (so simultaneously absorbed from other sources of energy, e.g. from a chemical reaction). The branch is defined when its voltage and/or current are themselves defined by an equation known as the **Branch Constitutive Equation**. This equation may be considered as a synonym for the branch.

Circuit theory deals with a variety of branches, as for instance the voltage source $v = 1\ \text{V}$, current source $i = 2\ \text{A}$, linear conductance $i = Gv$ or nonlinear resistance $v = 0.025\ \ln(1 + 10^{15} i)$. In general, both their voltage and current may be located on either the left or the right side of the constitutive equation; hence an implicit notation is needed to formalize a variety of branches. For instance $i - Gv = I$ defines a parallel combination of the conductance G with the current source I. A more general nonlinear branch may be drawn up as $f(i, v) = s$. This formula includes an excitation s (either voltage or current type). The examples presented demonstrate **independent branches** only, i.e. those whose equations are not dependent upon currents or voltages of other branches. Among the independent branches there are **independent sources**, that are constant or time-varying.

Branches may also be controlled by other voltages or currents, as for instance the linear Voltage-Controlled Current Source (VCCS) $i_a = \alpha v_b$ or the nonlinear Current-Controlled Voltage Source (CCVS) $v_a = \phi(i_c)$. An appropriate notation is now introduced with the intention of identifying the controlling mechanism of **controlled branches**. Branch variables, i and v, are given indices:

- a denotes the branch whose current/voltage is being described and is the dependent variable,
- b denotes another branch whose voltage is a controlling variable,
- c indicates another branch whose current is a controlling variable.

This convention will be used occasionally hereafter, when needed. Though a dependent branch may be controlled by any number of voltages and currents, formalism will be limited to cases in which there is only one controlling variable of either type.

Thus the interdependence of the branch's own and controlling signals may be expressed in a general implicit form

$$f\left(i_a, \frac{\mathrm{d}i_a}{\mathrm{d}t}, v_a, \frac{\mathrm{d}v_a}{\mathrm{d}t}, v_b, \frac{\mathrm{d}v_b}{\mathrm{d}t}, i_c, \frac{\mathrm{d}i_c}{\mathrm{d}t}, s(t)\right) = 0\,. \tag{1.7}$$

This equation includes a dependence on time through the excitation $s(t)$ and a dependence on derivatives, occurring in reactive branches, as for example the inductance $v_a - L(i_a)\,\mathrm{d}i_a/\mathrm{d}t = 0$, or the source $i_a - \gamma\,\mathrm{d}v_b/\mathrm{d}t = 0$ which is controlled by a derivative.

Branches are usually explicit functions of one of their branch variables. For example, a controlled voltage source is defined by the explicit voltage $v_a = \phi(v_b)$, while the expression for the conductance of a diode $i_a = I_s[\exp(v_a/V_T) - 1]$, gives rise to an explicit current. However, the latter might be transformed into the current to voltage function $v = V_T \ln(1 + i/I_s)$. Therefore we have to distinguish two **types of branches**, the current-defined and voltage-defined. They are drawn in Figure 1.4(b) and (c).

The branch is current-defined if its current is a function of its own voltage, controlling variables or their derivatives, namely

$$i_a = f_i\left(v_a, \frac{\mathrm{d}v_a}{\mathrm{d}t}, v_b, \frac{\mathrm{d}v_b}{\mathrm{d}t}, i_c, \frac{\mathrm{d}i_c}{\mathrm{d}t}, s(t)\right). \tag{1.8}$$

Conductances, capacitances, and current sources belong to the above type.

The branch is voltage-defined if its voltage is a function of its own current, controlling variables or their derivatives, namely

$$v_a = f_v\left(i_a, \frac{\mathrm{d}i_a}{\mathrm{d}t}, v_b, \frac{\mathrm{d}v_b}{\mathrm{d}t}, i_c, \frac{\mathrm{d}i_c}{\mathrm{d}t}, s(t)\right). \tag{1.9}$$

Resistances, inductances, and voltage sources are of this type.

A resistor is the simplest example of a branch that can be solved either for the current $i = v/R$ or for the voltage $v = Ri$ (excluding the case $R = 0$). Similarly the diode conductance is solvable for both its current and voltage. As can be seen from these examples the branch stands rather for an equation than for even an ideal element. Thus a resistance and a corresponding conductance are quite different branches. Some branches are transformable from one type to another (e.g. a nonzero resistance) while others are not.

Branches dependent on time derivatives of voltages or currents are known as reactive. They model two important physical phenomena: storage of energy in the electric field and in the magnetic field. If the frequency is not too high, these fields can be considered separately as if they were independent (although, in general they are not, according to Maxwell's Laws). The electric field is set up in capacitors as a result of a voltage difference. The magnetic field is set up in the medium surrounding a conductor due to the flow of current.

Let us consider the linear capacitance $i_a = C\,dv_a/dt$. On its positive plate the charge $q_a = C\,v_a$ is accumulated. The equation for current is set up by differentiating the charge with respect to time. This rule is valid not only in linear cases. From the physics of electricity it is known that the current is, in general, a derivative of the charge, namely

$$i_a = \frac{dq_a}{dt}. \tag{1.10}$$

The charge itself may be a nonlinear function of the voltage v_a, as shown in Figure 1.5(a). Provided that there is no interference of the electric field and external fields the voltage v_a results only in polarization of the charge on the plates. The sum of the charge on the plates is constant, independent of the voltage, and has no effect on the current; hence it may be set as zero (see Figure 1.5(a)). An intermediate differentiation of (1.10) with respect to the terminal voltage yields

$$i_a = \frac{dq_a}{dv_a}\frac{dv_a}{dt} = C(v_a)\frac{dv_a}{dt} \tag{1.11}$$

including the instantaneous capacitance (i.e. a derivative of the charge versus the voltage). Equation (1.11) differs from (1.10) in a differentiated (state) variable. We have two possible choices of differentiated variables: the charge or the voltage. The former appears to be a better choice as will be shown later on.

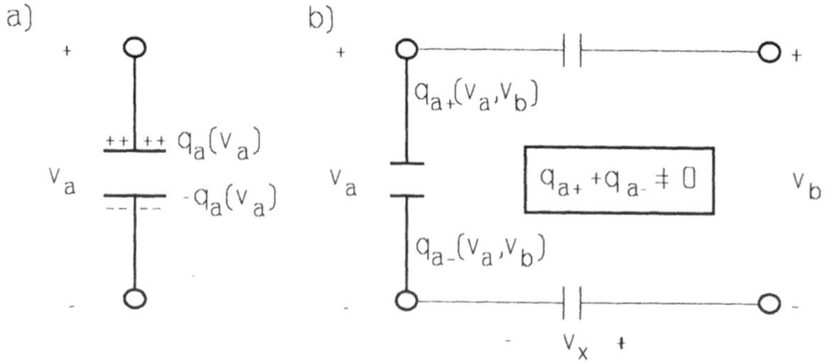

Figure 1.5 (a) Charge storage in an isolated capacitance and (b) the case of charge dependent on an external voltage

In general, the charge may be controlled by external signals, although there is no physical reason for considering the current as its controlling variable. Hence, let the charge stored in the *a*th branch be dependent on the local voltage v_a and the external voltage v_b, as shown in Figure 1.5(b)

$$q_a = q_a(v_a, v_b). \tag{1.12}$$

To our surprise, a notation like this is rather imprecise from the physical point of view. The case of the controlled charge appears to be quite complicated, anyway. To clarify this remark we have to remember that the plates of our capacitor may react with nodes of the *b*th branch and so two additional mutual capacitors, drawn with a thin line, arise. Obviously this complication does not occur in the case of systems composed of distinct capacitors, but is quite common in integrated devices (e.g. MOS). Charges q_{a+}, q_{a-} stored on the plates have now to be distinguished. Although seemingly associated with the *a*th branch, they are partly located also on the mutual capacitances. Therefore definition of the charge in terms of v_a, v_b with no regard to voltages across the mutual capacitors (v_x and $v_a - v_x - v_b$ in Figure 1.5(b)) is quite inadequate. The sum q_{a+}, q_{a-} is not constant. It depends on the mutual capacitances and voltages established between their terminals. That is why while taking into account an external control v_b one should not neglect electric reaction with the environment. However, if the mutual capacitances describing this reaction are assumed to be zero the behavior of the charge is the same as in the previous case shown in Figure 1.5(a). Moreover, if the sum of charges on all four terminals involved is taken, instead of the two only, the sum would be constant, independent of the voltages. Our consideration is subject to Gauss's Law. We will recall it again later on.

Our problem may be clearly demonstrated by combining Equation (1.10) with (1.12) to take the voltages as the differentiated variables. We assume our two charges q_{a+}, q_{a-} to be dependent on the voltages v_x and $v_a - v_x - v_b$. After differentiation of equations such as (1.12) we obtain

$$\begin{aligned} i_{a+} &= \frac{\mathrm{d}q_{a+}}{\mathrm{d}v_a}\frac{\mathrm{d}v_a}{\mathrm{d}t} + \frac{\mathrm{d}q_{a+}}{\mathrm{d}(v_a - v_b - v_x)}\frac{\mathrm{d}(v_a - v_b - v_x)}{\mathrm{d}t} \\ i_{a-} &= \frac{\mathrm{d}q_{a-}}{\mathrm{d}v_a}\frac{\mathrm{d}v_a}{\mathrm{d}t} + \frac{\mathrm{d}q_{a-}}{\mathrm{d}v_x}\frac{\mathrm{d}v_x}{\mathrm{d}t}. \end{aligned} \tag{1.13}$$

If the expressions for charges are physical, then the first terms in both equations (1.13) are instantaneous capacitances, that differ in the sign only. The two second terms stand for the mutual capacitances. The notation of (1.13) is meaningful only if it constitutes the physical reality, i.e. if it is derived from a certain function (1.12) for the charge, and all voltages influencing this charge are involved. Otherwise, the mutual terms in (1.13) are not physical. Hence, one must not choose arbitrarily a set of controlling variables and invent models directly in the form (1.13), because although the instantaneous capacitances may seem to be quite correct in fact they may not be derived from physical charge.

Magnetic effects in inductors need to be treated similarly. A magnetic flux induced by a current flowing through a conductor may be, in general, a nonlinear function $\psi_a = \psi_a(i_a)$ of this current. According to Faraday's Law, a voltage established across a conductor, equal to the time derivative of the flux, is the result of the reaction of that conductor to the flux

$$v_a = \frac{d\psi_a}{dt} \tag{1.14}$$

or alternatively $v_a = L(i_a)\, di_a/dt$ where $L(i_a) = d\psi_a/di_a$ is called the instantaneous inductance.

The flux may also be a common product of several branch currents. Each branch reacts with a portion of the flux. Hence, its voltage is the time derivative of this flux. Let us consider two coupled inductors in branches a and c. The ath branch is taken to react with the flux $\psi_a(i_a, i_c)$ produced by the two branch currents. Its terminal voltage is the time derivative of the flux, namely $v_a = d\psi_a(i_a, i_c)/dt$ or alternatively $v_a = L_{aa}(i_a, i_c)\,(di_a/dt) + L_{ac}(i_a, i_c)\,(di_c/dt)$ in terms of current derivatives. The quantity $L_{aa}(i_a, i_c) = \partial\psi_a(i_a, i_c)/\partial i_a$ denotes the instantaneous inductance of the ath branch, and $L_{ac}(i_a, i_c) = \partial\psi_a(i_a, i_c)/\partial i_c$ is the instantaneous mutual inductance. Of course, a similar inductive effect may arise in the cth branch due to the flux $\psi_c(i_a, i_c)$ that reacts with this branch. This flux may be different from the previous $\psi_a(i_a, i_c)$ if the coils are not 100% linked to each other. It also induces a voltage across the cth branch $v_c = d\psi_c(i_a, i_c)/dt$. This treatment clarifies the concept of mutual inductance between branches.

Summing up the characteristics of branches, we take our circuit to be composed of n_b branches. Hence it produces n_b branch equations that are, in general, dependent on branch voltages, currents, and possibly their derivatives or alternatively on derivatives of charges and fluxes. The latter are functions of voltages and currents respectively. A system of n_b equations is therefore suitable for a concise description of all the branches that are met in practice. Such a system can be written as follows:

$$\mathbf{f}\left[\mathbf{i}, \mathbf{v}, \frac{\mathrm{d}\mathbf{q}(\mathbf{v})}{\mathrm{d}t}, \frac{\mathrm{d}\boldsymbol{\psi}(\mathbf{i})}{\mathrm{d}t}, \mathbf{s}(t)\right] = \mathbf{0}. \tag{1.15}$$

It covers basic (1.7) - (1.9) and reactive (1.10), (1.14) branches.

A branch with a linear characteristic is called linear. The conductance $i_a = G v_a$, source $v_a = E$, controlled source $i_a = g_{ab} v_b$, and capacitance $i_a = C\,\mathrm{d}v_a/\mathrm{d}t$ are linear instances; the conductance $i_a = I_s[\exp(v_a/V_T) - 1]$, and capacitance $q_a = Q_o \exp(v_a/V_T)$ are nonlinear examples. Circuits consisting of linear branches are called linear circuits.

Note our system of graphical symbols for branches, which is shown in Table 1.1. Symbols for nonlinear resistors, conductors, capacitors and inductors are crossed.

Table 1.1 The system of graphical symbols for circuit elements

element	linear	nonlinear
resistor conductor		
independent and controlled vopltage source	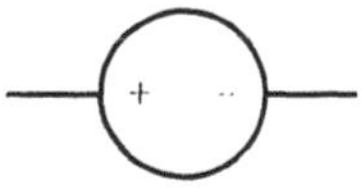	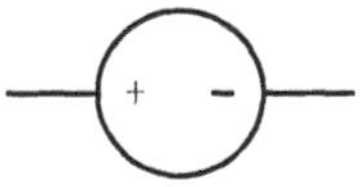
independent and controlled current source	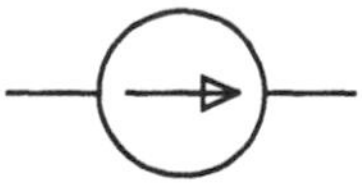	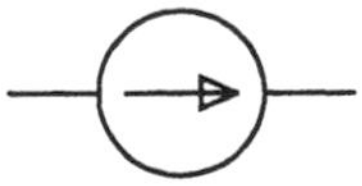
capacitor		
inductor		

1.2.2 Circuit topology

Electrical circuits are constructed by connecting together a number of branches with common terminals known as circuit nodes. Each branch has two terminals - an initial node and a final node, with the branch current flowing from the initial node to the final one. Each circuit node joins the initial and/or final terminals of a number of branches. Knowing which node each branch is connected to defines the circuit structure, also termed the circuit topology. Thus, to define the circuit two things are necessary: the characterization of individual branches and the topological information.

An obvious way of defining the topology consists of the unique identification of nodes (e.g. by means of numbers) and assigning a pair (*initial_node, final_node*) to each branch. This simple idea is implemented in input net-lists for general-purpose circuit simulators such as PSPICE. The net-list usually contains a standardized name of an element, two numbers identifying its terminal nodes and certain values of parameters. In Figure 1.6(a) we have a circuit consisting of 3 nodes numbered from 0 to 2 and 4 branches, whose PSPICE net-list is given below.

```
I1   0  1  1MAmp
R2   1  0  1KOhm
R3   1  2  1KOhm
R4   2  0  1KOhm
```

where I1 stands for the current source. The second and third column of the net-list define the topology by pairs of nodes. The same may be represented in a graphical manner by using directed graphs, whose arrows denote directions of currents, as is shown in Figure 1.6(b).

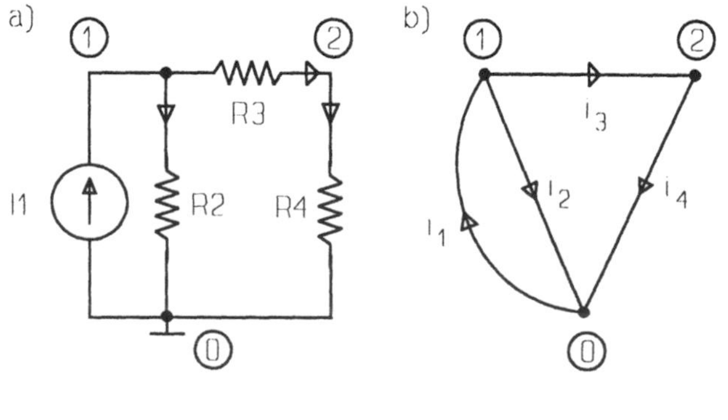

Figure 1.6 (a) A simple circuit and (b) a graph representing its topology

Let us consider node 1 where branches I1, R2 and R3 meet. Nodes are ideal points where no charge storage occurs, and hence the sum of the currents must be zero. Thus we have $-i_1+i_2+i_3=0$. This simple principle is known as **Kirchhoff's Current Law** (KCL). According to Figure 1.6(b) three KCL equations at nodes 1, 2 and 0 respectively may be written

$$\begin{aligned} -i_1 + i_2 + i_3 &= 0 \\ -i_3 + i_4 &= 0 \\ i_1 - i_2 - i_4 &= 0 \end{aligned} \quad . \qquad (1.16)$$

The KCL equations corresponding to all nodes of the circuit contain redundancy, because they are linearly dependent, so any one can be derived from the others. This observation leads to the concept of a reference node at which the KCL equation may be omitted. Taking node 0 as the reference (the number 0 is conventionally reserved for it) the KCL Equation (1.16), reduced by deletion of the third equation, may be written in the matrix form

$$\begin{bmatrix} -1 & 1 & 1 & 0 \\ 0 & 0 & -1 & 1 \end{bmatrix} \begin{bmatrix} i_1 \\ i_2 \\ i_3 \\ i_4 \end{bmatrix} = \mathbf{0} \qquad (1.17)$$

or in general

$$\mathbf{A}\ \mathbf{i} = \mathbf{0} \qquad (1.18)$$

where **i** is the vector of branch currents and **A** is known as the incidence matrix containing only coefficients 0 or +1, −1. Two quantities are of great importance in our consideration: the number of nodes (excluding the reference) n_n and the number of branches n_b. The incidence matrix is hence of dimension $n_n \times n_b$. Its element a_{ij} is 1 if the *i*th branch begins at the *j*th node, -1 if the *i*th branch ends at the *j*th node, and 0 otherwise.

In a certain assembly of the branches, some of them may form so-called loops, i.e. closed paths. In Figure 1.6(a) sets {I1, R2}, {I1, R3, R4} and {R2, R3, R4} are the loops that may be extracted from our circuit. Except at very high frequencies, where the concept of closed circuit breaks down, conservation of energy mean the work done around a loop is zero. So, the sum of voltages of the branches constituting the loop must be zero. Thus, in the loops of Figure 1.6(a) the branch voltages obey the **Kirchhoff's Voltage Law** (KVL)

$$\begin{aligned} v_1 + v_2 \qquad &= 0 \\ -v_2 + v_3 + v_4 &= 0\,. \\ v_1 + v_3 + v_4 &= 0 \end{aligned} \tag{1.19}$$

The above equations again contain redundancy, because each may be obtained from the others, just as each of the loops arises from a combination of two others. It can be proved that a topological circuit representation contains $n_b - n_n$ independent loops. In Figure 1.6(a) two independent, so-called fundamental, loops exist: {I1, R2} and {R2, R3, R4}. Therefore KVL may be reduced to these two loops, which can be written in a compact matrix form

$$\begin{bmatrix} 1 & 1 & 0 & 0 \\ 0 & -1 & 1 & 1 \end{bmatrix} \begin{bmatrix} v_1 \\ v_2 \\ v_3 \\ v_4 \end{bmatrix} = \mathbf{0} \tag{1.20}$$

where the matrix is known as the loop matrix **B** with dimensions $(n_b - n_n) \times n_b$. Following a consistent direction around each loop, the element b_{ij} of the matrix **B** is found to be 1 if the *j*th branch voltage is of the *i*th loop direction, −1 if it is in opposition to the *i*th loop direction, and finally 0 if it does not belong to the *i*th loop. The branch voltages may be collected in a column vector **v**. Hence, the KVL equations yield a formula

$$\mathbf{B}\,\mathbf{v} = \mathbf{0}\,. \tag{1.21}$$

It can be shown that topological matrices **A** and **B** both contain the same full information on the assembly of branches. For any circuit and loop we emphasize a rule relating the incidence of branches with respect to nodes and loops. The node may be (i) out of the loop or (ii) belong to the loop and hence incident with its two branches. In case (ii) these two branches are in a consistent (or alternatively reverse) orientation with respect to the node if and only if their orientation with respect to the loop is reverse (or alternatively consistent). This leads to the formula

$$\mathbf{B}\,\mathbf{A}^t = \mathbf{0}\,. \tag{1.22}$$

Thus the incidence matrix **A** may substitute for **B** and is of primary importance in our consideration because it contains the full topological information and its definition is very simple (see Equation (1.17) once more).

More extensive treatment of the circuit topology is covered in References [1.6 - 1.9].

1.2.3 Canonical equations and structural identity

In this book the term **Canonical Circuit Equations** (CCE) is introduced to obtain a unified treatment of the topics in circuit simulation.

Two Kirchhoff's Laws (1.18), (1.21) have given n_b equations based only on a circuit topology, and hence satisfied independently of types of branches involved. On the other hand, the full set of the branch variables (also referred to as the unknowns) includes $2n_b$ quantities: the currents **i** and the voltages **v**. To characterize the circuit uniquely n_b branch equations of the general form (1.15) have to be added. Finally, we obtain the set of so-called Canonical Circuit Equations (CCE) written briefly as

$$\begin{aligned} &\mathbf{A\,i} = \mathbf{0} \\ &\mathbf{B\,u} = \mathbf{0} \\ &\mathbf{f}\left[\mathbf{i}, \mathbf{v}, \frac{\mathrm{d}\mathbf{q}(\mathbf{v})}{\mathrm{d}t}, \frac{\mathrm{d}\boldsymbol{\psi}(\mathbf{i})}{\mathrm{d}t}, \mathbf{s}(t)\right] = \mathbf{0} \end{aligned} \quad . \tag{1.23}$$

The CCE term, seemingly based on very well-known concepts, introduces a surprising order to our treatment. It will therefore be a crucial point in our unified approach to circuit simulation methods.

To give an instructive example, let us recall the circuit given in Figure 1.1(b) whose branches have been characterized in Section 1.1.1. Let R_1 be the current-defined branch and i_D, C_D be treated separately (they may also be taken together as one current-defined branch). Furthermore, let the capacitance be described in terms of the derivative of the voltage. Assuming certain current directions and correspondingly opposite voltage directions CCE arise as follows

$$\begin{aligned} &i_{EG} + i_{RI} = 0 \\ &-i_{R1} + i_D + i_{CD} - i_I = 0 \\ &v_{VG} - v_{R1} - v_D = 0 \\ &v_D - v_{CD} = 0 \\ &v_{CD} + v_I = 0 \\ &v_{VG} = V_G(t) \\ &i_{R1} = v_{R1}/R_1 \\ &i_D = I_s[\exp(v_D/V_T) - 1] \\ &i_{CD} = C_{D0} * \exp(v_{CD}/V_T) * \mathrm{d}v_{CD}/\mathrm{d}t \\ &i_I = I \end{aligned} \tag{1.24}$$

where the first two lines represent KCL, the next three stand for KVL, and last five characterize branches.

The value of CCE consists in the compact, clear circuit characterization that distinguishes the structural (topological) and "material" aspect of the circuit. The methods of circuit analysis are somehow equivalent to the transformation of an original circuit into a substitute circuit which can then be manipulated. This corresponds to mathematical operations on the original canonical equations, that produce substitute canonical equations. The form of our canonical set easily shows whether the original and substitute circuits differ in structure (topology), or only in the physical meaning of the variables and the form of branches.

The topological part of CCE depends on the incidence and loop matrices. Due to their interdependence (1.22) the topological information contains redundancy. Nevertheless, to retain the set of $2n_b$ primary branch variables (currents and voltages) we do not eliminate the loop matrix **B** from our consideration.

Circuits of the same number of nodes and identically assembled branches, and of the same node and branch enumeration are called **structurally identical.** It is obvious that the identity of their **A** matrices is a necessary and sufficient condition of this property. The same **A** and **B** matrices occurring in several CCE sets certainly prove the structural identity of the circuits characterized by these sets. Thus topological identity can be found from CCE by simple inspection.

Let us consider, for instance, an extremely simple circuit: the parallel connection of the voltage source E and its load - the conductance G. The set of CCE takes the following form

$$\begin{aligned} i_V + i_G &= 0 \\ v_V - v_G &= 0 \\ v_V &= V \\ i_G &= G\, v_G \end{aligned} \quad . \tag{1.25}$$

If both sides of the above CCE are differentiated with respect to G at the point i_{V0}, i_{R0}, that satisfies (1.25), we obtain the set

$$\begin{aligned} \partial i_V/\partial G + \partial i_G/\partial G &= 0 \\ \partial v_V/\partial G - \partial v_G/\partial G &= 0 \\ \partial v_V/\partial G = \partial V/\partial G &= 0 \\ \partial i_G/\partial G &= G\,(\partial v_G/\partial G) + v_{GO} \end{aligned} \quad . \tag{1.26}$$

From its first and second line, this set stands for a certain circuit of the same topology as (1.25). However, its variables are derivatives of the original variables but its branches are quite different, derived from the original by differentiation. Thus (1.26) describes a substitute circuit, structurally identical to (1.25). It can, for example, be used to calculate derivatives for the purpose of sensitivity analysis. The canonical equations and the structural identity concept will be extensively used in the introductory discussion of most of simulation methods.

Subsection 1.2 has covered the basic constituent concepts in the mathematical characterization of circuits. To test our understanding we can try to work through the following problems:

- What do we mean by circuit topology and how may it be expressed in a matrix form? Explain the concept of the node and the loop.
- What are the physical backgrounds of Kirchhoff's Laws (KCL, KVL). How may KCL be expressed at a particular node and for the whole circuit? What do we mean by KVL around a particular loop and in a whole circuit?
- Explain the concepts of current- and voltage-defined branches.
- Explain what the Canonical Circuit Equations consist of.
- Give an example and explain the structural identity of circuits.

1.3 Circuit equations

A basic circuit description by means of the canonical equations has been introduced in Section 1.2.3. Such a characterization provides a simple approach to the derivation and understanding of simulation algorithms, though its direct application to computationally efficient algorithms is doubtful. Efficient equations do not contain branch variables. Another properly modified vector of the unknowns is introduced instead. After this so-called **transformation of circuit variables** the set may be reduced by eliminating redundant equations. Among a number of transformations known for many years the nodal transformation of voltages is unequalled. Therefore other transformations will only be mentioned briefly in what follows.

1.3.1 Transformation of circuit variables

In this section we turn to the nodal transformation of circuit voltages and show that this is an instance of the more general concept of transforming the circuit variables. Voltage, according to its nature, is uniquely defined only with respect to a reference point. Thus after a circuit reference node has been selected (see Section 1.2.2), n_n voltages at all other nodes may be determined with respect to that reference. They may be collected in the vector of node voltages $\mathbf{v}_n$ of the dimension n_n. Notice that the branch voltages satisfying KVL are all derivable from the nodal voltages, as a branch voltage is equal to the difference between the voltages at the positive and the negative terminals. For example Figure 1.7 presents the topology of the circuit given in Figure 1.6(a) with the branch voltages denoted. The following is the relation between the branch and nodal voltages

$$\begin{aligned} v_1 &= -v_{n1} \\ v_2 &= v_{n1} \\ v_3 &= v_{n1} - v_{n2} \\ v_4 &= v_{n2} \end{aligned} \tag{1.27}$$

Any choice of the nodal voltages is physical, that is all branch voltages evaluated from the nodal voltages automatically obey KVL, and thus all KVL equations are eliminated from further consideration. A general relation between branch and nodal voltages is derivable from a structure of connections embedded in the incidence matrix (see Section 1.2.2) so that we have

$$\mathbf{v} = \mathbf{A}^t \mathbf{v}_n \tag{1.28}$$

where superscript "t" stands for a matrix transposition. This so-called **nodal transformation** of voltages is quite general. Equation (1.27) is just a particular instance. Substitution of (1.28) in KVL (1.21) due to (1.22) yields an identity that proves KVL to be satisfied for any nodal voltages.

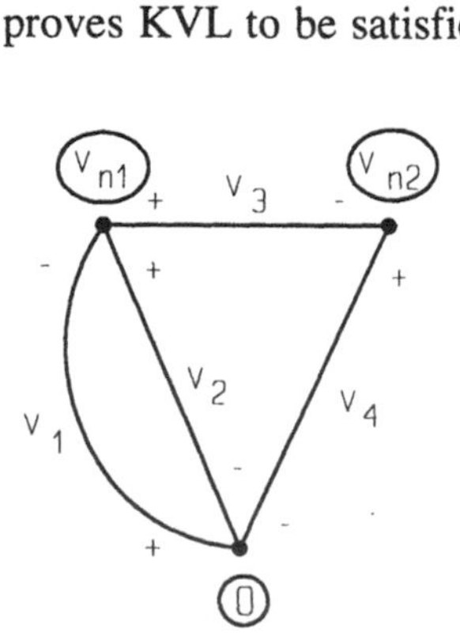

Figure 1.7 Graphical presentation of the relation between nodal and branch voltages

By using linear algebra we can discover that the branch voltages obeying KVL constitute the n_n-dimensional linear space over the set $\{-1,0,1\}$. The nodal voltages are one of its possible bases. Other selections of the base would also be available. Let us consider, for example, the circuit shown in Figure 1.6(a). After certain independent loops have been selected within the circuit, a set of n_n branches may be taken away from these loops in such a manner that the rest of the circuit will contain no loops. In graph theory [1.6 - 1.9] such a subgraph free of loops is called a tree; branches not belonging to the tree are called links. Voltages across branches of a tree also constitute the base of space of branch voltages. Hence, the voltages of links may be derived from tree voltages by virtue of the KVL balances around the loops. This would provide a transformation of voltages. Similarly, instead of branch voltages we can also transform branch currents. After a tree has been selected, currents flowing through links may be taken as the base of branch currents satisfying KCL. This so-called **mesh transformation** of currents takes the form $\mathbf{i} = \mathbf{B}^t \mathbf{i}_L$ parallel to the nodal transformation ($\mathbf{i}_L$ means currents of links). The above two transformations make use of a tree and thus need an automatic tree generation procedure. Except for state-space methods (of limited usefulness and so beyond the scope of our book), these transformations have almost been forgotten. The nodal transformation appears to prevail in virtually all applications.

1.3.2 Transformation of canonical circuit equations

The nodal transformation provides a simple means of eliminating branch voltages and KVL formulae from the set of canonical equations (1.23). Let us recall the example shown in Figure 1.1(b) and express the branch voltages by the nodal voltages: $v_{VG}=v_{n1}$, $v_{R1}=v_{n1}-v_{n2}$, $v_D=v_{n2}$, $v_{CD}=v_{n2}$, $v_I=-v_{n2}$. Substituting the above relations in (1.24), the KVL equations are eliminated and we obtain the transformed CCE

$$
\begin{aligned}
& i_{VG} + i_{RI} = 0 \\
& -i_{RI} + i_D + i_{CD} - i_I = 0 \\
& v_{n1} = V_G(t) \\
& i_{RI} = (v_{n1} - v_{n2})/R_1 \\
& i_D = I_s\,[\exp(v_{n2}/V_T) - 1] \\
& i_{CD} = C_{DO} * \exp(v_{n2}/V_T) * \mathrm{d}v_{n2}/\mathrm{d}t \\
& i_I = I
\end{aligned}
\qquad . \qquad (1.29)
$$

Equations of the above kind are, in general, derivable from the canonical set (1.23) by substituting the nodal transformation (1.28)

$$\begin{aligned} &\mathbf{A\,i} = \mathbf{0} \\ &\mathbf{f}\,[\mathbf{i}, \mathbf{A}^t\mathbf{v}_n, \frac{d\mathbf{q}(\mathbf{A}^t\mathbf{v}_n)}{dt}, \frac{d\boldsymbol{\psi}(\mathbf{i})}{dt}, \mathbf{s}(t)] = \mathbf{0}\,. \end{aligned} \tag{1.30}$$

In this way the number of circuit variables has been reduced to n_b+n_n and the redundant topological information embedded in the loop matrix has been eliminated. Several practical types of circuit characterization may be evaluated from this set.

1.3.3 Nodal equations

The types of circuit characterization depend on the character of the branches involved. Let us consider the simplest case where the branches are all current-defined and controlled only by voltages, that is having the general form

$$i_a = f_a[v_a, \frac{dv_a}{dt}, v_b, \frac{dv_b}{dt}, s(t)]\,. \tag{1.31}$$

Specifically they may be current excitations $s(t)$, linear or nonlinear conductances, current sources controlled by voltages or by their derivatives, capacitances or their combinations. If the branches are of the form (1.31), then the canonical circuit characterization (1.30) takes the form

$$\begin{aligned} &\mathbf{A\,i} = \mathbf{0} \\ &\mathbf{i} = \mathbf{f}\,[\mathbf{A}^t\mathbf{v}_n, \frac{d\mathbf{q}(\mathbf{A}^t\mathbf{v}_n)}{dt}, \mathbf{s}(t)] \end{aligned} \tag{1.32}$$

and consists of n_n KCL formulae and n_b branches expressed in terms of the nodal voltages only. Elimination of the currents from both KCL and branch equations (1.32) is possible because the branches are of the current-defined type and we then have

$$\mathbf{A}\;\;\mathbf{f}\,[\mathbf{A}^t\mathbf{v}_n, \frac{d\mathbf{q}(\mathbf{A}^t\mathbf{v}_n)}{dt}, \mathbf{s}(t)] = \mathbf{0}\,. \tag{1.33}$$

We have obtained n_n current balance equations in terms of branch functions. These so-called **Nodal Equations** (NE) are easily obtained by simple circuit inspection. For example, we may consider the nonlinear

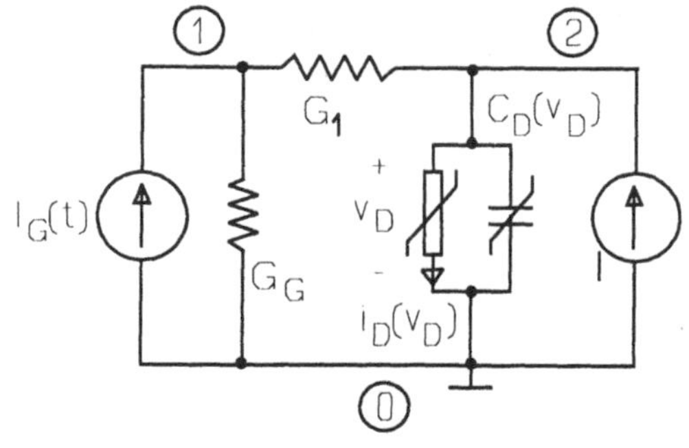

Figure 1.8 A circuit amenable to analysis by nodal approach

circuit in Figure 1.8 consisting of only the current-defined branches (resistors are modeled as conductors). NE have been created by formulating KCL at each node, assuming the conventional plus sign for currents flowing from a node and the minus sign for currents flowing towards the node. Hence, expressing each current in terms of the branch functions and the node voltages we have

$$
\begin{aligned}
&-I_G(t) + G_G v_{n1} + G_1 (v_{n1} - v_{n2}) = 0 \\
&-G_1 (v_{n1} - v_{n2}) + i_D(v_{n2}) + C_D(v_{n2}) \frac{dv_{n2}}{dt} - I = 0
\end{aligned}
\tag{1.34}
$$

where $i_D(u_D)$, $C_D(u_D)$ are known functions, e.g. as in Figure 1.1(b).

Nodal formulation is an extremely limited approach to circuit characterization. It needs to be extended to cover voltage defined elements.

1.3.4 Modified nodal equations

The nodal equations that we have introduced rely on eliminating all currents from the transformed canonical equations (1.30). Once some of the branches are voltage-defined such an elimination is impossible, even though the elimination of currents flowing through the current defined branches is still practicable. Thus, we may consider extracting a subset of the currents that will not be eliminated from the equations. These currents may not be removable from the system for two reasons: because they flow through the voltage defined branches or because they are controlling variables. Thus, while formulating the equations, branches with currents that are not supposed to be eliminated have to be especially selected. They may be of the current or voltage-defined type. The term class II has been

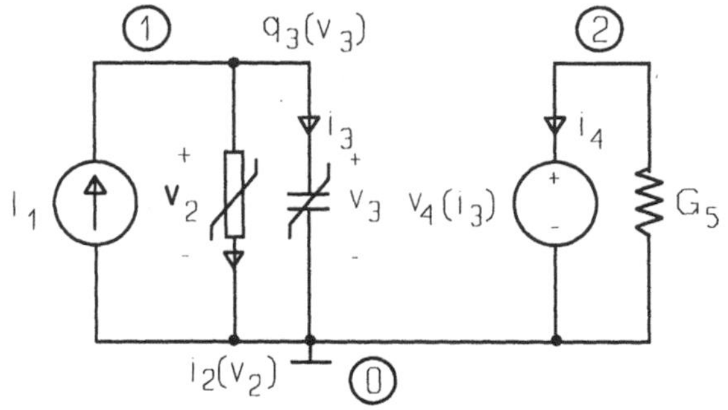

Figure 1.9 A circuit amenable to analysis by MNE

coined to identify them. The rest of the circuit branches are current defined and may be controlled by class II currents. We consider them to belong to class I, and their currents are to be disregarded from consideration.

Let us consider, for example, the circuit shown in Figure 1.9 comprising one voltage-defined branch $v_4(i_3)$ and one controlling current i_3 flowing through the capacitive branch $q_3(v_3)$. These two branches constitute the class II. The others, in particular a nonlinear conductance $i_2(v_2)$, belong to the class I. First, we write the KCL equations corresponding to our two nodes. Class I currents will be expressed by the branch equations and the node voltages while the two class II currents i_3, i_4 stay explicit:

$$\begin{aligned} -I_1 + i_2(v_{n1}) + i_3 &= 0 \\ i_4 + G_5 v_{n2} &= 0 \end{aligned} \quad . \tag{1.35}$$

In this way the branch formulae, except for the two belonging to class II, have been eliminated from the system. The two noneliminated yield

$$\begin{aligned} v_{n2} - v_4(i_3) &= 0 \\ i_3 - \frac{dq_3(v_{n1})}{dt} &= 0 \end{aligned} \quad . \tag{1.36}$$

By putting (1.35) and (1.36) together we arrive at a set of Modified Nodal Equations (MNE), which depend on $n_n + n_{bII}$ variables (nodes and branches of class II).

We split currents of the circuit (1.32) into the classes I and II. Then KCL yields

$$[\mathbf{A}_\mathrm{I} \;\; \mathbf{A}_\mathrm{II}] \begin{bmatrix} \mathbf{i}_\mathrm{I} \\ \mathbf{i}_\mathrm{II} \end{bmatrix} = \mathbf{0} \,. \tag{1.37}$$

After the branch equations have also been split into the two corresponding subsets, the canonical equations (1.32) take a new form

$$\begin{aligned} &\mathbf{A}_\mathrm{I}\,\mathbf{i}_\mathrm{I} + \mathbf{A}_\mathrm{II}\,\mathbf{i}_\mathrm{II} = \mathbf{0} \\ &\mathbf{i}_\mathrm{I} = \mathbf{f}_\mathrm{I}\left(\mathbf{A}^t\mathbf{v}_n, \frac{\mathrm{d}\mathbf{q}_\mathrm{I}(\mathbf{A}^t\mathbf{v}_n)}{\mathrm{d}t}, \mathbf{i}_\mathrm{II}, \frac{\mathrm{d}\boldsymbol{\psi}_\mathrm{I}(\mathbf{i}_\mathrm{II})}{\mathrm{d}t}, t\right) \\ &\mathbf{f}_\mathrm{II}\left(\mathbf{A}^t\mathbf{v}_n, \frac{\mathrm{d}\mathbf{q}_\mathrm{II}(\mathbf{A}^t\mathbf{v}_n)}{\mathrm{d}t}, \mathbf{i}_\mathrm{II}, \frac{\mathrm{d}\boldsymbol{\psi}_\mathrm{II}(\mathbf{i}_\mathrm{II})}{\mathrm{d}t}, t\right) = \mathbf{0} \end{aligned} \tag{1.38}$$

where the second subset stands for the branch equations of class I; it is current-defined and possibly controlled by the currents of class II. The third subset characterizes the branches of class II, that may be implicit, current or voltage-defined, and possibly controlled by the currents of class II. The first two equations of (1.38) can be combined to give the Modified Nodal Equations

$$\begin{aligned} &\mathbf{A}_\mathrm{I}\;\mathbf{f}_\mathrm{I}\left(\mathbf{A}^t\mathbf{v}_n, \frac{\mathrm{d}\mathbf{q}_\mathrm{I}(\mathbf{A}^t\mathbf{v}_n)}{\mathrm{d}t}, \mathbf{i}_\mathrm{II}, \frac{\mathrm{d}\boldsymbol{\psi}_\mathrm{I}(\mathbf{i}_\mathrm{II})}{\mathrm{d}t}, t\right) + \mathbf{A}_\mathrm{II}\,\mathbf{i}_\mathrm{II} = \mathbf{0} \\ &\mathbf{f}_\mathrm{II}\left(\mathbf{A}^t\mathbf{v}_n, \frac{\mathrm{d}\mathbf{q}_\mathrm{II}(\mathbf{A}^t\mathbf{v}_n)}{\mathrm{d}t}, \mathbf{i}_\mathrm{II}, \frac{\mathrm{d}\boldsymbol{\psi}_\mathrm{II}(\mathbf{i}_\mathrm{II})}{\mathrm{d}t}, t\right) = \mathbf{0} \,. \end{aligned} \tag{1.39}$$

This modified nodal type of circuit characterization well suits the practical needs and offers a variety of numerical advantages in real situations. After becoming familiar with its principles, we can proceed to the methods and algorithms of linear and nonlinear d.c./transient analysis covered in subsequent sections.

Subsection 1.3 has been dedicated to the styles of circuit characterization useful in deriving practical simulation methods. To test our understanding we can try to work through the following problems and answer the questions:

- What manipulation has been done to eliminate KVL from consideration? Explain the concept of the transformation of circuit variables.

- What do we mean by saying that the transformed canonical equations are a common root of practical circuit formalizations?
- What sorts of branches can be used in the Nodal Equations?
- Explain the two sorts of branches (class I and class II) involved in the formulation of Modified Nodal Equations.
- Explain the idea of Modified Nodal Equations as an extension of Nodal Equations.

References

1.1. Banzaf, W., *Computer-Aided Circuit Analysis Using SPICE*, Prentice-Hall International (1989).
1.2. Meares, L. G., *Spice Applications Handbook*, Intusoft, San Pedro, CA (1990).
1.3. Muller, K. H., *A Spice Cookbook*, Intusoft, San Pedro, CA (1991).
1.4. Tuinenga, P. W., *SPICE: A Guide to Circuit Simulation and Analysis Using PSPICE*, Prentice Hall Inc., Englewood Cliffs, NJ (1992).
1.5. *Circuit Analysis Reference Manual*, version 5.2, MicroSim Corporation, Irvine, CA (July 1992).
1.6. Seshu, S. and Reed, M. B., *Linear Graphs and Electrical Networks*, Addison-Wesley Publishing Company Inc., Reading, MA (1961).
1.7. Chen, W. K., *Applied Graph Theory*, North-Holland Publishing Co., Amsterdam (1971).
1.8. Johnson, D. R. and Johnson, J. R., *Graph Theory with Engineering Applications*, The Ronald Press Co. (1972).
1.9. Mayeda, W., *Graph Theory*, John Wiley and Sons Inc., New York (1972).

Chapter 2

Numerical Analysis of Linear Circuits

2.1 Formulation of algebraical circuit equations

2.1.1 General characterization of linear circuits

A linear circuit consists of linear branches, for instance the resistance $v_a = R\,i_a$, conductance $i_a = G\,v_a$, independent source $v_a = V$, controlled source $v_a = T_{ab}\,v_b$, capacitor $i_a = C\,\mathrm{d}v_a/\mathrm{d}t$. Hence, it includes these components described by linear algebraical or differential equations. Due to the possibility of voltage or current definition of the branches and to their current and/or voltage controlled character, they may be expressed in the following implicit form

$$L_a v_a + \hat{L}_a \frac{\mathrm{d}q_a}{\mathrm{d}t} + L_b v_b + K_a i_a + \hat{K}_a \frac{\mathrm{d}\psi_a}{\mathrm{d}t} + K_c i_c = s(t) \qquad (2.1)$$
$$q_a = q_{a0} + C_{aa} v_a + C_{ab} v_b\ , \quad \psi_a = \psi_{a0} + L_{aa} i_a + L_{ac} i_c$$

where the right-hand side $s(t)$ stands for the time domain excitation. Hence, according to the concepts introduced in Section 1.2.3, the linear circuit has the following canonical characterization

$$\begin{aligned} \mathbf{A\,i} &= \mathbf{0} \\ \mathbf{B\,v} &= \mathbf{0} \\ \mathbf{K\,i} + \hat{\mathbf{K}}\frac{\mathrm{d}\boldsymbol{\psi}(\mathbf{i})}{\mathrm{d}t} + \mathbf{L\,v} + \hat{\mathbf{L}}\frac{\mathrm{d}\mathbf{q}(\mathbf{v})}{\mathrm{d}t} &= \mathbf{s}(t) \\ \boldsymbol{\psi}(\mathbf{i}) = \boldsymbol{\psi}_0 + \mathbf{L\,i}\ , \quad \mathbf{q}(\mathbf{v}) &= \mathbf{q}_0 + \mathbf{C\,v} \end{aligned} \qquad (2.2)$$

which may be viewed as a special case of (1.23). The third row of (2.2) describes branches of all possible types, current or voltage controlled. Their excitations have been written on the right-hand side.

The numerical techniques dedicated to particular types of linear circuit analysis (e.g. d.c., a.c. or time domain) produce linear algebraical equations that may be solved with relative ease by means of general-purpose methods presented in Chapter 3. The algebraic-differential circuit characterization (2.2), after certain manipulations appropriate to the type of signals involved, becomes a purely algebraical, so-called substitute circuit description that corresponds to a particular analysis. There are three cases when substitute algebraical equations arise.

- **The direct current (d.c.) analysis of linear circuits** (a trivial case of rather little practical importance). A d.c. steady state under constant excitations $\mathbf{s}(t)=\mathbf{s}$ can be derived by open-circuiting all capacitors and short-circuiting inductors. This is equivalent to setting to zero all the differential components of the branch equations (2.1), (2.2) and hence yielding the algebraic canonical formulation (2.3).

- **The steady state in linear circuits subject to sinusoidal (a.c.) excitations** that may be described in terms of the phasors mentioned in Section 1.1.2. Phasors are complex numbers representing the amplitudes and phases of harmonic signals. If signals are expressed in terms of phasors their derivatives are also phasors and hence the circuit (2.2) is transformed into an immitance circuit (2.3) of the same structure, but including complex branch equations. This case will be treated later on.

- **Transient analysis,** mentioned in Section 1.1.2 and extensively discussed in the next chapters, relies on time discretization. At an instant t_n so-called differentiation formulae are used to obtain a local interpolation of waveforms, and therefore to replace the differential relation between signals and their derivatives by an algebraic relation satisfied by a certain class of interpolating functions (typically polynomials). We can observe, without further discussion, that, after the polynomial-type differentiation formula for an instantaneous derivative $dx/dt=\gamma(x-p_x)$ has been substituted in Equation (2.2), the purely algebraical companion circuit (2.3) is gained.

Thus in the above cases a substitute algebraic canonical characterization has been obtained, which is structurally identical with the original (2.2)

$$\begin{aligned} \mathbf{A\,i} \quad\quad\; &= \mathbf{0} \\ \mathbf{B\,u} \quad\quad\; &= \mathbf{0}\,. \\ \mathbf{K\,i} + \mathbf{L\,u} &= \mathbf{s} \end{aligned} \tag{2.3}$$

Notice that the detailed form of the coefficients **K**, **L**, and the excitation **s** is dependent on the type of analysis. Since we are investigating general techniques for creating efficient algebraic equations for linear circuits, consideration may be focused on an algebraic circuit (2.3) of real or complex coefficients. This general formalism will be manipulated below by means of circuit transformation and reduction tools developed in the previous chapter and actually applied to the linear case.

2.1.2 Nodal Equations

Nodal Equations (NE) are created by expressing branch voltages in terms of nodal voltages (see nodal transformation (1.28)) and then eliminating all branch currents due to the assumed current definition of branches. The nodal equations are a linear case of the general idea from Section 1.3.3. NE are constituted as current balance equations at all nodes excluding the reference one, and set up in such a way that each current is expressed in terms of branch constitutive function nodal voltages. Conventionally a current flowing away from a node is positive, while one flowing towards a node is negative. To show how nodal equations look, let us consider the circuit of Figure 2.1. The current balance at node 1 (numbers written in circles) yields the formula

$$G_2 v_{n1} - G_{54} v_{n2} = I_1 \tag{2.4}$$

where v_{n1}, v_{n2} are corresponding nodal voltages. Similarily at node 2

$$-G_{32} v_{n1} + (G_4 + G_{54}) v_{n2} = 0 \, . \tag{2.5}$$

This simple style of linear circuit description may be easily formalized in a general form. The existence of the nodal equations depends on whether all branches are current-defined and at most voltage-controlled. The general *a*th branch acceptable to NE, controlled by the *b*th branch voltage, is shown in Figure 2.2. It may be formalized as

$$i_a = G_{aa} v_a + G_{ab} v_b + I_a \, . \tag{2.6}$$

Provided that branches of the circuit (2.3) are all of the form (2.6) then

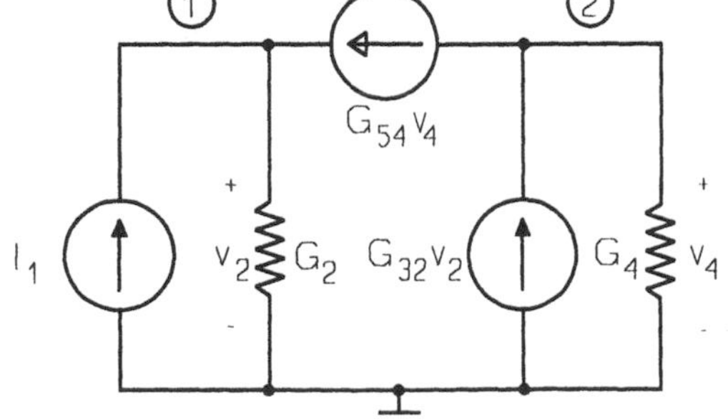

Figure 2.1 A circuit acceptable to nodal equations

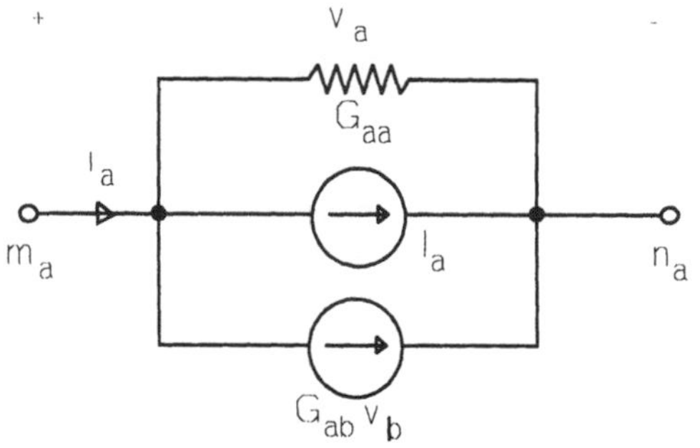

Figure 2.2 A general current-defined branch acceptable to nodal equations

the branch equations may be succinctly expressed as $\mathbf{i}=\mathbf{G}\,\mathbf{v}+\mathbf{I}$, where $\mathbf{G}$ is the matrix of branch conductances (including the parameters G_{aa} and G_{ab} of all branches), while $\mathbf{j}$ is the vector of branch current sources I_a. After the branch voltages have been expressed in terms of the nodal voltages (see (1.28)) the branch voltages and KVL disappear and (2.3) yields

$$\begin{aligned}&\mathbf{A}\,\mathbf{i}=\mathbf{0}\\&\mathbf{i}=\mathbf{G}\,\mathbf{A}^t\,\mathbf{v}_n+\mathbf{I}\,.\end{aligned} \tag{2.7}$$

By eliminating $\mathbf{i}$ between these equations we obtain the nodal equation

$$\mathbf{A}\,\mathbf{G}\,\mathbf{A}^t\,\mathbf{v}_n=-\mathbf{A}\,\mathbf{I} \tag{2.8}$$

where $\mathbf{A}\,\mathbf{G}\,\mathbf{A}^t$ is the nodal admittance matrix (denoted as $\mathbf{Y}$), while $-\mathbf{A}\,\mathbf{I}$ is the vector of nodal currents, further denoted as $\mathbf{b}$.

Formulation of the nodal equations by multiplying the matrices **A**, **G**, **I** is extremely impractical. We have shown them to be very friendly in creation. Now, it will be shown that they have nice properties when it comes to setting them up in the computer. To obtain an algorithm we should first notice that the branch matrices **G**, **I** are sums of elementary matrices corresponding to each branch ($\mathbf{G}=\sum_a \mathbf{G}^a$, $\mathbf{I}=\sum_a \mathbf{I}^a$). Hence, due to the linear dependence of $\mathbf{Y}$ and $\mathbf{b}$ on $\mathbf{G}$ and $\mathbf{I}$ respectively, the nodal matrices are also the sums of branch contributions: $\mathbf{Y}=\sum_a \mathbf{Y}^a$, $\mathbf{b}=\sum_a \mathbf{b}^a$. These contributions amount to $\mathbf{Y}^a=\mathbf{A}\mathbf{G}^a A^t$, $\mathbf{b}^a=\mathbf{A}\mathbf{I}^a$. They are very sparse matrices including only a few nonzero elements at certain positions. Usually we demonstrate them in a table form known as a stamp. These stamps have rows and columns enumerated by means of integers assigned to the circuit nodes. The reference node has the number

0 and so it does not appear in the stamp. The general *a*th branch, shown in Figure 2.2, is taken to be stretched between nodes numbered m_a and n_a, while the current flows away from the former towards the latter. This yields the following stamp:

	m_a	n_a	m_b	n_b	
m_a	G_{aa}	$-G_{aa}$	G_{ab}	$-G_{ab}$	$-I_a$
n_a	$-G_{aa}$	G_{aa}	$-G_{ab}$	G_{ab}	I_a
		$\mathbf{Y}^a$			$\mathbf{b}^a$

Let us notice that once a branch is incident to the reference node 0 the corresponding row and column of the stamp do not exist.

Of course, the branch of Figure 2.2 may be reduced with relative ease to a single element, namely: conductance, voltage controlled current source (VCCS) and independent current source. For instance the conductance arises when assuming the only nonzero coefficient to be G_{aa}. Hence, (2.6) transforms into $i_a = G_{aa} v_a$, while the stamp reduces to the four terms of the first two columns.

Moreover, the nodal equations accept a current-defined multi-pole described by the admittance matrix. For example, we consider the three-pole connected to nodes m_1, m_2, m_3 and shown in Figure 2.3. Its branch currents expressed in terms of nodal voltages take the form

$$\begin{aligned} i_1 &= Y_{11} v_{n1} + Y_{12} v_{n2} + Y_{13} v_{n3} \\ i_2 &= Y_{21} v_{n1} + Y_{22} v_{n2} + Y_{23} v_{n3} \, . \\ i_3 &= Y_{31} v_{n1} + Y_{32} v_{n2} + Y_{33} v_{n3} \end{aligned} \tag{2.9}$$

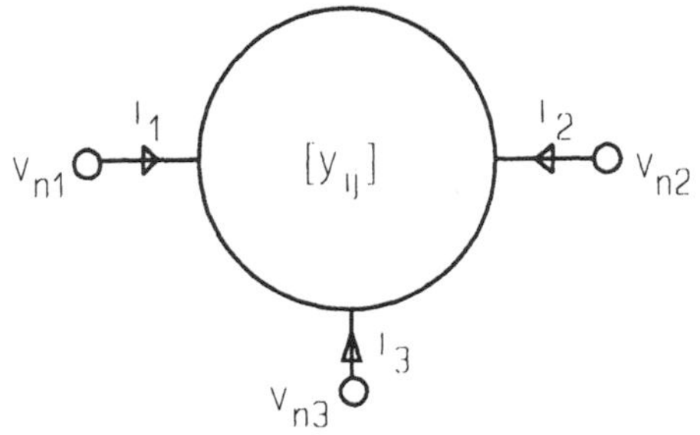

Figure 2.3 A three-pole

It can be readily shown that this element yields the stamp

	m_1	m_2	m_3
m_1	Y_{11}	Y_{12}	Y_{13}
m_2	Y_{21}	Y_{22}	Y_{23}
m_3	Y_{31}	Y_{32}	Y_{33}

Topics covered in this section can be found in books [2.2, 2.3, 2.5].

2.1.3 Modified nodal equations

2.1.3.1 Example

To begin the discussion let us recall the Modified Nodal Equations (MNE) form of nonlinear circuit description, introduced in Section 1.3.4. Further we concentrate on its linear case. MNE accept both voltage- and current-defined as well as voltage- and current-controlled branches. The two classes of branches have been distinguished: class II consisting of voltage-defined branches or branches including controlling currents and class I composed of the other branches. The set of unknowns includes the nodal voltages and currents of class II. To write MNE first we create KCL equations at each node, apart from the reference one. While setting up the current balance relation we explicitly take the currents of class II, i.e. those that belong to the set of unknowns. The other currents are expressed in terms of branch constitutive functions and the nodal volt-

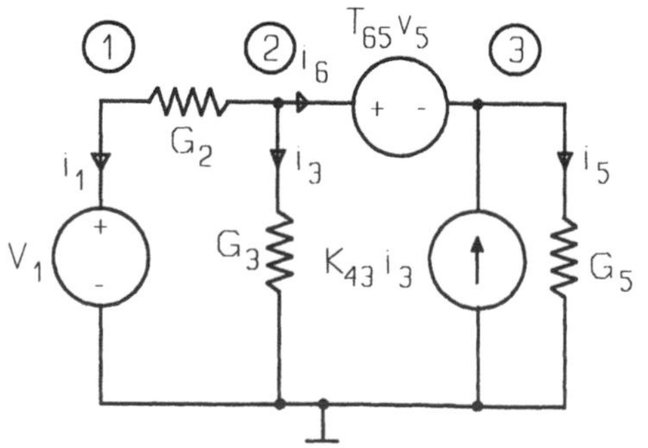

Figure 2.4 A circuit acceptable to modified nodal equations

ages exactly in the same manner as was carried out in the case of nodal equations in Section 2.1.2.

Let us consider the linear circuit of Figure 2.4 whose equations may not be formulated according to the nodal equation form, because they contain voltage-defined and current-controlled branches. To create the vector of unknowns the vector of the nodal voltages v_{n1}, v_{n2}, v_{n3} is taken first. Then the currents i_1, i_6 of the voltage-defined branches V_1, T_{65} are added. Moreover, the controlling current i_3 is introduced. Finally, to show that MNE may accept any current taken as unknown, the current i_5 is selected and, hence, the branch G_5 is chosen to be of class II. Thus, we write three KCL equations corresponding to the nodes

$$\begin{aligned} &G_2(v_{n1}-v_{n2})+i_1=0\\ &G_2(v_{n2}-v_{n1})+i_3+i_6=0\,.\\ &-K_{43}i_3+i_5-i_6=0 \end{aligned} \tag{2.10}$$

Then, the branch equations of class II are written in terms of nodal voltages and the currents selected to the vector of unknowns

$$\begin{aligned} &-v_{n1}=-V_1\\ &G_3v_{n2}-i_3=0\\ &G_5v_{n3}-i_5=0\\ &-v_{n2}+(T_{65}+1)v_{n3}=0\,. \end{aligned} \tag{2.11}$$

The final set of MNE may be presented in a matrix form

$$\begin{bmatrix} G_2 & -G_2 & 0 & 1 & 0 & 0 & 0\\ -G_2 & G_2 & 0 & 0 & 1 & 0 & 1\\ 0 & 0 & 0 & 0 & -K_{43} & 1 & -1\\ -1 & 0 & 0 & 0 & 0 & 0 & 0\\ 0 & G_3 & 0 & 0 & -1 & 0 & 0\\ 0 & 0 & G_5 & 0 & 0 & -1 & 0\\ 0 & -1 & T_{65}+1 & 0 & 0 & 0 & 0 \end{bmatrix} \begin{bmatrix} v_{n1}\\ v_{n2}\\ v_{n3}\\ i_1\\ i_3\\ i_5\\ i_6 \end{bmatrix} = \begin{bmatrix} 0\\ 0\\ 0\\ -V_1\\ 0\\ 0\\ 0 \end{bmatrix} \tag{2.12}$$

MNE for linear circuits have been primarily introduced by C. W. Ho, R. K. Brayton and F. G. Gustavson. [2.1]

2.1.3.2 General consideration of acceptable branches

Circuit description by modified nodal equations (MNE) has been considerably extended in comparison with the usual nodal equations (NE) by the inclusion of voltage-defined and current-controlled branches. We have two sorts of acceptable branches: current- and voltage-defined. A current-defined branch is similar to the one demonstrated in Figure 2.2 which is acceptable to the usual NE. It has been extended, however, by a current-controlled component, as shown in Figure 2.5(a). Moreover, the voltage-defined branch, as shown in Figure 2.5(b), is acceptable. It may be a voltage and/or current-controlled branch.

The current-defined *a*th branch presented in Figure 2.5(a), which is voltage-controlled by the *b*th branch and current-controlled by the *c*th branch, is of a form

$$i_a = G_{aa} v_a + G_{ab} v_b + K_{ac} i_c + I_a . \tag{2.13}$$

The corresponding voltage-defined branch of Figure 2.5(b) yields the characterization

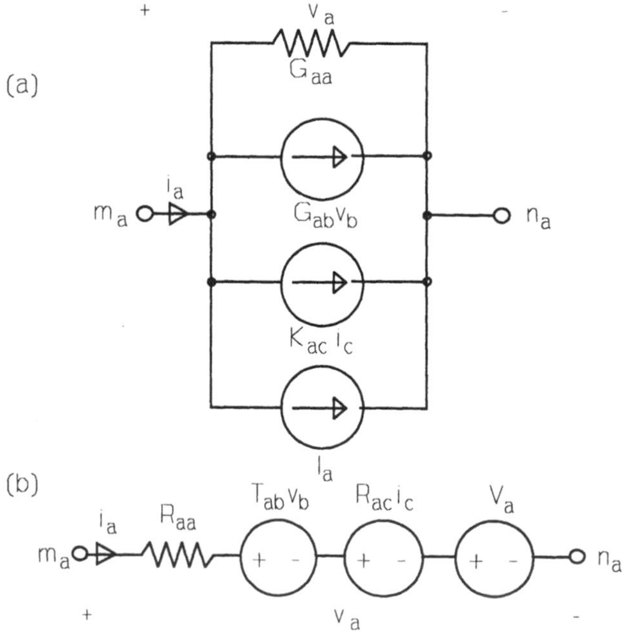

Figure 2.5 Generalized (a) current defined (b) voltage defined branches acceptable to modified nodal equations

$$v_a = Z_{aa} i_a + T_{ab} v_b + Z_{ac} i_c + V_a \,. \tag{2.14}$$

According to the MNE form we must select branches from class II. Their currents will not be eliminated from the set of equations and so will be circuit variables (also called unknowns). Hence, the crucial point is to properly choose the classes. Class II is composed of:

- all voltage-defined branches,
- all branches, whose currents are controlling variables,
- any other branches that have been chosen to add their currents to the vector of unknowns.

The other branches are taken to be of class I. All voltage-defined branches are hence included in class II. As for the current-defined branches they may have currents selected as circuit variables (class II) or their currents will be eliminated from the equations.

The vector of circuit variables (unknowns) of MNE is assumed to be composed of nodal voltages and currents of class II. Hence, it includes:

(i) nodal voltages (except the reference node),
(ii) currents of voltage-defined branches,
(iii) controlling currents,
(iiii) any other currents, if the user wants to add them.

Recalling Figure 2.4 it can be seen that the currents i_1, i_6 have been selected according to the voltage definition (ii), i_3 is a controlling current (iii) and i_5 has been added due to the possibility (iiii).

Rules of circuit variables selection in MNE are not unique. Many different specific implementations are possible. At least the nodal voltages, currents of voltage-defined branches and controlling currents have to be selected as unknowns. The commonly used simulator PSPICE also uses MNE, but the need for considering controlling currents separately is eliminated as the input language enables current controlling only from voltage sources.

In our book the reader is encouraged to make use of the ubiquitous simulator PSPICE and the simulator OPTIMA 3, included on Diskette B. OPTIMA, extensively exploited in our book, is especially dedicated to provide the reader with a deeper insight into simulation. A description of its input language and presentation of principles of elements modeling is given in TUTORIAL.DOC and the on-line help. Simulator OPTIMA opens to the reader the world of interesting experiments in simulation.

The specific OPTIMA implementation of MNE sets up an unknowns vector composed of: nodal voltages, currents flowing through voltage-defined branches (that is independent and controlled voltage sources, inductors, zero-valued and modeled resistors, operational amplifiers and ports of transmission lines), controlling currents (that may flow through the above mentioned elements and also through any resistor) and also charges and fluxes of nonlinear reactive elements. This extension connected with the time-domain analysis will be discussed later on.

2.1.3.3 General formulation of linear modified nodal equations

Let a linear circuit be given in a general canonical form (2.3). Branches can be divided into classes I and II. Class I includes only the current-defined branches (2.13). They may be controlled by currents of class II. Class II contains branches of any type (we express them in a general implicit form although in practice they may be implicit or explicit of the form (2.13) or (2.14)). Their currents $\mathbf{i}_{II}$ are selected as circuit variables. The currents of class I are denoted as $\mathbf{i}_I$. Thus, (2.3) after the transformation (1.28) yields

$$\begin{aligned} \mathbf{A}_I \mathbf{i}_I + \mathbf{A}_{II} \mathbf{i}_{II} &= \mathbf{0} \\ \mathbf{i}_I &= \mathbf{G}_I \mathbf{A}^t \mathbf{v}_n + \mathbf{K}_I \mathbf{i}_{II} + \mathbf{I}_I \\ \mathbf{K}_{II} \mathbf{i}_{II} + \mathbf{L}_{II} \mathbf{A}^t \mathbf{v}_n &= \mathbf{I}_{II} \end{aligned} \tag{2.15}$$

where the incidence matrix $\mathbf{A}$ has been divided into two parts according to the division of branches into classes I and II. The matrices $\mathbf{G}_I, \mathbf{K}_I$ stand for the respective coefficients of the branches (2.13). Identically we have the matrices $\mathbf{K}_{II}, \mathbf{L}_{II}$ consisting of the coefficients and $\mathbf{s}_{II}$ standing for the excitations of the branches (2.13) and (2.14) of class II.

After $\mathbf{i}_I$ has been eliminated from Equations (2.15) we have

$$\begin{bmatrix} \mathbf{A}_I \mathbf{G}_I \mathbf{A}^t & \mathbf{A}_I \mathbf{K}_I + \mathbf{A}_{II} \\ \mathbf{L}_{II} \mathbf{A}^t & \mathbf{K}_{II} \end{bmatrix} \begin{bmatrix} \mathbf{v}_n \\ \mathbf{i}_{II} \end{bmatrix} = \begin{bmatrix} -\mathbf{A}_I \mathbf{I}_I \\ \mathbf{s}_{II} \end{bmatrix} \tag{2.16}$$

known as Modified Nodal Equations (MNE). They are composed of n_n (the number of nodes excluding the reference) KCL equations expressed in terms of the directly involved currents of class II and of class I currents indirectly given through the class I branch formulae and the nodal voltages. This set is extended by n_{II} (the number of class II branches) branch equations written in terms of their currents and nodal voltages. We will succinctly formalize MNE as $\mathbf{Y}\,\mathbf{x} = \mathbf{b}$.

2.1.3.4 Computer formulation of modified nodal equations

As will be readily shown, the stamps, known from nodal equations, dedicated to algorithmic formulation of the equations are available also for linear modified nodal equations (MNE). Equation (2.16) proves that the four parts of the matrix and the two of the right-hand-side (RHS) are linear transformations (i.e. derived by matrix multiplications) of the branch matrices $\mathbf{G}_{\mathrm{I}}, \mathbf{K}_{\mathrm{I}}$, etc., in a manner similar to that which was proved in Section 2.1.2 while considering the nodal equations. The branch matrices are the superposition of components corresponding to each branch because each component constitutes a separate row of the branch equations. Hence, the matrix **Y** and RHS **b** is a sum of so-called stamps, i.e. elementary matrices $\mathbf{Y}^a$, $\mathbf{b}^a$ corresponding to each branch.

From the point of view of the stamps MNE accept three branch types:

- **type v**: the voltage-defined branch of the form (2.14),
- **type ix**: the current-defined branch of the form (2.13) whose current has been taken into the circuit variables (unknowns) vector **x**,
- **type i**: the current-defined branch of the form (2.13) whose current has not been taken into the circuit variables (unknowns) vector.

For the current-defined *a*th branch (2.13) shown in Figure 2.5(a) whose current is not a circuit variable we have the stamp

type i

	m_a	n_a	m_b	n_b	i_c	
m_a	G_{aa}	$-G_{aa}$	G_{ab}	$-G_{ab}$	K_{ac}	$-I_a$
n_a	$-G_{aa}$	G_{aa}	$-G_{ab}$	G_{ab}	$-K_{ac}$	I_a
			$\mathbf{Y}^a$			$\mathbf{b}^a$

The rows and columns of the matrix correspond to numbers of nodes and current variables. The nodes and the class II currents have to be numbered in one sequence using successive integers (where 0 stands for the reference node not represented in the stamp). The nodal voltages and currents may be mixed in order. At the border of the stamp we have symbols of nodes and currents that stand for numbers addressing row-column positions of the matrix. The stamp of the RHS vector appears on the right side of the double line.

In the case of the *a*th current-defined branch (2.13) whose current has been selected as a circuit variables the stamp

type ix

	m_a	n_a	m_b	n_b	i_c	i_a	
m_a						1	
n_a						-1	
i_a	G_{aa}	$-G_{aa}$	G_{ab}	$-G_{ab}$	K_{ac}	-1	$-I_a$

$\mathbf{Y}^a$ $\mathbf{b}^a$

appears to include a third row corresponding to the branch equation and the column labeled with i_a denoting a current variable generated by this branch.

The voltage-defined branch (2.14) of Figure 2.5(b) produces the stamp

type v

	m_a	n_a	m_b	n_b	i_c	i_a	
m_a						1	
n_a						-1	
i_a	-1	1	T_{ab}	$-T_{ab}$	Z_{ac}	Z_{aa}	$-V_a$

$\mathbf{Y}^a$ $\mathbf{b}^a$

While formulating circuit matrices by means of the stamps techniques certain branches appear to be incident with the reference node 0. Then, those rows and columns of the stamps that correspond to grounded terminals will be omitted as KCL at the reference node is obeyed automatically.

The stamps easily reduce to specific cases relating to particular elements (e.g. conductance provides the stamp of sort i or ix where $G_{ab}=K_{ac}=I_a=0$). MNE accept interesting branches that need more comment. The short-circuit is a voltage-defined branch: v_a=0 and may be treated as a zero voltage source or zero resistance. Its stamp arises from the type v. The open circuit is an empty element $i_a=0$ of an empty stamp i.

However, if we wanted its zero current to be taken as unknown, then a stamp would be easily obtained from the type ix.

The ideal operational amplifier is an asymptotic case of a voltage-controlled voltage source (VCVS) once its gain approaches infinity. The stamp for a VCVS derived from the sort v takes the form

	m_a	n_a	m_b	n_b	i_a
m_a					1
n_a					-1
i_a	-1	1	T_{ab}	$-T_{ab}$	

$\mathbf{Y}^a$

The third row includes the branch equation. It may be divided by T_{ab} to obtain

	m_a	n_a	m_b	n_b	i_a
m_a					1
n_a					-1
i_a	$-1/T_{ab}$	$1/T_{ab}$	1	-1	

$\mathbf{Y}^a$

As the voltage gain t_{ab} approaches infinity we obtain the stamp of an ideal operational amplifier

	m_b	n_b	i_a
m_a			1
n_a			-1
i_a	1	-1	

It represents the short-circuit v_b=0 with no current between nodes m_b and n_b and any current flowing from the node m_a to n_a.

MNE accept a hybrid characterization of the multi-pole of Figure 2.3, provided that terminals are divided into sections 1 and 2

$$\begin{bmatrix} \mathbf{i}_1 \\ \mathbf{v}_{n2} \end{bmatrix} = \begin{bmatrix} \mathbf{H}_{11} & \mathbf{H}_{12} \\ \mathbf{H}_{21} & \mathbf{H}_{22} \end{bmatrix} \begin{bmatrix} \mathbf{v}_{n1} \\ \mathbf{i}_2 \end{bmatrix}. \tag{2.17}$$

Thus the MNE need currents $\mathbf{i}_2$ of section 2 to be added to the variables vector. After collecting the nodal voltages of sections 1, 2 and currents of section 2 in three groups, labeled by $\mathbf{n}_1$, $\mathbf{n}_2$, $\mathbf{i}_2$, the block stamp takes the form

	$\mathbf{n}_1$	$\mathbf{n}_2$	$\mathbf{i}_2$
$\mathbf{n}_1$	$\mathbf{H}_{11}$		$\mathbf{H}_{12}$
$\mathbf{i}_2$	$\mathbf{H}_{21}$	**-1**	$\mathbf{H}_{22}$

$\mathbf{Y}^a$

2.1.3.5 How do the stamps generate equations? Example.

Let us consider the circuit in Figure 2.4. We begin with the stamp of G_2

$$\begin{array}{c} \\ 1 \\ 2 \\ 3 \\ i_1 \\ i_3 \\ i_5 \\ i_6 \end{array}
\begin{array}{c} \begin{array}{ccccccc} 1 & 2 & 3 & i_1 & i_3 & i_5 & i_6 \end{array} \\
\left[\begin{array}{ccccccc} G_2 & -G_2 & & & & & \\ -G_2 & G_2 & & & & & \\ & & & & & & \\ & & & & & & \\ & & & & & & \\ & & & & & & \\ & & & & & & \end{array}\right] \end{array}
\begin{array}{c} RHS \\ \left[\begin{array}{c} \\ \\ \\ \\ \\ \\ \\ \end{array}\right] \end{array} \tag{2.18}$$

After the stamps of V_1, G_3 and K_{43} have been added we obtain

$$
\begin{array}{c}
 \\ 1 \\ 2 \\ 3 \\ i_1 \\ i_3 \\ i_5 \\ i_6
\end{array}
\begin{array}{c}
\begin{array}{cccccccc} 1 & 2 & 3 & i_1 & i_3 & i_5 & i_6 \end{array} \\
\left[\begin{array}{ccccccc}
G_2 & -G_2 & & 1 & & & \\
-G_2 & G_2 & & & & & \\
 & & & & & & \\
-1 & & & & & & \\
 & & & & & & \\
 & & & & & & \\
 & & & & & &
\end{array}\right]
\end{array}
\begin{array}{c}
RHS \\
\left[\begin{array}{c}
 \\ \\ \\ -V_1 \\ \\ \\ \\
\end{array}\right]
\end{array}
\tag{2.19}
$$

$$
\begin{array}{c}
 \\ 1 \\ 2 \\ 3 \\ i_1 \\ i_3 \\ i_5 \\ i_6
\end{array}
\begin{array}{c}
\begin{array}{cccccccc} 1 & 2 & 3 & i_1 & i_3 & i_5 & i_6 \end{array} \\
\left[\begin{array}{ccccccc}
G_2 & -G_2 & & 1 & & & \\
-G_2 & G_2 & & & 1 & & \\
 & & & & & & \\
-1 & & & & & & \\
 & G_3 & & & -1 & & \\
 & & & & & & \\
 & & & & & &
\end{array}\right]
\end{array}
\begin{array}{c}
RHS \\
\left[\begin{array}{c}
 \\ \\ \\ -V_1 \\ \\ \\ \\
\end{array}\right]
\end{array}
\tag{2.20}
$$

$$
\begin{array}{c}
 \\ 1 \\ 2 \\ 3 \\ i_1 \\ i_3 \\ i_5 \\ i_6
\end{array}
\begin{array}{c}
\begin{array}{cccccccc} 1 & 2 & 3 & i_1 & i_3 & i_5 & i_6 \end{array} \\
\left[\begin{array}{ccccccc}
G_2 & -G_2 & & 1 & & & \\
-G_2 & G_2 & & & 1 & & \\
 & & & & -K_{43} & & \\
-1 & & & & & & \\
 & G_3 & & & -1 & & \\
 & & & & & & \\
 & & & & & &
\end{array}\right]
\end{array}
\begin{array}{c}
RHS \\
\left[\begin{array}{c}
 \\ \\ \\ -V_1 \\ \\ \\ \\
\end{array}\right]
\end{array}
\tag{2.21}
$$

After the element G_5 has been introduced the system takes the form

$$\begin{array}{c} \\ 1 \\ 2 \\ 3 \\ i_1 \\ i_3 \\ i_5 \\ i_6 \end{array}
\begin{array}{c}
\begin{array}{ccccccc} 1 & 2 & 3 & i_1 & i_3 & i_5 & i_6 \end{array} \\
\left[\begin{array}{ccccccc}
G_2 & -G_2 & & 1 & & & \\
-G_2 & G_2 & & & 1 & & \\
 & & & & -K_{43} & 1 & \\
-1 & & & & & & \\
 & G_3 & & & -1 & & \\
 & & G_5 & & & -1 & \\
 & & & & & &
\end{array}\right]
\end{array}
\begin{array}{c} RHS \\ \left[\begin{array}{c} \\ \\ \\ -V_1 \\ \\ \\ \\ \end{array}\right] \end{array}
\qquad (2.22)$$

Finally, the inclusion of T_{65} yields the complete matrix and RHS

$$\begin{array}{c} \\ 1 \\ 2 \\ 3 \\ i_1 \\ i_3 \\ i_5 \\ i_6 \end{array}
\begin{array}{c}
\begin{array}{ccccccc} 1 & 2 & 3 & i_1 & i_3 & i_5 & i_6 \end{array} \\
\left[\begin{array}{ccccccc}
G_2 & -G_2 & & 1 & & & \\
-G_2 & G_2 & & & 1 & & 1 \\
 & & & & -K_{43} & 1 & -1 \\
-1 & & & & & & \\
 & G_3 & & & -1 & & \\
 & & G_5 & & & -1 & \\
 & -1 & T_{65}+1 & & & &
\end{array}\right]
\end{array}
\begin{array}{c} RHS \\ \left[\begin{array}{c} \\ \\ \\ -V_1 \\ \\ \\ \\ \end{array}\right] \end{array}
\qquad (2.23)$$

obviously identical with those formulated in (2.12).

2.1.4 Two-graph modified nodal equations

J. Vlach and K. Singhal have developed a version of modified nodal equations [2.3] presenting a slightly changed style of the circuit topology characterization. Their idea is a consequence of concepts of current and voltage graphs which occur in the theory of graphs [1.5-1.9]. A simple idea of this, seemingly advanced, approach is the elimination from the modified nodal equations of unknown currents flowing through independent voltage sources and one of two nodes incident to each short-circuit. The justification for this is that the voltage of a short-circuit is obvious, while the current through a voltage source may be uninteresting. Hence, in circuits including a great number of voltage excitations in comparison to the number of nodes, considerable savings arise. Unfortunately this condition is not often met in practice, so the Vlach-Singhal modification is of much less importance then the primary modified nodal equations introduced in Section 2.1.3. The Vlach-Singhal method is known as that of the Two-graphs Modified Nodal Equations (TGMNE).

2.1.4.1 Two-graph modified characterization of circuit topology

In Section 1.2.2 the characterization of the circuit topology has been introduced. The topology (also known as a circuit structure) consists of branches and nodes properly connected. Each of these two things models two phenomena. In the case of the branch, it stands for a flowing current and simultaneously a voltage developed across its terminals (which is of the opposite direction). For the node, it represents a point where currents are consistent with KCL and moreover denotes a node voltage. Hence, two features of the circuit topology may be distinguished: (i) a structure of currents, i.e. the way of connecting currents to their joining points, (v) a structure of the branch voltages developed between the nodal voltages. The current structure (i) responsible for KCL is known as the **current-graph** (i-graph, in short), the voltage structure related to KVL is called the **voltage-graph** (v-graph, in short). Both graphs have been taken to be identical so far, characterized by the same incidence matrix **A**. We will show that they may also be different. A considerable saving in the number of circuit variables is available when not all branch voltages or currents are taken into KCL and KVL formulae. This idea will be explained below in detail.

First, let us turn to savings in KCL. We consider the i-graph whose branches are assigned to currents while nodes are points where currents join together. A part of a circuit including two nodes A and B is demonstrated in Figure 2.6(a). A current balance at node A takes the form: $i_1+i_2+i_a=0$ while at B: $i_3+i_4-i_a=0$. Once the current i_a is declared to be of no interest and need not be represented in circuit equations then it can be omitted in KCL by setting the balance: $i_1+i_2+i_3+i_4=0$ at the nodes A and B together. It means that while forming the i-graph the branch i_a is collapsed into one point AB as demonstrated in Figure 2.6(b).

In a similar way, if i_a is 0 (open-circuit) then the *a*th branch of the i-graph may be deleted and its current is eliminated from KCL, as in Fig-

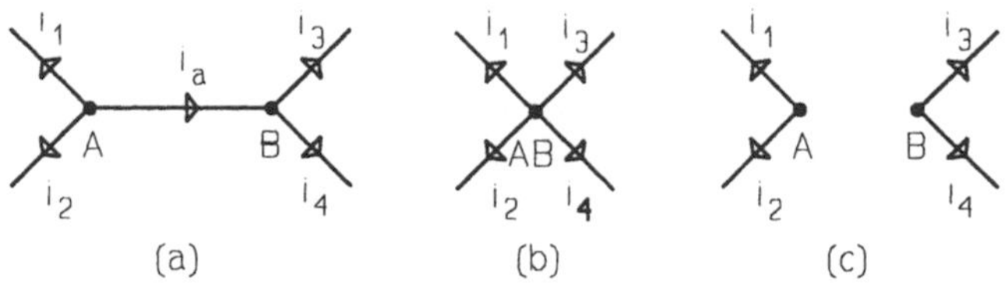

Figure 2.6 Creation of i-graphs: (a) an original structure, (b) collapsing two nodes while a current is of no interest, (c) deletion of a branch while a current is zero

ure 2.6(c). In other cases, except the two of Figures 2.6(b,c), the i-graph has a branch correspondent to each branch current. The i-graph constitutes the branch currents vector $\mathbf{i}_i$ resulting from $\mathbf{i}$ after certain currents have been cancelled due to the principles demonstrated in Figures 2.6(b,c). The incidence matrix $\mathbf{A}_i$ relating to the i-graph is constructed by the same rules as the usual incidence matrix (1.17) (see Section 1.2.2) but it applies to the i-graph instead of the full circuit topology. Hence, columns of $\mathbf{A}_i$ correspond to the currents $\mathbf{i}_i$ while rows are assigned to the possibly collapsed nodes. Finally, we obtain the reduced KCL including as many equations as the number of i-graph nodes (i.e. n_b minus a number of collapsed nodes)

$$\mathbf{A}_i \, \mathbf{i}_i = \mathbf{0} \,. \tag{2.24}$$

Now, we turn to KVL. As is known from Section 1.2.2 they are obeyed by all voltages established around loops of the circuit, as e.g. shown in Figure 2.7(a). The voltage-graph (v-graph) includes branches standing for branch voltages and nodes assigned to nodal voltages.

If there is a short-circuit between nodes A and B, then the voltage across it is 0 and hence in the v-graph the nodes A and B collapse into one node AB, as demonstrated in Figure 2.7(b). Thus, the KVL: $v_1+v_2+v_3+v_4=0$ reduces to: $v_2+v_3+v_4=0$ due to the short-circuit $v_1=0$.

In some cases also certain branch voltages may be of no interest (e.g. voltages across current sources) and so their elimination from equations is possible. Deletion of the branch v_1 (see Figure 2.7(c)) opens a loop including this branch and hence the KVL balance around this loop is disabled. Although, in this way the calculation of certain nodal voltages becomes impossible nevertheless the loop matrix $\mathbf{B}$ known from Section 1.2.2 (see (1.20), (1.21)) is reduced by a number of rows.

Thus, the v-graph arises by collapsing short-circuits and eliminating

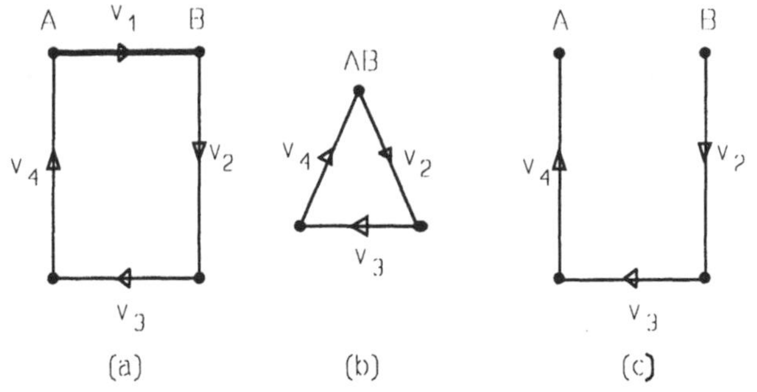

Figure 2.7 Creation of v-graphs: (a) an original structure, (b) collapsing two nodes while a voltage is zero, (c) deletion of a branch while a voltage is of no interest

those branches whose voltages are of no interest. It is characterized by the loop matrix $\mathbf{B}_\mathrm{v}$ and the incidence matrix $\mathbf{A}_\mathrm{v}$ similar to the matrices $\mathbf{B}$ and $\mathbf{A}$ known from Section 1.2.2 but related to the v-graph instead of the circuit topology. Branch voltages of the v-graph $\mathbf{v}_\mathrm{v}$ are close to $\mathbf{v}$ apart from positions eliminated due to the cases shown in Figures 2.7(b,c). They obey KVL

$$\mathbf{B}_\mathrm{v}\,\mathbf{v}_\mathrm{v} = \mathbf{0}\,. \tag{2.25}$$

The number of KVL equations amounts to the number of v-graph loops, that is to n_b minus the number of v-graph nodes (excluding the reference). Such KVL constitutes a v-graph nodal transformation

$$\mathbf{v}_\mathrm{v} = \mathbf{A}_\mathrm{v}^{t}\,\mathbf{v}_{n\mathrm{v}} \tag{2.26}$$

where $\mathbf{v}_{n\mathrm{v}}$ are the nodal voltages of the v-graph.

Application of the two-graph version of Kirchhoff's Laws (2.24), (2.25) slightly diminishes the number of equations and unknowns in canonical equations in comparison with the original number from Section 1.2.3. Instead of (1.23) the two-graph canonical equations may be written

$$\begin{aligned} &\mathbf{A}_\mathrm{i}\,\mathbf{i}_\mathrm{i} = \mathbf{0} \\ &\mathbf{B}_\mathrm{v}\,\mathbf{v}_\mathrm{v} = \mathbf{0} \\ &\mathbf{K}\,\mathbf{i}_\mathrm{i} + \mathbf{L}\,\mathbf{v}_\mathrm{v} = \mathbf{s} \end{aligned} \tag{2.27}$$

where the number of circuit variables is $2n_b$ minus a number of collapsed and eliminated branches in both graphs. The number of equations constituted by Kirchhoff's Laws is n_b minus a number of collapsed branches of the i-graph and eliminated branches of the v-graph. Hence, to obtain a unique canonical description, branch constitutive formulae for all branches except those collapsed in the v-graph and eliminated from the i-graph have been added. Savings are dependent on the number of collapsed and eliminated branches. After some algebra two-graph modified nodal equations may also be written "manually", like in Section 2.1.3.1.

2.1.4.2 Example of two-graph modified nodal equations

To demonstrate how the two-graph characterization of topology changes the modified nodal equations in practice, the circuit in Figure 2.8(a) has been considered. Two-graph modified nodal equations (TGMNE) are first composed of the balance of currents at the i-graph nodes. The i-graph is demonstrated in Figure 2.8(b). It arises by collapsing voltage sources V_1 and V_3, i.e. nodes 1 and 2 collapse into the reference 0. Hence, the i-graph will only include the nodes 3 and 4 where KCL balances have to be written.

To choose circuit variables the v-graph in Figure 2.8(c) has to be taken into account. It arises by collapsing the short-circuit between nodes 3 and 4 and by deleting the voltage across the current source K_{65} (assumed to be of no interest). Thus, we have a collapsed node 3+4 and hence three nodal voltages v_{n1}, v_{n2}, v_{n3+4} constitute the v-graph.

MNE have the unknowns vector extended with the additional currents of branches of class II (i.e. voltage defined and including controlling currents). In TGMNE these currents are selected from the i-graph. Thus we take the controlling current i_5, but neither i_1 nor i_3, collapsed in the i-graph, need be added anymore.

Finally, KCL at nodes 3, 4 of the i-graph is expressed in terms of the

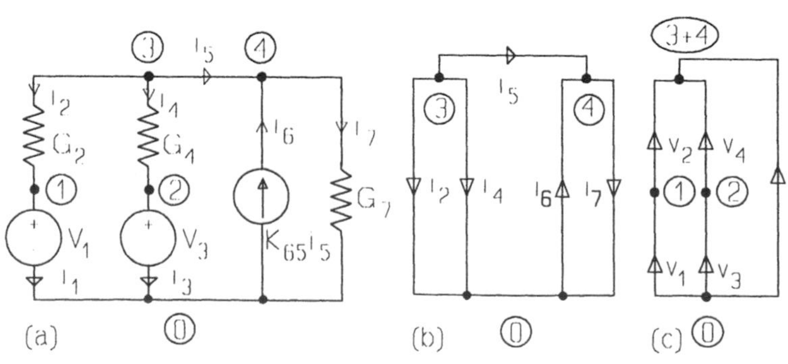

Figure 2.8 (a) A circuit acceptable by TGMNE, (b) its i-graph, (c) its v-graph

v-graph nodal voltages and the i-graph current variable i_5 of class II.

$$\begin{aligned} &G_2(v_{n1}-v_{n3+4})+G_4(v_{n2}-v_{n3+4})-i_5=0 \\ &-i_5-K_{65}i_5+G_7v_{n3+4}=0\,. \end{aligned} \tag{2.28}$$

Furthermore, certain branch equations of class II are added. All voltage defined branches and those including controlling currents are involved apart from the short-circuits (collapsed in the v-graphs) and possibly open circuits (deleted from the i-graphs). In our example we have the sources V_1 and V_3 only.

$$\begin{aligned} v_{n1}&=V_1 \\ v_{n2}&=V_3\,. \end{aligned} \tag{2.29}$$

As can be seen from the above discussion, if MNE were to be formulated a set would comprise 7 variables. Using TGMNE we have only 4 unknowns. Savings of one nodal voltage and the currents of V_1, V_3 have been achieved.

2.1.4.3 General formulation of two-graph modified nodal equations

The modified nodal equations introduced in Section 2.1.3 have been set up by dividing the branches into two classes. Class I consists of the current-defined branches whose currents had not been taken into the circuit variables vector due to their elimination from the equations. As for the voltage- or current-defined branches from class II their currents become the explicit unknowns in MNE for the circuit. The two-graph modified nodal equations (TGMNE) accept current-defined branches of the i-graph, that is $i_a=G_{aa}v_a+G_{ab}v_b+K_{ac}i_c+I_a$, where coefficients may not be zero simultaneously (open-circuits are excluded from the i-graphs). These branches constitute class I, and may be described altogether as follows

$$\mathbf{i}_{\mathrm{Ii}}=\mathbf{G}_{\mathrm{I}}\mathbf{v}_{\mathrm{v}}+\mathbf{K}_{\mathrm{I}}\mathbf{i}_{\mathrm{IIi}}+\mathbf{I}_{\mathrm{I}} \tag{2.30}$$

where the currents $\mathbf{i}_{\mathrm{Ii}}$ will not be selected as circuit variables. The above branches are possibly dependent on the v-graph branch voltages $\mathbf{v}_{\mathrm{v}}$ and the class II branch currents of the i-graph $\mathbf{i}_{\mathrm{IIi}}$.

Now, turning to the class II of TGMNE, it consists of voltage- and current-defined branches from the v-graph (hence short-circuits are excluded) selected due to their voltage-defined type, the flow of a controlling current or the user's will. Thus, voltage-defined branches

$v_a = R_{aa} i_a + T_{ab} v_b + R_{ac} i_c + V_a$ (where coefficients may not be simultaneously zero) or current-defined branches $i_a = G_{aa} i_a + G_{ab} v_b + K_{ac} i_c + I_a$ dependent on the v-graph voltages or the i-graph class II currents are acceptable. They may be written in the following implicit form

$$\mathbf{K}_{\text{IIv}} \mathbf{v}_{\text{v}} + \mathbf{L}_{\text{IIv}} \mathbf{i}_{\text{IIi}} = \mathbf{s}_{\text{IIv}} . \tag{2.31}$$

As for circuit variables they are composed of the v-graph branch voltages $\mathbf{v}_{\text{v}}$ (across all branches apart from collapsed short-circuits), and of the i-graph class II currents $\mathbf{i}_{\text{IIi}}$. The latter flow through the voltage-defined branches, except the collapsed voltage sources, or control other branches, or are taken to the class II due to the user's will (as mentioned in a description of class II).

Joining Kirchhoff's Laws (2.24), (2.25) with the branch sets (2.30), (2.31) the two-graph canonical equations (2.27) arise. Let $\mathbf{v}_{nv}$ be the v-graph nodal voltages. After KCL have been divided into two sections and the branch voltages $\mathbf{v}_{\text{v}}$ have been transformed to the nodal voltages by using the transformation (2.26) we obtain

$$\begin{aligned} &\mathbf{A}_{\text{Ii}} \mathbf{i}_{\text{Ii}} + \mathbf{A}_{\text{IIi}} \mathbf{i}_{\text{IIi}} = \mathbf{0} \\ &\mathbf{i}_{\text{Ii}} = \mathbf{G}_{\text{I}} \mathbf{A}_{\text{v}}^{t} \mathbf{v}_{nv} + \mathbf{K}_{\text{I}} \mathbf{i}_{\text{IIi}} + \mathbf{I}_{\text{I}} . \\ &\mathbf{K}_{\text{IIv}} \mathbf{A}_{\text{v}}^{t} \mathbf{v}_{nv} + \mathbf{L}_{\text{IIv}} \mathbf{i}_{\text{IIi}} = \mathbf{s}_{\text{IIv}} \end{aligned} \tag{2.32}$$

Elimination of $\mathbf{i}_{\text{Ii}}$ yields TGMNE in a succinct form

$$\begin{bmatrix} \mathbf{A}_{\text{Ii}} \mathbf{G}_{\text{I}} \mathbf{A}_{\text{v}}^{t} & \mathbf{A}_{\text{Ii}} \mathbf{K}_{\text{Ii}} + \mathbf{A}_{\text{IIi}} \\ \mathbf{K}_{\text{IIv}} \mathbf{A}_{\text{v}}^{t} & \mathbf{L}_{\text{IIv}} \end{bmatrix} \begin{bmatrix} \mathbf{v}_{nv} \\ \mathbf{i}_{\text{IIi}} \end{bmatrix} = \begin{bmatrix} -\mathbf{A}_{\text{Ii}} \mathbf{I}_{\text{Ii}} \\ \mathbf{s}_{\text{IIv}} \end{bmatrix} . \tag{2.33}$$

Let a circuit be composed of n_n nodes (except the reference) and n_{II} branches of class II (as involved in modified nodal equations). Within this set we assume n_V voltage sources to be collapsed and n_{sc} short-circuits. Hence, TGMNE are composed of $n_n - n_V$ KCL equations at the i-graph nodes expressed in terms of $n_n - n_{sc}$ v-graph nodal voltages (via branch equations) and also in terms of $n_{\text{II}} - n_V$ i-graph class II currents (explicitly). Furthermore, $n_{\text{II}} - n_{sc}$ equations of v-graph class II branches written in terms of the same variables defined above are added. The dimension of this set is equal to $n_n + n_{\text{II}} - n_V - n_{sc}$ and so the saving in comparison with MNE amounts to $n_V + n_{sc}$.

Let us stress that there is no exact correspondence between the KCL balance points and the nodal voltages. The currents taken to the vector of the unknowns are also not exactly the same as those that flow through branches whose equations have been added to the set. This feature causes an increased workload with the two-graph method in "manual" use.

2.1.4.4 Computer-aided formulation of two-graphs equations

The stamps techniques, as in the case of the modified nodal approach, may be applied for computer-aided generation of the two-graph modified nodal equations (TGMNE). Since there is not an exact correspondence between equations and variables of TGMNE (because of the separate topologies of the i-graph and v-graph) we have to introduce distinct enumerations of the equations and variables. Considering the numbering of variables let the reference voltage be obviously 0 and the nodal voltages $\mathbf{v}_{nv}$ of the v-graph and currents $\mathbf{i}_{\mathrm{IIi}}$ of the i-graph follow a continuous integer enumeration. These numbers will be column addresses to circuit matrix positions. In stamps we will denote them as $(\textit{node symbol})_{\mathrm{v}}$ and $(\textit{current symbol})_{\mathrm{i}}$. As for the enumeration of the equations, the KCL balances in the i-graph and the class II branch equations follow another continuous sequence of integer addresses. In the stamps those numbers are symbolically denoted as $(\textit{node symbol})_{\mathrm{i}}$ and $(\textit{current symbol})_{\mathrm{v}}$.

To demonstrate this way of numbering let us turn to the example of Figure 2.8(a). Circuit variables have to be chosen according to the following scheme: first nodal voltages of the v-graph v_{n1}, v_{n2}, v_{n3+4}, and then the class II currents of the i-graph i_5, take numbers ranging from 1 to 4. In terms of symbols used in the stamp borders these numbers have been denoted as: $(1)_{\mathrm{v}}$, $(2)_{\mathrm{v}}$, $(3{+}4)_{\mathrm{v}}$, $(\mathrm{i}_5)_{\mathrm{i}}$. As for the equations they are set up in the following order: two KCL balances at the i-graph's nodes 3 and 4 appear at positions 1 and 2, then two equations of the class II v-graph branches V_1, V_3 including the currents i_1, i_3 arise at the positions 3 and 4. Thus the symbols of positions in the stamps will be $(3)_{\mathrm{i}}$, $(4)_{\mathrm{i}}$, $(\mathrm{i}_1)_{\mathrm{v}}$, $(\mathrm{i}_3)_{\mathrm{v}}$.

The stamps of TGMNE corresponding to the branches of Figure 2.5 will be presented in what follows. Several types have been distinguished:

- **type i**: the current-defined branch whose current is not a circuit variable,
- **type ix**: the current-defined branch whose current is selected as a circuit variable,
- **type vr**: the voltage-defined branch of nonzero resistance,
- **type ve**: the voltage source (to be collapsed in the i-graph),
- **type vsc**: the short-circuit.

type i

	$(m_a)_v$	$(n_a)_v$	$(m_b)_v$	$(n_b)_v$	$(i_c)_i$	
$(m_a)_i$	G_{aa}	$-G_{aa}$	G_{ab}	$-G_{ab}$	K_{ac}	$-I_a$
$(n_a)_i$	$-G_{aa}$	G_{aa}	$-G_{ab}$	G_{ab}	$-K_{ac}$	I_a
			$\mathbf{Y}^a$			$\mathbf{b}^a$

type vsc

	$(i_a)_i$	
$(m_a)_i$	1	
$(n_a)_i$	-1	
	$\mathbf{Y}^a$	$\mathbf{b}^a$

type vr

	$(m_a)_v$	$(n_a)_v$	$(m_b)_v$	$(n_b)_v$	$(i_a)_i$	$(i_c)_i$	
$(m_a)_i$					1		
$(n_a)_i$					-1		
$(i_a)_v$	-1	1	T_{ab}	$-T_{ab}$	R_{aa}	R_{ac}	$-V_a$
				$\mathbf{Y}^a$			$\mathbf{b}^a$

type ve

	$(m_a)_v$	$(n_a)_v$	$(m_b)_v$	$(n_b)_v$	$(i_c)_i$	
$(i_a)_v$	-1	1	T_{ab}	$-T_{ab}$	R_{ac}	$-V_a$
			$\mathbf{Y}^a$			$\mathbf{b}_a$

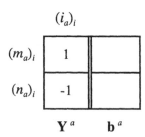

type ix

	$(m_a)_v$	$(n_a)_v$	$(m_b)_v$	$(n_b)_v$	$(i_a)_i$	$(i_c)_i$	
$(m_a)_i$					1		
$(n_a)_i$					-1		
$(i_a)_v$	G_{aa}	$-G_{aa}$	G_{ab}	$-G_{ab}$	-1	K_{ac}	$-I_a$
				$\mathbf{Y}^a$			$\mathbf{b}^a$

TGMNE can be found also in works of Singhal and Vlach. [2.2, 2.3]

2.1.5 Tableau equations

Tableau equations have been developed by G. D. Hachtel, R. K. Brayton, and F. G. Gustavson. [2.4] This style of circuit description is orientated towards specific, extremely efficient, sparse techniques for solving linear sets of equations. The sparse matrix technology, covered in Chapter 3, takes advantage of the existence of a low percentage of non-zero matrix elements. This feature considerably reduces a number of nontrivial (i.e. nonzero) operations during the solution process. Hence, a crucial point is to provide a rather small number of nonzero entries but not necessarily a small matrix size. In the case of tableau equations large matrix dimensions are even intentional. Provided a very economical sparsity-exploiting solution algorithm is used, advantage may be taken of the tableau equations, though they are considerably extended in size even in comparison with the canonical equations.

2.1.5.1 The essence of tableau equations. Example

First, let us consider an algebraic case, i.e. the linear circuit (2.3). The nodal transformation (1.28) forces branch voltages to obey KVL. Hence, by involvement of the nodal voltages $\mathbf{v}_n$ the KVL equations and the branch voltages $\mathbf{v}$ may be fully eliminated from (2.3) to obtain a so-called transformed canonical set for further reductions towards the NE or MNE forms. Such an approach has been applied so far, but another partial elimination is also possible. It consists of deleting the KVL equations but leaving the branch constitutive formulae unchanged. Hence, due to the explicit voltages $\mathbf{v}$ left in the branch expressions, the need arises for the transformation (1.28) which has to be added to the set of equations (for combining $\mathbf{v}$ with $\mathbf{v}_n$). Thus, instead of (2.3), the following circuit equations will be obtained

$$\begin{aligned} \mathbf{A}\,\mathbf{i} &= \mathbf{0} \\ \mathbf{A}^t\mathbf{v}_n - \mathbf{v} &= \mathbf{0}\,. \\ \mathbf{K}\,\mathbf{i} + \mathbf{L}\,\mathbf{v} &= \mathbf{s}(t) \end{aligned} \tag{2.34}$$

This formula is already very close to the tableau set. It consists of KCL balance equations at all nodes except the reference, the nodal transformation equations, and all the branch constitutive expressions. Turning to Figure 2.9 and recalling the above reasoning subsequent stages of the tableau system formulation will be demonstrated.

First, KCL yields the expressions

$$\begin{aligned} i_1 + i_2 &= 0 \\ -i_2 + i_3 + i_4 &= 0\,. \\ -i_4 + i_5 &= 0 \end{aligned} \tag{2.35}$$

Then, we add the nodal transformation formulae

$$\begin{aligned} v_1 &= v_{n1} \\ v_2 &= v_{n1} - v_{n2} \\ v_3 &= v_{n2} \\ v_4 &= v_{n2} - v_{n3} \\ v_5 &= v_{n3} \end{aligned} \quad . \tag{2.36}$$

Now, the branch equations expressed in terms of their branch variables can be written. For resistive branches we simply set up the formulae

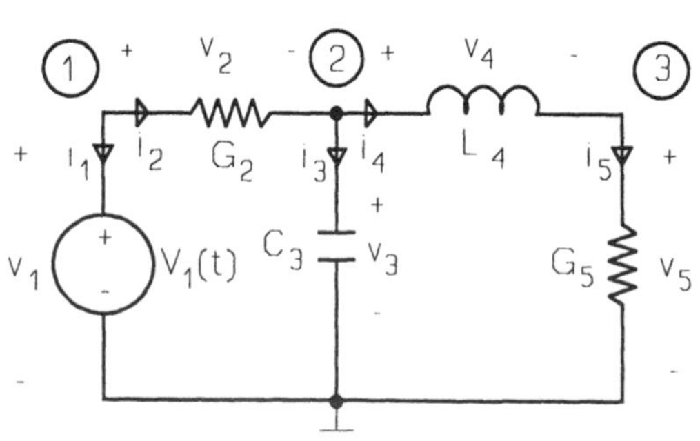

Figure 2.9 A simple linear circuit

$$\begin{aligned} &v_1 = V_1 \\ &i_2 - G_2 v_2 = 0 \; . \\ &i_5 - G_5 v_5 = 0 \end{aligned} \tag{2.37}$$

Once we turn to capacitive and inductive branches the tableau approach yields additional charge and flux type variables. Each branch description is therefore divided into two parts: a differential expression for the current or voltage and an algebraic one for the charge or flux. In the case of C_3 and L_4 the former is

$$i_3 - \frac{dq_3}{dt} = 0, \qquad v_4 - \frac{d\psi_4}{dt} = 0 \tag{2.38}$$

where the charge q_3 and the flux ψ_4 cen be formalized in an algebraical manner taking into account the initial values q_{30} and ψ_{40}

$$\begin{aligned} &q_3 - C_3 v_3 = q_{30} \\ &\psi_4 - L_4 i_4 = \psi_{40} \; . \end{aligned} \tag{2.39}$$

Although we are not familiar yet with time domain analysis, we have to use some topics relating to it. Such a need arises because the tableau formulation makes use of a differential form of branches in the transient analysis. Numerical techniques provided later on transform instantaneous derivatives into an algebraic form by means of, so-called, differentiation formulae. For example, the derivative of x at instant t may be substituted by the implicit Euler formula $dx/dt = (1/h)(x - x_0)$ where h is an integration step while x_0 is a value of the signal x at the previous instant $t - h$. Eliminating in this way the derivatives in (2.38) we obtain instead the algebraic equations:

$$\begin{aligned} &i_3 - (1/h) C_3 v_3 = -(1/h) q_{30} \\ &v_4 - (1/h) L_4 i_4 = -(1/h) \psi_{40} \quad . \end{aligned} \tag{2.40}$$

The equations (2.35)-(2.37), (2.39) and (2.40) taken together yield a tableau set rewritten below in matrix form

v_{n1}	v_{n2}	v_{n3}	i_1	i_2	i_3	i_4	i_5	v_1	v_2	v_3	v_4	v_5	q_3	ψ_4			
			1	1											v_{n1}		
				-1	1	1									v_{n2}		
						-1	1								v_{n3}		
-1								1							i_1		
-1	1								1						i_2		
	-1									1					i_3		
	-1	1									1				i_4		
		-1										1			i_5	=	
								1							v_1		V_1
				-1					G_2						v_2		
										C_3			-1		v_3		$-q_{30}$
						L_4								-1	v_4		$-\psi_{40}$
							-1					G_5			v_5		
					1								$-1/h$		q_3		$-q_{30}/h$
						1								$-1/h$	ψ_4		$-\psi_{40}/h$

The tableau equations, similarly to the previously introduced styles of circuit description, may be formalized in a general compact form. They are composed of four categories of equations:

(KCL) n_n Kirchhoff's Current Law expressions written in terms of the branch currents $\mathbf{i}$, namely $\mathbf{A}\mathbf{i}=\mathbf{0}$;

(NT) n_b nodal transformation formulae joining the branch and nodal voltages, namely $\mathbf{A}^t\mathbf{v}_n-\mathbf{v}=\mathbf{0}$;

(ABE) n_b algebraical branch equations consisting of (i) the voltage-defined (2.14) or current-defined (2.13) constitutive formulae for resistive branches as well as (ii) charge and flux constitutive expressions defining capacitors and inductors

$$\begin{aligned} q_a &= C_{aa}v_a + C_{ab}v_b + q_{a0} \\ \psi_a &= L_{aa}i_a + L_{ac}i_c + \psi_{a0}\,. \end{aligned} \tag{2.41}$$

Charges are in general voltage controlled while fluxes are current controlled. ABE have a matrix form $\mathbf{K}\mathbf{i}+\mathbf{L}\mathbf{v}+\mathbf{D}\mathbf{q}+\mathbf{P}\boldsymbol{\psi}=\mathbf{s}$.

(DE) Differential equations for capacitors $i_a=\mathrm{d}q_a/\mathrm{d}t$ and inductors $v_a=\mathrm{d}\psi_a/\mathrm{d}t$ that may be written altogether as $\mathbf{M}_C\mathbf{i}=\mathrm{d}\mathbf{q}/\mathrm{d}t$ and $\mathbf{M}_L\mathbf{v}=\mathrm{d}\boldsymbol{\psi}/\mathrm{d}t$, where zero-unit matrices involve selected capacitor currents and inductor voltages from full vectors of branch currents and voltages respectively: $\mathbf{i}_C=\mathbf{M}_C\mathbf{i}$, $\mathbf{v}_L=\mathbf{M}_L\mathbf{v}$. A proper time discretization at instant t and substitution of derivatives by means of the differentiation formulae (DF) transform branch differential equations into an algebraic form $\mathbf{M}_C\mathbf{i}-\gamma\mathbf{q}=-\gamma\mathbf{p}_q$, $\mathbf{M}_L\mathbf{v}-\gamma\boldsymbol{\psi}=-\gamma\mathbf{p}_\psi$. In general, operating on a variable x the linear DF (presented later on in detail) imposes an algebraic relation $\mathrm{d}x/\mathrm{d}t=\gamma(x-p_x)$ at instant t between a derivative and a variable by involving the quantity p_x representing the past history of x, i.e. a linear combination of values of x at a few instants preceding t). A number of DE amounts to a number of capacitors and inductors together (n_C+n_L).

Collecting these components of the tableau system the succinct formalization is yielded

$$\begin{array}{l}\text{KCL}\\ \text{NT}\\ \text{ABE}\\ \text{DE}_\text{C}\\ \text{DE}_\text{L}\end{array}\begin{bmatrix}\mathbf{0} & \mathbf{A} & \mathbf{0} & \mathbf{0} & \mathbf{0}\\ \mathbf{A}^t & \mathbf{0} & -\mathbf{1} & \mathbf{0} & \mathbf{0}\\ \mathbf{0} & \mathbf{K} & \mathbf{L} & \mathbf{D} & \mathbf{P}\\ \mathbf{0} & \mathbf{M}_C & \mathbf{0} & -\gamma\mathbf{1} & \mathbf{0}\\ \mathbf{0} & \mathbf{0} & \mathbf{M}_L & \mathbf{0} & -\gamma\mathbf{1}\end{bmatrix}\begin{bmatrix}\mathbf{v}_n\\ \mathbf{i}\\ \mathbf{v}\\ \mathbf{q}\\ \boldsymbol{\psi}\end{bmatrix}=\begin{bmatrix}\mathbf{0}\\ \mathbf{0}\\ \mathbf{s}\\ -\gamma\mathbf{p}_q\\ -\gamma\mathbf{p}_\psi\end{bmatrix} \qquad (2.42)$$

We may write it shortly as $\mathbf{Y}\,\mathbf{x}=\mathbf{b}$. Its size equals $n_n+2n_b+n_C+n_L$.

2.1.5.2 Stamps for computer formulation of tableau equations

As for previous nodal and modified nodal equations, the tableau matrices are superpositions of elementary branch contributions $\mathbf{Y}^a$, $\mathbf{b}^a$. Rules for formulating the tableau system yield four general stamps.

For the current-defined resistive branch in Figure 2.5(a) described by the formula (2.13) the following stamp may be written

		n_a	m_a	i_a	i_c	u_a	u_b	
KCL	n_a			1				
KCL	m_a			-1				
NT	b_a	1	-1			-1		
ABE	b_a			-1	K_{ac}	G_{aa}	G_{ab}	$-I_a$

$\mathbf{Y}^a$ $\qquad\qquad$ $\mathbf{b}^a$

The columns have been numbered by means of symbols of nodes according to Figure 2.5 and of current, voltage, charge or flux variables. The rows have symbolic numbers in terms of a name of the section (KCL, NT, ABE, DE) and a name of an equation within the section. KCL rows are identified by nodes, while rows of NT, ABE, DE are represented by branch symbols (e.g. b_a stands for the ath branch).

A resistive voltage defined branch of Figure 2.5(b) described by Equation (2.14) yields the stamp

		n_a	m_a	i_a	i_c	v_a	v_b	
KCL	n_a			1				
KCL	m_a			-1				
NT	b_a	1	-1			-1		
ABE	b_a			R_{aa}	R_{ac}	-1	T_{ab}	$-V_a$

The capacitive branch (2.41) results in the stamp

		n_a	m_a	i_a	v_a	v_b	q_a	
KCL	n_a			1				
KCL	m_a			-1				
NT	b_a	1	-1		-1			
ABE	b_a				C_{aa}	C_{ab}	-1	$-q_{a0}$
DE	b_a			1			$-\gamma$	$-\gamma p_{qa}$

Finally the inductive branch (2.41) yields

		n_a	m_a	i_a	v_a	i_c	ψ_a	
KCL	n_a			1				
KCL	m_a			-1				
NT	b_a	1	-1		-1			
ABE	b_a			L_{aa}		L_{ac}	-1	$-\psi_{a0}$
DE	b_a				1		$-\gamma$	$-\gamma p_{\psi a}$

Some advantages of the tableau description will be stressed now. Its variables, consisting of nodal and branch voltages, branch currents, charges and fluxes, are all available directly from the solution vector. If they were needed during the d.c. or transient analysis they could be quickly extracted from the solution with no additional calculations. The availability of charges and fluxes may be essential especially in certain implementations of time domain analysis.

The most important reason for developing the tableau method is the possibility of automatic selection between different orders of elimination of variables: equivalent in fact to the choice of circuit description form. As will be shown, the elimination order arises from automatic ordering techniques during the solution process. Common algorithms for deriving a solution to linear algebraic equations, known as gaussian elimination or LU method, are covered in Chapter 3. All arithmetic operations, whether or not the operands are zero, consume the same computer time. Thus, considerable savings may be achieved when sparse implementations of these algorithms are used that, in fact, consist of carrying out nonzero arithmetic operations only. For instance, the multiplications $0*a$ or subtractions a-0 may be easily neglected. To jump to topics in Chapter 3 notice that good numerical properties of the solution (low truncation error) occur when the proper reordering of rows and/or columns is involved. It consists of an intentional order of eliminating variables and equations from the system so as to optimize the magnitudes of, so-called, pivots and a number of fill-ins. The elimination relies on calculating a variable from one equation and substituting it into the others. After the few first variables have been eliminated the others yield a reduced system equivalent to a reduced circuit description. The whole reordering and elimination may be done in a fully numerical manner (using these pivotal and fill-ins criteria) or instead the introductory steps, when several variables are to be eliminated first, may be carried out analytically (hence gaining e.g. modified nodal equations). For instance, eliminating first the branch variables $\mathbf{v}$, $\mathbf{i}$, $\mathbf{q}$, $\boldsymbol{\psi}$ a reduced system results, drawn up in terms of the nodal voltages $\mathbf{v}_n$. It will be exactly the same as the well-known system of nodal equations (NE). Similarly, if all variables except class II currents and nodal voltages (see Section 2.1.3) are deleted the rest will form the modified nodal equations (MNE). The choice of a certain style of equations, in fact, for an intentional analytical elimination of some variables from the tableau system. The LU algorithm, including reordering applied to the tableau, performs such an elimination without any knowledge of the equation forms invented. It is only driven by the values of matrix elements and the numerical properties of a system. Hence, an efficient sparse procedure may select a "form" of circuit description dedicated exactly to data involved in a possibly better way than we could do for example by choosing MNE arbitrarily. The tableau approach, including an efficient sparse procedure, seems to be a generalized strategy for gaining an optimal style of circuit equations. Its efficiency lies only in an efficient algorithm for solving sparse systems.

In practice a weak point of existing sparse matrix methods is the need for a globally optimal reordering strategy. It is difficult, however, to predict, when the solution process is being started, what will be the best order from $n!$ possible permutations (n is the size of the system). Usually a decision due to the permutations is local at each step and so a local optimum only is yielded. In large systems arrived at by the tableau method the size of decision space for reordering is huge.

The number of nonzero elements and operations in tableau approach is not less than e.g. in MNE. Hence, to sum up our consideration, we can say that, in practice, the efficiency of the tableau approach is merely comparable to MNE and does not improve on it considerably. Certain examples may be shown where the tableau appears to be slightly better then MNE but MNE are more ubiquitous.

Other treatments of the tableau method of circuit description may be found in a paper by Hachtel *et al* [2.4] as well as in books [2.2, 2.3].

In addition, a trial two-graph tableau formulation has been proposed involving a reduction of certain equations thanks to i-u-graphs properties. [2.2, 2.3] This technique is of little practical importance however.

To sum up Section 2.1 and enable the reader to repeat its main topics the following problems can be formulated

- Compare sets of branches acceptable to NE, MNE, TGMNE and tableau forms of circuit equations.
- Try to characterize and compare sizes and sparsity of matrices produced by the above forms of circuit equations.
- What forms of equations are the easiest in "manual" formulation ?
- Invent a linear circuit composed of a variety of current- and voltage-defined branches and set up "manually" its MNE, TGMNE, tableau equations.
- Select from this circuit a small subcircuit built of a few elements. Set up a symbolic notation for nodes and current variables and draw up a stamp for this subcircuit by means of the equations created in the preceding item. Try to identify the types of the branches involved and compare your result with a combination of general stamps.
- Explain the sense of viewing the tableau method as a generalization of other forms of circuit equations where the problem of selecting the best form of equations has been moved to the pivoting stage.

2.2 Frequency domain a.c. circuit analysis

2.2.1 A.c. analysis of linear circuits

This section covers the responses of linear circuits subject to sinusoidal excitations at a single frequency. These circuits are composed of such branches as resistances, conductances, capacitors, inductors, controlled and independent sources, that may be characterized, in general, as linear branches (2.1). The algebraic-differential first order canonical equations (2.2) succinctly describe the circuits involved. Assuming sinusoidal excitations at an angular frequency ω and certain initial conditions (voltages across capacitors and currents through inductors or, in general, values of all variables being differentiated) a circuit response appears to be a sum of two components: transient and impressed. A transient component is an exponential response to the initial conditions when excitations are zero. An impressed component is a sinusoidal response to the excitations when zero initial conditions are taken. In nonautonomous (i.e. not generating) circuits a transient response dies away after some time. Thus, a steady state amounts to an impressed component which may be proved to be sinusoidal of the same frequency ω as the excitations but with a certain magnitude and phase shift (delay) with respect to them. This steady-state response to sinusoidal excitations is known as the **alternating current (a.c.) response**. Computer derivation of a.c. responses is called the **a.c. analysis**.

To demonstrate these properties of linear circuits let us consider the example of Figure 2.10. Its mathematical characterization is derived from KVL around the loop. Introducing an unknown current i, a voltage $L\,\mathrm{d}i/\mathrm{d}t$ across the inductor, and a voltage Ri across the resistor a KVL balance yields

$$L\frac{\mathrm{d}i}{\mathrm{d}t}+Ri=A_u\sin(\omega t)\,. \tag{2.43}$$

The reader familiar with linear differential equations could solve Equation (2.43) formally to obtain a sum of the exponential component (dependent on i_0, the initial value of i at instant zero) and the sinusoidal component

$$i(t)=i_0\exp(-\frac{R}{L}t)+\frac{E}{\sqrt{R^2+\omega^2L^2}}\sin(\omega t+\phi),\ \ \phi=-\arctan\frac{\omega L}{R} \tag{2.44}$$

The exponential transient signal is responsible for the unsteady state while the sinusoidal component is impressed by the source and is slightly shifted in phase. After several time constants $\tau=L/R$ the transient has died away and the purely sinusoidal signal, known as the a.c. response, is practically reached.

Usually, the a.c. analysis of linear circuits is carried out many times at distinct frequency points to obtain the amplitudes and phases of circuit outputs as functions of frequency. This, so-called, **frequency-domain a.c. analysis** yields frequency characteristics of amplitude and phase. Turning to the example of Figure 2.10 the amplitude takes the form

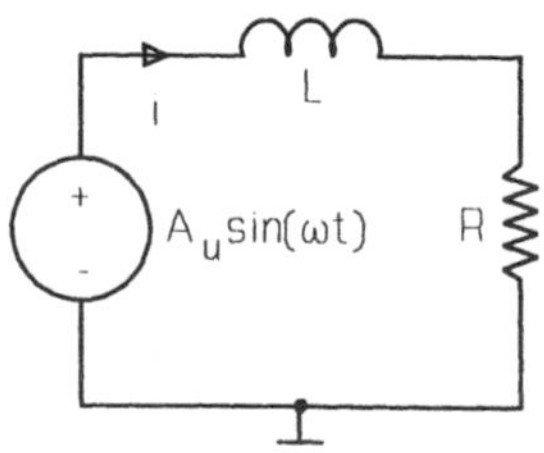

Figure 2.10 A simple RL circuit

$|i|=A_u/\sqrt{(R^2+\omega^2L^2)}$ while the phase $\phi=-\arctan(\omega L/R)$. This trivial analytical instance exhibits the connections of linear sinusoidally excited circuits with linear ordinary differential equations. In more complicated cases the differential approach is too time-consuming. Direct, well-known algebraic techniques for a.c. response derivation are used instead. They have been extensively applied in circuit theory for many years. Thanks to the sinusoidal character of a signal at a single frequency a phasor algebra could be developed. Once the original algebraic-differential equations have been transformed into algebraic phasor equations, the original circuit changes into a substitute immitance circuit. This technique is explained below.

2.2.1.1 Introduction to phasor algebra

In this section simple reasoning leads to fully algebraic techniques for calculating linear circuit responses to sinusoidal excitations. First of all let us notice that any linear differential equation having a sinusoidal right-hand side (i.e. excitation) implies an impressed solution component of a sinusoidal character. This fact arises from properties of the sine function whose derivatives all remain sine or cos functions. This mathematical observation reflects the physical properties of linear systems to exactly transmit such signals. Thus, provided circuit variables are sinusoidal then periodical factors appear to be negligible while the algebraic operations on amplitudes and phases remain. To explain the algebraic approach to a.c. response calculation let us recall the RL circuit of Figure 2.10. Assuming the excitation $A_u\sin(\omega t)$ the circuit response i may be predicted as $A_i\sin(\omega t+\phi)$. Placing this predictor in Equation (4.43) we obtain the formula

$$L\,A_i\omega\,\cos(\omega t+\phi)+R\,A_i\,\sin(\omega t+\phi)=A_u\sin(\omega t) \qquad (2.45)$$

satisfied for all t. Both sides of it can be differentiated to obtain

$$\begin{aligned}-L\,A_i\omega^2\,\sin(\omega t+\phi)+R\,A_i\omega\,\cos(\omega t+\phi)&=\\ =A_u\omega\cos(\omega t)&\end{aligned} \qquad (2.46)$$

also fulfilled at each instant t. Both (2.45) and (2.46) are satisfied if A_i and ϕ are the amplitude and phase of a signal that fulfil Equation (2.43). Hence, dividing (2.46) by ω and adding Equation (2.45) multiplied by imaginary unit j we obtain the complex equation

$$\begin{aligned} &L A_i \mathrm{j}\omega\,[\cos(\omega t+\phi)+\mathrm{j}\sin(\omega t+\phi)]+\\ &+R A_i\,[\cos(\omega t+\phi)+\mathrm{j}\sin(\omega t+\phi)]=\\ &=A_u\,[\cos(\omega t)+\mathrm{j}\sin(\omega t)]\,. \end{aligned} \tag{2.47}$$

As $\cos(\omega t+\phi)+\mathrm{j}\sin(\omega t+\phi)=\exp(\mathrm{j}\,\omega t)\exp(\mathrm{j}\,\phi)$ then (2.47) yields

$$\begin{aligned} &L A_i \mathrm{j}\,\omega \exp(\mathrm{j}\,\omega t)\exp(\mathrm{j}\,\phi)+\\ &+R A_i \exp(\mathrm{j}\,\omega t)\exp(\mathrm{j}\,\phi)=A_u \exp(\mathrm{j}\,\omega t)\,. \end{aligned} \tag{2.48}$$

Now a complex auxiliary variable $I=A_i\exp(\mathrm{j}\,\phi)$ can be introduced. After both sides have been divided by the quantity $\exp(\mathrm{j}\,\omega t)$ (nonzero) the time independent complex algebraic equation arises. In this the unknown variable I is a, so-called, phasor

$$(\mathrm{j}\,\omega L+R)\,I=A_u\exp(\mathrm{j}0)\,. \tag{2.49}$$

A phasor is a compact representation of a harmonic signal; its modulus stands for the amplitude of an oscillation, while its argument denotes a shift in phase with respect to an excitation. Thus, the simple algebraic equation (2.49) yields the solution $I=A_u/(\mathrm{j}\,\omega L+R)$, whose amplitude and phase are respectively: $|I|=A_u/\sqrt{(R^2+\omega^2L^2)}$ and $\phi=\mathrm{Arg}\,I=-\arctan(\omega L/R)$. They obviously equal the steady-state of (2.44) obtained from the differential approach.

The above consideration shows that due to the sinusoidal character of signals a phasor of a sinusoidal wave $x(t)$ can be seen as the linear transformation $[(\mathrm{d}x/\mathrm{d}t)/\omega+\mathrm{j}x]/\exp(\mathrm{j}\omega t)$. Its properties are as follows: the a.c. signal $x(t)=A_x\sin(\omega t+\phi)$ is mapped onto a phasor $X=A_x\exp(\mathrm{j}\,\phi)$, the linear expression $a\,x(t)+b\,y(t)$ is mapped onto $a\,X+b\,Y$ while moreover the derivative $\mathrm{d}x(t)/\mathrm{d}t$ transforms into $\mathrm{j}\omega X$. These rules constitute a phasor algebra.

Application of the phasor transformation to the circuit of Figure 2.10 results in Equation (2.49), i.e. maps the reactive circuit onto the algebraic one, the resistance $v=Ri$ changes into the phasor representation $V=RI$, while the inductance $v=\mathrm{d}\psi/\mathrm{d}t$, $\psi=Li$ yields the phasor model $V=\mathrm{j}\omega\Psi=\mathrm{j}\omega L I$ known as an impedance. Instead of handling the differential characterization (2.43) the phasor description (2.49) may be now considered. It is easily obtainable from KVL around the loop of Figure 2.10 using the phasor models of branches. This technique is a crucial point of the a.c. analysis.

2.2.1.2 Substitute immittance circuit

The a.c. analysis of linear circuits comprises a transformation of all circuit branches into phasor, that is immittance, models. This may be formally carried out by applying the phasor algebra constitutive rules, introduced in Section 2.2.1.1, to branch equations. It can readily be shown that a conductance $i=Gv$ transforms into the current defined branch $I=GV$. Similarly e.g. for CCVS $v_a=R_{ac}i_c$ yields $V_a=R_{ac}I_c$. For a capacitor $i=\mathrm{d}q/\mathrm{d}t$, $q=Cv$, and after the rule of phasor transformation of derivatives has been applied, the model $I=\mathrm{j}\omega Q=\mathrm{j}\omega CV$ is obtained, which is the admittance $I=YV$ with the coefficient $Y=\mathrm{j}\omega C$ representing the capacitance under a.c. signals. The sinusoidal voltage $v=A_u\sin(\omega t+\phi)$ or current $i=A_i\sin(\omega t+\phi)$ sources transform into constant phasor sources $V=A_u\exp(\mathrm{j}\phi)$ and $I=A_i\exp(\mathrm{j}\phi)$ respectively. After substitution of all branches of a linear circuit by proper immittance models a substitute immittance circuit arises.

To carry out the linear a.c. analysis, equations of a substitute immittance circuit have to be formulated first, for instance in a modified nodal equations (MNE) form. They have to be solved at each frequency point specified by a user. This will be demonstrated by means of the example of Figure 2.11(a). To generate an immittance circuit the voltage sinusoidal source transforms into the constant voltage source $V_1=A\exp(\mathrm{j}0)=A$, conductance G_4 is unchanged, inductor L_2 becomes the impedance $Z_2=\mathrm{j}\omega L_2$ and finally the capacitor C_3 yields the admittance $Y_3=\mathrm{j}\omega C_3$. Voltages and currents of the immittance circuit are phasors (complex numbers) corresponding to the original voltages and currents. The immittance circuit obtained is drawn in Figure 2.11(b). It is of the same topology as the original circuit. Complex MNE created according to Section 2.1.3 take the following form:

$$\begin{bmatrix} 0 & 0 & 1 & 1 \\ 0 & \mathrm{j}\omega C_3+G_4 & 0 & -1 \\ 1 & 0 & 0 & 0 \\ -1 & 1 & 0 & \mathrm{j}\omega L_2 \end{bmatrix} \begin{bmatrix} V_{n1} \\ V_{n2} \\ I_1 \\ I_2 \end{bmatrix} = \begin{bmatrix} 0 \\ 0 \\ V_1 \\ 0 \end{bmatrix}. \tag{2.50}$$

After they have been solved by means of methods described fully in Chapter 3 the complex unknowns are obtained. From each phasor signal the amplitude arises by evaluating the magnitude, while phase is derived from the argument of a complex number.

Techniques for generating the immittance equations may be drawn up

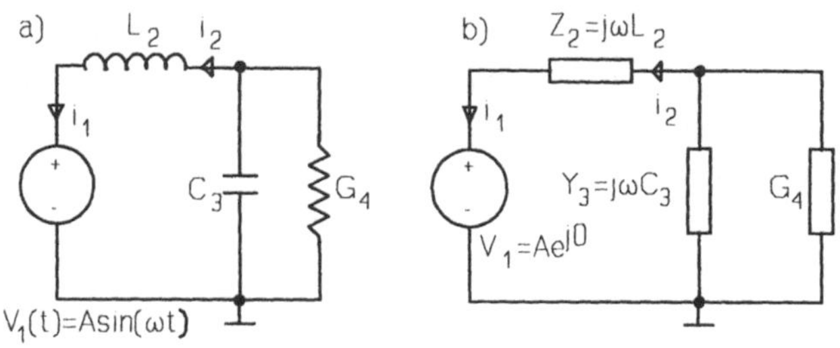

Figure 2.11 a) A linear circuit and b) its substitution immitance model for a.c. analysis

in a slightly generalized form. Let a linear circuit have the canonical equations (2.2), with zero initial charge and flux values $\mathbf{q}_0$, ψ_0 (because of the steady a.c. state). Applying the phasor transformation to both sides a set of linear complex equations yields

$$\begin{aligned} &\mathbf{A}\,\mathbf{I} = \mathbf{0} \\ &\mathbf{B}\,\mathbf{V} = \mathbf{0} \\ &(\mathbf{K}_1 + j\omega \mathbf{K}_2 \mathbf{L})\,\mathbf{I} + (\mathbf{L}_1 + j\omega \mathbf{L}_2 \mathbf{C})\,\mathbf{V} = \mathbf{S} \end{aligned} \tag{2.51}$$

with the phasor branch variables **I**, **V** involved. As can be seen, this system can be interpreted as the canonical description of a certain circuit, structurally identical with the original circuit. Its signals are all phasors of original signals while branches derive from the original branches by means of the phasor transformation. Hence, they are immittance counterparts of the differential originals. For the circuit of Figure 2.11(a) such an immittance substitute circuit has been shown in Figure 2.11(b). Proper formulation of the immittance models is the essence of a.c. analysis.

To show, in general, how immittance models are derived from circuit branches let us consider a linear circuit composed of current- or voltage-defined branches, independent, voltage or current controlled. The *a*th current-defined branch takes the general form (including conductances, capacitances, and current sources)

$$i_a = G_{aa} v_a + \hat{G}_{aa} \frac{\mathrm{d}v_a}{\mathrm{d}t} + G_{ab} v_b + \hat{G}_{ab} \frac{\mathrm{d}v_b}{\mathrm{d}t} + K_{ac} i_c + \hat{K}_{ac} \frac{\mathrm{d}i_c}{\mathrm{d}t} + j_a \tag{2.52}$$

while the *a*th voltage-defined branch is similar

$$v_a = R_{aa} i_a + \hat{R}_{aa} \frac{\mathrm{d}i_a}{\mathrm{d}t} + T_{ab} v_b + \hat{T}_{ab} \frac{\mathrm{d}v_b}{\mathrm{d}t} + R_{ac} i_c + \hat{R}_{ac} \frac{\mathrm{d}i_c}{\mathrm{d}t} + e_a \tag{2.53}$$

and covers resistances, inductors and voltage sources. After a phasor transformation has been applied to the current-defined branch its immittance counterpart is yielded

$$I_a = (G_{aa} + \mathrm{j}\omega \hat{G}_{aa}) V_a + (G_{ab} + \mathrm{j}\omega \hat{G}_{ab}) V_b + \\ + (K_{ac} + \mathrm{j}\omega K_{ac}) I_c + J_a . \tag{2.54}$$

Similarly, the voltage-defined branch transforms into

$$V_a = (R_{aa} + \mathrm{j}\omega \hat{R}_{aa}) I_a + (T_{ab} + \mathrm{j}\omega \hat{T}_{ab}) V_b + \\ + (R_{ac} + \mathrm{j}\omega \hat{R}_{ac}) I_c + E_a \tag{2.55}$$

where V_a, I_a, V_b, I_c are phasors of the signals v_a, i_a, v_b, i_c, while J_a, E_a are phasors of sinusoidal excitations j_a, e_a.

The immittance branches (2.54), (2.55) appear to be of the standard algebraic form (2.13), (2.14) accepted by MNE. Hence, equations may be formulated by means of the stamps introduced in Section 2.1.3.4. Further realization of the a.c. analysis is given below.

2.2.1.3 Algorithm for frequency domain analysis

The numerical a.c. analysis needs an immittance circuit to be formulated (e.g. in MNE style) at each discrete frequency ω and then solved by means of techniques introduced in Chapter 3. These two stages are simply repeated within a discretized frequency loop. The only problem specific to the a.c. analysis therefore is to have a proper substitute immittance model. The form of the equations (NE, MNE, tableau) as well as their solution techniques are not the essence of our analysis though they decide whether it is efficient and accurate or not. After the immittance equations have been solved at any particular frequency a complex (phasor) response vector **X** of all circuit variables is worked out. It is useful for evaluating a complex signal X_{out} at a user defined output that may be a voltage between two nodes $(X(node+) - X(node-))$, a current $(X(current\_variable\_number))$ selected as a circuit variable or any other current derived via branch voltages and branch constitutive equations. A user usually needs an output amplitude i.e. the modulus $|X_{\text{out}}|$ and the phase $\operatorname{Arg} X_{\text{out}}$ obtainable by means of a complex argument measured with respect to the zero phase of the excitations. The phase is within the range $(-\pi,\pi]$. If a single, unit excitation is provided the output response represents a complex transmittance. Its modulus is known as a gain while its argument is a shift in phase. The importance of maintaining an accurate representation of gain, its positive values, and a wide frequency range encourages us to adopt a logarithmic scale. For example, a passive filter may easily exhibit gains ranging from 10^{-6} to 1 over a frequency range in which we are interested. The logarithmic scale commonly employed is that of gain in decibels (dB), where $|X_{out}|_{dB} = 20\log_{10}|X_{out}|$. Phase is usually displayed on a linear scale, however.

The a.c. analysis is carried out at certain frequency points chosen within a user-defined range $[f_{min}, f_{max}]$ (where $f = \omega/(2\pi)$). This so-called frequency domain analysis is used to test how the amplification and phase shifting properties of linear circuits depend on frequency. The resulting frequency response characteristics play an important role in testing the bandwidth, cut-off frequencies, selectivity and quality factors of filters or attenuators. When checking these properties within narrow frequency ranges (e.g. in a resonant circuit case) a linear frequency scale is appropriate. To obtain a sequence of *np* linearly spaced discrete values f_i within the frequency range, the rule: $f_{i+1} = f_i + (f_{max} - f_{min})/(np-1)$ may be used. To yield a logarithmic scale the formula $f_{i+1} = f_i(f_{max}/f_{min})^{1/(np-1)}$ should be employed instead. After the circuit has been analyzed at the chosen frequency values a sequence of complex (phasor) responses is obtained and may be stored. Three systems of displaying the results of the frequency analysis are in use. In one, the amplitude and phase functions $|X_{out}(f)|$, $\mathrm{Arg}\, X_{out}(f)$ are known as the Bode plots. Another system displays the complex X_{out} response parametrized versus f on the complex ($\mathrm{Re}\, X_{out}$, $\mathrm{Im}\, X_{out}$) plane; this polar plot is known as a Nyquist diagram. Finally, a Nichols plot portrays the complex value X_{out} in a rectangular plane with coordinates of amplitude and phase. These results may easily be produced after the analysis has been terminated by means of a graphical postprocessor, that makes use of the complex results stored (e.g. in a real-imaginary part form) in a binary data disk file. An example of this is the use of PROBE to portray results from PSPICE simulations and NODE-VIEWER to portray results from OPTIMA simulations.

To gain some familiarity with practical aspects of the frequency domain analysis we can test these two simulators. In \SPICE directory of Diskette A we have several linear circuits: elliptic filter S2-ELIP.CIR, active filter S2-COR.CIR and quartz filter S2-PIEZO.CIR, etc. Their counterparts for OPTIMA are in \OPTIMA directory. Let us more extensively consider the elliptic 5th order RLC filter S2-ELIP.CIR demonstrated in Figure 2.12. Its net-lists for PSPICE and OPTIMA simulators are listed below. The net-list for OPTIMA is saved as O2-ELIP.CIR. Text and binary results S2-ELIP.OUT, S2-ELIP.DAT obtained by PSPICE and s2-ELIP.RES, S2-ELIP.DAT by OPTIMA are also included.

```
S2-ELIP.CIR
* FIFTH ORDER ELLIPTIC FILTER
VINP 1 0 AC 1

R1 1 2 10K
C1 2 0 29.7N
L1 2 3 1.43
C2 2 3 5.28N
C3 3 0 34.9N
L2 3 4 0.987
C4 3 4 15.5N
C5 4 0 23.3N
R2 4 0 10K

.PROBE
.AC DEC 100 100 10K
.END
```

```
O2-ELIP.CIR
# FIFTH ORDER ELLIPTIC FILTER
VINP 1 0 MINP MAGN 1
.MODEL MINP SRC
R1 1 2 10K
C1 2 0 29.7N
L1 2 3 1.43
C2 2 3 5.28N
C3 3 0 34.9N
L2 3 4 0.987
C4 3 4 15.5N
C5 4 0 23.3N
R2 4 0 10K
!
.PROBE AC V(4)
.AC DEC 100 100 10K
.END
```

Readers may carry out these simulations themselves or handle the existing results by means of available postprocessors. In Figures 2.13, 2.14 frequency characteristics of amplitude VDB(4) (in logarithmic scale) and of phase VP(4) have been displayed. We can easily observe low-pass properties with a cut-off frequency 1kHz and attenuation outside the band better then 50dB. The slope in the vicinity of 1kHz is about -134 dB per decade.

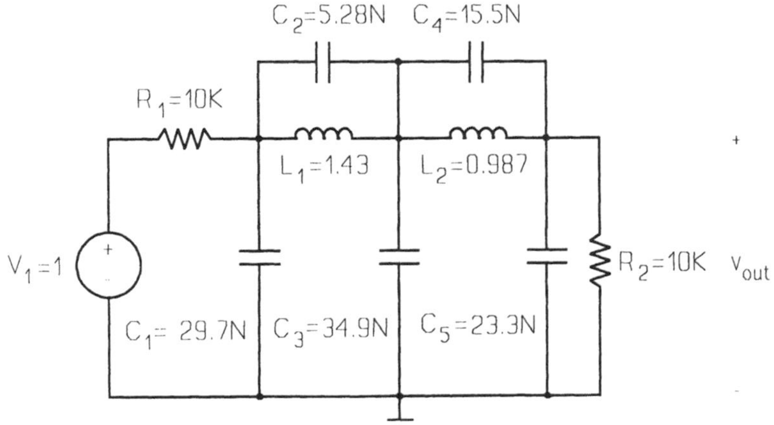

Figure 2.12 A 5th order elliptic filter

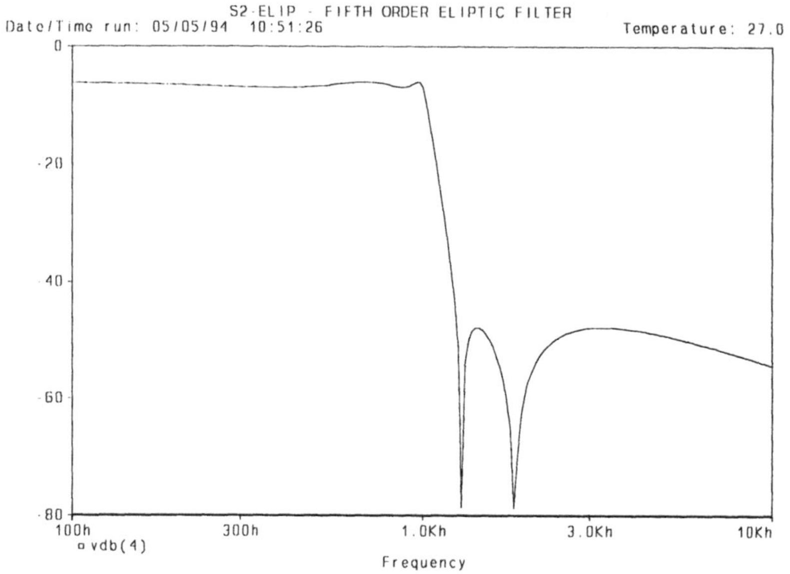

Figure 2.13 A Bode amplitude plot of the filter from Figure 2.12

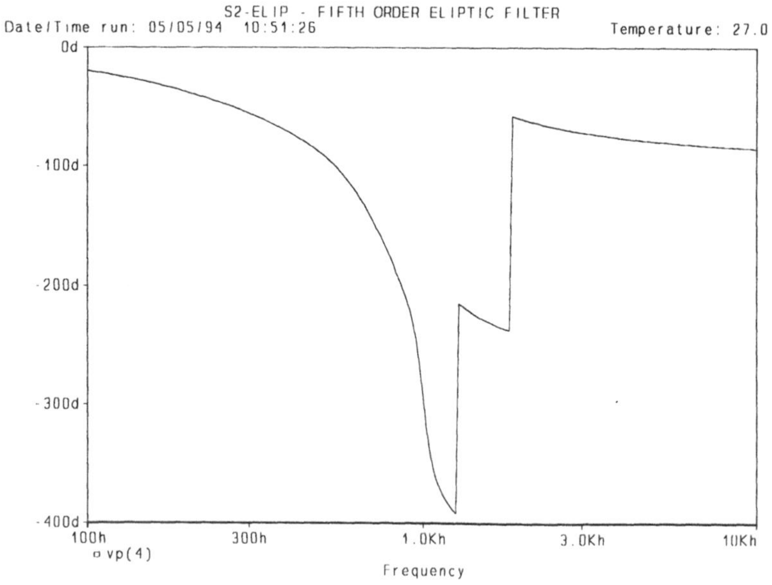

Figure 2.14 A Bode phase plot of the filter from Figure 2.12

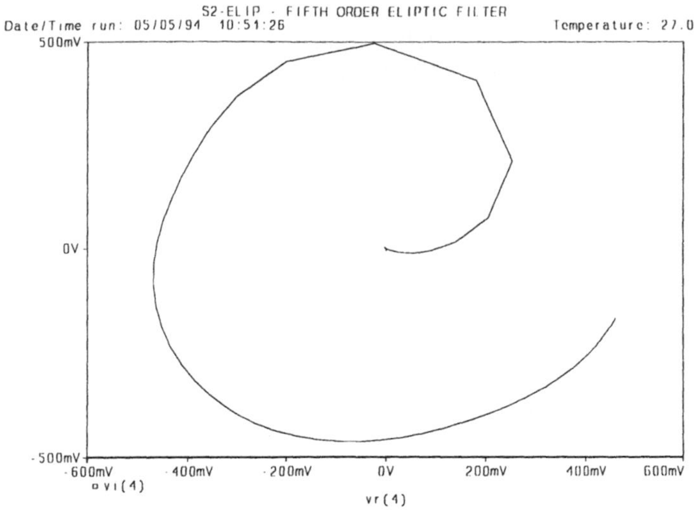

Figure 2.15 A Nyquist plot of the filter from Figure 2.12

Postprocessors may also be used to observe the other display styles. Assigning a real part VR(4) of the response V(4) to the X-axis and an imaginary part VI(4) to the Y-axis the Nyquist polar plot appears on the PROBE's screen shown in Figure 2.15. Similar effects are obtained from NODE-VIEWER while observing the quantities RE:V(4) and IM:V(4) on X and Y axes respectively. Selecting amplitude VM(4) on the Y-axis and

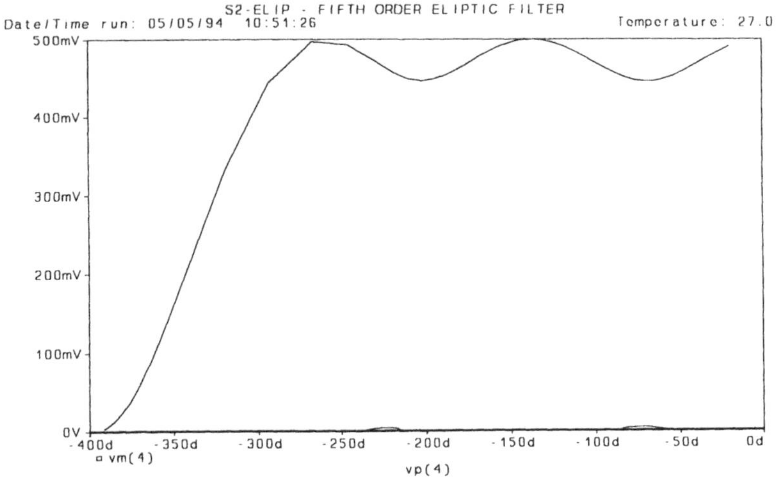

Figure 2.16 A Nichols plot of the filter from Figure 2.13

phase VP(4) on the X-axis the Nichols plot of Figure 2.16 arises. These examples demonstrate the common use of a.c. analysis in practice.

To get a better familiarity with frequency domain analysis let us finish this section with Algorithm 2.1 that outlines the computational structure.

Algorithm 2.1 Frequency domain analysis of linear circuits

> **Data**: *logarithmic* or *linear* frequency scale,
> *np* (≥2) points per range of analysis, i.e.
> from *freq1* to *freq2* (≥*freq1*);
> **if** *linear* **then** *aux* = (*freq2*-*freq1*)/(*np*-1);
> **else** *aux* = (*freq2*/*freq1*)^(1/(*np*-1));
> *freq* = *freq1*;
> **for** *i* = 1 **to** *np* **do**
> **begin**
> formulation of a complex (phasor) matrix and right-hand side **Y**, **b** of an immitance circuit at the radial frequency 2π**freq*;
> LU decomposition of **Y**;
> **if** (**Y** is singular) **then stop** {alarm exit};
> solution of **Y X** = **b** via forward and backward substitution;
> evaluation and storage of complex responses on circuit outputs from the complex circuit variables vector ***X***;
> **if** *linear* **then** *freq* = *freq*+*aux* **else** *freq* = *aux***freq*;
> **end**

The formulation of the complex immittance equations is realized in practice by means of the stamps techniques introduced in this chapter, according to the appropriate form of formulation (Nodal, Modified Nodal etc.). Stamps are obviously applied to immittance models (2.54), (2.55) of the original branches (2.52), (2.53). These models have the form (2.13), (2.14) accepted by MNE.

Frequency analysis may be found in many books (e.g. [2.3, 2.5]).

2.2.2 Small-signal frequency analysis of nonlinear circuits

2.2.2.1 The concept of a small-signal incremental analysis

Analog nonlinear circuits usually contain constant sources (say $\mathbf{s}_0$) known as direct current (d.c.) excitations. These impress a constant d.c. steady state called the d.c. response. In most cases the d.c. response is not the output signal of interest to the user. Circuit performance is related rather to changes in instantaneous signals subject to variable excitations.

The d.c. response only establishes the background to these changes. Assuming instantaneous branch signals $\mathbf{i}(t)$, $\mathbf{v}(t)$, and d.c. responses $\mathbf{i}_0$, $\mathbf{v}_0$ the quantities $\hat{\mathbf{i}}(t)=\mathbf{i}(t)-\mathbf{i}_0$, $\hat{\mathbf{v}}(t)=\mathbf{v}(t)-\mathbf{v}_0$ are called the incremental components of the responses. The incremental response is impressed by the incremental components of the excitations $\hat{\mathbf{s}}(t)=\mathbf{s}(t)-\mathbf{s}_0$ (the notation corresponds to the canonical description (2.2)). In practical circuits with continuous branch characteristics the smaller one makes the incremental excitations the smaller will be incremental responses. Once incremental signals are small, an instantaneous response slightly varies in a d.c. response vicinity. Thus arises the concept of piecewise linearization for incremental signals in the immediate vicinity of a particular d.c. response level.

It can be proved that for a continuous differentiable function $f(x)$, at any point x where $f'(x)\neq 0$, as the increment Δx is made progressively smaller so the differential increment $f'(x)\Delta x$ approaches closer to a true increment $\Delta f(x)=f(x+\Delta x)-f(x)$. In fact, while expanding the function as a Taylor series the relative discrepancy between true and differential increments (let us say a nonlinearity index) approaches zero as Δx is taken towards zero

$$\frac{\Delta f(x)-f'(x)\Delta x}{\Delta f(x)}=$$
$$=\frac{[f'(x)\Delta x+\frac{1}{2}f''(x)\Delta x^2+\frac{1}{6}f'''(x)\Delta x^3+\ldots]-f'(x)\Delta x}{f'(x)\Delta x+\frac{1}{2}f''(x)\Delta x^2+\frac{1}{6}f'''(x)\Delta x^3+\ldots}\underset{\Delta x\to 0}{\to}0. \tag{2.56}$$

Thus, the smaller are the incremental excitations the better are the linear approximations of branches at d.c. bias points, that is the better is the reflection of true signal increments. Many analog systems, e.g. amplifiers, rely in operation on processing small incremental signals. The influence of nonlinear branches appears insignificant and hence consideration of the asymptotic case $\Delta x\to 0$ is quite reasonable. The evaluation of incremental responses from this asymptotic state properly describes real systems for sufficiently small incremental excitations. Nonlinear circuits of nearly linear properties for incremental signals are known as quasi-linear circuits. In practical analysis these small incremental excitations are taken to be sinusoidal and therefore the a.c. steady-state in a linearized circuit is considered. Its solution is known as small-signal a.c. (or frequency domain) analysis. Provided a.c. incremental excitations are given, then amplitude and phase characteristics can be obtained by applying the phasor methods to the linearized circuit.

Some introductory concepts in the small-signal a.c. analysis of nonlinear circuits have been already demonstrated in Section 1.1.2. We could recall the circuit of Figure 1.2(a) and study once more its small-signal phasor characterization (1.5) yielding the response (1.6). To demonstrate more details of small-signal circuit operation let us consider the simple MOSFET amplifier of Figure 2.17(a). Applying the Schichman-Hodges model of n-channel MOSFET with polarization $V_{DS}>0$ and capacitances neglected

$$I_D = \begin{cases} \text{cut-off region:} \\ 0, \quad \text{for } V_{GS}<V_T \\ \\ \text{linear region:} \\ \dfrac{\gamma W}{L}(V_{GS}-V_T-\dfrac{V_{DS}}{2})\,V_{DS}, \quad \text{for } V_{GS}\geq V_T \text{ and } 0\leq V_{DS}\leq V_{GS}-V_T \\ \\ \text{saturation region:} \\ \dfrac{\gamma W}{2L}(V_{GS}-V_T)^2, \quad \text{for } V_{GS}\geq V_T \text{ and } V_{DS}>V_{GS}-V_T \end{cases} \tag{2.57}$$

we get the circuit of Figure 2.17(b) containing a controlled nonlinear current source. In Figure 2.18 a solution is shown in the $I_D(V_{DS})$ plane for the following parameters: γ is a gain factor (50 μA/V^2), V_T is a threshold voltage (1.72 V), W and L are channel width and length respectively (2 μm and 13 μm). The resistive load R_L=200 kΩ is represented by a straight line of slope $-1/R_L$. It crosses the MOSFET characteristic at the d.c. bias point P. In Figure 2.18 a dotted plot of the function $I_D=\gamma W(V_{DS}-V_T)^2/(2L)$ separates linear and saturation regions, i.e. corre-

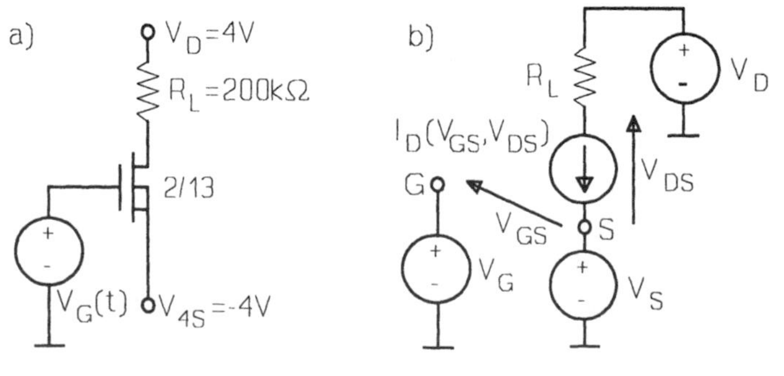

Figure 2.17 (a) A simple MOSFET amplifier and (b) its equivalent circuit

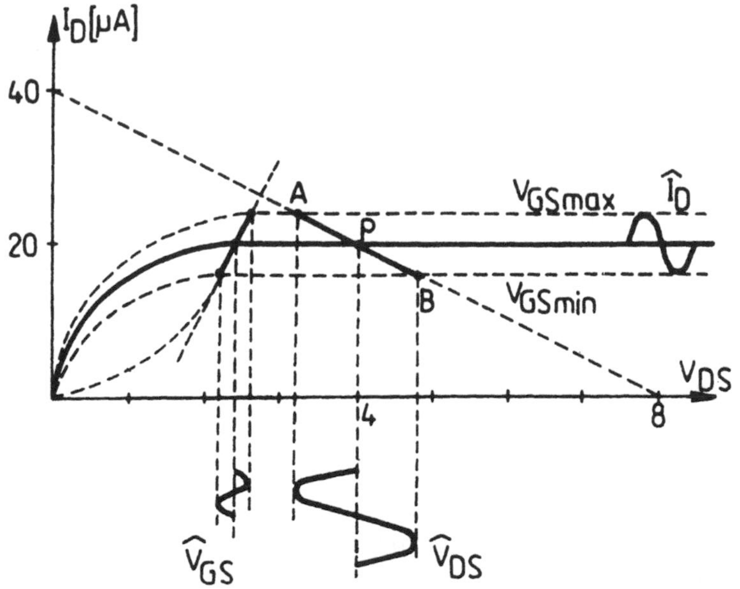

Figure 2.18 Graphical demonstration of small-signal operation of the MOSFET amplifier circuit from Figure 2.15(b)

sponds to the condition $V_{DS} = V_{GS} - V_T$. Since the d.c. gate voltage is zero, the gate-source bias is 4 V, while an a.c. gate voltage is $V_G(t) = A\sin(\omega t)$ If $A = 0.1$ V then the instantaneous MOSFET characteristic moves between two extreme positions corresponding to V_{GSmin}=3.9 V and V_{GSmax}=4.1 V as shown in Figure 2.18. Hence, the bias point moves within the segment AB. If the amplitude A (0.1 V) is sufficiently small the quadratic dependence $I_D(V_{GS})$ in the saturation region may be treated as linear in a vicinity of the bias. Taking the derivative of the drain current $G = \mathrm{d}I_D/\mathrm{d}V_{GS} = \gamma W(V_{GS0} - V_T)/L$ as the local slope of the MOSFET characteristic at the bias V_{GS0}=4 V the amplitude of I_D is Ag. Schematic sinusoids $\hat{V}_{GS}(t)$, $\hat{I}_D(t)$, $\hat{V}_{DS}(t)$ shown in Figure 2.18 explain how incremental signals arise from linearized mappings. Thus, the a.c. voltage across R_L has an amplitude $R_L A G$ and so the voltage gain is finally seen to be $R_L G = 3.5$. The PSPICE sample S2-RM-DC.cir provides a d.c. sweep analysis of the trajectory of a solution (*ID*, *VDS*) presented in Figure 2.18. A corresponding sample file S2-RM-WF.CIR gives waveforms of small incremental signals (using the command .TRAN) and demonstrates results of a small-signal a.c. analysis (for nonzero frequencies using the command .AC and for zero frequency using .TF).

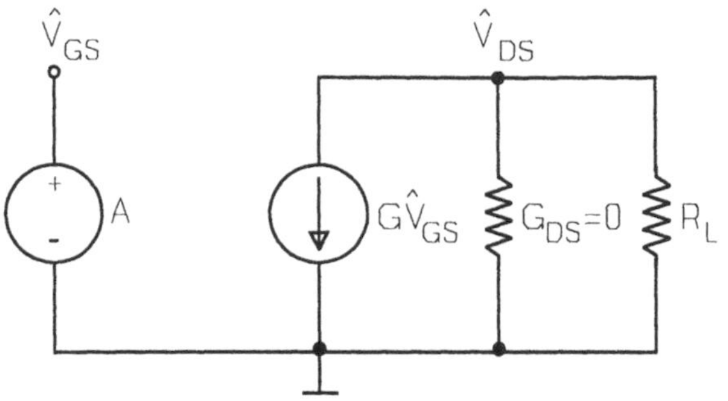

Figure 2.19 A substitution small-signal incremental circuit for a.c. analysis of the circuit from Figure 2.15(b)

In this small-signal analysis the nonlinear controlled current source of the Schichman and Hodges model of the MOS transistor has been replaced by the linear one arising from linearization at the bias point. This source operates on incremental signals, namely $I_D - I_{D0} = G(V_{GS} - V_{GS0})$. The resistive load remains unchanged. Sources have been substituted by sources of sinusoidal incremental signals. The gate source, due to its zero d.c. value, has been left the same, while the drain and source constant supply sources have been short-circuited. We have obtained a substitute linear circuit called a small-signal incremental circuit, drawn in Figure 2.19. Hence, the small-signal a.c. analysis is reduced to linear analysis of circuits like that of Figure 2.19. This appears to be the general conclusion.

Similar observation of small-signal operation of nonlinear elements can be carried out using other PSPICE examples. For instance, S2-RB-DC.CIR accompanied by S2-RB-WF.CIR demonstrate a bjt with a resistive load. Samples S2-CM-DC.CIR and S2-CM-WF.CIR show operation of a CMOS pair. Tasks S2-DF-DC.CIR and S2-DF-WF.CIR provide a more complicated case of a bjt with a bjt load. Each sample *-DC.CIR provides location of a dynamic bias point, and shows how it follows changes of an excitation. Corresponding samples *-WF.CIR provide observation a.c. characteristics and the distorted waveforms that arise because the amplitudes of real excitations are not infinitesimally small.

2.2.2.2 A substitute small-signal incremental immitance circuit

The small-signal a.c. analysis of the example of Figure 2.17(b) is equivalent to the solution of the incremental circuit of Figure 2.19 obtained by linearization in a d.c. response vicinity. To generalize this observation let us rewrite the general purpose canonical equations (1.23)

$$\begin{aligned} \mathbf{A}\mathbf{i} &= \mathbf{0} \\ \mathbf{B}\mathbf{v} &= \mathbf{0} \\ \mathbf{f}\left[\mathbf{i}, \mathbf{v}, \frac{\mathrm{d}\mathbf{q}(\mathbf{v})}{\mathrm{d}t}, \frac{\mathrm{d}\psi(\mathbf{i})}{\mathrm{d}t}, \mathbf{s}(t)\right] &= \mathbf{0}. \end{aligned} \tag{2.58}$$

Notice, that a steady-state d.c. response is a solution of equations

$$\begin{aligned} \mathbf{A}\mathbf{i}_0 &= \mathbf{0} \\ \mathbf{B}\mathbf{v}_0 &= \mathbf{0} \\ \mathbf{f}(\mathbf{i}_0, \mathbf{v}_0, \mathbf{0}, \mathbf{0}, \mathbf{s}_0) &= \mathbf{0} \end{aligned} \tag{2.59}$$

where the derivatives are substituted by zeros (due to the constant state expected), while the constant d.c. components $\mathbf{s}_0$ have been taken instead of the original excitations $\mathbf{s}(t)$. The small-signal operation relies on approximating the canonical equations by means of the first order Taylor series expansion in the vicinity of the d.c. solution. After $\mathbf{q}, \psi$ has been linearized, the left-hand sides of (2.58) may be treated as functions of branch variables, derivatives and excitations, and hence expanded in the vicinity of the bias $(\mathbf{i}_0, \mathbf{v}_0, \mathbf{0}, \mathbf{0}, \mathbf{s}_0)$

$$\begin{aligned} &\mathbf{A}\mathbf{i}_0 + \mathbf{A}(\mathbf{i} - \mathbf{i}_0) = \mathbf{0} \\ &\mathbf{B}\mathbf{v}_0 + \mathbf{B}(\mathbf{v} - \mathbf{v}_0) = \mathbf{0} \\ &\mathbf{f}(\mathbf{i}_0, \mathbf{v}_0, \mathbf{0}, \mathbf{0}, \mathbf{s}_0) + \frac{\partial \mathbf{f}}{\partial \mathbf{i}}(\mathbf{i} - \mathbf{i}_0) + \frac{\partial \mathbf{f}}{\partial \mathbf{v}}(\mathbf{v} - \mathbf{v}_0) + \\ &\quad + \frac{\partial \mathbf{f}}{\partial \dot{\psi}} \frac{\mathrm{d}\left[\psi_0 + \frac{\mathrm{d}\psi(\mathbf{i}_0)}{\mathrm{d}\mathbf{i}}(\mathbf{i} - \mathbf{i}_0)\right]}{\mathrm{d}t} + \frac{\partial \mathbf{f}}{\partial \dot{\mathbf{q}}} \frac{\mathrm{d}\left[\mathbf{q}_0 + \frac{\mathrm{d}\mathbf{q}(\mathbf{v}_0)}{\mathrm{d}\mathbf{v}}(\mathbf{v} - \mathbf{v}_0)\right]}{\mathrm{d}t} + \\ &\quad + \frac{\partial \mathbf{f}}{\partial \mathbf{s}}[\mathbf{s}(t) - \mathbf{s}_0] = \mathbf{0} \end{aligned} \tag{2.60}$$

where $\partial\mathbf{f}/\partial\mathbf{i}, \partial\mathbf{f}/\partial\mathbf{v}, \partial\mathbf{f}/\partial\dot{\psi}, \partial\mathbf{f}/\partial\dot{\mathbf{q}}, \partial\mathbf{f}/\partial\mathbf{s}$ are Jacobi matrices of the function $\mathbf{f}$ with respect to $\mathbf{i}, \mathbf{v}, \mathrm{d}\psi/\mathrm{d}t, \mathrm{d}\mathbf{q}/\mathrm{d}t, \mathbf{s}$ respectively at the point $(\mathbf{i}_0, \mathbf{v}_0, \mathbf{0}, \mathbf{0}, \mathbf{s}_0)$. In Equation (2.60) the zero order components are identical with the left-hand side of Equation (2.59) and hence may be substituted by zeros. After the hats have been introduced for denoting the incremental components

$$\begin{aligned}
&\mathbf{A}\hat{\mathbf{i}} = \mathbf{0}\\
&\mathbf{B}\hat{\mathbf{v}} = \mathbf{0}\\
&\frac{\partial \mathbf{f}}{\partial \mathbf{i}}\hat{\mathbf{i}} + \frac{\partial \mathbf{f}}{\partial \mathbf{v}}\hat{\mathbf{v}} + \frac{\partial \mathbf{f}}{\partial \dot{\psi}}\frac{\mathrm{d}[\psi_0 + \frac{\mathrm{d}\psi(\mathbf{i}_0)}{\mathrm{d}\mathbf{i}}\hat{\mathbf{i}}]}{\mathrm{d}t} + \\
&\qquad + \frac{\partial \mathbf{f}}{\partial \dot{\mathbf{q}}}\frac{\mathrm{d}[\mathbf{q}_0 + \frac{\mathrm{d}\mathbf{q}(\mathbf{v}_0)}{\mathrm{d}v}\hat{\mathbf{v}}]}{\mathrm{d}t} + \frac{\partial \mathbf{f}}{\partial \mathbf{s}}\hat{\mathbf{s}}(t) = \mathbf{0}\,.
\end{aligned} \tag{2.61}$$

This can be easily compared with the linear canonical set (2.2). By simple inspection it can be noticed that the system (2.61) stands for a canonical description of a linear circuit of the same topology as the original (2.58). Its branches are derived from the original branches by linearization at the d.c. solution point and by substituting for the circuit variables their corresponding incremental components. Thus, (2.61) identifies the linear, so-called, small-signal incremental circuit.

This circuit may be easily transformed into its immittance substitute suitable for a.c. analysis. By eliminating unimportant initial conditions $\mathbf{q}_0, \psi_0$ and applying the phasor transformation (see Section 2.2.1) to (2.61) we obtain

$$\begin{aligned}
&\mathbf{A}\hat{\mathbf{I}} = \mathbf{0}\\
&\mathbf{B}\hat{\mathbf{V}} = \mathbf{0}\\
&(\frac{\partial \mathbf{f}}{\partial \mathbf{i}} + \mathrm{j}\omega\frac{\partial \mathbf{f}}{\partial \dot{\psi}}\frac{\mathrm{d}\psi}{\mathrm{d}\mathbf{i}})\hat{\mathbf{I}} + (\frac{\partial \mathbf{f}}{\partial \mathbf{v}} + \mathrm{j}\omega\frac{\partial \mathbf{f}}{\partial \dot{\mathbf{q}}}\frac{\mathrm{d}\mathbf{q}}{\mathrm{d}\mathbf{v}})\hat{\mathbf{V}} = -\frac{\partial \mathbf{f}}{\partial \mathbf{s}}\hat{\mathbf{S}}\,.
\end{aligned} \tag{2.62}$$

This circuit is structurally identical with the original one (2.58) though it has branches not only linearized but also transformed into the jω domain. Variables of the immittance circuit (2.62) are complex phasors carrying the information of signals amplitudes and phases. These phasors have been denoted by capital letters with hats.

Thus, the essence of the small-signal a.c. analysis relies on creation of a small signal immittance circuit by linearizing and transforming into the phasor domain all branches and assembling them according to the original topology. In practice, for a given topology, all branches have to be subsequently transformed. Linear branches do not change their form at all, only terminal signals become increments. For example a conductance $i = Gv$ in a d.c. state fulfils $i_0 = Gv_0$ and hence, subtracting both formulae we have the incremental branch $\hat{i}=G\hat{v}$ of practically the same form. For any nonlinear resistive branch, for instance $i_a=f_a(v_b)$, the Taylor linearization at the bias point yields $i_a=f_a(v_{b0})+G_{ab0}(v_b-v_{b0})$. Hence, in the incremental immittance case we have $\hat{I}_a=G_{ab0}\hat{V}_b$ where G_{ab0} is a derivative of f_a at the bias point v_{b0}. Reactive elements are handled similarly. A nonlinear charge $q(v)$ after linearization gives $q_0+C_0(v-v_0)$. Hence, after differentiation, the incremental branch $\hat{i}=i-i_0=i-0=C_0\,\mathrm{d}\hat{v}/\mathrm{d}t$ arises. In the phasor domain it takes an admittance form $\hat{I}=\mathrm{j}\,\omega\,C_0\,\hat{V}$, where C_0 is an instantaneous (in other words differential) capacitance $\mathrm{d}q(v_0)/\mathrm{d}v$ at the bias point. For nonlinear inductors a similar consideration gives the impedance model $\hat{V}=\mathrm{j}\omega L_0\,\hat{I}$ with a differential inductance at the bias point.

To finally demonstrate the techniques of generating the small-signal immittance circuit let us consider the nonlinear circuit of Figure 2.20(a). The following nonlinear branches are involved there:

$$i_2(v_2)=I_s[\exp(v_2/V_T)-1],\ C_3(v_3)=C_0\exp(v_3/V_T),\ v_4(i_2)=V_T(i_2/I_s)^3.$$

Let the index 0 denote, as usually, the bias components. Linear branches have the immittance models: $\hat{V}_5=Z_5\hat{I}_5$, where $Z_5=\mathrm{j}\omega L_5$ and $\hat{I}_6=G_6\hat{V}_6$. Excitation takes the form $\hat{I}_1=A\exp(\mathrm{j}0)$. Nonlinear conductances transform into the linear ones $\hat{I}_2=G_{20}\,\hat{V}_2$, where the coefficient can be written as $G_{20}=\partial i_2(v_{20})/\partial v_2=(I_s/V_T)\exp(v_{20}/V_T)$. The controlled source yields $\hat{V}_4=R_{420}\hat{I}_2$, $R_{420}=\partial v_4(i_{20})/\partial i_2=(3V_T/I_s)(i_{20}/I_s)^2$, being its linear counterpart. Finally, a nonlinear capacitance transforms into a linear admittance $\hat{I}_3=Y_{30}\,\hat{V}_3$, where $Y_{30}=\mathrm{j}\omega C_0\exp(v_{30}/V_T)$. A resultant substitute immittance circuit is demonstrated in Figure 2.20(b). Its MNE can be easily written.

2.2.2.3 Specifics of the implementation

The small-signal a.c. analysis of nonlinear circuits can be implemented according to the same Algorithm 2.1 as the linear a.c. analysis. The only difference lies in another way of formulating substitute immitance equations. In a linear case they are formulated as e.g. MNE of the immitance circuit derived directly from the original linear circuit. In the small-signal case the original circuit is transformed into a small-signal incremental form, then its immittance counterpart is created from which MNE can be generated by means of the stamps technique. The substitute immittance circuit is only an abstract auxiliary concept. In practice, the linearized immitance equations are designated directly by built-in nonlinear models where appropriate formulae for derivatives are incorporated and by in-

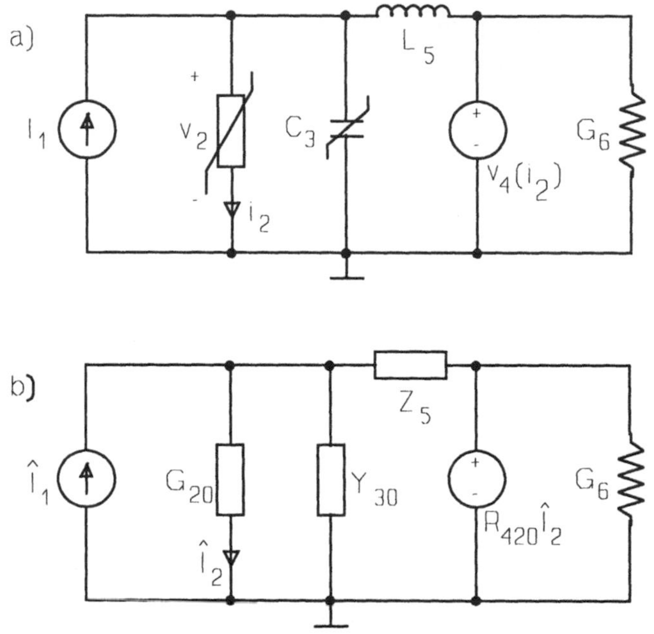

Figure 2.20 (a) A nonlinear circuit and (b) its small-signal immitance counterpart dedicated to small-signal AC analysis

variant topological information using the stamps technique. As nonlinear models are built of expressions for branch variables and their partial derivatives embedded in program code, the coefficients of the immittance branches derived from these derivatives can be easily calculated. They are next introduced at appropriate matrix positions.

To demonstrate the possibility of creating immittance stamps for the automated formulation of small-signal immittance equations let us assume a circuit that is composed of the following unified current or voltage-defined branches

$$\begin{aligned} i_a &= f_{ia}\left[v_a, v_b, i_c, \frac{\mathrm{d}q_a(v_a, v_b, i_c)}{\mathrm{d}t}, s(t)\right] \\ v_a &= f_{va}\left[i_a, v_b, i_c, \frac{\mathrm{d}\psi_a(i_a, v_b, i_c)}{\mathrm{d}t}, s(t)\right] \end{aligned} \tag{2.65}$$

These two formulae cover independent sources, resistive elements as well as charge- or flux-defined controlled components.

Let us assume that the d.c. nonlinear analysis was carried out prior to small-signal a.c. calculations. Hence, the d.c. response is available. Now, the branches (2.65) can be linearized at this bias point and transformed into a phasor form. Expanding (2.65) in a first order Taylor series, introducing incremental components and carrying out a phasor transformation we get linear immittance counterparts, also current or voltage-defined respectively

$$\begin{aligned} \hat{I}_a &= G_{aa}\hat{V}_a + G_{ab}\hat{V}_b + K_{ac}\hat{I}_c + \hat{J}_a \\ \hat{V}_a &= R_{aa}\hat{I}_a + T_{ab}\hat{V}_b + R_{ac}\hat{I}_c + \hat{E}_a\,. \end{aligned} \tag{2.66}$$

They are exactly the same as branches (2.13), (2.14), acceptable to MNE, and hence MNE stamps from Section 2.1.3.4 can be applied to obtain immitance equations. After some algebra, coefficients of (2.66) take the form including derivatives calculated at the bias point

$$\begin{aligned} G_{aa0} &= \frac{\partial f_i}{\partial v_a} + \mathrm{j}\omega\frac{\partial f_i}{\partial \dot{q}}\frac{\partial q}{\partial v_a} & G_{ab0} &= \frac{\partial f_i}{\partial v_b} + \mathrm{j}\omega\frac{\partial f_i}{\partial \dot{q}}\frac{\partial q}{\partial v_b} \\ K_{ac0} &= \frac{\partial f_i}{\partial i_c} + \mathrm{j}\omega\frac{\partial f_i}{\partial \dot{q}}\frac{\partial q}{\partial i_c} & \hat{J}_a &= -\frac{\partial f_i}{\partial s}\hat{S} \\ R_{aa0} &= \frac{\partial f_v}{\partial i_a} + \mathrm{j}\omega\frac{\partial f_v}{\partial \dot{\psi}}\frac{\partial \psi}{\partial i_a} & T_{ab0} &= \frac{\partial f_v}{\partial v_b} + \mathrm{j}\omega\frac{\partial f_v}{\partial \dot{\psi}}\frac{\partial \psi}{\partial v_b} \\ R_{ac0} &= \frac{\partial f_v}{\partial i_c} + \mathrm{j}\omega\frac{\partial f_v}{\partial \dot{\psi}}\frac{\partial \psi}{\partial i_c} & \hat{E}_a &= -\frac{\partial f_v}{\partial s}\hat{S}\,. \end{aligned} \tag{2.67}$$

This consideration proves that circuits composed of a variety of branches covered by the formulae (2.65) may be analyzed using the stamps for MNE.

In the particular cases of semiconductor device models, specialized stamps are usually created according to the general rules above. A.c. models implemented in general purpose simulators work out the stamp components and then fill the matrix with them for each discrete frequency. As a practical example we consider an outline of a physical MOSFET model appearing in both PSPICE and OPTIMA simulators and shown in Figure 2.21. This model has 6 inter-terminal nonlinear capacitors, parasitic drain-substrate and source-substrate p-n junctions and a controlled source standing for a channel current. The model involves 6 nodal variables connected with terminal nodes: *ng* (gate), *nd* (drain), *ns* (source), *nb* (substrate) and 2 internal nodes *nsi* (internal source), *ndi* (internal drain). The stamp for the small-signal a.c. analysis indicates which complex matrix components have to be added to the matrix to represent the MOSFET. It takes the form shown in Table 2.1. The capacitances written in the stamp mean differential capacitances at the bias point derived from the nonlinear capacitors shown in Figure 2.21. The conductances G_{D1} and G_{D2} are derivatives of the current characteristics of diodes D1 and D2 at bias points. Furthermore, small "g" denotes differential conductances of the source I_D, namely: $g_d = \partial I_D / \partial V_{DB}$, $g_s = \partial I_D / \partial V_{SB}$, $g_g = \partial I_D / \partial V_{GB}$ calculated at the bias point. Once certain expressions for capacitors, I_D and the currents of D1, D2 are provided we are able to evaluate the corresponding immittance model coefficients and substitute them in place of symbols in the stamp given in Table 2.1.

The information regarding small-signal a.c. analysis may be found in many books. (e.g. [2.2, 2.3, 2.5])

The following questions and problems can help us to repeat topics treated in Section 2.2 and to check out comprehension.

- Explain the concept of a phasor and explain the interpretation of its magnitude and phase.
- Using the circuit shown in Figure 2.12 as an example explain the concept of the substitute immittance counterpart of a linear circuit.
- Explain the computational structure of a.c. analysis.
- What types of characteristics can we obtain from the frequency-domain a.c. analysis of linear circuits and what is their practical

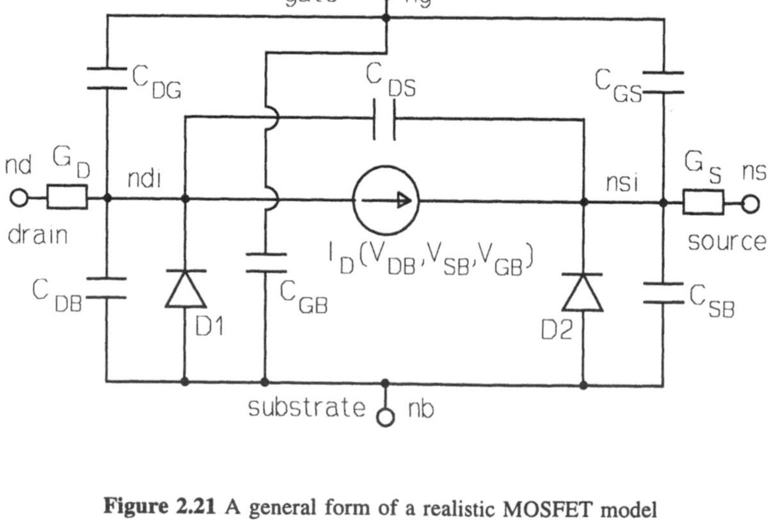

Figure 2.21 A general form of a realistic MOSFET model

Table 2.1 The stamp of the MOSFET model for a.c. analysis

	ng	*ndi*	*nsi*	*nb*	*nd*	*ns*
ng	$j\omega(C_{DG}+C_{GB}+C_{GS})$	$-j\omega C_{DG}$	$-j\omega C_{GS}$	$-j\omega C_{GB}$		
ndi	$-j\omega C_{DG}+g_G$	$j\omega(C_{DB}+C_{DG}+C_{DS}+G_{D1}+G_D+g_D$	$-j\omega C_{DS}+g_S$	$-j\omega C_{DB}+-G_{D1}-g_D+-g_S-g_G$	$-G_D$	
nsi	$-j\omega C_{GS}-g_G$	$-j\omega CC_{DS}-g_D$	$j\omega(C_{GS}+C_{SB}+C_{DS})+G_{D2}+G_S-g_S$	$-j\omega C_{SB}+-G_{D2}+g_D++g_S+g_G$		$-G_S$
nb	$-j\omega C_{GB}$	$-j\omega C_{DB}-G_{D1}$	$-j\omega C_{SB}-G_{D2}$	$j\omega(C_{DB}+C_{GB}+C_{SB})+G_{D1}+G_{D2}$		
nd		$-G_D$			G_D	
ns			$-G_S$			G_S

usfulness?

- What do we mean by incremental components of signals and by small signals?
- Explain the concept of a linearized immitance circuit for small-signal. a.c. analysis. Give examples of small-signal immittance models for such nonlinear elements as conductance, VCVS, capacitor, inductor, etc. Write stamps for these models.

References

2.1. Ho, C. W., Ruehli, A. E., Brennan, P. A., *The Modified Nodal Approach to Network Analysis*, IEEE Transactions on Circuits and Systems, vol. 22, no. 6, p. 504-509 (1975).

2.2. Singhal, K., Vlach, J., *Formulation of Circuit Equations*, in: *Circuit Analysis Simulation and Design, Part 1*, Ruehli, A. E. (Ed.), North Holland, Amsterdam (1986).

2.3. Vlach, J., Singhal, K., *Computer Methods for Circuit Analysis and Design*, Van Nostrand Reinhold, New York (1983).

2.4. Hachtel, G. D., Brayton, F. K., Gustavson, F. G., *The Sparse Tableau Approach to Network Analysis and Design*, IEEE Transactions

on Circuit Theory, vol. 18, no. 1, p. 101-113 (1971).
2.5. Mc Calla, W.J., *Fundamentals of Computer-Aided Circuit Simulation*, Kluwer Academic Publishers, Boston (1987).

Chapter 3

Numerical Solution of Linear Algebraic Equations

This chapter presents common numerical methods of solving algebraic linear equations dedicated to circuit simulation. It has been shown in previous chapters that computer methods of linear circuit analysis rely on two factors: formulating linear algebraic equations by means of stamps and then solving these equations. The latter is a considerably sophisticated numerical procedure that needs extremely efficient implementation as it is usually repeated many times within loops of the program. From the computational point of view the analyzer of linear equations is the heart of any simulation program, except for third generation relaxation simulators discussed in Chapter 9. In preparing its efficient implementation it is essential to take advantage of all possible tools and programming tricks to speed up peformance.

Topics related to solution of systems of linear equations may be considered quite separately, without any knowledge of a circuit. This chapter is dedicated to solution methods known as **finite.** They achieve a result in a limited number of operations, i.e. with no iterations. **Iterative** (so called relaxation) **methods**, where an accurate solution is theoretically not reached in a finite number of iterations, will be discussed in Chapter 9. In practice, finite methods need an economical implementation based on the exploitation of matrix sparsity. A considerable part of a circuit matrix is usually zero, and hence essential savings arise when only non-zero operations are carried out. This, so-called, sparse matrix technique is presented in Section 3.3. The topics mentioned are extensively investigated in a variety of books on numerical methods for linear algebra [3.3].

3.1 Introduction to simultaneous linear algebraic equations

The problem known as a system of m simultaneous linear algebraic equations with n unknowns is one of the general topics in linear algebra. It is closely related to the concept of linear mapping in a linear space. A system of n ordered real numbers can be arranged in the n-dimensional vector succinctly as $\mathbf{x} = [x_1, x_2, \ldots, x_n]^t$. A set of all vectors, like this obeying certain rules constituting their algebra, is known as the n-dimensional linear space R^n. A function $\mathbf{b}=\mathbf{f}(\mathbf{x})$ assigning a vector $\mathbf{b}\in R^m$ from the m-dimensional space to the vector $\mathbf{x}\in R^n$ is called a linear mapping. It may be written in terms of matrix operations

$$\begin{bmatrix} a_{11} & a_{12} & a_{13} & \dots & a_{1n} \\ a_{21} & a_{22} & a_{23} & \dots & a_{2n} \\ a_{31} & a_{32} & a_{33} & \dots & a_{3n} \\ \dots & \dots & \dots & \dots & \dots \\ a_{m1} & a_{m2} & a_{m3} & \dots & a_{mn} \end{bmatrix} \begin{bmatrix} x_1 \\ x_2 \\ x_3 \\ \dots \\ x_n \end{bmatrix} = \begin{bmatrix} b_1 \\ b_2 \\ b_3 \\ \dots \\ b_m \end{bmatrix} \tag{3.1}$$

or succinctly $\mathbf{A}\,\mathbf{x}=\mathbf{b}$. Once the mapping **A** and the vector **b** are known there is a mathematical problem whether and when there exists a proper **x** obeying (3.1) and a practical problem how to derive it. Equation (3.1) constitutes a system of linear equations. Linear dependence is ubiquitous and so the need for solving linear equations often arises also in electrical engineering.

3.1.1 Solvability of sets of linear equations

The concept of the linear combination $\alpha_1\mathbf{x}_1+\alpha_2\mathbf{x}_2+\ldots+\alpha_i\mathbf{x}_i$ of a number of vectors, where $\alpha_1,\ldots,\alpha_i$ are real weights, is of great importance in linear algebra. Certain vectors are known as linearly independent if none of them can be expressed as a linear combination of the others. They are linearly dependent otherwise. The linear mapping **A** may have linearly independent columns or some of them may be linear combinations of the others. A maximum number of linearly independent rows and columns of the matrix is known as its order, namely $\mathrm{ord}(\mathbf{A})$. Among n vectors of a dimension m, at most m and obviously not more then n are linearly dependent. Hence $\mathrm{ord}(\mathbf{A})\le\min(m,n)$. Notice that if the system (3.1) has a solution then the right-hand side (RHS) **b** is a linear combination of columns of **A**. Hence, the number of linearly independent columns of **A** is the same as the number of linearly independent columns of the matrix **A|b** (**A** extended with **b**) only when **b** is a linear combination of columns of **A** (that is when a solution of the equations exists). This leads to the **Kronecker-Cappelli Theorem**:

> *A solution of the system of simultaneous algebraic linear equations (3.1) exists if and only if* $\mathrm{ord}(\boldsymbol{A})=\mathrm{ord}(\boldsymbol{A}\,|\boldsymbol{b})$. *Moreover the solution is unique once* $\mathrm{ord}(\boldsymbol{A})=n$ *and ambiguous if* $\mathrm{ord}(\boldsymbol{A})<n$.

The nonsingularity of any quadratic matrix (i.e. $\det \mathbf{A} \neq 0$) denotes, in fact, the linear independence of its columns (and so rows). The order of any rectangular matrix is, hence, equal to the dimension of its largest nonsingular quadratic submatrix. Thus the uniqueness of the solution of (3.1) is equivalent to the nonsingularity of a quadratic submatrix of dimension n. Once $m<n$ the system is known as underdefined. It is at most of order m and may be inconsistent or ambiguous. Systems obeying $m=n$ are quadratic of order not greater than n, while ones satisfying $m>n$ are overdefined of order at most n. Quadratic and overdefined cases imply inconsistent, unique or ambiguous systems.

In electrical engineering RHS vectors $\mathbf{b}$ stand for stimulations and so, in practice, there are systems having solutions for any stimuli. These occur when $\mathbf{A}$ maps R^n onto the whole R^m, i.e. has m linearily independent columns so that its order is m. This case may occur in ambiguous underdefined or quadratic nonsingular mappings. The latter is obviously the most common.

For quadratic systems they may be, in general, unique once the matrix is nonsingular. Then, the zero solution appears only for the zero stimulation ($\mathbf{b}=\mathbf{0}$). Otherwise the matrix may be singular, and hence a system inconsistent or ambiguous according to stimuli.

3.1.2 Ill-conditioned systems

3.1.2.1 Estimation of a solution perturbation

Any mathematical problem is known as being ill-conditioned when its results are strongly dependent on variations in its input data. So the system (3.1) is ill-conditioned when small perturbations in elements of $\mathbf{A}$ and $\mathbf{b}$ imply large changes in the solution $\mathbf{x}$. To gain worst-case estimates of these changes the aparatus of norms in linear spaces is useful.

A norm of the n-dimensional vector $\mathbf{x}$, denoted as $\|\mathbf{x}\|$, is a measure of the distance between $\mathbf{x}$ and the origin $\mathbf{0}$ in the space R^n. The norm is positive or zero (the latter only for the vector $\mathbf{0}$), satisfies $\|\mathbf{x}_1+\mathbf{x}_2\| \leq \|\mathbf{x}_1\| + \|\mathbf{x}_2\|$ (known as the triangle law) and $\|c\mathbf{x}\| = |c| \, \|\mathbf{x}\|$ (the rule of uniformity), where c is a real number while $|...|$ stands for an absolute value. The most important Euclidean norm of vectors is defined as $\|\mathbf{x}\| = (x_1^2+x_2^2+...+x_n^2)^{1/2}$ and all further consideration will be related to this norm.

Linear algebra also contains a concept of norms of linear transformation. It is an upper bound of a maximum (over the space) transformation of a vector norm. A norm of the mapping (or equivalently matrix) $\mathbf{A}$ arises from the maximization

$$\|\mathbf{A}\| = \max_{\mathbf{x}\neq\mathbf{0},\,\mathbf{x}\in R^n} \frac{\|\mathbf{A}\,\mathbf{x}\|}{\|\mathbf{x}\|} \tag{3.2}$$

and, after utilizing the vector Euclidean norm, definition (3.2) appears to be the, so-called, spectral norm, that is the square root of the maximum eigenvalue of the symmetric matrix $\mathbf{A}\,\mathbf{A}^t$, i.e. $\|\mathbf{A}\| = (\max_i \lambda_i(\mathbf{A}\,\mathbf{A}^t))^{1/2}$.

Definition (3.2) shows the usefulness of these norms for such algebraic estimations as $\|\mathbf{A}\,\mathbf{x}\| \le \|\mathbf{A}\|\,\|\mathbf{x}\|$ and also $\|\mathbf{A}_1\,\mathbf{A}_2\| \le \|\mathbf{A}_1\|\,\|\mathbf{A}_2\|$. These rules will be used for estimating changes in the solution of linear equations due to perturbations of coefficients. The simplest case arises when the right-hand side only is perturbed. Let $\mathbf{d}_b$ be the additive perturbation of the right-hand side. Then, an implied change in the solution is $\mathbf{d}_x = \mathbf{A}^{-1}(\mathbf{b}+\mathbf{d}_b) - \mathbf{A}^{-1}\mathbf{b} = \mathbf{A}^{-1}\mathbf{d}_b$. Hence, the norm of the change may be estimated as $\|\mathbf{d}_x\| \le \|\mathbf{A}^{-1}\|\,\|\mathbf{d}_b\|$. The norm of the inverse matrix is an absolute measure of how perturbations of the stimuli are reflected in the solution. To obtain the relative change notice that from (3.1) the estimations $\|\mathbf{b}\| \le \|\mathbf{A}\|\,\|\mathbf{x}\|$ and $1/\|\mathbf{x}\| \le \|\mathbf{A}\|/\|\mathbf{b}\|$ arise. Thus, multiplying the left and right sides of the two estimations recently obtained we have an important specification

$$\frac{\|\mathbf{d}_x\|}{\|\mathbf{x}\|} \le \|\mathbf{A}^{-1}\|\,\|\mathbf{A}\|\,\frac{\|\mathbf{d}_b\|}{\|\mathbf{b}\|} \tag{3.3}$$

showing that the quantity $\mathrm{cond}(\mathbf{A}) = \|\mathbf{A}^{-1}\|\,\|\mathbf{A}\|$, known as the condition number, is a measure of the propagation of a relative perturbation.

It can be shown, that this factor is of great importance in the more general case of perturbations in both matrix and RHS coefficients. To formalize this problem let $\mathbf{D}_A$ be a perturbation of **A**. If a perturbated system is nonsingular let its solution be $\tilde{\mathbf{x}} = \mathbf{x} + \mathbf{d}_x$. Subtracting (3.1) from the perturbed system $(\mathbf{A}+\mathbf{D}_A)(\mathbf{x}+\mathbf{d}_x) = \mathbf{b}+\mathbf{d}_b$ the change in solution takes the form

$$\mathbf{d}_x = (\mathbf{1}+\mathbf{A}^{-1}\mathbf{D}_A)^{-1}\,\mathbf{A}^{-1}\,(\mathbf{d}_b - \mathbf{D}_A\,\mathbf{x}) \tag{3.4}$$

where $\mathbf{C} = 1 + \mathbf{A}^{-1}\,\mathbf{D}_A$ has been assumed to be nonsingular. Perturbations that do not yield the matrix singularity are called small. To introduce a sufficient condition that this case occurs let us consider the auxiliary system $\mathbf{C}\,\mathbf{x} = \mathbf{0}$. Estimation of its left side needs the following mathematical development to be pursued $\|\mathbf{x}\| = \|\mathbf{x}+\mathbf{A}^{-1}\mathbf{D}_A\mathbf{x} - \mathbf{A}^{-1}\mathbf{D}_A\mathbf{x}\| \le \|\mathbf{x}+\mathbf{A}^{-1}\mathbf{D}_A\mathbf{x}\| + \|\mathbf{A}^{-1}\mathbf{D}_A\|\,\|\mathbf{x}\|$ to prove $\|\mathbf{C}\mathbf{x}\| = \|\mathbf{x}+\mathbf{A}^{-1}\mathbf{D}_A\mathbf{x}\| \ge \|\mathbf{x}\| - \|\mathbf{A}^{-1}\mathbf{D}_A\|\,\|\mathbf{x}\| = \|\mathbf{x}\|(1 - \|\mathbf{A}^{-1}\mathbf{D}_A\|)$. For the matrix perturbations obeying $\|\mathbf{A}^{-1}\mathbf{D}_A\| \le 0$ the right side of the given $\|\mathbf{Cx}\|$ estimation is positive for all nonzero **x** and hence $\|\mathbf{Cx}\|$ is positive for all $\mathbf{x}=\mathbf{0}$. Thus, the equation $\mathbf{Cx}=\mathbf{0}$ has a unique solution $\mathbf{x}=\mathbf{0}$ and its matrix **C** must be nonsingular. In practice, perturbations satisfying the above sufficient condition are called small perturbations.

To obtain an estimation of $\mathbf{C}^{-1}$ we notice that due to the definition of C the equation $\mathbf{C}^{-1} = \mathbf{1} - \mathbf{C}^{-1}\mathbf{A}^{-1}\mathbf{D}_A$ can be written and after norms have been involved, the estimation $\|\mathbf{C}^{-1}\| \le 1 + \|\mathbf{C}^{-1}\|\,\|\mathbf{A}^{-1}\|\,\|\mathbf{D}_A\|$ proves that

$$\|\mathbf{C}^{-1}\| \le \frac{1}{1 - \|\mathbf{A}^{-1}\|\,\|\mathbf{D}_A\|}. \tag{3.5}$$

To start the final estimation of a solution from (3.4) we gain

$$\|\mathbf{d}_x\| \le \|(\mathbf{1}+\mathbf{A}^{-1}\mathbf{D}_A)^{-1}\|\,\|\mathbf{A}^{-1}\|\,(\|\mathbf{d}_b\| + \|\mathbf{D}_A\|\,\|\mathbf{x}\|) \tag{3.6}$$

and taking into account (3.5), the relative perturbations take the form

$$\frac{\|\mathbf{d}_x\|}{\|\mathbf{x}\|} \le \frac{\|\mathbf{A}^{-1}\|\,\|\mathbf{A}\| \left(\dfrac{\|\mathbf{d}_b\|}{\|\mathbf{b}\|} + \dfrac{\|\mathbf{D}_A\|}{\|\mathbf{A}\|} \right)}{1 - \|\mathbf{A}^{-1}\|\,\|\mathbf{A}\| \dfrac{\|\mathbf{D}_A\|}{\|\mathbf{A}\|}}. \tag{3.7}$$

As can be seen, the essence of the propagation of perturbations lies in the condition number involved. For small perturbations the propagation is linear. According to the spectral norm definition $\mathrm{cond}(\mathbf{A})$ is equal to a square root of the ratio of the maximum and minimum eigenvalues of $\mathbf{A}\mathbf{A}^t$. Ill-conditioned systems have a large discrepancy between eigenvalues. The best $\mathrm{cond}(\mathbf{A}) = 1$ occurs for unitary matrices. Let the relative error of the floating point data representation be ε. It corresponds to the number of its mantissa bits *m*. The IEEE 754 norm assumes that double precision numbers are of $m=52$. Hence, $\varepsilon=2^{-m-1}=1.11\ 10^{-16}$. Small perturbations occur when $\varepsilon\,\mathrm{cond}(\mathbf{A})<1$. Hence well-conditioned systems should have a condition number smaller then $1/\varepsilon$. They are ill-conditioned otherwise.

3.1.2.2 Example of an ill-conditioned system

Now, we will demonstrate how ill-conditioned matrices may appear in even very simple cases. Consider the circuit in Figure 3.1 whose equations are as follows

$$\begin{bmatrix} 1+\alpha & -\dfrac{1}{1+\alpha} \\ -\alpha & \dfrac{1}{\alpha+1} \end{bmatrix} \begin{bmatrix} v_1 \\ v_2 \end{bmatrix} = \begin{bmatrix} 1 \\ 1 \end{bmatrix}. \tag{3.8}$$

Let their matrix be further denoted as **A**. For $\alpha = 99 = 10^2 - 1$ it yields

$$\boldsymbol{A} = \begin{bmatrix} 100 & -0.01 \\ -99 & 0.01 \end{bmatrix} \tag{3.9}$$

and then an exact solution of (3.8) is $v_1 = 2$, $v_2 = 19\,900$. To calculate the condition number let us create the auxiliary matrix

$$\mathbf{A}\mathbf{A}^t = \begin{bmatrix} 10000.0001 & -9900.0001 \\ -9900.0001 & -9800.9999 \end{bmatrix} \tag{3.10}$$

whose eigenvalues obtained from the characteristic equation

$$\begin{vmatrix} 10000.0001 - \lambda & -9900.0001 \\ -9900.0001 & -9800.9999 - \lambda \end{vmatrix} = 0 \tag{3.11}$$

are $\lambda_1 = 19801$, $\lambda_2 = D/\lambda_1$, where $D = \det(\mathbf{A}\mathbf{A}^t) = 10^{-4}$. Now the spectral norm of **A** is equal to $[\lambda_{max}(\mathbf{A}\mathbf{A}^t)]^{1/2} = \lambda_1$, while that of $\mathbf{A}^{-1}$ is

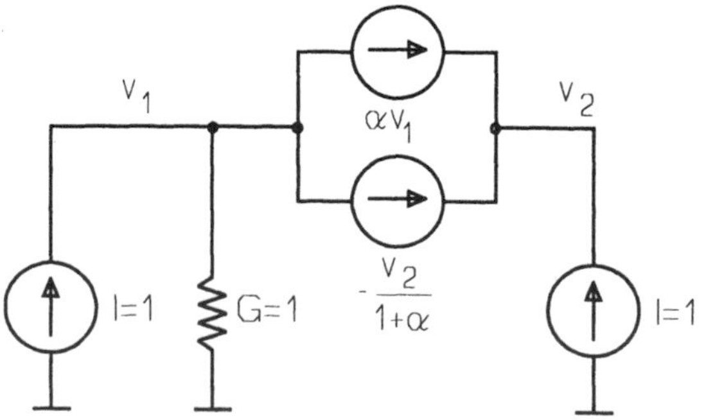

Figure 3.1 A simple ill-conditioned circuit

$[\lambda_{max}(\mathbf{A}^{-1}(\mathbf{A}^{-1})^t)]^{1/2}$. The latter is equal to $[\lambda_{max}((\mathbf{A}^t\mathbf{A})^{-1})]^{1/2} = 1/[\lambda_{min}(\mathbf{A}\mathbf{A}^t)]^{1/2} = 1/\lambda_2$ that is to the inverse of the minimum eigenvalue of the auxiliary matrix (3.10). Finally, the condition number is their product: $\mathrm{cond}(A) = \lambda_1/\sqrt{D} = 1\,980\,200$. Taking the relative level of the computer arithmetic $\varepsilon = 1.11\ 10^{-16}$ it can be seen that the relative error of the solution is something like $2\ 10^{-10}$. Hence, it is quite infinitesimal from the PSPICE printout. We may confirm this by testing the sample S3-TEST.CIR for PSPICE included on Diskette A.

Taking α as a variable the general solution $v_1 = 2$, $v_2 = (2\alpha+1)(\alpha+1)$, and after some algebra $\mathrm{cond}(A) \approx 2\alpha(\alpha+1)^2$ are obtained. For several large values of α approaching the limitations of the computer algebra, simulations of the sample S3-TEST.CIR have been carried out by means of PSPICE. Results are collected in Table 3.1. The exact solutions and the theoretical inaccuracy $\varepsilon\,\mathrm{cond}(\mathbf{A})$ have been written in the upper row for each value of α. The results derived from PSPICE and the inaccuracy $Er = \|\Delta\mathbf{v}\|/\|\mathbf{v}\| \approx |\Delta v_2|/|v_2|$ estimated from the printout are given in the lower row for each α. Until something like $\alpha = 999$ the PSPICE arithmetic appears to be quite sufficient. For bigger α, the extremely large condition number starts to imply an inaccuracy visible in the PSPICE printout. Table 3.1 shows the results for $\alpha = 39999$ and above. This value yields $v_1 = 1.9928$ instead of 2 and $v_2 = 7.174\ 10^9$ instead of the exact solution $7.19994\ 10^9$. Thus the actual error is 0.36% while the theoretical

Table 3.1 SPICE examination of the circuit from Figure 3.1

α	v_1	v_2	ε cond(A)	cond(A)
	PSPICE	PSPICE	Er	
39999	2	$3.19996\ 10^9$	$14.1\ 10^{-3}$	$1.28\ 10^{14}$
	1.9968	$3.195\ 10^9$	$1.56\ 10^{-3}$	
49999	2	$4.99995\ 10^9$	$27.5\ 10^{-3}$	$2.5\ 10^{14}$
	2.9950	$4.987\ 10^9$	$2.59\ 10^{-3}$	
59999	2	$7.19994\ 10^9$	$47.5\ 10^{-3}$	$4.32\ 10^{14}$
	1.9928	$7.174\ 10^9$	$3.6\ 10^{-3}$	
69999	2	$9.79993\ 10^9$	$75.5\ 10^{-3}$	$6.86\ 10^{14}$
	1.9902	$9.752\ 10^9$	$4.9\ 10^{-3}$	

is something like 4.75%. Numerical experiments collected in Table 3.1 show that theoretical errors arising from the worst-case norms based estimations are about one order bigger then actual inaccuracies observed. This effect is a result of rather pessimistic overestimated upper bounds used in the algebra of norms. From S3-TEST.CIR we may also discover that once $\alpha=99999$ the ill-conditioning effects exceed the capabilities of a computer and PSPICE fails to find a solution. This example proves that, while applying norms and the condition numbers, attention has to be paid to overestimated upper bounds.

To sum up this section and enable us to check our familiarity with its scope the following questions may be answered:

- What are the necessary and sufficient conditions to be fulfilled when the solution of a system of linear equations exists? Distiguish the general rectangular case and then the quadratic case.
- Explain the concept of the ill-conditioned system of equations and a definition of the condition number. Can a high precision in computer arithmetic override the ill-condition ?

3.2 Finite methods of solving linear algebraic equations

Finite methods of solving the system of simultaneous linear algebraic equations (3.1) theoretically (i.e. neglecting the arithmetic inaccuracy) can yield an exact solution in a finite number of operations such as additions, subtractions, multiplications, and divisions. The most common and dominant in practical applications are methods that derive from the, so-called, Gaussian elimination (GE). The ubiquitous success of GE is due to its simple implementation, good numerical properties and one of the lowest levels of computational complexity among a variety of known finite methods. This section is dedicated to a class of common finite methods arising from GE. They are equivalent to certain matrix triangular factorizations and are known as LU factorization methods. The original GE appears to be one of them. Subsection 3.2.1 explains GE and shows how it can be viewed as LU factorization. Subsection 3.2.2 covers a variety of possible LU factorization strategies while 3.2.3 treats the numerical properties of these methods. Their sparse implementations will be presented in Section 3.3.

3.2.1 Gaussian elimination

3.2.1.1 The idea of Gaussian elimination

To demonstrate the strategy of Gaussian elimination (GE) let us consider a simple example. As GE consists of subsequent modifications of each equation let a notation r_i^k be introduced to refer to the ith equation after the kth modification. Coefficients of these equations will be a_{ij}^k and b_i^k for right-hand sides (RHS).

$$\begin{aligned} r_1^0&: \quad 2x_1+3x_2+5x_3=10 \\ r_2^0&: \quad 3x_1+2x_2+x_3=6 \\ r_3^0&: \quad 5x_1+3x_2+2x_3=10\,. \end{aligned} \tag{3.12}$$

The first modification relies on eliminating x_1 from the 2nd and 3rd equations by subtracting the equation r_1^0 multiplied by a_{i1}^0/a_{11}^0 (i=2,3) from the equations r_i^0. This makes the coefficients of x_1 zero.

$$\begin{aligned} r_1^0&: \quad 2x_1+3x_2+5x_3=10 \\ r_2^1&: \quad -2.5x_2-6.5x_3=-9\;. \\ r_3^1&: \quad -4.5x_2-10.5x_3=-15 \end{aligned} \tag{3.13}$$

Once modified, the 2nd and 3rd equations have obtained the superscript 1.

Now, the second modification is applied relying on eliminating x_2 from the equation r_i^1 (i=3). This needs the equation r_2^1 to be multiplied by a_{i2}^1/a_{22}^1 and then subtracted from r_i^1. Finally, the triangular system is gained, namely

$$\begin{aligned} r_1^0&: \quad 2x_1+3x_2+5x_3=10 \\ r_2^1&: \quad -2.5x_2-6.5x_3=-9\quad. \\ r_3^2&: \quad 1.2x_3=1.2 \end{aligned} \tag{3.14}$$

The stage just finished is called the elimination. From the third equation of (3.14) x_3 can be easily obtained. Then, x_2 is obtainable from the second equation by substituting value of x_3. Finally, x_1 can be obtained from the first equation by means of the calculated x_3, x_2. This is known as backward substitution:

$$x_3 = \frac{1.2}{1.2} = 1$$
$$x_2 = \frac{-9-(-6.5)x_3}{-2.5} = \frac{-2.5}{-2.5} = 1 \, . \tag{3.15}$$
$$x_1 = \frac{10-5x_3-3x_2}{2} = \frac{2}{2} = 1$$

This GE strategy consisted of subsequent elimination of coefficients from the first column (related to x_1), then from the second column, and the third, etc. as shown in Figure 3.2(a). This is the column-wise order of eliminating. It may be formalized in the form of Algorithm 3.1.[1]

Algorithm 3.1 The principle of Gaussian elimination

> **for** k=1 **to** n-1 **do** {modifications loop}
>   **for** i=k+1 **to** n **do** {equations loop}
>   create r_i^k by subtracting the equation r_k^{k-1} multiplied by $a_{ik}^{k-1}/a_{kk}^{k-1}$ from the equation r_i^{k-1};

Instead of the column-wise the row-wise order may be developed. It relies on eliminating the first position of the second row, then the first and second positions of the third row, and so on, according to Figure 3.2(b). This approach is mathematically equivalent to the column-wise approach but differs in the order of eliminating components.

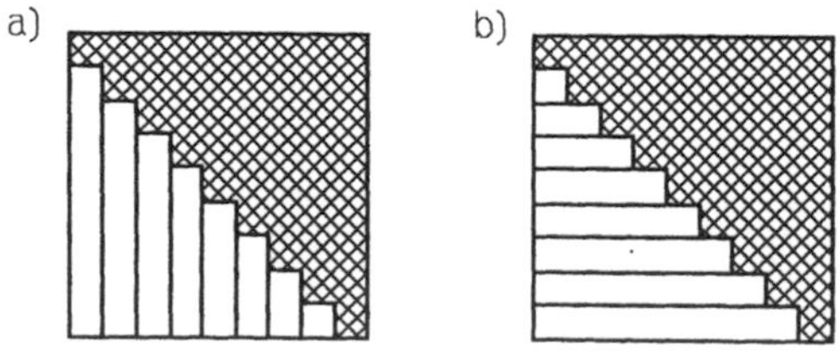

Figure 3.2 (a) Column-wise and (b) row-wise strategies of Gaussian elimination

---

[1] In the book we use a Pascal-like algorithmic language including rather obvious keywords, braces { } surrounding comments and operational braces **begin end**.

Algorithm 3.2 A principle of row-wise Gaussian elimination

> **for** i=2 **to** n **do** {equations loop}
>   **for** k=1 **to** i-1 **do** {modifications loop}
>   create r_i^k by subtracting the equation r_k^{k-1} multiplied
>   by $a_{ik}^{k-1}/a_{kk}^{k-1}$ from the equation r_i^{k-1};

Both versions are equivalent even on the score of computational complexity, although, in sparse implementations, the order of storing the matrix in memory may imply a different data access time. The column-wise algorithm will be more efficient with column-wise sparse matrix storage. Further discussion will be focused on the column-wise Algorithm 3.1. In terms of operations on coefficients (except the component x_k to be reduced) this algorithm takes the detailed form of Algorithm 3.3.

Algorithm 3.3 An extended row-wise Gaussian elimination scheme

> **for** k=1 **to** n-1 **do** {modifications loop}
>   **for** i=k+1 **to** n **do** {rows loop}
>   **begin**
> $a_{ik}^{*k-1} = a_{ik}^{k-1} / a_{kk}^{k-1}$; {auxiliary variables}
>   **for** j=k+1 **to** n **do** {columns loop}
> $a_{ij}^{k} = a_{ij}^{k-1} - a_{kj}^{k-1}\, a_{ik}^{*k-1}$; {matrix modification}
> $b_i^{k} = b_i^{k-1} - b_k^{k-1}\, a_{ik}^{*k-1}$; {right-hand sides modification}
>   **end**

The quantities a_{ik}^{*k-1} may be temporarily treated as auxiliary factors to be multiplied by both sides of the equations to gain, in the kth modification, the elimination of a variable x_k. Notice, that GE contains separate operations on the matrix and on the RHS, but carried out simultaneously.

Algorithm 3.4 Gaussian elimination dedicated row-wise backward substitution

> **for** i=n **downto** 1 **do**
> **begin**
>   **for** j=n **downto** $i+1$ **do**
> $b_i = b_i - a_{ij}\, x_j$;
> $x_i = b_i / a_{ii}$;
> **end**

After the elimination has been finished, according to Algorithm 3.3, we have to carry out the backward substitution (as in (3.15)). It is formalized in Algorithm 3.4 (superscripts of the modifications may be omitted as after GE they are uniquely defined as a row number minus 1). As can be seen, this algorithm applies the row-wise access to coefficients. The similar, column-wise approach is shown in Algorithm 3.5.

Algorithm 3.5 Gaussian elimination dedicated column-wise backward substitution

```
for j=n downto 1 do
begin
   for i=1 to j-1 do
      b_i = b_i - a_ij x_j;
   x_j = b_j / a_jj;
end
```

3.2.1.2 Gaussian elimination leads to a triangular decomposition in fact

As is known from the previous consideration, Algorithm 3.3 has transformed the system $\mathbf{A}\mathbf{x}=\mathbf{b}$ into the triangular form $\mathbf{U}\mathbf{x}=\bar{\mathbf{b}}$, as e.g. (3.14). The matrix $\mathbf{A}$ has been changed into the upper triangular matrix $\mathbf{U}$ and the vector $\mathbf{b}$ into the modified vector $\bar{\mathbf{b}}$. Hence the Gaussian elimination, in terms of the linear algebra, is equivalent to a certain linear mapping $\mathbf{M}$ applied to both sides of the equations. The LHS has been transformed into $\mathbf{MA}$ equal to $\mathbf{U}$ while the RHS transforms into $\mathbf{Mb}$, equal to $\bar{\mathbf{b}}$.

In order to better understand what operations are behind the mapping $\mathbf{M}$ let us observe the transformation process of a 3x3 system.

$$\begin{bmatrix} a_{11}^0 & a_{12}^0 & a_{13}^0 \\ a_{21}^0 & a_{22}^0 & a_{23}^0 \\ a_{31}^0 & a_{32}^0 & a_{33}^0 \end{bmatrix} \begin{bmatrix} x_1 \\ x_2 \\ x_3 \end{bmatrix} = \begin{bmatrix} b_1^0 \\ b_2^0 \\ b_3^0 \end{bmatrix}. \tag{3.16}$$

After the first modification, consisting of subtracting from the 2nd and the 3rd row the first row multiplied by their respective $a_{i0}^{*1}=a_{i1}^0/a_{11}^0$, $(i=2,3)$, we get

$$\begin{bmatrix} a_{11}^0 & a_{12}^0 & a_{13}^0 \\ & a_{22}^1 & a_{23}^1 \\ & a_{32}^1 & a_{33}^1 \end{bmatrix} \begin{bmatrix} x_1 \\ x_2 \\ x_3 \end{bmatrix} = \begin{bmatrix} b_1^0 \\ b_2^0 - a_{21}^{*0}\, b_1^0 \\ b_3^0 - a_{31}^{*0}\, b_1^0 \end{bmatrix}. \tag{3.17}$$

Notice that a_{ij}^{*k} are also auxiliary variables involved in Algorithm 3.3. After the second modification, that relies on subtracting from the 3rd row the 2nd row multiplied by $a_{32}^{*1}=a_{32}^{1}/a_{22}^{1}$, we get

$$\begin{bmatrix} a_{11}^0 & a_{12}^0 & a_{13}^0 \\ & a_{22}^1 & a_{23}^1 \\ & & a_{33}^2 \end{bmatrix}\begin{bmatrix} x_1 \\ x_2 \\ x_3 \end{bmatrix}=\begin{bmatrix} b_1^0 \\ b_2^0-a_{21}^{*0}b_1^0 \\ b_3^0-a_{31}^{*0}b_1^0-(b_2^0-a_{21}^{*0}b_1^0)a_{32}^{*1} \end{bmatrix}. \tag{3.18}$$

As can be seen, from the above right-hand side, GE applies the mapping **M** which is of a lower triangular form. A notation b_1^0, b_2^1, b_3^2 has been introduced in Algorithm 3.3 for the right-hand sides $\bar{\mathbf{b}}$ derived from GE. Hence, we may express the right-hand sides as follows

$$\begin{bmatrix} 1 & & \\ -a_{21}^{*0} & 1 & \\ -a_{31}^{*0}+a_{32}^{*1}a_{21}^{*0} & -a_{32}^{*1} & 1 \end{bmatrix}\begin{bmatrix} b_1^0 \\ b_2^0 \\ b_3^0 \end{bmatrix}=\begin{bmatrix} b_1^0 \\ b_2^1 \\ b_3^2 \end{bmatrix}. \tag{3.19}$$

This formula is an example of the general one $\mathbf{M}\mathbf{b}=\bar{\mathbf{b}}$ From (3.19) an inverse relation may be easily obtained. From the 2nd row, b_2^0 is yielded as a function of b_2^1 and b_1^0. From the 3rd row, b_3^0 is yielded as a function of b_1^0, b_2^1, b_3^2. After certain reductions the inverse dependence takes an amazingly simple form

$$\begin{bmatrix} 1 & & \\ a_{21}^{*0} & 1 & \\ a_{31}^{*0} & a_{32}^{*1} & 1 \end{bmatrix}\begin{bmatrix} b_1^0 \\ b_2^0 \\ b_3^0 \end{bmatrix}=\begin{bmatrix} b_1^0 \\ b_2^1 \\ b_3^2 \end{bmatrix}. \tag{3.20}$$

This can be easily extended to any number of equations. The inverse of GE mapping of the RHS $\mathbf{M}^{-1}\bar{\mathbf{b}}=\mathbf{b}$ takes the general form

$$\begin{bmatrix} 1 & & & & \\ a_{21}^{*0} & 1 & & & \\ a_{31}^{*0} & a_{32}^{*1} & 1 & & \\ \cdots & \cdots & \cdots & \cdots & \\ a_{n1}^{*0} & a_{n2}^{*1} & a_{n3}^{*2} & \cdots & 1 \end{bmatrix}\bar{\mathbf{b}}=\mathbf{b} \tag{3.21}$$

where the matrix (further denoted as **L** instead of $\mathbf{M}^{-1}$) is lower triangular composed of the auxiliary elements a_{ik}^{*k-1} which are calculated by Algorithm 3.3 during elimination. Hence, instead of carring out operations on the RHS in Algorithm 3.3 one may store the elements of **L** and solve (3.21) separately. This would stand for exactly the same operations but extracted from the elimination procedure, thanks to the storage of **L**.

Due to the relation $\mathbf{M}\,\mathbf{A}=\mathbf{U}$ the matrix **A** is equal to $\mathbf{M}^{-1}\mathbf{U}=\mathbf{L}\mathbf{U}$. Thus, provided that **L** exists, **A** has been factorized into a product of two triangular matrices, which will be discussed in detail further on.

During the Gaussian elimination, subdiagonal zero elements, arising at each *k*th modification from the reduction of the variable x_k, should not be stored. The elements of **U** are the only ones to be written down at upper triangular positions of **A**. Hence the lower triangular part remains free and may be filled with all subdiagonal elements of **L**. Thus, the GE is the same as calculation and storage of both matrices **L** and **U**, then derivation of the auxiliary vector $\bar{\mathbf{b}}$ from (3.21) and finally the backward substitution. This is much more convenient then the original Gaussian elimination Algorithm 3.3. Such an approach is called the LU factorization method. It will be discussed in the subsections that follow.

3.2.2 LU factorization methods

3.2.2.1 Two alternatives of LU normalization

In the Gaussian elimination (GE) so far, the resulting **U** factor includes nonnormalized (i.e. not unit) diagonal elements. The factor **L**, however, contains normalized columns and unit diagonal elements. This is related to a structure of algorithm whose elimination stage has been taken to be free of divisions by diagonal elements. Instead of this simple version of GE a dual approach may be developed. In this, at its elimination stage each row of the upper triangular system $\mathbf{U}\,\mathbf{x}=\bar{\mathbf{b}}$ is divided by the diagonal elements of **U**. Recalling the example (3.12), instead of the upper triangular system (3.14), gained previously, we additionally divide both sides of the first equation by 2, of the second by -2.5 and of the third by 1.2. Hence, a system with a normalized upper triangular matrix is obtained

$$
\begin{aligned}
r_1^{*0}:&\quad x_1+1.5x_2+2.5x_3= 5\\
r_2^{*1}:&\quad x_2+2.6x_3=3.6\\
r_3^{*2}:&\quad x_3=1.0
\end{aligned}
\qquad (3.22)
$$

where the ith equation after the kth modification and the additional normalization is referred to as r_i^{*k}. This dual version of GE may be formalized as Algorithm 3.6.

Algorithm 3.6 Gaussian elimination scheme with normalized rows

> **for** k=1 **to** n **do** {modifications loop}
> **begin**
>   create r_k^{*k-1} by dividing both sides of r_k^{k-1}
>   by a_{kk}^{k-1}; {normalization of equation}
>   **for** i=k+1 **to** n **do** {equations loop}
>   create r_i^{k} by subtracting the equation r_k^{*k-1} multiplied by
> a_{ik}^{k-1} from the equation r_i^{k-1};
> **end**

Each, kth modification is preceded by the division of the kth equation by the coefficient a_{kk}^{k-1} to yield a normalized equation with unit elements on the diagonal. Hence, the matrix **U** with diagonal units is produced.

Algorithm 3.6 can be easily developed to the form of Algorithm 3.7.

Algorithm 3.7 Gaussian elimination with a normalized **U** matrix

> **for** k=1 **to** n-1 **do**
> **begin**
>   **for** j=k+1 **to** n **do**
> $a_{kj}^{*k-1} = a_{kj}^{k-1} \,/\, a_{kk}^{k-1}$;
> $b_k^{*k-1} = b_k^{k-1} \,/\, a_{kk}^{k-1}$;
>   **for** i=k+1 **to** n **do**
>   **begin**
>   **for** j=k+1 **to** n **do**
> $a_{ij}^{k} = a_{ij}^{k-1} - a_{kj}^{*k-1}\, a_{ik}^{k-1}$;
> $b_i^{k} = b_i^{k-1} - b_k^{*k-1}\, a_{ik}^{k-1}$;
>   **end**
> **end**

The mapping **M** after some algebra, of the type already shown in Equation (3.19), appears to be lower triangular but with the nonnormalized diagonal. Its inverse can be proved to be

$$\boldsymbol{L} = \begin{bmatrix} a_{11}^{0} & & & \\ a_{21}^{0} & a_{22}^{1} & & \\ \cdots\cdots & & & \\ a_{n1}^{0} & a_{n2}^{1} & \cdots\cdots & a_{nn}^{n-1} \end{bmatrix} \tag{3.23}$$

instead of the normalized form (3.21), known so far. As can be seen, in this version divisions by diagonal elements have been carried from the lower to the upper part of a matrix.

Thus, in operating on the normalized matrix **U** the backward substitution Algorithm 3.4 or 3.5 has to be free of divisions by diagonal elements. Suitable changes in these algorithms may be done by the reader.

The approach that makes use of the normalized matrix **L** is also known as of Doolittle's type, while the dual approach with the normalized **U** is sometimes called the Crout's version.

3.2.2.2 When is LU factorization realizable?

By recalling the example (3.12) we see that the Gaussian elimination (GE), that is the LU factorization, is successful only if all diagonal elements obtained during the course of the algorithm are nonzero. We should notice that the values of the diagonal terms of the original system before calculations are of no importance at all. A nonzero value for the modified diagonal elements, arising during the elimination, is the only essential thing. The diagonal elements a_{ii}^{i-1} $(i = 1, ..., n)$ being dynamically worked out are known as the **pivots**. Let us consider a simple example of a system whose matrix is nonsingular

$$\begin{aligned} x_1 + 2x_2 + 3x_3 &= 6 \\ x_1 + 2x_2 + x_3 &= 4 \; . \\ 2x_1 + x_2 + 3x_3 &= 6 \end{aligned} \tag{3.24}$$

After the first modification the second pivot becomes zero

$$\begin{aligned} x_1 + 2x_2 + 3x_3 &= 6 \\ 0x_2 - 2x_3 &= -2 \\ -3x_2 - 3x_3 &= -6 \end{aligned} \tag{3.25}$$

and hence a second modification appears to be impossible. Thus, there is a question whether GE is of any use in this case. Fortunately we see that if we interchange the order of the 2nd and the 3rd equations in (3.24)

$$\begin{aligned} x_1 + 2x_2 + 3x_3 &= 6 \\ 2x_1 + \ \ x_2 + 3x_3 &= 6 \\ x_1 + 2x_2 + \ \ x_3 &= 4 \end{aligned} \tag{3.26}$$

we obtain a quite different result after the 1st modification

$$\begin{aligned} x_1 + 2x_2 + 3x_3 &= \ \ 6 \\ -3x_2 - 3x_3 &= -6 \\ -0x_2 - 2x_3 &= -2 \end{aligned} \tag{3.27}$$

accidentally in this case the same as after the 2nd modification. Hence, the final backward substitution yields the solution $x_1 = x_2 = x_3 = 1$. A similar transformation into the solvable case might be achieved by permuting the order of unknowns in the equations. Notice that permuations of equations (matrix and right-hand side rows) and unknowns (matrix columns) are, for a mathematician, only an unimportant manipulation with no implications for the solution. However, this is not unimportant in numerical analysis, where permutations decide whether GE will be successful or not. We may state in general that GE is successful only if all the resulting pivots are nonzero. This never happens in the singular matrix case which may have at most an ambiguous solution. Hence, GE is not useful in singular cases; moreover it may numerically fail even for nonsingular systems if implemented without permutations of row or column order.

There is a strong connection between the pivots and the determinants of certain submatrices. Let $\mathbf{A}_k$ be the matrix arising from $\mathbf{A}$ by deleting rows and columns of a number greater then k. It can be proved that the pivot a_{kk}^{k-1}, $k=1, ..., n$ exists if and only if det $\mathbf{A}^{k-1} \neq 0$ (i.e. the smaller of two lined submatrices in Figure 3.3 is nonsingular), and hence can be expressed as a ratio $a_{kk}^{k-1} = \det \mathbf{A}_k / \det \mathbf{A}_{k-1}$, of the determinants of two neighbouring submatrices. For the first pivot it is simply equal to a_{11}, that is $\det \mathbf{A}_0 = 1$ has to be taken in the above rule.

From this rule of expressing pivots in terms of determinants it can be seen that all pivots are nonzero if all submatrices $\mathbf{A}_k$ are nonsingular. Once the matrix $\mathbf{A}$ is nonsingular there exists the order of rows and columns such that all submatrices $\mathbf{A}_k$ are nonsingular and hence GE may be carried out. Unfortunately, examination of the determinants is too costly a way of choosing a proper ordering. On the other hand the reordering itself appears to be an indispensable part of the procedure.

Another important conclusion is that if the pivots have been found then all of them may be nonzero, which indicates the matrix nonsingularity (once the last pivot is nonzero) or singularity (otherwise). In such a case we have the following constructive formula for the determinant

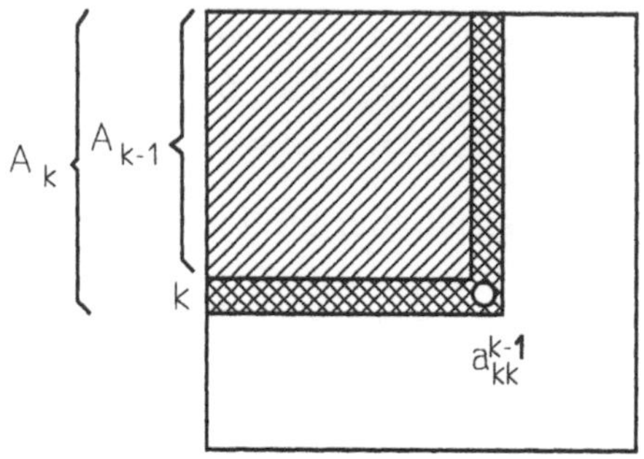

Figure 3.3 Interpretation of pivots by means of determinants of submatrices

$$\det \mathbf{A} = \det \mathbf{A}_n = \frac{\det \mathbf{A}_n}{\det \mathbf{A}_{n-1}} \frac{\det \mathbf{A}_{n-1}}{\det \mathbf{A}_{n-2}} \cdots \frac{\det \mathbf{A}_2}{\det \mathbf{A}_1} \frac{\det \mathbf{A}_1}{\det \mathbf{A}_0} = a_{nn}^{n-1} \ldots a_{11}^{0} \quad (3.28)$$

as a product of pivots. Once reordering has been implemented the rule (3.28) needs a modification: each permutation of rows or columns changes the sign of the determinant. Hence, the sign of (3.28) has to be changed if the number of permutations is odd.

If the GE fails after some steps, and hence not all pivots have been calculated, the issue of matrix singularity remains undecided.

Efficient reordering methods will be discussed in the next subsections.

3.2.2.3 A variety of LU factorization schemes

In Section 3.2.1.2 the Gaussian elimination has been shown to yield the LU factorization of a matrix. Four schemes of this factorization have been introduced: the column-wise and row-wise scheme as well as a scheme with a normalized upper or lower triangular factor. In what follows we turn to a variety of LU factorizations without viewing them as Gaussian eliminations any more. The four approaches mentioned differ in the location of divisions and order of entering and modifying matrix elements. Moreover, we will show a few other structures of calculations.

Once the manner of normalization is decided, the order of the calculations does not imply differences in factorization effects. The factorization obeys the equation $\mathbf{L}\,\mathbf{U} = \mathbf{A}$ that generates n^2 constraints imposed on n^2+n unknown elements of triangle factors. Once the n diagonal elements of $\mathbf{L}$ or $\mathbf{U}$ are taken to be unity then the factorization becomes unique irrespective of the order of calculations. Assuming a general form of triangular factors, containing components l_{ij}, u_{ij}, the product of factors may be formalized in the form

$$\sum_{k=1}^{\min(i,j)} l_{ik} u_{kj} = a_{ij}, \quad i,j=1, ..., n \tag{3.29}$$

which is satisfied by any factorization algorithm. Due to two possible normalizations, in further consideration we will focus on the case of the **L** matrix having the unit diagonal (l_{ii}=1, i=1,...,n). Then, the available factorization algorithms will differ in the organization of calculations but not in the result; that is always the same, satisfying (3.29), namely

$$\begin{aligned} u_{ij} &= a_{ij} - \sum_{k=1}^{i-1} l_{ik} u_{kj}, \quad j \geq i \\ l_{ij} &= \left(a_{ij} - \sum_{k=1}^{j-1} l_{ik} u_{kj}\right) / u_{jj}, \quad j<i \end{aligned} \tag{3.30}$$

When choosing the organization of calculations, it is essential to store all intermediate results in the same original matrix area, so the new data has to replace the previous data. Hence, the memory requirement is limited to n^2 numbers, **U** is located in the upper and diagonal part of the quadratic table, while subdiagonal elements of **L** are located in the lower part. The diagonal units of **L** need not be stored at all, obviously.

Four practical versions of the LU factorization (3.30) have been invented: submatrix-wise, column-wise, row-wise and mixed row-column-wise. They differ in the order of elements entered, according to Figure 3.4. The submatrix-wise approach outlined in Figure 3.4(a) is the same as the Gaussian elimination Algorithm 3.3. At the kth step it consists of dividing a hatched portion of the kth column by a pivot marked out with a circle, and then of modifying all elements of the submatrix including rows and columns with numbers ranging from k+1 to n. This version is formalized in Algorithm 3.8.

Algorithm 3.8 A submatrix-wise LU factorization

```
for k=1 to n-1 do {modifications loop}
   for i=k+1 to n do {rows loop}
   begin
      a_ik = a_ik / a_kk; {column normalization}
      for j=k+1 to n do {columns loop}
         a_ij = a_ij - a_kj a_ik; {submatrix modifications}
   end
```

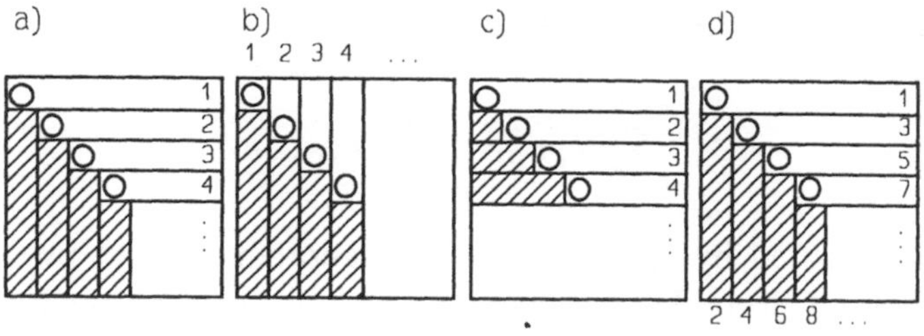

Figure 3.4 Four schemes of LU factorization strategy: (a) submatrix-wise, (b) column-wise, (c) row-wise, (d) mixed row-column-wise

This algorithm clearly exhibits the principle of the elementary modification step performed for subsequent values of k. To modify the position (i, j) of a matrix two border elements (i, k) and (k, j) are taken, multiplied and subtracted from a value at this position, as drawn in Figure 3.5. In different LU factorization algorithms this elementary step is applied in different sequences.

A short LU factorization, according to the above mnemotechnic rule, will be presented. Let us take the matrix

5	3	6
1	1	2
2	2	3

The first row needs no modification, 5 is the first pivot, and the subdiagonal portion of the first column needs division by this pivot. This produces the matrix whose 2×2 submatrix needs the first modification (for k=1)

5	3	6
0.2	1	2
0.4	2	3

Once k=1, the 2×2 submatrix is being modified now by means of border elements from the first row and column: 1-3*0.2=0.4, 2-0.2*6=0.8, 2-0.4*3=0.8, 3-0.4*6=0.6

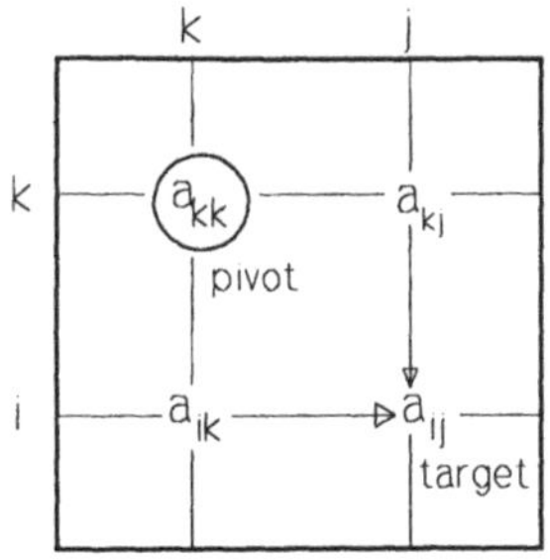

Figure 3.5 The principle of elementary modification

5	3	6
0.2	0.4	0.8
0.4	0.8	0.6

The second pivot is 0.4 and next the subdiagonal portion of the second column is divided by this. Now the 1×1 submatrix is ready for the second modification (for k=2)

5	3	6
0.2	0.4	0.8
0.4	2	0.6

The second modification manipulates only one element, making use of border elements taken from the second row and column: $0.6 - 2 \cdot 0.8 = -1$. We gain the final result

5	3	6
0.2	0.4	0.8
0.4	2	-1

The above table incorporates the part **U** in the diagonal and upper area

5	3	6
	0.4	0.8
		-1

and the **L** part including the subdiagonal elements and the diagonal units that are predefined

1		
0.2	1	
0.4	2	1

The resulting triangular factors may be used for further solution of the system as will be shown later on.

Now, we turn to other LU strategies. In the column-wise Algorithm 3.9 columns are sequentially entered and processed as shown in Figure 3.4(b).

Algorithm 3.9 A column-wise LU factorization

```
for j=1 to n do {columns loop}
begin
   for k=1 to j-1 do {modifications loop}
      for i=k+1 to n do {rows loop}
         a_ij = a_ij - a_kj a_ik; {column modifications}
   for i=j+1 to n do {rows loop once more}
      a_ij = a_ij / a_jj; {column normalization}
end
```

A practical factorization procedure may also be implemented according to the mixed row-column scheme of Figure 3.4(d). Then, Algorithm 3.10 has been obtained where the proper portions of rows and columns are produced alternately, and then a column portion is normalized by a pivot.

The row-wise scheme of Figure 3.4(c) has been left for the reader's invention.

Algorithm 3.10 A mixed row-column-wise LU factorization

```
for i=2 to n do
begin
  for j=i to n do {loop along a portion of i-th row}
    for k=1 to i-1 do
      a_ij = a_ij - a_kj a_ik; {modifications}
  for j=i+1 to n do {loop along a portion of i-th column}
  begin
    for k=1 to i-1 do
      a_ji = a_ji - a_ki a_jk; {modifications}
    a_ji = a_ji / a_ii; {column normalization}
  end
end
```

3.2.2.4 Solution of a factorized system

As we have already seen, the linear system $\mathbf{A}\,\mathbf{x}=\mathbf{b}$ after the matrix factorization $\mathbf{A}=\mathbf{L}\,\mathbf{U}$ has been transformed into the form $\mathbf{L}\,\mathbf{U}\,\mathbf{x}=\mathbf{b}$. The inside term $\mathbf{U}\,\mathbf{x}$ may be denoted as $\bar{\mathbf{b}}$ and treated as an auxiliary variable. Thus, we gain a triangular system $\mathbf{L}\,\bar{\mathbf{b}}=\mathbf{b}$ exactly the same as (3.21) achieved when Gaussian elimination was discussed. To solve this system, $\bar{b}_1$ should be derived from the first row and substituted into the second equation, and this should be subsequently solved for $\bar{b}_2$, etc. Let us recall the example from Section 3.2.2.3 and assume $\mathbf{b}'=[14,4,7]$.

$$\begin{bmatrix} 1 & & \\ 0.2 & 1 & \\ 0.4 & 2 & 1 \end{bmatrix} \begin{bmatrix} \bar{b}_1 \\ \bar{b}_2 \\ \bar{b}_3 \end{bmatrix} = \begin{bmatrix} 14 \\ 4 \\ 7 \end{bmatrix}$$

The following intermediate solution arises

$$\begin{aligned} \bar{b}_1 &= 14 \\ \bar{b}_2 &= 4-0.2\bar{b}_1 = 4-0.2*14 = 1.2 \\ \bar{b}_3 &= 7-0.4\bar{b}_1-2\bar{b}_2 = 7-0.4*14-2*1.2 = -1 \end{aligned} \tag{3.31}$$

Due to the order of deriving the unknowns this procedure is called the **forward substitution**. It may be expressed in general as

$$\bar{b}_i = -\Big(\sum_{k=1}^{i-1} l_{ik}\bar{b}_k\Big) + b_i, \quad i=1, ..., n \tag{3.32}$$

Implementation of the forward substitution may be row or column-wise. In the row-wise version, elements of **L** are entered in a row-wise manner. Hence, the rule (3.32) can be implemented as Algorithm 3.11 where the intermediate solution $\bar{\mathbf{b}}$ is located subsequently in places of **b**.

Algorithm 3.11 A row-wise forward substitution

```
for i=2 to n do
   for k=1 to i-1 do
      b_i = b_i - a_ik b_k;
```

After the forward substitution has been carried out the system $\mathbf{U}\,\mathbf{x} = \bar{\mathbf{b}}$ is solved. Turning to our example we gain the system

$$\begin{bmatrix} 5 & 3 & 6 \\ & 0.4 & 0.8 \\ & & -1 \end{bmatrix} \begin{bmatrix} x_1 \\ x_2 \\ x_3 \end{bmatrix} = \begin{bmatrix} 14 \\ 1.2 \\ -1 \end{bmatrix}$$

which needs the known **backward substitution**, which consists of deriving the last unknown from the last equation, substituting it in the previous equation, solving it for the previous unknown, etc.

Thus, we have

$$\begin{aligned} x_3 &= \bar{b}_3/(-1) = 1 \\ x_2 &= (\bar{b}_2 - 0.8x_3)/0.4 = 1 \\ x_1 &= (\bar{b}_1 - 6x_3 - 3x_2)/5 = 1 \end{aligned} \tag{3.33}$$

This stage may be in general drawn up as

$$x_i = \Big[-\Big(\sum_{k=n}^{i+1} u_{ik}x_k\Big) + \bar{b}_i\Big] \,/\, u_{ii}, \quad i=n, ..., 1 \tag{3.34}$$

but its implementation may be similarly row-wise or column-wise. The row-wise implementation takes the form of Algorithm 3.12.

Algorithm 3.12 A row-wise backward substitution

```
for i=n downto 1 do
begin
  for k=i+1 to n do
    b_i = b_i - a_ik b_k;
  b_i = b_i / a_ii;
end
```

3.2.2.5 Numerical complexity of the LU method

Numerical complexity is understood as the number of floating-point arithmetic operations during the course of an algorithm. Let a system be composed of n equations. After some algebra it can be shown that for all variants of the LU factorization, irrespective of which factor is normalized and what is the order of modifications, it requires $n^2/2 - n/2$ divisions, $n^3/3 - n^2/2 + n/6$ multiplications and the same number of subtractions. The numerical complexity of the forward and backward substitutions, also irrespective of which version is involved, is determined by n divisions, $n^2 - n$ multiplications and the same number of subtractions.

In very rough estimations of the computational effort a unit known as long operation has been used. The long operation is equivalent to division alone or multiplication and subtraction taken together, as they need approximately the same computational effort. Hence, the LU factorization requires $n^3/3 - n/3$, while forward-backward substitutions need n^2 long operations. Such an estimation of numerical complexity is not valid in sparse matrix implementations. In the full matrix approach it gives a good approximate measure of computational effort.

Consider a system described by the matrix $\mathbf{A}$ which has to be solved m times with different right-hand sides $\mathbf{b}_j$, $j=1, ..., m$. Then, the most economical approach needs only one LU factorization and m forward and backward substitutions. Hence, the full matrix computations require $n^3/3 - n/3 + mn^2$ long operations, much less than $m(n^3/3 - n/3 + n^2)$, arising if the LU factorization and forward-backward substitutions were, pointlessly, carried out m times. Total separation of factorization and substitutions is one of the essential advantages of the LU method.

3.2.3 Numerical difficulties in the LU method

3.2.3.1 How the order of rows and columns affects a solution inaccuracy

It has been shown in previous sections, that the order of rows and columns quite often decides whether LU factorization is successful or not, even in a nonsingular case. Moreover once the limited accuracy of machine arithmetic is taken into account this order can cause considerable error in a numerical solution. We will consider an instructive example. Let the machine arithmetic be of two accurate decimal digits, correctly rounded (that is not chopped). This degree of precision results in a representation error of any floating-point number x: $|(\tilde{x}-x)/x| \le \varepsilon = 0.05$. We consider the system

$$\begin{bmatrix} 0.031 & 3.2 \\ 72 & 100 \end{bmatrix} \begin{bmatrix} x_1 \\ x_2 \end{bmatrix} = \begin{bmatrix} 6.3 \\ 7300 \end{bmatrix} \tag{3.35}$$

whose exact solution is $x_1 = 100, x_2 = 1$. Its matrix, factorized taking into account the assumed precision, takes the form

0.031	3.2
72/0.031=2300	100-2300*3.2=100-7400=-7300

Hence, after forward substitution the following variables are obtained

$$\begin{aligned} \bar{b}_1 &= 6.3 \\ \bar{b}_2 &= 7300 - 2300 * 6.3 = 7300 - 14000 = -6700 \end{aligned} \tag{3.36}$$

while after backward substitution the following final result arises

$$\begin{aligned} x_2 &= (-6700)/(-7300) = 0.92 \\ x_1 &= (6.3 - 3.2 * 0.92)/0.031 = (6.3 - 2.9)/0.031 = 110 \end{aligned} \tag{3.37}$$

It exhibits something like a 10 per cent difference in comparison with the accurate solution. The reader might view this as the inevitable effect of so poor an arithmetic, but then he would be wrong. To demonstrate that the same arithmetic can give a much better solution let us solve this system with rows exchanged: the first row becoming the second while the second becomes the first. The same arithmetic gives a factorized matrix:

72	100
0.031/72=4.3 10^{-4}	3.2-4.3 10^{-4}*100=3.2-0.043=3.2

After forward substitution (including also reordered right-hand sides) we have

$$\begin{aligned} \bar{b}_1 &= 7300 \\ \bar{b}_2 &= 6.3 - 4.3\ 10^{-4} * 7300 = 6.3 - 3.1 = 3.2 \end{aligned} \tag{3.38}$$

while after backward substitution the quite accurate result

$$\begin{aligned} x_2 &= 3.2/3.2 = 1 \\ x_1 &= (7300 - 100)/72 = 100 \end{aligned} \tag{3.39}$$

is gained irrespective of the poor arithmetic.

To see why the order implies such a difference in numerical errors let us observe the factorized matrix. The worst error arises in element $u_{22} = 100 - 2300 \cdot 3.2$ due to the innacuracy of $l_{21} = 2300$, though the latter's relative perturbation does not exceed ε. In fact, the absolute error of u_{22} obeys $|\Delta u_{22}| = |u_{12} \Delta l_{21}| = |3.2 \cdot 2300 \cdot \Delta l_{21}/l_{21}| \leq 7300\varepsilon$. Thus the relative error may be estimated as

$$|\Delta u_{22}/u_{22}| \leq \frac{7400}{7300}\,\varepsilon \tag{3.40}$$

This inequality has shown that the error of l_{21} is dangerously propagated to the element u_{22} and moreover will reflect on the backward substitutions.

On the other hand, if rows are interchanged then the same component takes the form: $u_{22} = 3.2 - 4.3\,10^{-4}\,100 = 3.2$. Thus the innacuracy of l_{21} produces a successfully tuned error of u_{22}

$$|\Delta u_{22}/u_{22}| \leq \frac{4.3\,10^{-4}\,100}{3.2 - 4.3\,10^{-4}\,100}\,\varepsilon = 0.013\,\varepsilon \tag{3.41}$$

Comparison of (3.40) and (3.41) reveals that at the *k*th modification of the LU factorization the weak point is the error tuning ratio

$$\frac{|l_{ik} u_{kj}|}{|a_{ij} - l_{ik} u_{kj}|} \tag{3.42}$$

A similar effect arises in the forward substitution (let us compare (3.36) and (3.38)) where the error of the intermediate solution appears to be

$$\left|\Delta\bar{b}_i/\bar{b}_i\right| \le \frac{|l_{ik}\bar{b}_k|}{|\bar{b}_i - l_{ik}\bar{b}_k|}\,\varepsilon \tag{3.43}$$

A small value of the tuning ratio (3.42) is gained once $|l_{ik}u_{kj}| \ll |a_{ij}|$, that is if l_{ik} is small. In the first case discussed we have: $l_{21}=2300$, while in the second: $l_{21}=4.3\ 10^{-4}$. This observation explains our troubles with numerical errors. The algorithm including the lower matrix normalization consisted of dividing subdiagonal elements by pivots. If pivots are big numbers in comparison with the subdiagonal terms then the components of **L** are small.

To examine this problem in detail let us consider the submatrix-wise LU factorization Algorithm 3.8. The state of the matrix, before the kth modification is outlined in Figure 3.6. At this moment processing of rows and columns numbered from 1 to k-1 has just been finished. The submatrix including rows and columns ranging from k to n (lined in the figure) needs further manipulations. This involves dividing the subdiagonal (doubly lined) portion of the kth column by the pivot a_{kk}, and then modifying the submatrix incorporating the rows and columns ranging from k+1 to n. To understand the pivoting idea notice that, before pivotal divisions and modifications, any element (say a_{ij}) of the marked submatrix may become the pivot by selecting the jth column and interchanging it with the kth, and by selecting the ith row and permutating it with the kth. This strategy is known as **full pivoting**. Once reordering of rows is the only way acceptable, a pivot may be chosen from the doubly

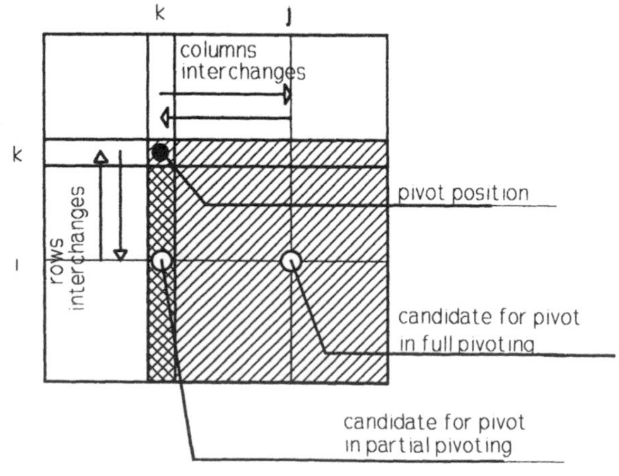

Figure 3.6 Graphical representation of partial and full pivoting

lined (in Figure 3.6) diagonal and subdiagonal portion of the *k*th column. Then, the selected row, say the *i*th, is interchanged with the *k*th. This strategy is known as **partial pivoting**. In both approaches, once a pivot of a maximum magnitude is selected from candidates then, at each modification step of the LU factorization (and so of the forward substitution) the error tuning ratio (3.42) will be the smallest, as the resulting elements of the lower triangular matrix appear to be not greater then unity. Thus, selecting the maximum pivot at the *k*th step we minimize modification errors at this step. However, we are not sure how large pivots will arise in later steps and whether they would not be more advantageous if another choice of pivot were made at the *k*th step. Hence, the pivoting involved is like a local optimization, i.e. we cannot find out such an order of rows and columns that guarantees the global minimum of the solution error. In practice, the partial pivoting is usually quite sufficient. If the matrix is primarily of rather dominating diagonal terms, then the choice of pivots from the diagonal, known as **diagonal pivoting**, is satisfactory. It consists of identical permutations of rows and columns. This strategy, due to its essential advantage, is often used in sparse implementations, for instance incorporated in PSPICE and OPTIMA.

In the above pivoting strategies a convenient detection of matrix singularity can be proved. Once a pivot maximized over a set of candidates appears to be zero (or very small) then a matrix is surely singular or very ill-conditioned.

3.2.3.2 Practical algorithms incorporating pivoting strategies

To approach a practical algorithm let the partial pivoting be added to each step of Algorithm 3.8, which leads to Algorithm 3.13. In practice the *k*th step of Algorithm 3.13 can be carried out by physically interchanging in memory elements of the *k*th and *imax*th row of the matrix and RHS vector. Usually this is an inefficient approach, except for the sparse implementations where interchanges consist of once performed data structure manipulations. Alternatively indirect addressing is available.

However, before we proceed to these more advanced implementations, the reader may study an instructive example of the LU method with partial pivoting. Let the reordering be viewed as something like physical interchanging. The matrix, RHS and enumeration of rows (in parantheses) are drawn up at three stages (for k=1,2,3) of Algorithm 3.13. We start from the following original system

Algorithm 3.13 LU factorization including a schematic partial pivoting

```
for k=1 to n-1 do
   begin
      search the kth column from the kth to nth position for an
         element of the maximum magnitude. Let it be found in the
         row of the number imax and have the magnitude amax ;
      if (amax ≤ minimum_pivot_level) stop {matrix singularity}
      if (k≠imax) interchange the kth row with the imaxth and
         do so with the right-hand sides;
      for i=k+1 to n do
      begin
         a_ik = a_ik / a_kk; {divisions by the pivot chosen}
         for j=k+1 to n do {modifications of the submatrix}
            a_ij = a_ij - a_kj a_ik;
      end
   end
```

0.1	10	0.9	101	(1)
10	1	10	30	(2)
100	1	20	130	(3)

At the first modification, the first pivot is chosen from the first column. As 100 is the element of greatest magnitude, rows (1) and (3) should be interchanged. Then, we divide subdiagonal elements of the first column by the pivot and modify the submatrix marked off

100	1	20	130	(3)
0.1	0.9	8	30	(2)
0.001	9.999	0.88	101	(1)

At the second step a pivot is chosen from the diagonal and subdiagonal portion of the second column. As 9.999 is of the maximum magnitude, we interchange rows (2) and (1). The division by the pivot and modification of the next submatrix yield

100	1	20	130	(3)
0.001	9.999	0.88	101	(1)
0.1	1/1111	7 1023/1111	30	(2)

Now, we turn to the forward substitution. It involves right-hand sides whose permutations are consistent with the matrix permutations. After some algebra, the intermediate unknowns: $\bar{b}_1=130$, $\bar{b}_2=100.87$, $\bar{b}_3=7\ 1023/1111$ are obtained. Then the backward substitution yields $x_1=1, x_2=10, x_3=1$, which are located in places of $b_3=130$, $b_2=30$, $b_1=101$. respectively (as can be seen from the reordering involved). If rows were physically reordered, then the solutions would not be permutated at all, e.g. b_3 would be located at the first physical position.

Other implementations make use of indirect addressing. Let $\mathbf{r}$ be a vector of n integers ranging from 1 to n. We may enter the first logical row by referring to it as to the r_1th row in the physical storage. For instance, once $r_1=2$ and $r_2=1$, then the first row is at the second physical position while the second row is at the first physical position, that is the rows 1 and 2 are interchanged. Once $r_1=1$ then the first row is obviously not reordered. If rows are not reordered, then addresses have their values $r_i=i$ (i=1,...,n). To interchange the pth and the qth logical rows the addresses r_p and r_q have to be interchanged, irrespective of their previous values. Manipulations of the addresses $\mathbf{r}$ can be added to Algorithms 3.13, 3.11 and 3.12. Then, we obtain LU factorization Algorithm 3.14 followed by the forward substitution Algorithm 3.15 (a counterpart of 3.11), and then, the backward substitution Algorithm 3.16 (arising from 3.12). To simplify the notation, indices in brackets have been introduced instead of subscripts.

The indirect addressing yields a "parasitic" reordering of the solution. In the case of partial pivoting x_i arises at the physical position $b[r[i]]$. Much more complicated is the case of full pivoting where, once column and row interchanges have been stored in vectors $\mathbf{c}$ and $\mathbf{r}$ respectively, then an order of solution arises, in which $x[c[i]]=b[r[i]]$ (i=1, ..., n). Thus, an *a posteriori* reordering of the solution is necessary. If symmetric permutations, i.e. diagonal pivoting is used then there is a very convenient situation of no solution reordering.

Finally, we notice that if reordering is used, the determinant changes its sign as many times as there are row and column interchanges.

Algorithm 3.14 The LU factorization by means of the indirect addressing techniques

```
for k=1 to n do r[k] = k; {initialization of pointers}
for k=1 to n-1 do
  begin {choice of a pivot from a portion of the kth column}
    amax = 0;
    for i=k to n do if |a[r[i],k]| >amax then
      begin
        amax = |a_{iw_i,k}|; imax = i;
      end
    if amax ≤ atol then
      begin {unsuccessful LU factorization}
        r[n] = 0;
        stop
      end
    {interchange of the kth row with imaxth}
    hlp = r[k] ; r[k] = r[imax] ; r[imax] = hlp ;
    for i=k+1 to n do {modification loop}
      begin
        a[r[i],k] = a[r[i],k] / a[r[k],k] ;
        for j=k+1 to n do
          a[r[i],j] = a[r[i],j] - a[r[i],k] a[r[k],j] ;
      end
  end
```

Algorithm 3.15 The row-wise forward substitution compatible with Algorithm 3.14

```
for i=1 to n do
  for k=1 to i-1 do
    b[r[i]] = b[r[i]] - a[r[i],k] b[r[k]] ;
```

Algorithm 3.16 The row-wise backward substitution compatible with Algorithm 3.15

```
for i=n downto 1 do
begin
  for k=i+1 to n do
    b[r[i]] = b[r[i]] - a[r[i],k]] b[k] ;
  b[r[i]] = b[r[i]] / a[r[i],i] ;
end
```

3.2.3.3 Numerical error propagation in an algorithm utilizing pivoting

Partial and full pivoting provide the best numerical quality of LU method implementations. It can be shown that thanks to pivoting, and the related normalization to unity of the lower triangular factor (hence, all its components do not exceed 1), the perturbated solution $\hat{\mathbf{x}}$ of a system may be viewed as the exact solution of a certain perturbated system whose matrix is $\mathbf{A}+\Delta\mathbf{A}$, while the norm of perturbation is not too large. To formalize this, so-called, superior numerical quality problem, let us introduce the following definition.

> An algorithm for solving the system $\mathbf{A}\mathbf{x}=\mathbf{b}$ is of the superior numerical quality in a certain class of matrices $\mathbf{A}$, if a real positive constant K (called the cumulation index) exists such that for all systems of this class the solution $\tilde{\mathbf{x}}$ derived by means of this algorithm, in floating-point arithmetic, may be expressed as the exact solution of the system $(\mathbf{A}+\Delta\mathbf{A})\tilde{\mathbf{x}}=\mathbf{b}$, where (taking a linear approximation of errors) the relation $\|\Delta\mathbf{A}\| \le K\varepsilon\|A\|$ is satisfied.

The above numerical property means that the errors of all arithmetic operations may be expressed as the result of a certain (say substitute) error in the matrix representation propagated through the ideal algorithm (i.e. one free from arithmetic operation errors). The level of this substitute error is expressed by means of the cumulation index, while the condition number $\mathrm{cond}(\mathbf{A})$ may be interpreted as the gain of propagation. Due to this observation, the worst-case estimation of the solution error takes the following form

$$\frac{\|\tilde{\mathbf{x}}-\mathbf{x}\|}{\|\mathbf{x}\|} \le \frac{K\varepsilon\,\mathrm{cond}(\mathbf{A})}{1-K\varepsilon\,\mathrm{cond}(\mathbf{A})}, \tag{3.44}$$

originating from the formula (3.7), where K is the cumulation index. In the formula (4.13) $\mathrm{cond}(\mathbf{A})=\|\mathbf{A}\|\,\|\mathbf{A}^{-1}\|$ is the condition number, while ε is the level of floating-point representation error also known from Section 3.1.2.1.

To present the worst-case estimation of the cumulation index let us view a real inaccurate solution as exactly worked out from the perturbed system $(\mathbf{A}+\Delta\mathbf{A})\hat{\mathbf{x}}=\mathbf{b}$ or identically from the system$(\mathbf{L}+\Delta\mathbf{L})$ $(\mathbf{U}+\Delta\mathbf{U})\hat{\mathbf{x}}=\mathbf{b}$ whose perturbation has been incorporated into triangular factors. Combining the two formulae and considering $\mathbf{A}=\mathbf{L}\mathbf{U}$ we obtain a matrix perturbation of the form $\Delta\mathbf{A}=\mathbf{L}\,\Delta\mathbf{U}+\Delta\mathbf{L}\,\mathbf{U}$. To get convenient tools for estimation we turn to the norms introduced in Section 3.1.2.1, even though, because of the resulting simplicity of calculation, the maximum norms (l^∞) are used instead of the spectral norms. In terms of maximum norms the distance in a linear space between a point (vector) and an origin is the maximum absolute value of the vector components, where maximization over a space dimension is involved. The maximum norm of a vector and correspondent norm of a matrix induced by l^∞ can be proved to be as shown below

$$\|\mathbf{x}\| = \max_{i=1,\ldots,n} |x_i|, \qquad \|\mathbf{A}\| = \max_{i=1,\ldots,n} \sum_{j=1}^{n} |a_{ij}| \tag{3.45}$$

Using these norms an estimation $\|\Delta\mathbf{A}\| \leq \|\boldsymbol{L}\| \, \|\Delta\mathbf{U}\| + \|\Delta\mathbf{L}\| \, \|\mathbf{U}\|$ is obtained. Due to pivoting norms of triangular factors it can be shown that they satisfy the estimations $\|\mathbf{L}\| \leq \mathrm{O}(n)$, $\|\mathbf{U}\| \leq \mathrm{O}(n^2)\|\mathbf{A}\|$ (the latter is practical but not very exact). Identically, after some algebra (outside the scope of our book) we obtain $\|\Delta\mathbf{L}\| \leq \mathrm{O}(n^2)\,\varepsilon$, $\|\Delta\mathbf{U}\| \leq \mathrm{O}(n^3)\,\|\mathbf{A}\|\,\varepsilon$. Finally, the estimation $\|\Delta\mathbf{A}\| \leq \mathrm{O}(n^4)\|\mathbf{A}\|\,\varepsilon$ arises, that yields the cumulation index of the order of the fourth power of dimension n. In practice, this result is often too pessimistic - about ten times overestimated.

To check familiarity with the topics of this section we may answer the following questions and work through the following problems:

- Invent a simple 3×3 system and try to carry out the four Gaussian elimination strategies: row-wise and column-wise with and without the normalization of the diagonal.
- Explain why GE is equivalent to matrix factorization into two triangular factors.
- Demonstrate the different strategies of LU factorization: submatrix-wise, column-wise and alternate row/column-wise with normalization of lower or upper factors. Explain the row and column-wise strategy of the forward and backward substitution.
- What is the goal of reordering? Explain two main disadvantages of the LU method without reordering: execution failure and too large a numerical error. Can we solve the optimal reordering problem, i.e. find an order suitable for any nonsingular system which minimizes an overall numerical error?
- Explain the practical means of implementing the reordering: physical copying and indirect addressing as well as partial and full pivoting. What is the influence of these approaches on the order of the solution obtained?

3.3 Sparse matrix techniques

Modern electronic technology enables us to produce integrated circuits (ICs) to an extremely large scale of integration. Thus we need to use software tools for simulating circuits composed of many hundreds of thousands of elements. After all the devices and parasitic elements have been modeled we obtain a circuit which can be characterized by many hundreds or thousands of variables (nodal voltages and some currents). Thus, d.c., a.c. and time-domain analyses, after the appropriate transformations presented in earlier chapters (discretization, linearization), produce huge systems of algebraic linear equations. Their solution using the algorithms described in this chapter would be a problem of extreme numerical complexity, time-consumption and memory requirement. For instance, if a circuit is represented by a double precision (8 bytes) MNE matrix of dimension 1000, then the storage of this matrix consumes 8MB of operating memory. This is definitely too large even for modern workstations.

However, not all the coefficients of linear equations solved during the course of d.c, a.c. and transient analysis are essential. Most of them appears to be always zero, and this feature, called matrix sparsity, can be exploited to simplify the solution process. As we know, arithmetical computer operations on zero operands are executed in the same manner as on nonzero floating-point numbers, and hence, they are equally time-consuming. On the other hand, LU decomposition and forward/backward substitutions can be organized in such a way that operations are performed only on nonzero elements of the matrix and RHS vector. This gives considerable savings in the number of operations and so in time. Moreover, we need only store the values of the nonzero elements of the matrix. This provides considerable savings in memory. Such a method of organizing the process of solution of linear equations is called the sparse matrix technique. It will be discussed in this section.

3.3.1 Introductory notes on sparse matrix techniques

Modified nodal and tableau equations, introduced in Chapter 2, are the most common systems for the formulation of linearized circuit equations. Matrices and RHS vectors set up according to these approaches are very sparse. If we define a matrix sparsity index as the ratio of the number of nonzero elements to the number of all its elements, then in larger practical equations it achieves a fraction of several per cent. Since in MNE the number of nonzero elements is approximately proportional to the number

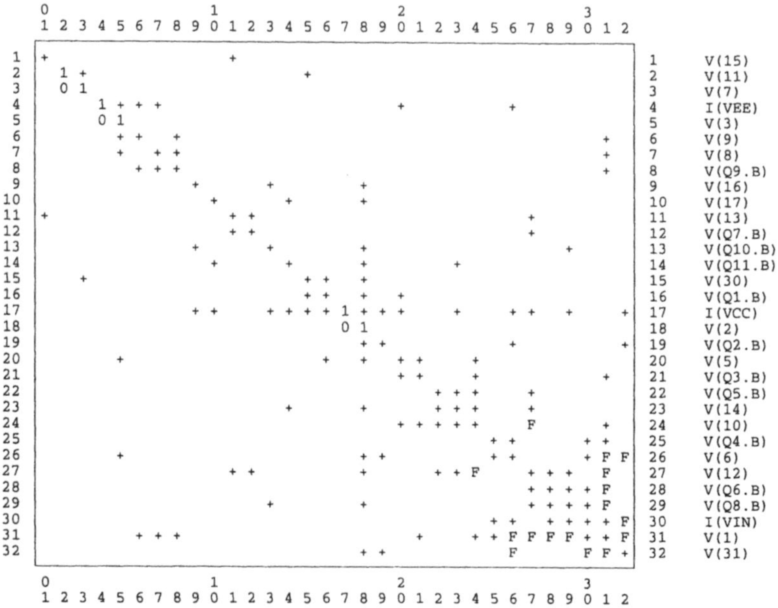

Figure 3.7 The sparsity structure of the matrix of example O3-SPAR3.CIR obtained from OPTIMA by means of the option MATVIEW

of elements, and can be estimated as $n_{nzer} \approx 5.5\, n_{elem}$, while the matrix dimension is equal to the number of circuit variables n, then the average sparsity index can be written as $s \approx 5.5 n_{elem}/n^2$. In large circuits, the number of elements is approximately proportional to the number of circuit variables (nodes), and hence, the sparsity index is inversely proportional to the circuit dimension n. For very large circuits, this sparsity index can be even smaller than 1 per cent.

On Diskette A we have sample circuits O3-SPAR1.CIR, O3-SPAR2.CIR and O3-SPAR3.CIR, where the sparse matrix of linearized extended MNE implemented in OPTIMA can be observed. The sparsity structure obtained for the bipolar example O3-SPAR1.CIR is shown in Figure 3.7. To generate the diagram of this matrix structure in the results text file we have used the option MATVIEW. Only nonzero elements are displayed. Elements varing in NR iterations are drawn with +, structural unities with 1, structural zeros with 0 and fill-ins with F. This option reports the statistics of sparsity. The list of quantities characterizing the matrices of these examples has been given in Table 3.2.

Table 3.2 Statistics of sparsity for three examples

example	dimension n	number of nonzero elements: *nnzer*	sparsity index
O3-SPAR1.CIR	61	315	8.47%
O3-SPAR2.CIR	50	212	8.48%
O3-SPAR3.CIR	32	145	14.16%

In the tableau matrix case, sparsity indexes are even smaller, since the elements introduce only a few times more nonzero elements than MNE, while the dimension of the matrix is about six times bigger. Thus, the sparsity index is about ten times smaller.

In circuit simulators such as PSPICE, OPTIMA, and many others, exploitation of sparsity is an important factor improving the efficiency of circuit analysis. However, yield is dependent on good organisation of data structures and the quality of programming.

Sparse matrix techniques for solving systems of linear circuit equations should provide the following three main features:

(i) Economical management of memory, storing only nonzero elements and their positions in a matrix, and efficiency entering these elements during the solution process. This results in memory consumption approximately proportional to the number of nonzero matrix elements.

(ii) Performing operations only on nonzero elements of the matrix and RHS vector. This considerably reduces the numerical complexity of the algorithm. In conventional methods, described in Section 3.2, the numerical complexity was about $n^3/3$, while in sparse techniques it is estimated to be $n^{1.1 \div 1.5}$. Thus savings in efficiency can be considerable.

(iii) Efficient selection of pivots controlling the maximum acceptable level of numerical error and simultaneously maintaining sparsity for **L** and **U** factors of the original matrix.

These features are achieved by the proper system of representing nonzero coefficients in data structures, and by efficient pivoting and appropriate programming of LU decomposition and forward/backward substitutions. These problems are discussed below.

3.3.2 Data structures for sparse matrix techniques

Appropriate data structures for storing nonzero elements of a circuit matrix and RHS vector are the foundation of any efficient algorithm of sparse matrix techniques. Data structures should not only store nonzero coefficients of equations, but provide fast access to each of them in the row-wise and column-wise order as well. The most ubiquitious is a system of orthogonal linked lists, recently applied in the package SPARSE [3.9], implemented in SPICE 3. A similar system has been used in the HOBO package implemented in OPTIMA. Many other sparse matrix technique packages are known [3.13]. In older FORTRAN implementations linked lists were carried out by means of arrays and indexes. Now, in more efficient C implementations, we use lists of structures and pointers for linking them.

The orthogonal linked list is a two-dimensional array of structures linked in two orthogonal directions. Each structure stands for one nonzero element of a matrix. If its dimension is n, then the RHS vector is stored as the $n+1$th column of a matrix. Each structure has two links, pointing to the next element in a row and the next element in a column. The following data are stored in one structure:

value	value of nonzero element (type double precision),
row	number of row (type integer),
col	number of column (type integer),
next_in_row	pointer to the same structure representing the next nonzero element in the same row,
next_in_col	pointer to the same structure representing the next nonzero element in the same column.

Thanks to pointers, we can proceed during calculations from one element to the next in a row and in a column. However, to find the first elements in rows and columns we require two additional arrays of pointers:

first_in_row[n]	contains at the ith position ($i=1, ..., n$) a pointer to the structure storing the first nonzero element in the ith row,
first_in_col[n]	contains at the ith position ($i=1, ..., n$) a pointer to the structure storing the first nonzero element in a column.

An additional table of pointers *diag*[n] points to structures representing diagonal matrix elements. A zero pointer means that there is no next element.

For example, the matrix below

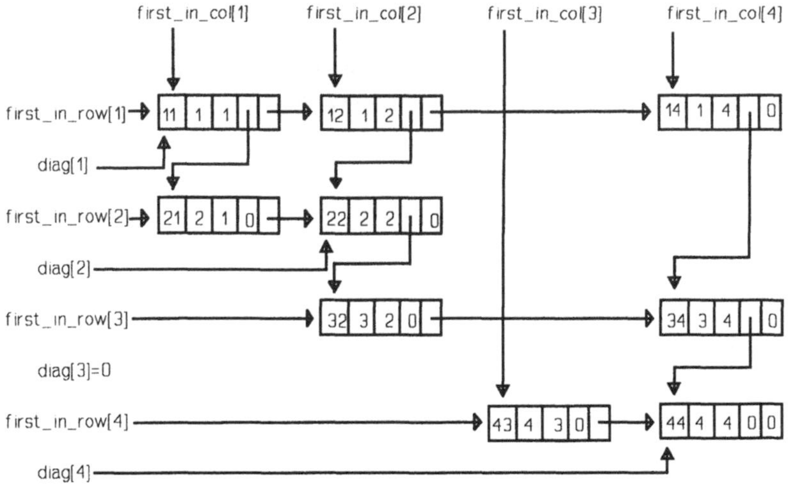

Figure 3.8 Example of a linked list storing the matrix given by Equation (3.46)

$$\begin{bmatrix} 11 & 12 & 0 & 14 \\ 21 & 22 & 0 & 0 \\ 0 & 32 & 0 & 34 \\ 0 & 0 & 43 & 44 \end{bmatrix} \begin{bmatrix} x_1 \\ x_2 \\ x_3 \\ x_4 \end{bmatrix} = \begin{bmatrix} 37 \\ 43 \\ 66 \\ 87 \end{bmatrix} \tag{3.46}$$

has the representation shown in Figure 3.8, where arrows denote links by means of pointers. These lists do not incorporate the RHS vector.

3.3.3 The problem of fill-ins and reordering

The whole process of solution of linear equations (LU factorization and forward/backward substitutions) consists of manipulating nonzero elements stored in linked lists. As it is known from Section 3.2, LU factorization changes elements of the matrix $\mathbf{Y}$, transforming it into $\mathbf{Q} = \mathbf{L} + \mathbf{U} - \mathbf{1}$, where $\mathbf{L}\,\mathbf{U} = \mathbf{Y}$. This transformation does not necessarily maintain the zero-nonzero element structure of the matrix $\mathbf{Y}$. New nonzero elements of $\mathbf{Q}$ may appear at positions which have not been represented in the linked list, since in $\mathbf{Y}$ these positions contained zeros. To store new positions, linked lists have to be extended by allocating new structures. This increases the number of nonzero elements, and thus reduces the efficiency of the sparse matrix techniques.

Matrices of MNE and tableau equations incorporate many positions never filled by the equation formulating procedure. They remain zero. We call them structural zeros. The rest of the elements are structurally nonzero. Among them we have positions filled by elements ±1, known as structural units. Other nonzero elements depend on time (excitations or parametric elements) or on signals (terms arising from linearization). They can vary in a wide range. These elements can be zero, but usually they are nonzero and, in general, have to be considered as nonzero.

If, during the LU decomposition a structural zero is replaced by a nonzero number, then this new nonzero element we call the fill-in. Fill-ins are additional nonzero elements, which have to be allocated in linked lists, and treated as nonzero terms in all further calculations. That is why economical management of memory and time requires as small a number of fill-ins as possible. Their minimization is the main criterion in reordering the rows and columns of a matrix. A proper reordering can provide a relatively small number of fill-ins, for instance not more than 5 per cent of matrix elements. Numerical experiments show that when minimization of fill-ins is not implemented, execution time can be extended by more than 100 per cent.

The number of fill-ins is strongly dependent on the order of rows and columns. As a classical example we demonstrate below a sparse matrix and give its LU factorization result according to Algorithm 3.8:

$$\begin{bmatrix} + & + & + & + & + \\ + & + & 0 & 0 & 0 \\ + & 0 & + & 0 & 0 \\ + & 0 & 0 & + & 0 \\ + & 0 & 0 & 0 & + \end{bmatrix} \underset{\text{LU}}{\Rightarrow} \begin{bmatrix} + & + & + & + & + \\ + & + & + & + & + \\ + & + & + & + & + \\ + & + & + & + & + \\ + & + & + & + & + \end{bmatrix}. \tag{3.47}$$

We notice that fill-ins appear at all structural zero positions. However, if an inverse order of rows and columns is taken, then no fill-ins arise:

$$\begin{bmatrix} + & 0 & 0 & 0 & + \\ 0 & + & 0 & 0 & + \\ 0 & 0 & + & 0 & + \\ 0 & 0 & 0 & + & + \\ + & + & + & + & + \end{bmatrix} \underset{\text{LU}}{\Rightarrow} \begin{bmatrix} + & 0 & 0 & 0 & + \\ 0 & + & 0 & 0 & + \\ 0 & 0 & + & 0 & + \\ 0 & 0 & 0 & + & + \\ + & + & + & + & + \end{bmatrix}. \tag{3.48}$$

Thus there is a way of enumeration that provides the minimum number of fill-ins. However, this strict minimization cannot be efficiently achieved. In practice, we use approximate, nonoptimal reordering strategies. To introduce a method of worst-case recognition of fill-ins, let us rewrite the main modifying formula (3.30), performed repeatedly in LU factorization and forward/backward substitutions: $a'_{ij} = a_{ij} - a_{ik}a_{kj}$. If the position (i, j) is a structural zero, while both (i, k) and (k, j) are structural nonzero elements, then in the most pessimistic case the modifying formula will generate a nonzero element (i, j). This happens always, except in those cases, where (i, j) is a nonzero element or at least one of the two structurally nonzero elements involved is actually zero. Such an outcome is possible since, the matrix contains a few per cent of nonzero elements while among structurally nonzero coefficients, there are terms dependent on signals, and they can become zero in some cases. Thus, the generation of a fill-in is not sure, but happens in the worst, though very probable case. The principle explained recognizes all fill-ins with a small excess, and gives a realistic upper bound for the number of fill-ins.

The problem of fill-in recognition and optimization has been extensively studied [3.1, 3.2, 3.5-8, 3.10-16]. The most conscientious method but also the slowest in operation is Berry's method [3.2]. The roughest but also the quickest is Markowitz method [3.11]. Surprisingly, Berry's method gives only about 5 per cent fewer fill-ins than Markowtz method. After years of studies, sparsity exploiting packages for linear circuits use mainly Markowitz method [3.11].

The Markowitz reordering method is based on the previously mentioned rule of fill-in generation during the modification: $a'_{ij} = a_{ij} - a_{ik}a_{kj}$.

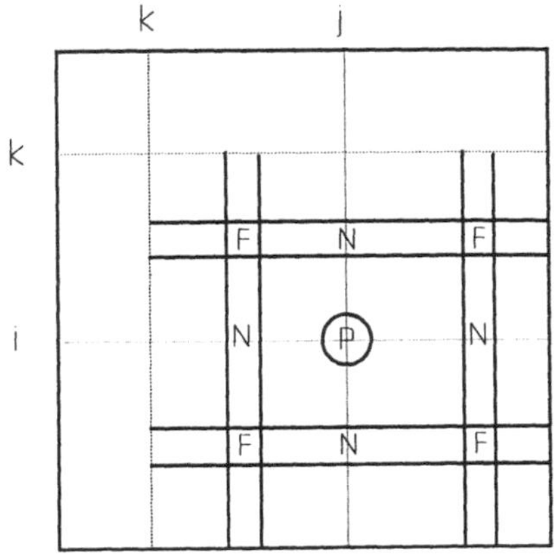

Figure 3.9 A mechanism of fill-in generation

Let us consider the kth step of the LU factorization Algorithm 3.8 with row and column reordering. The problem is drawn up in Figure 3.9. We operate only within the submatrix $\mathbf{Y}_k$, incorporating rows and columns from k to n. First, the element a_{ij} denoted as P is chosen as the pivot. This element will be treated as the kth pivot. Then, all elements of the ith row (within the submatrix $\mathbf{Y}_k$), except for the jth, will be divided by the pivot. Finally, standard modifications will be applied to all elements of the submatrix $\mathbf{Y}_k$, except for the ith row and the jth column. Fill-ins F are caused by four nonzero elements N of $\mathbf{Y}_k$. Two are in the ith row, and two others in the jth column. In the case shown, the number of fill-ins is at most 4. In general, it can be upper-bounded by the product $M_{ij}=r_i c_j$, where r_i is the number of structural nonzero elements in the ith row, while c_j is the number of structural nonzero elements in the jth column, both within the submatrix $\mathbf{Y}_k$ and without taking into account the element a_{ij}. The number M_{ij} we call the Markowitz measure corresponding to the position (i, j). At the kth step of LU factorization, Markowitz measures can be calculated with little computational effort for all elements of the submatrix $\mathbf{Y}_k$. For instance, at the first step of LU factorization of the matrix $\mathbf{Y}$, below, we can determine the array of Markowitz measures $\mathbf{M}_k$ for the whole matrix

$$\mathbf{Y}=\begin{bmatrix} + & 0 & + & 0 \\ 0 & + & 0 & 0 \\ 0 & 0 & + & + \\ + & 0 & + & + \end{bmatrix}, \qquad \mathbf{M}_1=\begin{bmatrix} 1 & . & 2 & . \\ . & 0 & . & . \\ . & . & 2 & 1 \\ 2 & . & 4 & 2 \end{bmatrix}. \tag{3.49}$$

These values of the Markowitz measure are taken into account in the selection of an optimal pivot which minimizes this measure. In the example (3.49), the pivot from the position (2, 2) provides no fill-ins, and hence should be chosen. The first row and column have to be permutated with the second row and column. Notice, that we have started from pivots which are one in a row or a column. They generate no fill-ins and are called singletons. Now, after the first modification, we obtain the matrix whose 3×3 submatrix of interest has the Markowitz measures:

$$\begin{bmatrix} . & . & . & . \\ . & x & x & 0 \\ . & 0 & x & x \\ . & x & x & x \end{bmatrix}, \quad \mathbf{M}_2=\begin{bmatrix} . & . & . & . \\ . & 1 & 2 & . \\ . & . & 2 & 1 \\ . & 2 & 4 & 2 \end{bmatrix}. \tag{3.50}$$

Using them we can select the second pivot and continue the procedure. However, we notice that two elements have the same measure 1, and we do not know which one to choose.

In the case of equal Markowitz measures, we select the element with the smallest value r_i, since it minimizes the number of divisions by a pivot. If this criterion also gives an ambiguous choice, then we select arbitrarily or, if numerical values are available, use a numerical criterion.

This Markowitz method of fill-in minimization considerably reduces numerical complexity. In fact, we can prove that the total number of arithmetic operations in a solution process can be described by the following formulae:

$$\begin{aligned} &\sum_{k=1}^{n-1} r_k && \text{divisions} \\ &\sum_{k=1}^{n-1} r_k c_k && \text{multiplications} \\ &\sum_{k=1}^{n-1} (r_k c_k - f_k) && \text{subtractions} \end{aligned} \tag{3.51}$$

where f_k indicates the number of true fill-ins in the submatrix $\mathbf{Y}_k$ (fill-in do not require subtraction).

In practice, reordering of both rows and columns is very inconvenient, since it destroys the satisfactory diagonal dominance produced by the matrix formulation procedure. Thanks to this diagonal dominance, in many practical cases we can even disregard the numerical criterion of pivoting and concentrate on the structural criterion, selecting pivots only from diagonal terms. Such a reordering is symmetrical and only one trace vector is sufficient for storing interchanges carried out. Also, we have to calculate, update, and store only the vector of measures r_k and c_k for diagonal elements, i.e. $k=1, ..., n$. If the reordering is implemented by updating links of structures in lists, then it has the same effect as in the case of physical element interchange (see Section 3.2.3.2) and the reordered solutions require reverse sorting based on the trace vector.

The above reordering can be performed using only structural information about a matrix. No numerical values are required. We arrange a loop over k from 1 to $n-1$, and at each step calculate r_k, c_k, and the Markowitz measure for all candidates at diagonal positions from k to n. Next, we select a pivot, perform the Markowitz worst-case prediction of fill-ins, allocate memory for them, and start treating them as structural nonzero elements in the next step. Then we proceed to the next k. To save time Markowitz measures are calculated once at the very beginning, and then they are updated for each k. This updating is based on updating the numbers r_i and c_j. They are increased when new fill-ins arise, and decreased when nonzero elements already of no interest leave the processed submatrix. We start diagonal pivoting from a determination of all singletons. Then all elements having the Markowitz measure equal to unity are taken. Next, candidates for pivots, whose measure is greater then one, are sought. During the search we minimize the Markowitz measure, and count how many elements have been found with the same measure as the previously determined minimum. If their number exceeds the assumed level, we break off the search, even when the end of a diagonal has not yet been reached.

In general-purpose simulators, this procedure is performed before analysis, when a circuit has been read in and compiled. In SPICE and similarily in OPTIMA, this is realized by the SETUP procedure. SETUP is preceded by an initial nonsymmetrical reordering, described in the next section. Next, linked-lists are allocated, Markowitz indexes are counted and diagonal reordering is generated. To save memory, SETUP should be repeated before each analysis. Between analyses memory is free.

3.3.4 Initial nonsymmetrical reordering

In the preceding section we have introduced a very efficient version of reordering, where pivots are selected from a diagonal. This requires good numerical properties of the diagonal. In MNE and extended MNE, introduced in Section 5.2.1.5, problems with the diagonal are caused only by voltage-defined elements, which introduce inconvenient diagonal terms. If a circuit contains a cut-set, consisting only of voltage-defined branches dividing the circuit into distinct parts, and if at least one of these parts is a nongrounded subcircuit built of resistors and conductances, then during LU factorization some pivots will become zero, as has been studied by I. Hajj *et al* [3.4]. For instance, the simple circuit in Figure 3.10(a) described by the equations

$$\begin{bmatrix} G_2 & -G_2 & 1 & 0 \\ -G_2 & G_2 & 0 & 1 \\ 1 & 0 & 0 & 0 \\ 0 & 1 & 0 & 0 \end{bmatrix} \begin{bmatrix} v_{n1} \\ v_{n2} \\ i_1 \\ i_3 \end{bmatrix} = \begin{bmatrix} 0 \\ 0 \\ V_1 \\ V_3 \end{bmatrix}. \tag{3.52}$$

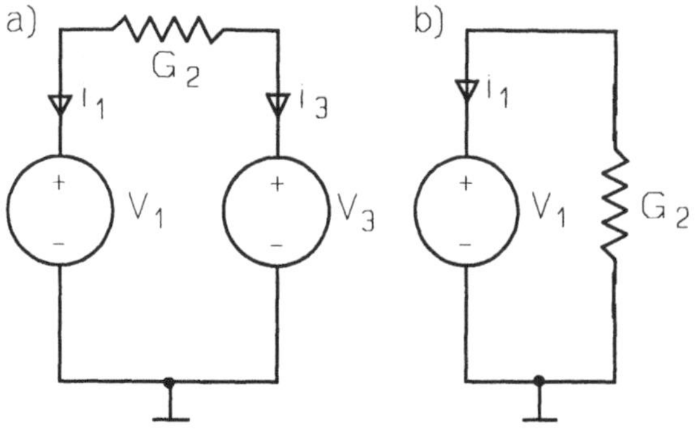

Figure 3.10 Simple circuits including: a) a cut-set of voltage sources, b) a grounded voltage source

incorporates the term G_2 at four symmetrical positions. After the first step of LU factorization, we obtain zero at position (2, 2). If we interchange the first row with the 3rd, and the 2nd with the 4th, then this inconvenient effect vanishes. All diagonal elements become unity:

$$\begin{bmatrix} 1 & 0 & 0 & 0 \\ 0 & 1 & 0 & 0 \\ G_2 & -G_2 & 1 & 0 \\ -G_2 & G_2 & 0 & 1 \end{bmatrix} \begin{bmatrix} v_{n1} \\ v_{n2} \\ i_1 \\ i_3 \end{bmatrix} = \begin{bmatrix} V_1 \\ V_3 \\ 0 \\ 0 \end{bmatrix}. \tag{3.53}$$

Another problem is caused by diagonal zero elements in Equation (3.52). During LU factorization they can become nonzero, but their value will be dependent on other matrix components, and hence can remain zero in some cases. Especially in linearized circuits it can be zero in one iteration while becoming nonzero in another. For the trivial circuit in Figure 3.10(b) we obtain the equations

$$\begin{bmatrix} G_2 & 1 \\ \mathbf{1} & 0 \end{bmatrix} \begin{bmatrix} v_{n1} \\ i_1 \end{bmatrix} = \begin{bmatrix} 0 \\ V_1 \end{bmatrix}, \tag{3.54}$$

where during LU factorization we obtain the first pivot G_2 and the second $-1/G_2$. If G_2 is very large or small in comparison with unity, then a large discrepancy between pivots arises, causing numerical difficulties. However, if we interchange rows, then both pivots become unity and no fill-ins arise. Those two examples are available in sample tasks O3-SPAR4.CIR and O3-SPAR5.CIR. Using option MATVIEW we can observe the structure of the matrix after rows reordering.

If we applied no reordering, then the analyser would break calculations with the report: "Structural singularity of the matrix".

Proceeding to the more general discussion, we have to consider a voltage-defined branch as shown in Figure 2.5(b). Let us rewrite below its MNE stamp:

	m_a	n_a	i_a	m_b	n_b	i_c	RHS
m_a			**1**				
n_a			-1				
i_a	**−1**	1	$\boldsymbol{R_{aa}}$	T_{ab}	$-T_{ab}$	R_{ac}	$-V_a$

where the three elements in bold type are of special interest. In different applications, the diagonal element $\boldsymbol{R_{aa}}$ can be altered over a wide range. In particular, it is zero for voltage sources, short-circuits, and d.c. analysis of inductors. Thus we do not want to take it as a pivot. However, we notice that the stamp has two bold-face unities located at symmetrical positions. Interchanging rows m_a (or equivalently n_a) with i_a we put these unities at diagonal positions, obtaining very good numerical behavior during LU factorization:

	m_a	n_a	i_a	m_b	n_b	i_c	RHS
m_a	**−1**	1	$\boldsymbol{R_{aa}}$	T_{ab}	$-T_{ab}$	R_{ac}	$-V_a$
n_a			-1				
i_a			**1**				

An identical row interchange should be performed for flux-defined branches described according to Extended MNE. Notice that in this interchange we can choose between two terminal nodes of the branch. This reordering eliminates many numerical pivoting problems.

Reordering should be applied to all voltage-defined branches. We start from grounded branches, where interchange of i_a with only the one terminal node is available. During this process we mark the node used. Then we proceed to floating voltage-defined branches. Since we have two terminal nodes, we interchange i_a with that node which has not been marked so far. If neither node is marked we have a degree of freedom to select that node p from $\{m_a, n_a\}$ which has a structural zero or a signal-dependent value at the position (p, p) (i.e. no conductance or a nonlinear element is connected to p respectively. This trick discards a pivot which is a structural zero or a variable, signal dependent element which can become zero for some signals. To experiment with this problem we can turn to examples S3-NPIV.CIR and O3-NPIV.CIR, where this effect can be tested.

The presented methods of row reordering, discussed in [3.4], provide good reordering in most cases except for loops of voltage-defined branches, where interchanges required by a few branches can be inconsistent. Another problem arises in the case of the ideal operation amplifier. This is a degenerate case where the voltage-defined branch has the stamp

	m_a	n_a	i_a	m_b	n_b
m_a	**0**		**1**		
n_a		**0**	**-1**		
i_a	**0**	**0**	**0**	1	-1

where no row interchanges eliminate all structurally zero pivots.

Algorithm 3.17 Nonsymmetrical reordering of the matrix

for k=1 **to** n **do if** ((k, k) is a structural zero and the kth row and column has one nonzero element)
  for j=1 **to** n **do**
    if ((k, j) and (j, k) are structural unities) interchange of the jth and the kth row;
for k=1 **to** n **do if** ((k, k) is a structural zero)
begin
  int=0;
  for j=1 **to** n **do if** ((k, j) and (j, k) are structural unities and (j, j) is a structural zero) **then**
  begin
    interchange of jth and kth row; *int*=1; **break**;
  end
  if(*int*=0) **for** j=1 **to** n **do if** ((k, j) and (j, k) are structural unities and (j, j) is a signal-dependent variable) **then**
  begin
    interchange of jth and kth row; **break**;
  end
end

The introduced reordering is summarized in Algorithm 3.17. We check the variability type of diagonal elements and look for structural zeros. In the first pass we select all zeros (p, p) which have one pair of symmetrical nonzero elements (k, p), (p, k) and interchange the row p and k. In the second pass we find all remaining structurally zero diagonal elements (p, p) having more pairs of symmetrical nonzero elements (k, p), (p, k), and select such a k that the interchange of the row p and k eliminates a structurally zero or signal-dependent pivot.

After all possible row interchanges produced by voltage-defined branches in Algorithm 3.17, the diagonal elements have to be once more checked for the existence of structurally zero pivots. Now we check the whole matrix, whether there is some other row interchange eliminating the structurally zero pivot and if not the structural singularity is reported. To increase the robustness of the structural pivoting the same auxiliary pivoting can be carried out also before Algorithm 3.17.

In general-purpose simulators this reordering is implemented for both MNE and tableau equations, after the circuit read-in and creation of internal numbering of circuit variables. From this numbering, row and column matrix indices arise. Due to the rows reordering, a trace of interchanges is generated and used as indirect addresses to modify row indices during matrix pointer generation.

3.3.5 Implementation of the solution procedure

3.3.5.1 Sparse LU factorization

In simulators, speed of execution depends mainly on quick formulation of equations and their efficient solution. The former is obtained by efficient computerization of device models and by storing pointers to structures representing nonzero matrix elements. These pointers are generated once in SETUP for all circuit elements, and then, during solution, they are used for quickly filling the matrix with the stamps of these elements. Matrix pointers are necessary, though they consume a lot of memory.

As for the sparse matrix solver, it is built of three distinct parts. The first one, located in SETUP, was described in the preceding sections. It prepares data structures, performs initial reordering, and diagonal Markowitz pivoting called **standard pivoting**. The second, DECOMP, is dedicated to LU factorization, while the third, SOLV, comprises forward/backward substitutions. The main purpose of DECOMP and SOLV is to minimize execution time. This requires perfect computerization.

As an example, we demonstrate an outline of the DECOMP procedure. Assuming that reordering and fill-in allocation has been done, we only have to perform the nonzero operations of the LU factorization. We will use the version of the factorization with normalized rows (known as Crout's) introduced in Algorithms 3.6 and 3.7 (distinction between the two versions of the LU factorization has been introduced in Section 3.2.2.1). The resulting procedure in C language takes the form of Algorithm 3.18. We only have to use pointers to refer from one element to the next in a row and in a column.

Algorithm 3.18 An outline of Crout's sparse LU factorization

```
typedef struct {  double value;
                  int    row;
                  int    col;
                  elem   *next_in_row ;
                  elem   *next_in_col }  elem;
elem *next_r, *next_c, *pointer, **diag;
int k,n;
double level; {minimum pivot level}
for (k=0 ; k < n ; k++)
/*------------------------------------------------------------------
   next_r     pointer to element a_kj
   next_c     pointer to element a_ik
   pointer    pointer to element a_ij
------------------------------------------------------------------*/
{ if ((diag[k] -> value)<=level) return (-1); {too small pivot}
  next_r = diag[k] -> next_in_row ;
  next_c = diag[k] -> next_in_col ;
  while (next_c != NUL)
  { next_c -> value = (next_c -> value)/(diag[k] -> value);
    pointer = next_c -> next_in_row;
    while (next_r != NUL)
    { if ((pointer -> col)==(next_r -> col))
      { pointer -> value = (pointer -> value)- \
            (next_c -> value)*(next_r -> value);
        next_r = next_r -> next_in_row;
      }
      pointer = pointer -> next_in_row;
    }
    next_c = next_c -> next_in_col;
  }
}
```

During LU factorization we require a numerical criterion for the pivots. This criterion can consist of checking whether $|a_{kk}| \leq \varepsilon\, a_{max} + \delta$, where, for example $\varepsilon = 10^{-6}$, $\delta = 10^{-10}$, while a_{max} is the maximum magnitude of pivot, met so far. In PSPICE and OPTIMA ε, δ can be set by the options PIVREL and PIVTOL respectively. Due to the initial reordering of rows, the matrix is diagonally dominant and hence theoretically the pivots should be satisfactory. However in nonlinear circuits many of them are dependent on linearized parameters and for specific signals can become zero. Then the numerical criterion will detect "Unsatisfactory LU decomposition".

To overcome this problem, an additional nonsymmetrical reordering (full pivoting) can be used. It takes into account the current submatrix (see Figure 3.6) and minimizes the Markowitz measure and the magnitude of the potential pivot simultaneously. Each candidate for being the pivot is checked to see that its selection will not destroy other diagonal elements. This reordering is called the **nonstandard** one. It is slightly inconvenient, because it increases execution time and can produce additional fill-ins. Nonstandard reordering is required, especially in d.c. analysis, because, due to the elimination of reactances and linearization, matrix elements can vary widely. In a.c. and transient analyses standard reordering in SETUP is more often sufficient. In OPTIMA standard pivoting is the default (also available by setting the option PIV to 0). To enable the nonstandard one we can set the option PIV to 2.

Moreover, PIV set to 1 provides matrix scaling. Scaling is very advantageous, since it does not change the matrix structure and number of fill-ins, and only increases pivots by an appropriate scaling factor. Since we employ Crout's approach, where rows of the matrix are divided by the pivot, the scaling of columns only is applicable. In such a scaling, if the kth pivot is too small, then we multiply the kth column of the matrix by a scaling factor s to obtain a satisfactory high level of a scaled pivot $s\, a_{kk}^{k-1}$. Then we have to rescale the solution as a product of s and an x_k obtained with scaling. To reduce errors we select s in the form 2^c.

3.3.5.2 Exploitation of the RHS vector sparsity

Kundert has shown in SPARSE [3.9], that exploitation of sparsity of the RHS vector gives considerable savings in forward substitution. We have implemented this idea also in the SOLV procedure. In many cases, especially in a.c. analysis, the RHS vector has a great number of zero elements. In data structures it is stored as a sparse $(n+1)$th column of a matrix. In this section we show how to use this sparsity.

If during the calculations of Markowitz measures we take into account also the RHS vector elements, then there is a trend to locate nonzero RHS elements in the ending section of the RHS vector. For instance, if this vector has m zeros at the very beginning, then the solution of $\mathbf{L}\,\mathbf{y}=\mathbf{b}$ (forward substitutions) will have also m beginning zeros, and hence we can start substitution from the $(m+1)$th position. In such a case, it is reasonable to use a version of forward substitutions including division by pivots, since this will save the divisions-time. This requires the Crout's version of LU factorization with normalized diagonal elements of $\mathbf{U}$, instead of the Doolittle version with the normalized diagonal of $\mathbf{L}$ (see Section 3.2.2.1). This explains why we have used Crout's algorithm in the discussion so far. Unfortunately, no savings are available in backward substitutions; hence we prefer to use the version that is free of divisions by pivots.

3.3.5.3 Sparse packages with code generation

In the approach discussed so far, the code of the solving procedure (e.g. that shown in Algorithm 3.18) was written in a high-level language and incorporated into the simulator. This code is written by means of loops and jumps. In fact, it is a sequence of operations $a'=a-b\cdot c$ (modifications) and $a'=a/b$ (divisions by pivots). If we know a particular matrix structure, then we can expand the solution procedure into a sequence of such instructions. It is called the linear code.

In many simulators the concept of generating a linear code for a particular circuit has been implemented. It consists of replacing Algorithm 3.18 by a similar one, where in places of numerical operations $a'=a-b\cdot c$ and $a'=a/b$ we introduce instructions for the generation of appropriate codes describing the type of operation and variables involved. This code generator can be incorporated into SETUP and run only once.

Two types of code are met in practice: compilable and interpretable. The compilable linear code can be generated in an assembler or, for portabilty onto different platforms, in a high-level language (e.g. C). It is compiled and linked with the simulator each time we analyze a new circuit. For up-to-date computers on a UNIX platform this approach is very convenient. Code generation techniques can be used also in other parts of a program, e.g. for matrix formulation. The compilable code is the quickest one.

The interpretable code is created by SETUP, but interpreted by a dedicated procedure incorporated into DECOMP and SOLV. Interpretable code contains instructions including the type of operation and pointers to operands. For example, we can have the instruction mod (a, b, c) which orders execution of the operation $(*a)=(*a)-(*b)*(*c)$ and the instruction div(a, b) for execution of $(*a) = (*a)/(*b)$. These instructions can be easily coded using an appropriate binary format including a bit of instruction and two or three pointers.

The weakest part of the code generation method is its huge size for large circuits, and hence large consumption of operating memory.

3.3.6 The sparse matrix technique dedicated to the tableau equations

The tableau method, discussed in Section 2.1.5, gives matrices about ten times sparser and a few times bigger. They require an extremely efficient sparse matrix solver. By appropriate reordering, the tableau equations can be reduced to MNE or another hybrid method. Thus, if an automatic reordering can be achieved which is optimal, then the resultant efficiency can be better than for any method where a reduction is presumed (e.g. MNE). The best implementations of this technique give slightly better efficiency than MNE.

A serious problem in tableau equations is the huge number of nonzero elements that vary in wide ranges. This requires full submatrix pivoting (not diagonal) with numerical criterion control. To make a solution efficient a method of grouping nonzero elements into classes, called types of variability, has been proposed [3.12]. The following types have been distinguished:

type 0) structural zeros,
type 1) structural unities,
type 2) other constants,
type 3) time-dependent elements,
type 4) elements dependent on time and/or signals from the last Newton-Raphson iteration.

We assign to each row and column of equations a variability type equal to the maximum type of their elements. Then, we reorder rows and columns according to their type of variability. Thus, the 1st are rows and columns having only structural unities, the 2nd are rows and columns dependent on constants and unities, the 3rd are rows and columns having time-dependent and constant coefficients, the 4th are rows and columns whose coefficients are dependent on signals (iterative branch coefficients) and/or on time (excitations or parametric coefficients).

To clarify this ordering concept let us consider the simple circuit shown in Figure 3.11. Its tableau equations have the form

$$\begin{array}{r} \text{KVL1:} \\ \text{KVL2:} \\ \text{BE1:} \\ \text{BE2:} \\ \text{KCL:} \end{array} \begin{bmatrix} 1 & 0 & 0 & 0 & -1 \\ 0 & 1 & 0 & 0 & 1 \\ 0 & 0 & 1 & 0 & 0 \\ 0 & G_2 & 0 & -1 & 0 \\ 0 & 0 & -1 & 0 & 0 \end{bmatrix} \begin{bmatrix} v_1 \\ v_2 \\ i_1 \\ i_2 \\ v_{n1} \end{bmatrix} = \begin{bmatrix} 0 \\ 0 \\ I_1 \\ 0 \\ 0 \end{bmatrix} \tag{3.55}$$

where we assume that I_1, G_2 are of the variability type 4. Thus, in the above equations the 1st, 2nd and 5th rows and the 1st, 4th and 5th columns are of type 1, while the 3rd, 4th rows and 2nd, 3rd columns are of type 4. After reordering, according to the above classification, we obtain

$$\begin{array}{r} \text{KVL1:} \\ \text{KVL2:} \\ \text{KCL:} \\ \text{--} \\ \text{BE1:} \\ \text{BE2:} \end{array} \left[\begin{array}{ccc|cc} 1 & -1 & 0 & 0 & 0 \\ 0 & 1 & 0 & 1 & 0 \\ 0 & 0 & 1 & 0 & -1 \\ \hline 0 & 0 & 0 & 0 & 1 \\ 0 & 0 & -1 & G_2 & 0 \end{array}\right] \left[\begin{array}{c} v_1 \\ v_{n1} \\ i_2 \\ \hline v_2 \\ i_1 \end{array}\right] = \left[\begin{array}{c} 0 \\ 0 \\ 0 \\ \hline I_1 \\ 0 \end{array}\right] \tag{3.56}$$

where, within the group of rows and columns of the same variability type, ordering is carried out using the fill-ins and numerical criteria.

Now we proceed with the solution process by eliminating unknowns starting from v_1, v_{n2}, etc. It is equivalent to Gaussian elimination, i.e. to the LU method. However, due to the partition into two groups of rows and columns, we can also partition the Gaussian elimination into two

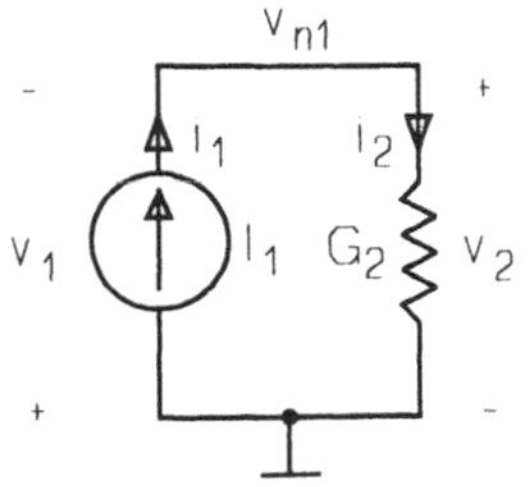

Figure 3.11 A simple linear circuit

stages. At the first stage, we perform three steps of elimination and stop before the final two steps. Such an operation is called the partial LU factorization. After it we obtain a circuit description reduced to two equations with unknowns v_2 and i_1. They contain terms of variability type 4. If these elements are subject to change in time and in NR loops (as in d.c. and transient analyses), then in these loops we only have to solve a system of two equations, instead of five.

To explain this idea in a more general way, let us write the reordered and partitioned tableau equations (3.56) in the following form:

$$\begin{bmatrix} \mathbf{Y}_{11} & \mathbf{Y}_{12} \\ \mathbf{Y}_{21} & \mathbf{Y}_{22} \end{bmatrix} \begin{bmatrix} \mathbf{x}_1 \\ \mathbf{x}_2 \end{bmatrix} = \begin{bmatrix} \mathbf{b}_1 \\ \mathbf{b}_2 \end{bmatrix}. \tag{3.57}$$

where the number of variables $\mathbf{x}_1$ is n_1. Thus we perform only n_1 steps of LU factorization Algorithm 3.8. This transformation, called the partial LU factorization, can be considered as a multiplication of both sides of tableau equations (3.57) by the matrix

$$\bar{\mathbf{L}}^{-1} = \begin{bmatrix} \mathbf{L}_{11}^{-1} & \mathbf{0} \\ -\mathbf{Y}_{21}\mathbf{Y}_{11}^{-1} & \mathbf{1} \end{bmatrix}. \tag{3.58}$$

It gives the partly reduced system

$$\begin{bmatrix} \mathbf{U}_{11} & \mathbf{L}_{11}^{-1}\mathbf{Y}_{12} \\ \mathbf{0} & \mathbf{Y}_{22}-\mathbf{Y}_{21}\mathbf{Y}_{11}^{-1}\mathbf{Y}_{12} \end{bmatrix} \begin{bmatrix} \mathbf{x}_1 \\ \mathbf{x}_2 \end{bmatrix} = \begin{bmatrix} \mathbf{b}_1' \\ \mathbf{b}_2' \end{bmatrix}. \tag{3.59}$$

The above transformation can be viewed as a factorization $\mathbf{Y} = \bar{\mathbf{L}}\,\bar{\mathbf{U}}$, where the lower matrix $\bar{\mathbf{L}}$, of the form

$$\bar{\mathbf{L}} = \begin{bmatrix} \mathbf{L}_{11} & \mathbf{0} \\ \mathbf{Y}_{21}\mathbf{U}_{11}^{-1} & \mathbf{1} \end{bmatrix} \tag{3.60}$$

automatically appears in the lower part of an array during partial factorization. The modified RHS, due to (3.60), can be treated as a solution of $\mathbf{L}_{11}\,\mathbf{b}_1' = \mathbf{b}_1$, followed by $\mathbf{b}_2' = \mathbf{b}_2 - \mathbf{Y}_{21}\mathbf{U}_{11}^{-1}\mathbf{b}_1'$.

This partial LU factorization algorithm fully modifies only the area of $\mathbf{Y}_{11}$, $\mathbf{Y}_{12}$, $\mathbf{Y}_{21}$ (i.e. that of variability type 1). The area $\mathbf{Y}_{22}$ remains partly unmodified, which corresponds to the fact that the equations are still not solved for unknowns $\mathbf{x}_2$.

After the partial LU factorization is complete and we have circuit equations in the form (3.59), we can terminate the solution process by solving the equations arising from (3.59), namely:

$$(\mathbf{Y}_{22} - \mathbf{Y}_{21}\mathbf{U}_{11}^{-1}\mathbf{L}_{11}^{-1}\mathbf{Y}_{12})\,\mathbf{x}_2 = \mathbf{b}_2' \tag{3.61}$$

We notice that if we have distinguished among $\mathbf{x}_2$ a certain two groups, e.g. of variability types 3 and 4, then we can continue the solution process, by partial elimination of the type 3 variables from Equation (3.61).

A great advantage of this approach is that the solution for variables corresponding to a higher type of variability can be performed by means of the reduced, i.e. smaller, submatrix. If calculations are repeated within the NR loop, then we repeat only the solution of the reduced equation (3.61). This requires a surprisingly small computational effort, since $\mathbf{Y}_{22}$ is subject to change in NR iterations, while $\mathbf{Y}_{11}$, $\mathbf{Y}_{12}$, $\mathbf{Y}_{21}$ are only of variability types 1 or 2, and hence they remain constant. From partial factorization we obtain results $\mathbf{L}_{11}$, $\mathbf{b}_1'$, $\mathbf{Y}_{21}\mathbf{U}_{11}^{-1}$, $\mathbf{L}_{11}^{-1}\mathbf{Y}_{12}$. Therefore, in the NR loop we need only calculate $\mathbf{b}_2' = \mathbf{b}_2 - \mathbf{Y}_{21}\mathbf{U}_{11}^{-1}\mathbf{b}_1'$, update $\mathbf{Y}_{22}$, and solve (3.61). If we require also $\mathbf{x}_1$, we have to perform only the backward substitution $\mathbf{U}_{11}\mathbf{x}_1 = \mathbf{b}_1 - \mathbf{L}_{11}^{-1}\mathbf{Y}_{12}\mathbf{x}_2$.

To sum up, we notice that in time loops of transient analysis we repeat the solution of (3.61) reduced to the variables of types 3 and 4. In the NR loop we can repeat the solution of the equations (3.61) reduced to the variables of type 4. In fact, the partial LU factorization technique is a circuit decomposition technique. It stands for extracting nested subcircuits corresponding to successive types of variability. Decomposition problems will be discussed in Chapter 8.

To check our comprehension we can work through the following problems and answer the questions posed.

- What are the advantages of the sparse matrix technique as regards memory consumption and numerical complexity (time consumption)? Compare with the full matrix technique.
- Explain what data structures are used to save memory and how savings in numerical complexity have been achieved.
- What do we mean by the fill-in? Invent a simple example to explain a mechanism for its production.

- How can we construct a worst-case prediction of fill-ins? Explain the concept of matrix reordering for the minimization of their number.
- What elements and topological configurations cause problems with providing a dominant character to a diagonal in MNA? Explain a non-symmetrical reordering which can overcome them.
- Explain the concept of an equations solver implemented using the fixed code and the compiled or interpreted code generated by the analyzer.
- What variability types can be distinguished in a sparse matrix? Invent a simple matrix having elements of two variability types only: the constant one, and the one varying in NR iterations. Explain how we should reorder this matrix and how to exploit in d.c. analysis the advantage of partial LU factorization.

References

3.1 Abdel-Malek, H.L., Elbardissy, S.A., *The Independent Variable Technique for Sparse Linear Equations,* Proc. IEEE Int. Symp. Circ. Syst., p. 527-530 (1989).

3.2 Berry, R.D., *An Optimal Ordering of Electronic Circuit Equations for a Sparse Matrix Solution*, IEEE Trans. Circ. Theory, vol. 18, no. 1, p. 40-50 (1971).

3.3 Forsythe, G. E., Moler, C. E., *Computer solution of linear algebraic systems*, Prentice-Hall, Englewood Cliffs, N. J. (1967).

3.4 Hajj, I.N., Yang, P., Trick, T.N., *Avoiding Zero Pivots in the Modified Nodal Approach*, IEEE Trans. Circ. Syst., vol. 28, no. 4, p. 271-279 (1981).

3.5 Hsieh, H.Y., *Pivoting-Order Computation Method for Large Random Sparse Systems*, IEEE Trans. Circ. Theory, no. 4, p. 225-230 (1974).

3.6 Hsieh, H.Y., *Fill-in Comparison Between Gauss-Jordan and Gaussian Eliminations*, IEEE Trans Circ. Syst., no. 4, p. 231-233 (1974).

3.7 Hsieh, H.Y., Ghaussi M.S., *An Optimal Pivoting Algorithm in Sparse Matrices.*, IEEE Trans. Circ. Theory, no. 1, p. 93-96 (1972).

3.8 Hsieh, H.Y., Ghaussi, M.S., *A Probabilistic Approach to Optimal Pivoting and Prediction of Fill-In for Random Sparse Matrices,* IEEE Trans. Circ. Theory, vol. 19, no. 7, p. 329-336 (1972).

3.9 Kundert, K.S., *Sparse Matrix Techniques*. In: *Circuit Analysis Simulation and Design,* Ed. A.E. Ruehli, Amsterdam-New York-Oxford-Tokyo, Elsevier (North Holland) (1986).

3.10 Nakhla, M., Vlach, J., *An Optimal Pivoting Order for the Solution of Sparse Systems of Equations*, IEEE Trans. Circ. Syst., no. 4, p. 222-225 (1974).

3.11 Markowitz, H.M. *The Elimination Form of the Inverse and its Application to Linear Programming,* Management Science, vol. 3, p. 255-269 (1957).

3.12 Norin, R.S., Pottle, Ch, *Effective Ordering of Sparce Matrices Arising from Nonlinear Electrical Networks,* IEEE Trans. Circ. Theory, vol. 18, no. 1, p. 139-145 (1971).

3.13 *Sparse matrix computations*, Ed. J.R. Bunch, D.J. Rose, Academic Press (1976).

3.14 Sangiovanni-Vincentelli, A. *A Note on Bipartite Graphs and Pivot Selection in Sparse Matrices*, IEEE Trans. Circ. Syst., no. 12, p. 817-821 (1976).

3.15 Sangiovanni - Vincentelli, A., *Circuit Simulation.* In: *Computer Design Aids for VLSI Circuits,* Ed. P. Antognetti, D.O. Pederson, H. DeMan H. Sijthof & Noordhoff (1981).

3.16 Tinney, W.F., Walker, J.W., *Direct Solutions of Sparse Network Equations by Optimally Ordered Triangular Factorization,* Proc. IEEE, vol. 55, no. 11, p. 181-1809 (1967).

Chapter 4

D.c. Analysis of Nonlinear Circuits

4.1 Introduction to d.c. analysis

4.1.1 Importance of d.c. analysis

In most practical electronic circuits a constant steady state resulting from d.c. excitations is the most commonly used simulation mode. If system performance relies on instantaneous signal variations then usually there is a reference signal impressed by the energy sources. This reference signal provides a background for the time-varying signals. In the case of small-signal a.c. operation (see Section 1.1.2) signals vary slightly in the vicinity of the bias. Therefore, if small incremental signals are considered, the behavior of a nonlinear circuit depends not only on its topology and the character of branches but also on the bias impressed. Thus, an efficient way of calculating of the bias is essential for a system designer. The design process for analog integrated circuits always starts with the d.c. analysis and verification of d.c. signals.

Similarily, systems operating with time-varying signals of a large dynamic range covering a wide range of nonlinear characteristics of the system elements, require a sequence of d.c. simulations with an excitation changing gradually. This so-called d.c. sweep or multipoint d.c. analysis examines the dependence of responses on the excitations, assuming that the latter vary slowly enough to neglect reactive effects. Such an analysis is useful for analog and digital (logic) circuits with large signals.

Thus the d.c. analysis is an introduction to other types of circuit analyses, such as small-signal a.c. (discussed in Section 3.2), time-domain and sensitivity. The d.c. analysis appears to be the most common in circuit simulation.

4.1.2 Substitute d.c. nonlinear circuits

The response of a dynamic electrical circuit to d.c. excitations tends towards the d.c. steady state, which is theoretically reached only at infinity but in practice occurs as early as in a few time constants. Signals corresponding to the d.c. steady state are constant, and hence their derivatives reach zero. There are no currents flowing through capacitors and zero voltages are reached across inductors. Thus, the former reduce to open circuits, and the latter to short circuits. This state is maintained until information bearing variable signals are switched on. To recall the essence of the d.c. analysis we may turn to Section 1.1.2 where the instructive d.c. response has been derived from the substitute nonlinear d.c. circuit of Figure 1.3(a). This circuit was obtained by open-circuiting capacitors and short-circuiting inductors in the circuit of Figure 1.1(b). This is equivalent to zeroing all the derivatives in the branch equations. Recalling canonical circuit equations (CCE) (1.23), assuming a constant d.c. excitation $\mathbf{s}_0$ and zeroing derivatives, CCEs transform into algebraic d.c. equations

$$\begin{aligned} &\mathbf{A}\,\mathbf{i} = \mathbf{0} \\ &\mathbf{B}\,\mathbf{v} = \mathbf{0} \\ &\mathbf{f}(\mathbf{i}, \mathbf{v}, \mathbf{0}, \mathbf{0}, \mathbf{s}_0) = \mathbf{0}\,. \end{aligned} \tag{4.1}$$

These equations describe a resistive nonlinear substitute circuit, structurally identical (see Section 1.2.3) with the original (1.23). The d.c. substitute circuit arises when all reactive branches are replaced by their resistive counterparts derived by setting zero for derivatives in branch constitutive formulae, as demonstrated in the transformation of the circuit of Figure 1.1(b) into the circuit of Figure 1.3(a). In particular, capacitors become open-circuits while inductors transform into short-circuits. For simplicity CCE (4.1) of the substitute resistive circuit will be written shortly as

$$\begin{aligned} &\mathbf{A}\,\mathbf{i} = \mathbf{0} \\ &\mathbf{B}\,\mathbf{v} = \mathbf{0} \\ &\mathbf{f}(\mathbf{i}, \mathbf{v}) = \mathbf{s}_0\,. \end{aligned} \tag{4.2}$$

Therefore, d.c. analysis is reduced to the problem of calculating the branch responses $\mathbf{i}$, $\mathbf{v}$ of the subsitute d.c. circuit. This, in general, involves solving a nonlinear algebraic set of equations and is an iterative numerical problem of significant complexity. If we can predict a good initial approximation to the d.c. signals of circuit elements (e.g. base-emitter and base-collector voltages of bipolar junction transistors) then, a second order locally convergent numerical method of solving algebraic equations, e.g. Newton–Raphson (NR), is quite suitable. Due to its simplicity and good convergence rate in a solution vicinity the NR method is common in professional simulators. However, usually it is implemented not in a fundamental version but with considerable improvements.

In some cases, the substitute d.c. circuit described by CCE (4.1) is not appropriate for calculating d.c. responses although the orginal circuit

(1.23) was quite physical and unique. Equations (4.1) may be singular so that its solution may be indefinite or ambiguous. The first case means that a circuit does not reach the d.c. steady state, while the second denotes that an infinite number of different d.c. responses is possible and all depends on the transient states preceding them. To discuss these singularities precisely we introduce the voltage $\mathbf{v}=\mathbf{A}^t\mathbf{v}_n$ and mesh $\mathbf{i}=\mathbf{B}^t\mathbf{i}_L$ transformation (see Section 1.3.1). They provide a set of independent basic voltages and currents that satisfy both KCL and KVL (this can be proved by substituting the transformations in KCL, KVL and considering interdependencies like (1.22)). Currents $\mathbf{i}_L$ may be viewed as link currents or any base of the linear space of currents satisfying KCL. Thus, (4.1) transforms into

$$\mathbf{f}(\mathbf{A}^t\mathbf{v}_n, \mathbf{B}^t\mathbf{i}_L, \mathbf{0}, \mathbf{0}) = \mathbf{s}_0 . \tag{4.3}$$

The total number of the variables $\mathbf{v}_n$ and $\mathbf{i}_L$ is theoretically equal n_b. However, some of the variables $\mathbf{v}_n$ and $\mathbf{i}_L$ may not appear in (4.3). This happens when in the original algebraic-differential branch functions some components of $\mathbf{v}_n$ and $\mathbf{i}_L$ appear only in the differentiated form. Then system (4.3) has less variables than equations and hence can be singular. The missing variables are inconsistent or ambiguous, not available from the substitute d.c. circuit described by (4.1). This singularity is caused by two critical circuit primitives: the floating node and the short loop.

- The floating node is a node incident only with a number of such branches as capacitors, independent current sources, open circuits, or, in general, any CD branches that after setting zero to derivatives transform into the form $i=\text{const}$ as shown in Figure 4.1(a). In a dynamic case, KCL at this node gives

$$I_1 + C_2\frac{\mathrm{d}(v_n - v_{n2})}{\mathrm{d}t} + C_3\frac{\mathrm{d}(v_n - v_{n3})}{\mathrm{d}t} = 0 . \tag{4.4}$$

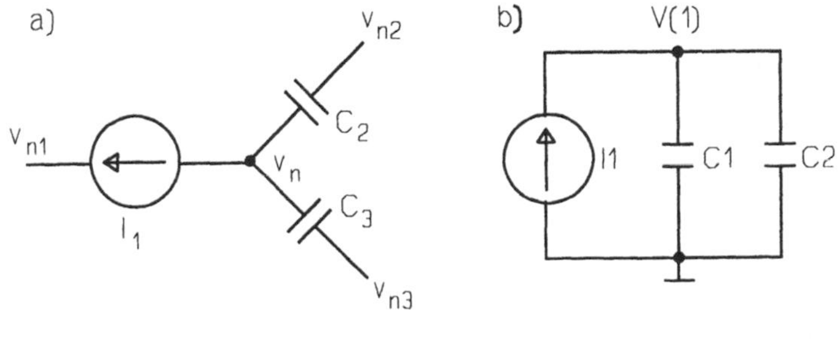

Figure 4.1 (a) Floating node and (b) circuit that exhibits floating node effect

When a d.c. steady state is gained, then the above derivatives reach zero, the source reaches its constant value and KCL takes the following form

$$I_{1DC} = 0 \tag{4.5}$$

which proves that the voltage of the floating node has no influence on KCL. Hence, substitute d.c. equations of any circuit incorporating the floating node are singular and during the solution process, due to a zero pivot, the LU factorization fails. Once $I_{1DC} \neq 0$, (4.5) is inconsistent, i.e. the d.c. steady state is never reached and we have the endless charging of capacitors by d.c. current sources. Otherwise, (4.5) is ambiguous, i.e. KCL is satisfied by any d.c. voltage v_n. However, this ambiguity does not happen in practice when we take into account the time-profile of the transient response. For a particular transient a unique asymptotic v_n is always reached. However, different transients (due to different reactive elements or initial conditions) would give different d.c. steady states.

The above discussion is demonstrated via the circuit of Figure 4.1(b). In the file O4-1.CIR of directory \OPTIMA on Diskette A we have its description for the simulator OPTIMA. The source $I_1(t)$ that varies in time exponentially from *INIT* to *I1DC* has been defined by means of OPTIMA tools for creating external user-defined models. To familiarize ourselves with OPTIMA modeling we can turn to the TUTORIAL.DOC on Diskette A. However, the basic information needed to understand the extensions available in OPTIMA, compared with PSPICE is available from examples in this book. In our example the d.c. response resulting from the excitation *I1DC* will be an asymptotic state derived from a transient state impressed by the source $I_1(t)$ starting from value *INIT* and by zero initial conditions on capacitors. O4-1.CIR provides several singular cases. First the case of $I1DC = 1$mA simulates the absence of a d.c. steady state when the nodal voltage $V(1)$ endlessly increases with asymptotic slope 500 V/s. The second, third and fourth cases, with $I1DC = 0$ and *INIT* taking three distinct values, result in three different transients that reach three different d.c. steady-state responses. In the fifth case we try to obtain a d.c. response using common d.c. analysis methods, that is by solving algebraic equations. Due to our previous consideration we are not surprised that the simulator fails during the LU factorization. To protect the user from such numerical problems PSPICE verifies net-lists and disables analysis in floating node cases, as demonstrated by the S4-1.CIR file in \SPICE directory of Diskette A. Such a verification has not been implemented in OPTIMA to leave the user with wider simulation possibilities.

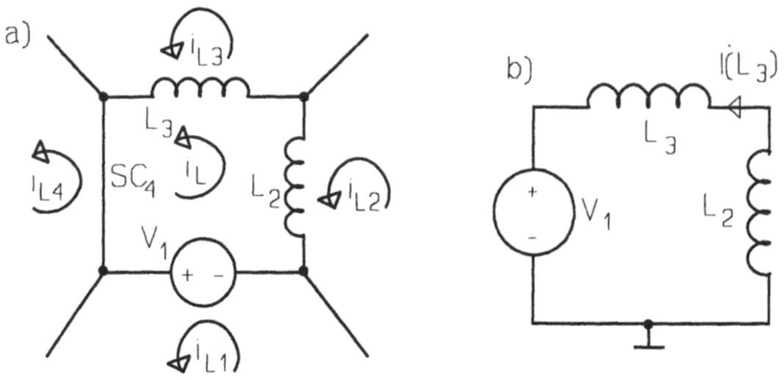

Figure 4.2 (a) A singular circuit loop and (b) example of a circuit including a singular loop

Identical problems arise in the case of a loop which comprises independent voltage sources, inductors, short circuits, or other VD branches that after setting zero to derivatives transform into $v = \text{const}$, as shown in Figure 4.2(a). The KVL around this loop gives the equation:

$$V_1 + L_3 \frac{d(i_L - i_{L3})}{dt} + L_2 \frac{d(i_L - i_{L2})}{dt} = 0. \tag{4.6}$$

In the d.c. steady state it turns into the singular form:

$$V_{1DC} = 0 \tag{4.7}$$

which cannot be solved for the mesh current i_L. Once $V_{1DC} \neq 0$, the inconsistence of (4.7) shows that no steady state of the mesh current is reached. Otherwise, the d.c. mesh current is indefinite. The sample circuit including a singular loop is shown in Figure 4.2(b). Handling the OPTIMA file O4-2.CIR similar effects as for the O4-1.CIR are obtained.

4.1.3 Outline of iterative methods used in practical simulators

Common implementations of d.c. analysis assume that no singularities occur and apply d.c. substitute circuits to work out d.c. responses by solving its system of simultaneous algebraic nonlinear equations. A number of iterative methods for solving nonlinear algebraic systems are known. In circuit simulation, the Newton-Raphson (NR) method is of the greatest importance. Furthermore, in rare cases the multidimensional secant (SEC) method is useful. In general purpose simulators, the NR method is usually accompanied by auxiliary step modification techniques. The following essential advantages of the NR method have to be emphasized:

- a good (quadratic) local convergence in the solution vicinity,
- reliability when implemented with a step-limiting mechanism in cases of circuits including typical, monotonic and regular nonlinear characteristics when iterations start from points not too far from the solution,
- simplicity in practical implementation.

Some people stress the weak points of this method:

- need for explicit formulae for derivatives,
- weak convergence when starting from a bad approximation to a solution,
- weak convergence in the cases of nonlinearities of rapidly changing slopes, negative slopes, hysteretic ranges, discontinuities of derivatives,
- weak convergence in strong feedback and large gain circuits.

Difficulties with convergence may be especially hard if the simulator utilizes user-defined nonlinear models, whose properties are not predictable in general so that the step limiting procedure cannot take advantage of the known properties of nonlinear characteristics.

In extremely hard tasks, to override numerical difficulties, continuation methods have been developed to extend the range of acceptable starting points. These methods may be viewed as transformation of the static (algebraic) case into the dynamic one.

This chapter is not an overview of the whole variety of methods suitable for solving nonlinear algebraic problems. A deep insight into these problems is provided by the well-known monograph [4.10]. Besides the basic NR method a variety of advanced modifications is known. They, however, often appear to be too costly or efficient only in specific cases (dedicated to some classes of nonlinear elements), and hence they have not proved to be satisfactory in general-purpose simulators. Circuit simulation needs methods that are of small computational complexity per iteration and well-convergent for a wide class of circuits met in practice. That is why further consideration is focused only on the most common methods. The next section is dedicated to the basic NR method. Then we will proceed to the modified NR algorithm used in popular simulators.

To test our understanding of the above topics we can work through the following problems.

- Explain the concept of d.c. analysis as a steady state, subject to constant excitations.

- What do we mean by the d.c. substitute nonlinear circuit?
- Explain two types of numerical singularities appearing in d.c. substitute circuits. Describe the physical behavior of the d.c. steady state in these singular cases. Demonstrate it by means of an instructive OPTIMA simulation.

4.2 A basic Newton-Raphson method

4.2.1 The case of a single nonlinear equation

4.2.1.1 Iterative formula and its local properties

At the beginning of our discussion let us consider the instructive example, a single equation with one unknown $f(v)=0$. Let f be twice differentiable within a closed interval $X \subseteq R$, while $v^{(p)}$ is a point on this interval. Using a Taylor series expansion our function may be expressed as the sum of a linear component and a residuum, which is a small value of the same order as the squared increment of v. Utilizing Taylor's theorem we can state that for all v, there is a point $\tilde{v}=v^*+\tau(v-v^*)$, $0 \leq \tau \leq 1$, such that

$$f(v)=\underbrace{f(v^{(p)})+f'(v^{(p)})(v-v^{(0)})}_{\text{linear component}}+\underbrace{\frac{1}{2}f''(\tilde{v})(v-v^{(p)})^2}_{\text{residuum}} \tag{4.8}$$

where f', f'' are the first and the second derivatives versus v. Let v^* be the theoretical solution of our problem $f(v^*)=0$. In general, it is not a solution of the linear part of the expansion, though, if the point $v^{(p)}$ is located close to x^*, then the residuum is relatively small and the solution of the linear component is even closer to v^* then $v^{(p)}$. Hence, the idea is to work out a next approximation $v^{(p+1)}$ as a solution of the linear component

$$f(v^{(p)})+f'(v^{(p)})(v^{(p+1)}-v^{(p)})=0\,. \tag{4.9}$$

When an iterative process starts from a given $v^{(0)}$ it may generate a sequence of approximate solutions $\{v^{(p)}\}_{p=0}^{\infty}$. From Equation (4.9), it can be seen that if the sequence is convergent, then $\lim v^{(p)} = \lim v^{(p+1)}$. Hence, the component including the derivative has to tend to zero. This proves that the limit of $f(v^{(p)})$ must be zero, and hence, the limit of the sequence has to fulfill our equation. Thus, once the convergence is ascertained we are sure to reach one of the solutions of our equation. Features of this method, in the vicinity of a theoretical solution, are known as properties of local convergence. If an appropriate starting point $v^{(0)}$ has been selected, the recursion

$$v^{(p+1)} = v^{(p)} - \frac{f(v^{(p)})}{f'(v^{(p)})} \tag{4.10}$$

called the iterative formula, may give a solution to our equation.

Unfortunately, the iterative formula is not always convergent, and even not always executable. To be executable it needs a nonzero derivative (located in the denominator) at all points arising during the iterative process. The examination of criteria when this condition is satisfied is the problem of so-called global properties of convergence, mentioned in Section 4.2.1.4.

At present, we limit our consideration to local convergence and try to estimate its rate. Subtracting both sides of (4.9) from (4.8) we obtain the expression that joins the solution errors in two subsequent iterations

$$f'(v^{(p)})(v^{(p+1)} - v^{*}) = \frac{1}{2} f''(\bar{v})\,(v^{(p)} - x^{*})^2 . \tag{4.11}$$

If in the whole range X the second derivative has an upper bound P, that is: $|f''(v)| \le P < +\infty$, while the first derivative has a lower bound D (succinctly $|f'(v)| \ge D > 0$), then from (4.11) we have a current error estimation

$$|v^{(p+1)} - v^{*})| \le \frac{P}{2D} |v^{(p)} - v^{*}|^2 . \tag{4.12}$$

If the above assumptions regarding the derivatives hold, then the solution error decreases in a quadratic manner at each iteration. An iterative process convergent in such a way is known as quadratically convergent. Quadratic convergence is possible in a solution vicinity only when at the solution itself the equation has a nonzero first derivative. The coefficient on the left side of (4.12) is called the convergence ratio. The bigger its value, the slower is the quadratic convergence and it approaches linear convergence. Small values of the ratio give quick convergence. As has been shown, this behavior occurs in the case of a function whose absolute value of first derivative is large while its second derivative is small in the vicinity of the solution. Such a function has a considerable slope and weak nonlinearity.

When the first derivative at the theoretical solution is zero, the sequence $f'(v^{(p)})$ decreases to zero. Hence, dividing both sides of (4.11) by small components of this sequence we observe the convergence slowing down. In this case, the error estimation (4.12) is quite impossible. To obtain another estimation we expand $f(v)$ at the point v^* in Taylor series with a second order residual term

$$f(v)=\frac{1}{2}f''(\tilde{v})\,(v-v^*)^2,\quad \tilde{v}=v^*+\tau(v-v^*),\ \ \tau\in[0,1]\,. \tag{4.13}$$

The derivative of (4.13) takes the form $f'(v)=f''(\tilde{v})\,(v-v^*)$. Substituting (4.13) and its derivative into Equation (4.10) we obtain the relation $|v^{(p+1)}-v^*|=|v^{(p)}-v^*|/2$, that proves a linear decrease in error, known as linear convergence.

4.2.1.2 Instructive example of the iterative NR process

The single equation case is instructive, although it is of limited practical importance. However, it helps us to become familiar with the NR iterations. Let us examine the circuit of Figure 4.3, whose only equation is of the form $f(v)-I=0$, where $i=f(v)$ is a nonlinear conductance. To demonstrate in detail the course of iterations, this circuit will be simulated by OPTIMA, where we are able to switch off the more advanced step limiting mechanism (for all external models by the option LIMU=0 or for a particular external model by is option LIMIT=OFF) and so obtain pure NR iterations. Moreover, OPTIMA provides that any nonlinear model may be defined and assigned to any element. In what follows a nonlinear conductance is to be represented by an external model.

In the following discussion we aim at comparing quadratic and linear convergence. Let the conductance of the circuit shown in Figure 4.3 be of the quadratic form $f(v)=\mathrm{sign}(v)\,v^2$. We are going to solve the circuit for v while it is excited by $I=4$. In fact, the equation is of the form $v^2-4=0$ (the function sign may be neglected due to positive values of v). Hence, it has the accurate solution $v=2$. Now, the iterative formula (4.10) takes the form

$$v^{(p+1)}=v^{(p)}-\frac{(v^{(p)})^2-4}{2v^{(p)}}\,. \tag{4.14}$$

Another example arises when we select the nonlinear conductance $f(v)=-v^2+4v$. It results in the circuit equation $v^2-4v+4=0$ of the same

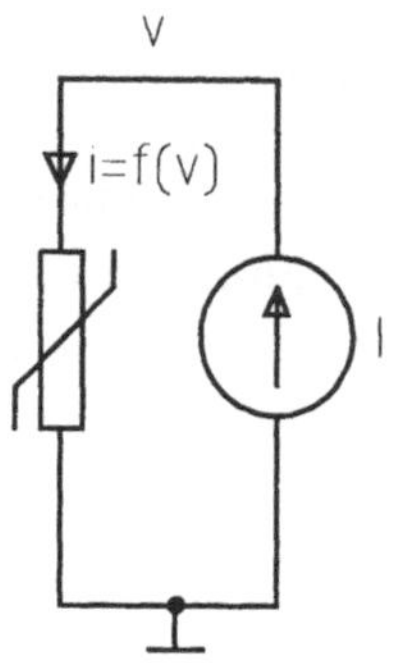

Figure 4.3 Circuit described by a single equation

exact solution $v=2$. Unfortunately, the first derivative at the solution is now zero and convergence is much slower. Writing the iterative formula

$$v^{(p+1)}=v^{(p)}-\frac{-(v^{(p)})^2+4v^{(p)}-4}{-2v^{(p)}+4}=\frac{v^{(p)}+2}{2} \tag{4.15}$$

we observe only linear convergence.

Assuming a starting point $v^{(0)}=1$, two sequences of approximate solutions could be derived from (4.14) and (4.15), but we prefer to demonstrate it letting OPTIMA handle the file O4-3.CIR including a description of the circuit shown in Figure 4.3. This circuit contains a nonlinear conductance driven by a current source. Programming this nonlinear conductance in OPTIMA exploits the External Model Description Language (EMDL) [4.16] which is extensively described in TUTORIAL.DOC. Its main idea will be explained in what follows.

The circuit description file O4-3.CIR provides after the title line, within braces { ... } an external model called QUAD. It depends on a common parameter A (default value 0) which is a coefficient of the conductance

COMMON A=0

and an individual parameter IFL (default 0) whose values 1 and 0 select between the two nonlinearities of the conductance involved

INDIVIDUAL IFL=0

Values of common parameters are common for all elements calling the same model line .MODEl. Individual parameters are individual for each element.

The model returns an output variable F whose type is I(NA,NB), i.e. the controlled current connected from the terminal NA to NB

OUTPUT I(NA,NB):F

where symbols NA, NB are declared as terminal nodes of the model

BLOCK_NODES NA,NB

The current F is controlled by the internal controlling variable U of the type V(NA,NB), i.e. it is a voltage across terminals NA and NB. A starting value of this voltage for NR iterations is 1V.

CONTROL V(NA,NB):U=1

The terminal nodes introduced are formal. Actual circuit nodes are assigned in an element line at the moment when the model is called by the element.

Before we proceed to comment further, the listing of O4-3.CIR is given.

```
TESTING OF NR ITERATIONS IN CIRCUIT OF FIGURE 4.3
# this test is in detail explained in Section 4.2.1.2 of the book
{
  NONLINEAR MODEL QUAD
  BLOCK_NODES NA,NB
  COMMON A=0
  INDIVIDUAL IFL=1
#   IFL=1  f=a*v*v  (default)
#   IFL=0  f=a*v*(4-v)
  CONTROL V(NA,NB):U=1
# Starting value U=1 has been set
  OUTPUTS G(NA,NB):F
     IF(IFL!=0&&IFL!=1)EXIT(1)
  ENDPP
  F=IFL*A*U*U+(1-IFL)*A*U*(4-U)
  F:U=IFL*2*A*U+(1-IFL)*(-2*U+4)
  ENDMDL
}
RFUN 1 0 MODL IFL 1
.MODEL MODL QUAD A 1
I 0 1 4
# this source provides the reference voltage 2V
# error of V(1) can be calculated as V(1,2)
V 2 0 2
```

```
!
.OPTI RELT 1E-8 LIMU 0
.PRINT OP V(1),V(1,2)
# In this case IFL=1, i.e. equation v^2-4=0 is solved
# We observe the quadratic convergence
.OP *
# We set IFL to 0, i.e. equation -v^2+4v-4=0 is solved
.ALT RFUN IFL 0
# We observe the linear convergence
.OP *
.END
```

Table 4.1 Subsequent NR iterations in quadratic and linear convergence cases

no.	quadratic convergence		linear convergence	
	V(1)	V(1)-V(2)	V(1)	V(1)-V(2)
1	2.5000	5.0000E-01	1.7500	-2.5000E-1
2	2.0500	5.0000E-02	1.8750	-1.2500E-1
3	2.0006	6.0976E-04	1.9375	-1.6250E-2
4	2.0000	9.2922E-04	1.9688	-3.125E-3
5	2.0000	2.2204E-15	1.9844	-1.5625E-2
6	2.0000	<1E-15	1.9922	-7.8125E-3
7	2.0000	<1E-15	1.9961	-3.9063E-3
8	-	-	1.9980	-1.9531E-3
9	-	-	1.9990	-9.7656E-4
10	-	-	1.9995	-4.8828E-4
11	-	-	1.9998	-2.4414E-4
12	-	-	1.9999	-1.2207E-4
13	-	-	1.9999	-6.1035E-5
14	-	-	1.9999	-3.0517E-5

The output variable F and its derivative F:U with respect to the controlling variable U are described by the expressions given.

After the brace } ending the model definitions the net-list begins with the nonlinear resistance RFUN connected between nodes 1 and 0, and hence the assignment NA=1, NB=0 occurs. This resistance is modeled by the model QUAD, and hence stands for the conductance $i=f(v)$ involved. In fact, this conductance is described by a current source controlled by its own voltage V(1,0). RFUN calls the .MODEL line MODL to which the external model QUAD is assigned. In the .MODEL line also the parameter A obtains its final value 1 (so far it had a default 0V only). The individual parameter IFL obtains its final value in the RFUN line. Moreover, as well as RFUN, the current source $I=4\,\mathrm{A}$ and the reference voltage source $V=2\,\mathrm{V}$ is introduced. After ! we have the option RELTOL=1E-8 defining the level of relative accuracy and LIMU=0 switching off modifications to the NR method for all external models). Command .PRINT with OP specifier provides printouts at each NR iteration. Command .OP starts the operating point analysis. The results of simulation and errors have been collected in V(1) and V(1)-V(2) columns of Table 4.1.

In the case of IFL=1 7 iterations have given accuracy to more than nine 15. At each iteration, a number of accurate digits increases twice. This is the advantage of the quadratic convergence.

In the case of IFL=0 OPTIMA reaches an accuracy of 5 digits in as many as 14 iterations, as is demonstrated in the third and fourth columns of Table 4.1. The error decreases only twice per iteration. A difference between rates of convergence in these two cases is really impressive.

4.2.1.3 Graphical representation of NR iterations

In Figure 4.4 a function $f(v)$ has been plotted. It crosses the v-axis at the ideal solution v^* of $f(v)=0$. Let $v^{(p)}$ be the pth approximate solution. A tangent formed at this solution stands for the linearized equation (4.9). The next, i.e. $p+1$th, approximate solution is obtained at a point where the tangent crosses the v-axis. Vertical dotted lines with arrows show a manner of mapping the v coordinates onto the curve $f(v)$ to find points where tangents touch the curve. The figure demonstrates how two subsequent approximate solutions have been derived by means of the NR iterative process starting from a given point. Their convergence to the ideal solution x^* is easily visible from the figure.

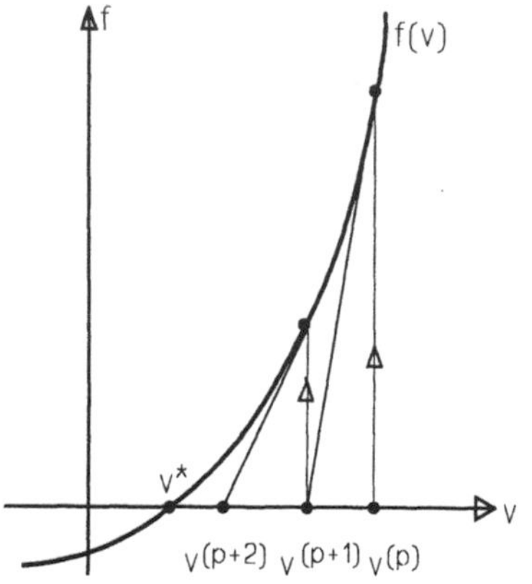

Figure 4.4 Graphical representation of NR iterations: thin continues lines are tangents, dotted lines indicate tangency points

4.2.1.4 Some remarks relating to global convergence properties

Global convergence theory tries to state precisely what assumptions, relating to the nonlinear function $f(v)$ and the starting point $v^{(0)}$, have to be made to gain convergence of the NR method. In general, this is a hard problem. Mathematicians can give some sufficient conditions of convergence, but it is not useful to predict whether a particular numerical task will be convergent or not. We confine the discussion here to some introductory aspects of global covergence in the case of a single equation. A detailed study of this problem can be found in the most comprehensive monograph dedicated to algebraic nonlinear equations [4.10].

The main concept of global convergence theory utilizes a so-called **convergence region**. It is such a maximum vicinity around the ideal solution v^* (that is a vicinity of the longest radius r) that all its points v (i.e. points satisfying $|v-v^*|\leq r$) are successful starting points for NR iterations (4.9). The radius of the convergence region is dependent on features of the function $f(v)$. Usually, we are only able to check whether a given point is within the region and sometimes calculate an upper bound for the distance from the given point to an unknown exact solution. Details of the global convergence estimation are outside the scope of our consideration due to their being of little practical importance in studying and developing practical d.c. algorithms. They may be neglected as long as we are not interested in precise mathematical proofs of convergence.

However, to show what features of the function $f(v)$ have repercussions on the rate of convergence or even on possible nonconvergence, we are going to present a theorem. Nevertheless, it will not help us at all to predict a good approximation to the exact solution (which might be used as a good starting point for NR process). For a single nonlinear equation we can formulate the following global convergence theorem.

Provided:

(i) $f(v)$ is defined and once differentiable in a closed segment $X \in R$ and has a root $v^* \in X$,

(ii) its derivative $f'(x)$ is within X of the Lipschitz property with a Lipschitz constant L, that is for any two $v_a, v_b \in X$ the following inequality $|f'(v_a) - f'(v_b)| \leq L\,|v_a - v_b|$ holds,

(iii) there is such a point $v^{(0)} \in X$ that $f'(v^{(0)}) \neq 0$ and

$$\frac{2\,L f(v^{(0)})}{[f'(v^{(0)})]^2} \leq 1 \tag{4.16}$$

and moreover the segment inclusion $[v^{(0)} - 2\Delta, v^{(0)} + 2\Delta] \subset X$ holds, where $\Delta = |f(v^{(0)})/f'(v^{(0)})|$,

then

(i) all derivatives $f'(v^{(p)})$ at points generated by the recursion (4.10) starting from $v^{(0)}$ remain nonzero, and thus the infinite sequence of these points may be produced,

(ii) this sequence converges to the solution v^*,

(iii) this solution it unique within $[v^{(0)} - 2\Delta, v^{(0)} + 2\Delta]$,

(iiii) for all p an error of the pth approximation to the solution can be bounded as $|v^{(p)} - v^*| \leq \Delta\, 2^{-p}$.

The above theorem enables us to estimate how far from the ideal solution lies a current point, but it offers no help in planning a strategy to select the starting point. We are only able to state that successful convergence, from a given point, depends on the first derivative and Lipschitz constant (or equivalently second derivatives for functions twice differentiable in X) within the set X. The general trend is similar to that in the local convergence case: the greater are the first derivatives and the smaller the Lipschitz constants, the better is the global convergence within X. Functions of considerable slope and weak nonlinearity are less sensitive to the starting point choice, and hence easily provide convergence.

To sum up we emphasize that the above conditions are only sufficient, that is there are functions and starting points that do not fulfil them but still provide successful NR process. The global conditions are often too pessimistic.

4.2.2 NR method for a system of algebraic equations

4.2.2.1 Iterative formulae

In circuit simulation we usually deal with systems of equations, not with a single one. The NR method is applicable also in this case. Let $\mathbf{f}(\mathbf{v}) = \mathbf{0}$ be a system of n algebraical equations with n unknowns. Given $\mathbf{v}^{(p)}$ as a starting point and taking a Taylor linearization of the vector function at this point (performing a first order expansion in the series) the following system of n algebraic linear equations arises, which is a multidimensional counterpart of the linear case (4.9)

$$\frac{\partial \mathbf{f}(\mathbf{v}^{(p)})}{\partial \mathbf{v}} \mathbf{v}^{(p+1)} = -\mathbf{f}(\mathbf{v}^{(p)}) + \frac{\partial \mathbf{f}(\mathbf{v}^{(p)})}{\partial \mathbf{v}} \mathbf{v}^{(p)}. \tag{4.17}$$

It can be solved only if the derivatives matrix $\partial \mathbf{f}/\partial \mathbf{v}$ (known as the Jacobi matrix) is nonsingular. With no additional comment we state that in the multidimensional case the NR method is also quadratically convergent (that is $\|\mathbf{v}^{(p+1)} - \mathbf{v}^*\| \le \alpha \|\mathbf{x}^{(p)} - \mathbf{v}^*\|^2$) in all cases, apart from the case of the Jacobi matrix singularity $\partial \mathbf{f}(\mathbf{v}^*)/\partial \mathbf{v}$ at the solution. In such a case linear convergence $\|\mathbf{v}^{(p+1)} - \mathbf{v}^*\| \le \alpha \|\mathbf{v}^{(p)} - \mathbf{v}^*\|$ occurs. A global convergence theorem is available also in this vector case. [4.3, 4.10]

4.2.2.2 Acceptable innacuracy in NR iterative formulae

Let the iterative formula (4.17) be convergent. While its iterative solutions tend to a limit, the term $\mathbf{v}^{(p+1)}$ approaches $\mathbf{v}^{(p)}$. Then, the two components of the Jacobi matrix, located on both sides of (4.17), become equal to each other and therefore can be eliminated. Hence, (4.17) turns into $-\mathbf{f}(\mathbf{v}^{(p)}) = \mathbf{0}$. This leads to the following conclusions:

(i) The numerical error of the solution is determined only by the calculation of the function $\mathbf{f}(\mathbf{v})$.

(ii) Inaccuracy in the Jacobi matrix affects only convergence or even divergence of the NR process, but never causes a solution error.

These conclusions enables us to purposely carry out inaccurate calculations of derivatives. In circuit simulators the greatest computational effort is spent on the calculation of nonlinear expressions of branch functions and their derivatives. Considerable savings can be achieved by simplifying the calculation of derivatives. These simplifications may consist of using derivatives obtained from earlier iterations (so-called bypassing) or approximate formulae for the Jacobi matrix.

Table 4.2. Relation between derivatives perturbation and number of iterations obtained from O4-4.cir

GAM	number of iterations	
	O4-4.CIR	O4-6.CIR
6	89	200
5	74	196
4	60	166
3	46	143
2	32	112
1.2	20	41
1.1	18	44
1	17	24
0.9	16	22
0.8	16	19
0.7	17	17
0.6	22	20
0.55	37	40
0.52	77	114

Inaccuracy in the calculation of derivatives affects the rate of local convergence in the solution vicinity and the convergence range. It can be shown that the convergence is no longer quadratic. However, a considerably small inaccuracy infinitesimally influences or even slightly decreases a number of required NR iterations. We demonstrate this effect by means of artificial perturbations of derivatives in external models for OPTIMA. Let us handle the sample O4-4.CIR. This file incorporates a circuit consisting of three diodes with exponential d.c. characteristics $i = I_s[\exp(v/V_T) - 1]$, common in bipolar circuits, and causing problems in the NR method. These diodes have been modeled by means of the external model DEXP incorporating a formula for current (denoted as CURR) and its derivative with respect to a terminal voltage (denoted as

CURR:VS). To modify the derivative, the latter has been multiplied by the adjusting factor GAM. If its value is 1, this provides the exact derivative, while other values stand for perturbations that yield a loss of correspondence between functions and derivatives.

After running the file O4-4.CIR we can form Table 4.1 where the relation between the number of iterations and GAM has been presented. From these examples we can find out that perturbations ranging from 1.2 to 0.8 weakly affect convergence or even slightly decrease the number of iterations. Moreover, we can observe that our examples are considerably resistant to perturbations ranging from 6 to 0.52, for a maximum number of 100 NR iterations imposed. In this range the convergence is maintained, though in a gradually inceasing number of iterations. Perturbations greater than 1, i.e. the increase of derivatives, are less critical, e.g. even for 600 per cent perturbations the iterations remain convergent, though after several dozen iterations.

There are examples where inaccurate derivatives infinitesimally affect the convergence (run the sample O4-5.CIR). In circuits with many nonlinear elements such as the bipolar amplifier O4-6.CIR shown in Table 4.2 the influence of perturbations is much more critical.

In the examples presented we have introduced in fact a scaling factor 1/GAM to all circuit variables. This can be extended to a general problem of NR step limiting. We take at each iteration, instead of the basic NR solution $\mathbf{v}^{(p+1)}$, a slightly modified point $\hat{\mathbf{v}}^{(p+1)} = \hat{\mathbf{v}}^{(p)} + \Gamma^{(p)}(\mathbf{v}^{(p+1)} - \hat{\mathbf{v}}^{(p)})$, where $\hat{\mathbf{v}}^{(p)}$ is an identically modified solution of the preceding iteration, $\mathbf{v}^{(p+1)}$ is a nonmodified solution of the current iteration, while the diagonal matrix $\Gamma^{(p)} = \mathrm{diag}\,[\gamma_1^{(p)},...,\gamma_n^{(p)}]$ comprises limiting factors for all components of $\mathbf{v}$. Therefore the ith component is limited in step according to the formula: $\hat{v}_i^{(p+1)} = \hat{v}_i^{(p)} + \gamma_i^{(p)}(v_i^{(p+1)} - \hat{v}_i^{(p)})$. Application of the NR step limiting formula to (4.17) gives the iterative equation

$$\frac{\partial \mathbf{f}(\mathbf{v}^{(p)})}{\partial \mathbf{v}}[\Gamma^{(p)}]^{-1}\,\mathbf{v}^{(p+1)} = -\mathbf{f}(\mathbf{v}^{(p)}) + \frac{\partial \mathbf{f}(\mathbf{v}^{(p)})}{\partial \mathbf{v}}[\Gamma^{(p)}]^{-1}\,\mathbf{v}^{(p)}. \tag{4.18}$$

From the above it is seen that step limiting is equivalent to a dynamic scaling of the Jacobi matrix by means of a diagonal matrix whose elements are reciprocals of limiting factors. Each column of the Jacobi matrix is divided by its limiting factor smaller than unity. This increases the matrix elements, and hence all derivatives are perturbated by a factor greater than unity.

We have demonstrated that large perturbations are not advantageous in the imediate vicinity of the solution. However, a dynamic scaling can be useful in initial iterations when we are far from the solution and the basic NR process is weakly convergent. This scaling can enlarge a convergence region and improve a convergence rate. These problems will be dicussed later on in the section dedicated to practical quasi-NR algorithms.

4.2.3 Methods for automatic formulation of iterative equations

4.2.3.1 Substitute iterative circuit

In Section 4.2.2 the process of solving a system of simultaneous nonlinear algebraic equations has been transformed into an iteratively repeated solution of systems of linear algebraic equations (4.17). In this section we take into account not nonlinear equations but rather nonlinear circuits which can be described in the d.c. steady state by d.c. canonical equations (4.2). To obtain iterative equations (4.17) corresponding to a d.c. nonlinear circuit (4.2) we could try to formulate nonlinear equations of (4.2) (e.g. by means of the modified nodal transformation of Section 1.3.4), and then carry out their Taylor linearization (see Section 4.2.2). In that way, the iterative equations (4.17) can be formulated. However, computer formulation of nonlinear equations from branch constitutive and from topological information is quite inefficient. We are able to formulate the iterative equations (4.17) directly from the input net-list (using branch definitions and topology) instead. This is the method of so-called substitute iterative circuits.

Let us start from the substitute d.c. nonlinear circuit (4.2). After Taylor linearization at the point $\mathbf{i}^{(p)}, \mathbf{v}^{(p)}$ we obtain

$$\begin{aligned}
&\mathbf{A}\,\mathbf{i}^{(p)} + \mathbf{A}\,(\mathbf{i}^{(p+1)} - \mathbf{i}^{(p)}) = \mathbf{0}\\
&\mathbf{B}\,\mathbf{v}^{(p)} + \mathbf{B}\,(\mathbf{v}^{(p+1)} - \mathbf{v}^{(p)}) = \mathbf{0}\\
&\mathbf{f}(\mathbf{i}^{(p)}, \mathbf{v}^{(p)}) + \frac{\partial \mathbf{f}(\mathbf{i}^{(p)}, \mathbf{v}^{(p)})}{\partial \mathbf{i}}(\mathbf{i}^{(p+1)} - \mathbf{i}^{(p)}) +\\
&\quad + \frac{\partial \mathbf{f}(\mathbf{i}^{(p)}, \mathbf{v}^{(p)})}{\partial \mathbf{v}}(\mathbf{v}^{(p+1)} - \mathbf{v}^{(p)}) = \mathbf{s}_0
\end{aligned} \tag{4.19}$$

where functions and derivatives are calculated at the linearization point derived from the preceding iteration. Hence, (4.19) can be rewritten as

$$\begin{aligned}
&\mathbf{A}\,\mathbf{i}^{(p+1)} = \mathbf{0}\\
&\mathbf{B}\,\mathbf{v}^{(p+1)} = \mathbf{0}\\
&\frac{\partial \mathbf{f}}{\partial \mathbf{i}}\,\mathbf{i}^{(p+1)} + \frac{\partial \mathbf{f}}{\partial \mathbf{v}}\,\mathbf{v}^{(p+1)} = \mathbf{s}\,.
\end{aligned} \tag{4.20}$$

where

$$\mathbf{s} = \mathbf{s}_0 - \mathbf{f}(\mathbf{i}^{(p)}, \mathbf{v}^{(p)}) + \frac{\partial \mathbf{f}}{\partial \mathbf{i}} \mathbf{i}^{(p)} + \frac{\partial \mathbf{f}}{\partial \mathbf{v}} \mathbf{v}^{(p)} . \qquad (4.21)$$

Comparing equations (4.20) with the canonical equations of linear circuit (2.3) it can be seen that the former is a canonical representation of a linear circuit structurally identical with the original circuit (1.23) (see Section 1.2.3) as well as with the substitute d.c. circuit (4.2). The branch variables of the circuit (4.20) have the values of the original branch variables at the p+1 NR iteration. Branches of the circuit (4.20) are linear counterparts of original nonlinear branches. The coefficients of these linear branches are expressed as partial derivatives of nonlinear functions **f** while excitations tak the form (4.21) instead of the original excitations. The circuit described by equations (4.20) is called **the substitute iterative circuit** dedicated to NR d.c. analysis. To obtain this circuit we have placed linearized counterparts in place of original branches. Linearized branches are called iterative models of original branches. Their construction will be shown in the next section.

When iterative models of all circuit branches are known, we are able to formulate circuit iterative equations (4.17). For example, they can be written as MNE (see Section 2.1.3) of the introduced substitute iterative circuit whose equations have the form (4.20). The iterative equations can be solved by means of LU methods described in Chapter 3.

4.2.3.2 Iterative models of nonlinear branches

Branch iterative models are obtained from a Taylor lineariztion of branch equations in the vicinity of the working point of the preceding iteration (denoted as pth). For example, let us consider the conductance $i=f(v)$. Its linearization at $v^{(p)}$ gives the iterative model $i^{(p+1)}=f(v^{(p)})+[\partial f(v^{(p)})/\partial v](v^{(p+1)}-v^{(p)})$ incorporating the linear conductance $G^{(p)}=\partial f(v^{(p)})/\partial i$ in parallel with the independent current source $I^{(p)}=f(v^{(p)})-G^{(p)}v^{(p)}$. Similarly, we can obtain an iterative model of the nonlinear CCVS $v=f(i)$. It consists of $v_a^{(p+1)}=R_{ac}^{(p)}i_c^{(p+1)}$, $R_{ac}^{(p)}=\partial f(i_c^{(p)})/\partial v$ (the linear CCVS) connected in series with $V^{(p)}=f(v^{(p)})-R_{ac}^{(p)}i_c^{(p)}$ (the voltage source). Of course linear branches after Taylor linearization remain unchanged, e.g. the conductance $i=Gv$ takes the form $i^{(p+1)}=Gv^{(p)}+G(v^{(p+1)}-v^{(p)})$ i.e. $i^{(p+1)}=G\,v^{(p+1)}$.

All the above iterative models are acceptable to MNE and can be easily combined with stamps for matrix formulation. To explain this approach let us consider the resistive circuit of Figure 4.5. Its nonlinear

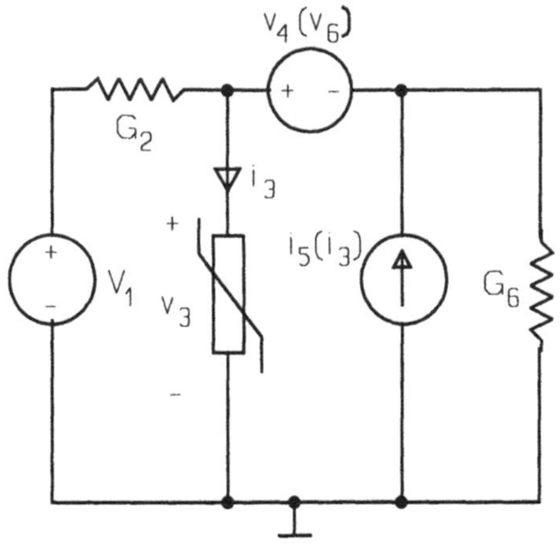

Figure 4.5 A resistive nonlinear circuit dedicated to demonstration of substitute iterative circuit

functions are: $v_3 = V_T \ln(i_3/i_s + 1)$, $v_4 = V_0 \arctan(v_6/V_0)$, $i_5 = \alpha(i_3/i_0)^3$. Their iterative models derived by Taylor linearization are as follows:

$$
\begin{aligned}
& v_3^{(p+1)} = R_3^{(p)}\, i_3^{(p+1)} + V_3^{(p)} \text{ where} \\
& R_3^{(p)} = \frac{V_T}{i_3^{(p)} + I_s}, \quad V_3^{(p)} = V_T \ln\left(\frac{i_3^{(p)}}{I_s} + 1\right) - R_3^{(p)}\, i_3^{(p)} \\
& v_4^{(p+1)} = T_{46}^{(p)}\, v_6^{(p+1)} + V_4^{(p)} \text{ where} \\
& T_{46}^{(p)} = \frac{1}{1 + (v_6^{(p)}/V_0)^2} \\
& V_4^{(p)} = V_0 \arctan(v_6^{(p)}/V_0) - T_{46}^{(p)}\, v_6^{(p)} \\
& i_5^{(p+1)} = K_{53}^{(p)}\, i_3^{(p+1)} + I_5^{(p)} \text{ where} \\
& K_{53}^{(p)} = \frac{3\alpha(i_3^{(p)})^2}{I_0^3}, \quad I_5^{(p)} = \alpha(i_3^{(p)}/I_0)^3 - K_{53}^{(p)}\, i_3^{(p)}
\end{aligned}
\tag{4.22}
$$

Linear branches remain unchanged, independent sources have their d.c. values. We obtain the iterative circuit shown in Figure 4.6. To find the p+1th iterative solution MNE have to be formulated

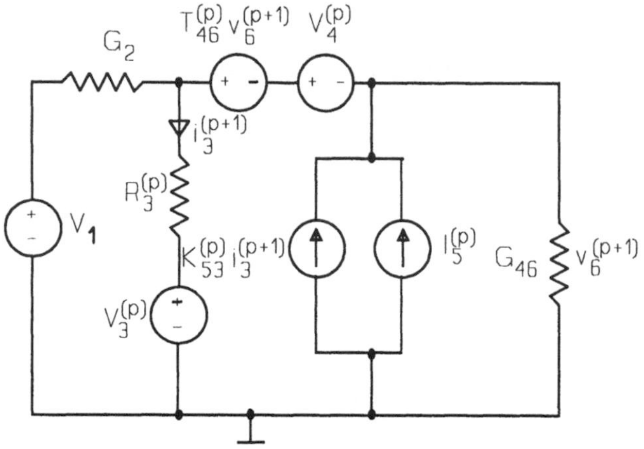

Figure 4.6 Substitute iterative circuit derived from the circuit of Figure 4.5

$$\begin{bmatrix} G_2 & -G_2 & 0 & 1 & 0 & 0 \\ -G_2 & G_2 & 0 & 0 & 1 & 1 \\ 0 & 0 & G_6 & 0 & -K_{53}^{(p)} & -1 \\ -1 & 0 & 0 & 0 & 0 & 0 \\ 0 & -1 & 0 & 0 & R_3^{(p)} & 0 \\ 0 & -1 & T_{46}^{(p)}+1 & 0 & 0 & 0 \end{bmatrix} \begin{bmatrix} v_{n1}^{(p+1)} \\ v_{n2}^{(p+1)} \\ v_{n3}^{(p+1)} \\ i_1^{(p+1)} \\ i_3^{(p+1)} \\ i_4^{(p+1)} \end{bmatrix} = \begin{bmatrix} 0 \\ 0 \\ I_5^{(p)} \\ -V_1 \\ -V_3^{(p)} \\ -V_4^{(p)} \end{bmatrix} \tag{4.23}$$

Assuming certain starting values $i_3^{(0)}$, $v_6^{(0)}$ of the controlling variables (for p=0) and repeating the solution of (4.23) within a loop controlled by p=1,2,... the d.c. analysis of circuit of Figure 4.5 can be obtained.

To introduce general rules for the creation of iterative circuits let us introduce standard nonlinear branches, current or voltage-defined

$$\begin{aligned} i_a &= \phi_i (v_a, v_b, i_c) \\ v_a &= \phi_v (i_a, v_b, i_c) \end{aligned} . \tag{4.24}$$

Their linearization gives corresponding linear branches whose coefficients are derivatives calculated at the last solution. The iterative model of the current-defined branch (4.24) is

$$
\begin{aligned}
i_a^{(p+1)} &= G_{aa}^{(p)} v_a^{(p+1)} + G_{ab}^{(p)} v_b^{(p+1)} + K_{ac}^{(p)} i_c^{(p+1)} + I_a^{(p)} \\
G_{aa}^{(p)} &= \frac{\partial \phi_i (v_a^{(p)}, v_b^{(p)}, i_c^{(p)})}{\partial v_a} \\
G_{ab}^{(p)} &= \frac{\partial \phi_i (v_a^{(p)}, v_b^{(p)}, i_c^{(p)})}{\partial v_b} \\
K_{ac}^{(p)} &= \frac{\partial \phi_i (v_a^{(p)}, v_b^{(p)}, i_c^{(p)})}{\partial i_c} \\
I_a^{(p)} &= \phi_i (v_a^{(p)}, v_b^{(p)}, i_c^{(p)}) - G_{aa}^{(p)} v_a^{(p)} - G_{ab}^{(p)} v_b^{(p)} - K_{ac}^{(p)} i_c^{(p)}
\end{aligned}
\tag{4.25}
$$

while that of the voltage-defined branch is

$$
\begin{aligned}
v_a^{(p+1)} &= R_{aa}^{(p)} i_a^{(p+1)} + T_{ab}^{(p)} v_b^{(p+1)} + R_{ac}^{(p)} i_c^{(p+1)} + V_a^{(p)} \\
R_{aa}^{(p)} &= \frac{\partial \phi_v (i_a^{(p)}, v_b^{(p)}, i_c^{(p)})}{\partial i_a} \\
T_{ab}^{(p)} &= \frac{\partial \phi_v (i_a^{(p)}, v_b^{(p)}, i_c^{(p)})}{\partial v_b} \\
R_{ac}^{(p)} &= \frac{\partial \phi_v (i_a^{(p)}, v_b^{(p)}, i_c^{(p)})}{\partial i_c} \\
V_a^{(p)} &= \phi_v (i_a^{(p)}, v_b^{(p)}, i_c^{(p)}) - R_{aa}^{(p)} i_a^{(p)} - T_{ab}^{(p)} v_b^{(p)} - R_{ac}^{(p)} i_c^{(p)}
\end{aligned}
\tag{4.26}
$$

To use the above formulae we have to know the expressions for branch functions ϕ_i, ϕ_v and, moreover, find expressions for partial derivatives. Thus, in general-purpose circuit simulators iterative models of nonlinear devices consist of encoded expressions. For example, SPICE code incorporates a MOSFET model composed of expressions for terminal currents and their partial derivatives symbolically derived once and built into the program code. Such models we call built-in models of devices. In the case of user-defined nonlinear branches, read from a simulator input file, derivatives may also be derived and given by the user in the symbolic form (as in our program OPTIMA). Alternatively, they can be automatically worked out by the program itself via numerical differentiation (as in OPTIMA, when the user omits symbolic expressions for derivatives) or by symbolic differentiation of model expressions.

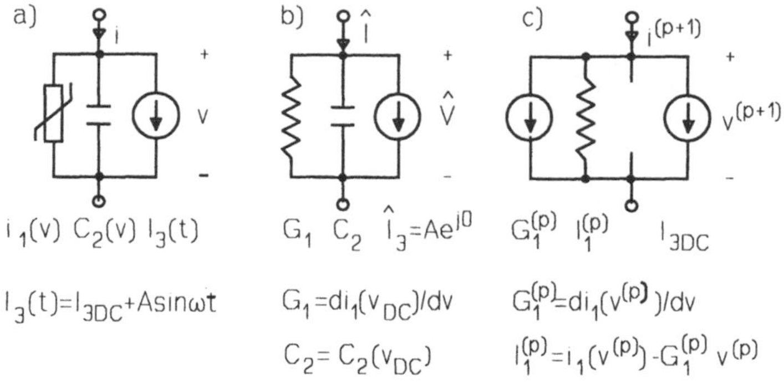

Figure 4.7 Comparison of substitute models of branches: (a) original branch, (b) small-signal a.c. immitance model, (c) iterative d.c. NR model

Coming back to iterative models (4.25) and (4.26) we observe that they are of the generalized linear form of Figure 2.5 dedicated to MNE. Their coefficients are derivatives calculated at a point where the branch independent variables (arguments) are from the last iteration. Moreover, an iterative branch incorporates a substitute source consisting of d.c. excitation (if any exists in our branch) and a substitute iterative component resulting from the function nonlinearity. For example, a nonlinear conductance in parallel with a current source $i=I_s[\exp(v/V_T)-1]+I_0$, after the Taylor linearization yields the current-defined branch $i^{(p+1)}=G^{(p)}v^{(p+1)}+I^{(p)}$. It is composed of the linear conductance $G^{(p)}=(I_s/V_T)\exp(v^{(p)}/V_T)$, and the source $I^{(p)}$ whose value is composed of the d.c source I_0 and an iterative component $I_s\exp(v^{(p)}/V_T)-G^{(p)}v^{(p)}$.

Due to the standard form of iterative branches (4.25), (4.26) compatible with MNE we are able to construct their stamps. They will be of types i, ix, v, given in Section 2.1.3.4, but with coefficients (4.25), (4.26) involved. These stamps may be viewed as MNE stamps of general nonlinear branches (4.24) dedicated to nonlinear d.c. NR analysis.

Comparing the d.c. iterative branches (4.25), (4.26) with a.c. substitute branches (2.66), (2.67) it is clear that for resistive branches there is a considerable similarity between both substitute representations. The coefficients are obtained from the same partial derivatives. The first difference lies in the point where the derivatives are calculated: the a.c. model uses the bias point (that is a point obtained after d.c. iterations have converged), while the d.c. model uses a point designated in the last iteration. The second difference is that in the a.c. case there is no source, while in the d.c. case we have the substitute auxiliary source.

For sources, in the a.c. case we have their substitute incremental phasor values, while in the d.c. case their d.c. components.

Finally, for reactive branches we obtain certain linearized reactive counterparts in the a.c. case, while only appropriate short-circuits or open-circuits (see Section 4.1) in the d.c. case. For example, let us take the branch of Figure 4.7(a), whose small-signal a.c. immitance counterpart is shown in Figure 4.7(b) while its iterative d.c. model is given in Figure 4.7(c)

4.2.3.3 Incremental iterations and incremental substitute circuit

Instead of the iterative NR equations (4.17) the so-called incremental iterative equations can be written

$$(\partial \mathbf{f}/\partial \mathbf{v})\,\Delta \mathbf{v}^{(p+1)} = -\mathbf{f}(\mathbf{v}^{(p)}) \tag{4.27}$$

where increments of circuit variables between the pth and $p+1$ iteration constitute the vector of unknowns. Then, a current solution is obtained from the sum $\mathbf{v}^{(p+1)} = \mathbf{v}^{(p)} + \Delta\mathbf{v}^{(p+1)}$. From a mathematical point of view this approach is equivalent to that previously presented. However, there are two advantages of the incremental approach: (i) RHS vector components are smaller and equibalanced and so a slightly better numerical accuracy in forward/backward substitution (see Chapter 3) can be achieved, (ii) the substitute iterative source has the advantageous property of approaching zero when iterations converge.

Canonical equations of the iterative circuit (4.19) may be easily rewritten in a form incorporating increments of branch variables. Taking incremental branch currents and voltages $\Delta\mathbf{i}^{(p+1)} = \mathbf{i}^{(p+1)} - \mathbf{i}^{(p)}$, $\Delta\mathbf{v}^{(p+1)} = \mathbf{v}^{(p+1)} - \mathbf{v}^{(p)}$ we obtain an iterative circuit described by the following canonical equations

$$\begin{aligned} &\mathbf{A}\,\Delta\mathbf{i}^{(p+1)} = \mathbf{0} \\ &\mathbf{B}\,\Delta\mathbf{v}^{(p+1)} = \mathbf{0} \\ &\frac{\partial \mathbf{f}}{\partial \mathbf{i}}\Delta\mathbf{i}^{(p+1)} + \frac{\partial \mathbf{f}}{\partial \mathbf{v}}\Delta\mathbf{v}^{(p+1)} = -\mathbf{f} \end{aligned} \tag{4.28}$$

This circuit is structurally identical with the original, but its branch variables are increments of the respective branch signals from the pth to $p+1$th iteration. The circuit (4.28) consists of incremental iterative branches that differ from nonincremental branches only in their excitations. These excitations are much simpler: they use values of branch functions $\mathbf{f}$ calculated at the last iteration. The latter is something like an error index and decreases towards zero while the process converges. Compared with the nonincremental case, this modification causes even constant excitations to change their models. However, the change is limited to RHS, while the contents of the matrix remain the same.

Now we will create incremental iterative models of standard nonlinear branches (4.24). For the CD branch this model is obtained by subtracting the current $i_a^{(p)}$ from both sides of (4.25). Similarly, for the VD branch we subtract $v_a^{(p)}$ from both sides of (4.26). After increments have been introduced we obtain

$$\Delta i_a^{(p+1)} = G_{aa}^{(p)} \Delta v_a^{(p+1)} + G_{ab}^{(p)} \Delta v_b^{(p+1)} + K_{ac}^{(p)} \Delta i_c^{(p+1)} + I_a^{(p)}$$

$$G_{aa}^{(p)} = \frac{\partial \phi_i(v_a^{(p)}, v_b^{(p)}, i_c^{(p)})}{\partial v_a}, \quad G_{ab}^{(p)} = \frac{\partial \phi_i(v_a^{(p)}, v_b^{(p)}, i_c^{(p)})}{\partial v_b}$$

$$K_{ac}^{(p)} = \frac{\partial \phi_i(v_a^{(p)}, v_b^{(p)}, i_c^{(p)})}{\partial i_c}$$

$$I_a^{(p)} = \phi_i(v_a^{(p)}, v_b^{(p)}, i_c^{(p)}) - i_a^{(p)} \tag{4.29}$$

$$\Delta v_a^{(p+1)} = R_{aa}^{(p)} \Delta i_a^{(p+1)} + T_{ab}^{(p)} \Delta v_b^{(p+1)} + R_{ac}^{(p)} \Delta i_c^{(p+1)} + V_a^{(p)}$$

$$R_{aa}^{(p)} = \frac{\partial \phi_v(i_a^{(p)}, v_b^{(p)}, i_c^{(p)})}{\partial i_a}$$

$$T_{ab}^{(p)} = \frac{\partial \phi_v(i_a^{(p)}, v_b^{(p)}, i_c^{(p)})}{\partial v_b} \tag{4.30}$$

$$R_{ac}^{(p)} = \frac{\partial \phi_v(i_a^{(p)}, v_b^{(p)}, i_c^{(p)})}{\partial i_c}$$

$$V_a^{(p)} = \phi_v(i_a^{(p)}, v_b^{(p)}, i_c^{(p)}) - v_a^{(p)}$$

Notice that in the above equations an excitation is calculated in the preceding iteration, as a difference between the value of branch variable derived from the branch function and the one from the iterative circuit. For each nonlinear branch the latter has to be derived from the iterative model at the preceding iteration

$$
\begin{aligned}
i_a^{(p)} &= i_a^{(p-1)} + G_{aa}^{(p-1)} \Delta v_a^{(p)} + G_{ab}^{(p-1)} \Delta v_b^{(p)} + K_{ac}^{(p-1)} \Delta i_c^{(p)} + I_a^{(p-1)} \\
v_a^{(p)} &= v_a^{(p-1)} + R_{aa}^{(p-1)} \Delta i_a^{(p)} + T_{ab}^{(p-1)} \Delta v_b^{(p)} + R_{ac}^{(p-1)} \Delta i_c^{(p)} + V_a^{(p-1)}
\end{aligned}
\tag{4.31}
$$

and stored for use in the next iteration.

Now, we will examine some examples. After linearization of a diode equation $i = I_s[\exp(v/V_T) - 1]$ at the pth point we obtain the nonincremental model $i^{(p+1)} = I_s\,[\exp(v^{(p)}/V_T) - 1] + [(I_s/V_T)\exp(v^{(p)}/V_T)] \cdot (v^{(p+1)} - v^{(p)})$. Then, subtracting the current $i^{(p)}$ from both sides, we obtain the incremental model: $\Delta i^{(p+1)} = G^{(p)}\Delta v^{(p+1)} + I^{(p)}$, where $G^{(p)} = (I_s/V_T)\exp(v^{(p)}/V_T)$ is the same as in the nonincremental case, while the RHS exhibits the error index, namely $I^{(p)} = I_s[\exp(v^{(p)}/V_T) - 1] - i^{(p)}$, specific to the incremental approach.

Another interesting example is related to a constant voltage source $v = V$. After the voltage $v^{(p)}$ has been subtracted from both sides we get the iterative model $\Delta v^{(p+1)} = V - v^{(p)}$. Usually, as for example in MNE, at each iteration the voltage has a constant value V (this results from the manner of formulating equations - see the stamp of type v). Hence, the incremental model of a voltage source is a short-circuit. Only in the first iteration could the model be a nonzero voltage source, due to a possible difference between V and the starting value $v^{(0)}$.

4.2.3.4 Graphical representation of iterative models

The demonstration of iterative models on a plane is possible only for two-dimensional problems. Let us consider the circuit shown in Figure 4.8 composed of a nonlinear conductance $i = f(v)$ and an element $i = I - v/R$ (the current source I connected to a resistance R). The solution problem may be discussed in the (v, i) plane. An exact d.c. response is located at the point (v_0, i_0), where both linear and nonlinear characteristics cross each other. An iterative NR solution consists of replacing the nonlinear characteristic by its first order Taylor expansion. The latter is represented by a tangent to the nonlinear characteristic set up at the point where the expansion has been taken. A current solution appears to be a point where the tangent crosses the line representing the driver IR. Let us consider the p–1 iteration represented by a tangent set up for $v^{(p-1)}$. The pth solution $(v^{(p)}, i^{(p)})$ appears at the crossing of the two straight lines, the tangent and IR. It is essential to observe that this point does not lie on the nonlinear characteristic. If it were located there, this would mean that the final solution had been reached. The point $(v^{(p)}, f(v^{(p)}))$ is located on the nonlinear characteristic. Its current coordinate is different

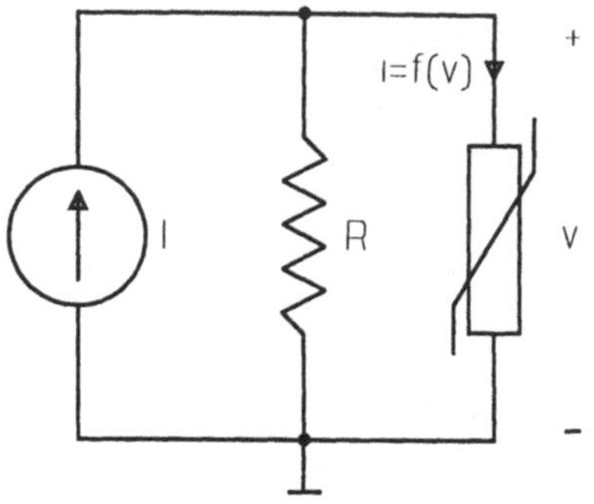

Figure 4.8 A simple nonlinear circuit whose iterative solution is demonstrated in Figure 4.9

from the NR solution $i^{(p)}$. The difference $f(v^{(p)}) - i^{(p)}$ defines the error of the pth iteration. To continue the iterative process we have to create a new tangent, relating to $v^{(p)}$ (drawn as the p+1 tangent in Figure 4.9).

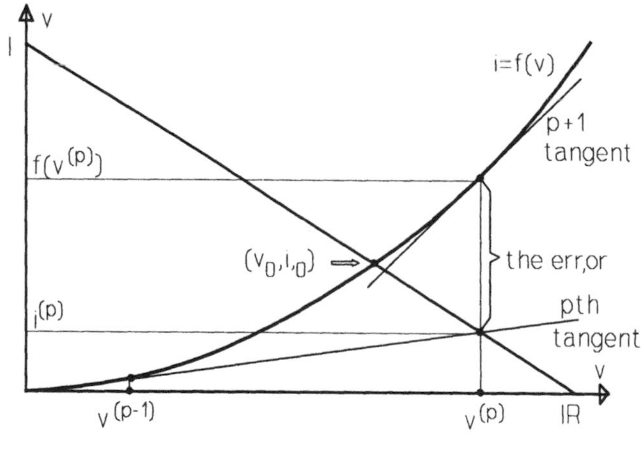

Figure 4.9 Graphical representation of iterative solution of circuit shown in Figure 4.8

4.2.3.5 Secant iterations

The secant method is the only practical alternative to the NR approach. It may be viewed as a modification of the latter, where derivatives have been replaced by finite differences. However, due to a slightly worse convergence and numerical complexity, the secant method is of limited application and our discussion will be brief.

Let the function $y=f(x)$, $y, x \in R$ constitute an equation $f(x)=0$ and at two points $x^{(p)}, x^{(p-1)}$ take values $f(x^{(p)})$, $f(x^{(p-1)})$ respectively. The straight line passing through these points is known as the secant, as it is demonstrated in Figure 4.10. Assuming that the equation for it $y=ax+b$ satisfies the given points, we construct the incremental formula $f(x^{(p)})-f(x^{(p-1)})=a(x^{(p)}-x^{(p-1)})$, which finally yields the secant

$$y=\frac{f(x^{(p)})-f(x^{(p-1)})}{x^{(p)}-x^{(p-1)}}(x-x^{(p)})+f(x^{(p)}). \tag{4.32}$$

The point where the secant crosses the X-axis is taken to be the next solution $x^{(p+1)}$, and it can be obtained from

$$\frac{f(x^{(p)})-f(x^{(p-1)})}{x^{(p)}-x^{(p-1)}}(x^{(p+1)}-x^{(p)})=-f(x^{(p)}). \tag{4.33}$$

The sequence $\{x^{(p)}\}_{p=1}^{\infty}$ that can be derived from the recursion (4.33) may be convergent only to the solution of our equation. The procedure of generating this sequence utilizes only the function $f(x)$. No derivatives are needed. Two calculations of the nonlinear function appear, instead of the calculation of a function and one of its derivatives in the NR case. Notice, that the secant formula may be viewed as a quasi-NR formula with a considerably inaccurate derivative. Its convergence may be proved to be slower than quadratic but faster than linear. The convergence exponent can be proved to be $(1+\sqrt{5})/2\approx 1.6$.

In circuit simulation this scalar case is not sufficient. Fortunately, the secant method may be easily extended to multivariate functions. Let us proceed to systems of n algebraic nonlinear equations with n unknowns. Each of these equations may be expressed as $y=f(\mathbf{x})=0$. The function f is of n variables, and hence can be approximated by a hyperplane $y=\mathbf{a}^t\mathbf{x}+b$

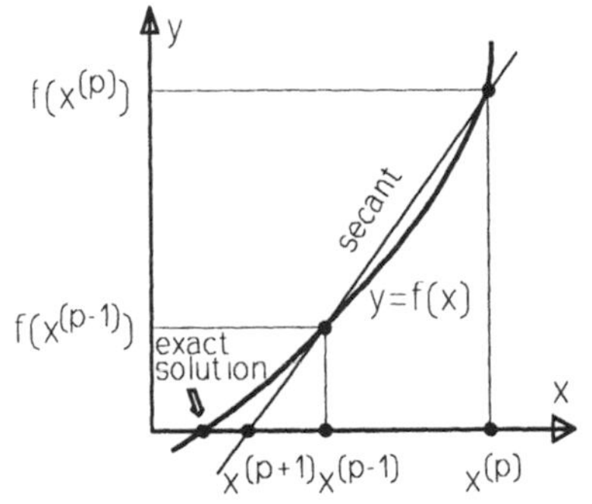

Figure 4.10 Secant linearization of the nonlinear function and its application to secant iteration

passing through n+1 points $(\mathbf{x}^{(i)}, f(\mathbf{x}^{(i)}))$, $i=p, p-1, ..., p-n$. The intersection of n such hyperplanes (corresponding to n equations) constitutes a subsequent p+1 iterative solution. To set up the secant we must determine the coefficients $\mathbf{a}^t, b$. As the hyperplane satisfies the given points we can extract the coefficients by substituting the coordinates of the points and subtracting side-wise the resulting equations with the unknowns $\mathbf{a}^t, b$. We obtain n incremental equations

$$\begin{aligned} f(\mathbf{x}^{(p)})-f(\mathbf{x}^{(p-1)}) &= \mathbf{a}^t(\mathbf{x}^{(p)}-\mathbf{x}^{(p-1)}) \\ f(\mathbf{x}^{(p-1)})-f(\mathbf{x}^{(p-2)}) &= \mathbf{a}^t(\mathbf{x}^{(p-1)}-\mathbf{x}^{(p-2)}) \\ &\dots \\ f(\mathbf{x}^{(p+1-n)})-f(\mathbf{x}^{(p-n)}) &= \mathbf{a}^t(\mathbf{x}^{(p+1-n)}-\mathbf{x}^{(p-n)}) \end{aligned} \quad . \tag{4.34}$$

Collecting, side by side, the left and the right sides of (4.34) we obtain the vector $\Delta\mathbf{f}$ and matrix $\Delta\mathbf{X}$

$$\Delta f = \begin{bmatrix} f(\mathbf{x}^{(p)})-f(\mathbf{x}^{(p-1)}) \\ f(\mathbf{x}^{(p-1)})-f(\mathbf{x}^{(p-2)}) \\ \dots \\ f(\mathbf{x}^{(p+1-n)})-f(\mathbf{x}^{(p-n)}) \end{bmatrix}, \quad \Delta\mathbf{X} = \left[\, \mathbf{x}^{(p)}-\mathbf{x}^{(p-1)} \,|\, \dots \,|\, \mathbf{x}^{(p+1-n)}-\mathbf{x}^{(p-n)} \,\right]. \tag{4.35}$$

Hence, (4.34) transforms into

$$\Delta\mathbf{X}^t\,\mathbf{a} = \Delta\mathbf{f}\,. \tag{4.36}$$

After solution of (4.36) for **a** we have the secant in the form

$$y = \mathbf{a}^t(\mathbf{x}-\mathbf{x}^{(\mathrm{p})}) + f(\mathbf{x}^{(p)})\,. \tag{4.37}$$

The transformation of $y=f(\mathbf{x})$ into (4.37) we call the secant linearization. The special case, when increments are not taken with respect to a preceding point but are taken with respect to a chosen reference point, the same for all increments, is known as the *regula falsi* approach.

It can be shown that the secant linearization transforms a linear function into the same linear function. Thus, transformation of circuit canonical equations yields the same KCL, KVL linear relations and substitute linear branches derived by the secant transformation of original branches. Hence, the secant method is easily implementable by means of a substitute iterative circuit consisting of secant iterative models.

To demonstrate secant iterative models let us consider a nonlinear conductance $i=f(v)$. We obtain the secant $i=f(v^{(p)})+a(v^{(p+1)}-v^{(p)})$, with the coefficient $a=[f(v^{(p)})-f(v^{(p-1)})]/(v^{(p)}-v^{(p-1)})$. This result stands for the well-known NR like branch $i^{(p+1)}=G^{(p)}v^{(p+1)}+I^{(p)}$, with a conductance and current source of values $G^{(p)}=a$, $I^{(p)}=f(v^{(p)})-G^{(p)}v^{(p)}$.

Much more complicated models arise for multivariate branch functions, e.g. for the general current-defined branch $i_a=\phi(v_a, v_b, i_c)$. In this case (4.37) takes the specific form

$$i_a^{(p+1)}=G_{aa}^{(p)}v_a^{(k+1)}+G_{ab}^{(p)}v_b^{(p+1)}+k_{ac}^{(p)}i_c^{(p+1)}+I_a^{(p)} \tag{4.38}$$

where coefficients $G_{aa}^{(p)}, G_{ab}^{(p)}, K_{ac}^{(p)}$ are components of the vector **a**. They can be derived from Equation (4.36) which in this case takes the following form

$$\begin{bmatrix} v_a^{(p)}-v_a^{(p-1)} & v_b^{(p)}-v_b^{(p-1)} & i_c^{(p)}-i_c^{(k-1)} \\ v_a^{(p-1)}-v_a^{(p-2)} & v_b^{(p-1)}-v_b^{(p-2)} & i_c^{(p-1)}-i_c^{(p-2)} \\ v_a^{(p-2)}-v_a^{(p-3)} & v_b^{(p-2)}-v_b^{(p-3)} & i_c^{(p-2)}-i_c^{(p-3)} \end{bmatrix}\begin{bmatrix} G_{aa}^{(p)} \\ G_{ab}^{(p)} \\ K_{ac}^{(p)} \end{bmatrix}= \tag{4.39}$$

$$=\begin{bmatrix} \phi(v_a^{(p)},v_b^{(p)},i_c^{(p)})-\phi(v_a^{(p-1)},v_b^{(p-1)},i_c^{(p-1)}) \\ \phi(v_a^{(p-1)},v_b^{(p-1)},i_c^{(p-1)})-\phi(v_a^{(p-2)},v_b^{(p-2)},i_c^{(p-2)}) \\ \phi(v_a^{(p-2)},v_b^{(p-2)},i_c^{(p-2)})-\phi(v_a^{(p-3)},v_b^{(p-3)},i_c^{(p-3)}) \end{bmatrix}.$$

From the above we observe that for multivariate branches the secant method is of considerably greater numerical complexity then the NR approach. It requires the storage of many previous solutions and each branch at each iteration requires the solution of (4.39). In general, due to these disadvantages and its slightly weaker rate of convergence the secant method appears to be worse than the NR approach.

However, in special cases when most of the circuit branches are subject to the NR linearization, a few others whose derivatives are not available may be linearized in a secant manner, to omit the differentiation embedded. This mixed NR-secant linearization usually worsens the convergence only infinitesimally.

4.2.4 Realization of the basic d.c. analysis algorithm

4.2.4.1 Tests for convergence

The basic methods of d.c. analysis introduced so far may be realized in the form of iterative algorithms. We break the NR iterative loop when the required accuracy has been reached. The user imposes a maximum allowable error level in the form of a relative accuracy index ε, and absolute accuracy index δ (often separate for currents δ_i (e.g. 1 pA) and for voltages δ_u (e.g. 1 μV). To explain the meaning of these indices let us consider two numbers a and b. The distance between them satisfies the accuracy indices involved when the inequality $|a-b| \le \varepsilon \max\{|a|, |b|\} + \delta$ holds. Similarly, we examine the distance between vectors $\mathbf{a}$ and $\mathbf{b}$ checking the above inequality for all pairs a_i, b_i of components.

In simple d.c. analysis programs, a test for convergence usually examines the number of iterations and the distance between circuit variables in two successive iterations. This criterion applied to a circuit variables vector takes the form

break after the $p+1$ solution has been obtained if:

$$p+1 \ge \max\_number\_of\_iterations; \ \{\text{practical nonconvergence}\}$$

or for all components x_i of $\mathbf{x}$ (4.40)

$$|x_i^{(p+1)} - x_i^{(p)}| \le \varepsilon \max\{|x_i^{(p+1)}|, |x_i^{(p)}|\} + \delta; \ \{\text{convergence}\}$$

In a more advanced case, independent variables y and dependent variables z_i of all circuit branches $y(\mathbf{z})$ should be taken into account instead of x_i, as the criterion (4.40) misrepresents the distance between an iterative and an exact solution. The ideal test for convergence should check the distance $\|\mathbf{x}^{(p+1)} - \mathbf{x}^*\|$, while $\|\mathbf{x}^{(p+1)} - \mathbf{x}^{(p)}\|$ is only a substitute measure. Considering that $\mathbf{x}^{(p+1)} - \mathbf{x}^{(p)} = (\mathbf{x}^{(p+1)} - \mathbf{x}^*) - (\mathbf{x}^{(p)} - \mathbf{x}^*)$, taking norms of both sides, and applying the rule $\|\mathbf{x}-\mathbf{y}\| \ge |\|\mathbf{x}\| - \|\mathbf{y}\||$, the following formula arises: $\|\mathbf{x}^{(p+1)} - \mathbf{x}^{(p)}\| \ge |\|\mathbf{x}^{(p+1)} - \mathbf{x}^*\| - \|x^{(p)} - \mathbf{x}^*\||$. This estimation shows that even if the LHS is small we are only sure that the RHS is also small. However, both components of the latter (denoting the absolute value of the distance between a current iterative solution and an asymptotic one) may be quite large. This occurs when the iterative process is slowly convergent.

A more advantageous convergence test uses an error index shown in Figure 4.9 which stands for the distance between a branch independent variable derived from the nonlinear characteristic and one derived from the iterative circuit. It is equal to zero when an exact solution at a nonlinear branch has been reached. Moreover, it is always equal to zero for linear branches. Hence, once the error indices approach zero, for all nonlinear branches, then the d.c. circuit solution is reached. To check the test the stopping flag *stop* is introduced and initialized to 1 (which means: break the iterative loop). Suppose the calculations of the pth iterative solution have been acomplished and we have just entered into the $p+1$ iteration. Then, before the iterative equations have been solved, a loop over all nonlinear branches is constructed (it may be realized simultaneously by setting up the circuit matrix at this iteration). Details of the test will now be explained by means of the sample branch $i=\phi(v)$ shown in Figure 4.9. Within the loop over branches a branch constitutive variable $i^{(p)}$ is first derived using a stored iterative model of the preceding iteration $i^{(p)}=\phi(v^{(p-1)})+[\partial\phi(v^{(p-1)})/\partial v](v^{(p)}-v^{(p-1)})$. It is represented in Figure 4.9 by a tangent at the point $v^{(p-1)}$. Then, $\phi(v^{(p)})$ is calculated from the branch function and the distance is checked. Since the function $i(v)$ can be very flat, a distance between two recent independent variables has to be checked simultaneously. This gives the test for convergence

$$\begin{aligned} &|\phi(v^{(p)})-i^{(p)}|\geq \varepsilon \max\{|i^{(p)}|,|\phi(v^{(p)})|\}+\delta_i \quad \text{or} \\ &|v^{(p)}-v^{(p-1)}|\geq \varepsilon \max\{|v^{(p)}|,|v^{(p-1)}|\}+\delta_v \end{aligned} \quad . \tag{4.41}$$

Fulfilment of (4.41) means convergence at the current branch has not been reached, and then we set the flag *stop* to 0. A current iterative model and iterative solution are stored and we proceed to the next branch. After all branches have been taken, the flag *stop* is checked: *stop* = 1 means that convergence criteria are fulfilled for all branches and the NR procedure may be finished. Otherwise, at least one branch does not satisfy the criterion. If the number of performed iterations is greater than *max_number_of_iterations*, iterations are terminated. Otherwise, the $p+1$ iteration should be continued by solving linear equations and proceeding to the next NR iteration.

Once the incremental iterations approach is applied (Section 4.2.3.3), general nonlinear branches (4.24) give iterative models (4.29), (4.30). In these models, sources equal to the respective error indices discussed above are introduced into the RHS vector of (4.27). Hence, the RHS is the residual vector while the error indices are its components. The test (4.41) has checked these components separately. Another possible test (used in CAzM [4.4]) examines overall the norm $\|\mathbf{f}(\mathbf{x}^{(p)})\|<\delta$ and moreover $\|\mathbf{x}^{(p+1)}-\mathbf{x}^{(p)}\|/\|\mathbf{x}^{(p)}\|\leq\varepsilon$.

4.2.4.2 An outline of a practical basic d.c. analysis algorithm

The NR algorithm for calculating d.c. responses of nonlinear circuits has been outlined in Algorithm 4.1. This algorithm is a basic realization of the NR method and does not contain auxiliary mechanisms for efficiency improvement. The latter will be discussed in Section 4.3 and will be added to this outline later on. The algorithm discussed here consists of a main iterative NR loop **for**. The number of its iterations is at most *max_number_of_iterations*. At each iteration we have: (i) for each branch of the circuit: the NR dedicated linearization (if it is nonlinear) and formulation of iterative equations (this block also incorporates the test for convergence), (ii) solution of algebraic linear equations. The latter is a standard procedure based on the algorithms of Chapter 3, usually implemented in sparse matrix techniques. The solution obtained is then contributed to (i) as a new point for circuit linearization and the process is repeated. To start the NR loop an initial point has to be assumed at the very beginning.

The most important part of the algorithm for formulating linearized iterative equations and testing convergence according to (4.41) is step (i). We need an appropriate data structure *struct* where each nonlinear branch has its own segment in memory. Assuming that all circuit branches are of the general form $y=y(\mathbf{z})$ (where y is a branch independent variable current or voltage, while $\mathbf{z}$ is a vector of independent variables including: controlling variables and the current of a voltage-defined branch and the voltage of a current-defined branch, e.g. $i_a=f(v_a, v_b)$ gives $z=[v_a, v_b]^t$) the memory belonging to a particular branch should store:

$y^{(0)}$ (branch constitutive variable y at last but one iteration),

$\mathbf{g}^{(0)} \equiv \partial y(\mathbf{z}^{(0)})/\partial \mathbf{z}$ (the row vector of coefficients of the iterative model at the last but one iteration),

$\mathbf{z}^{(0)}$ (independent variables at last but one iteration),

$\mathbf{z}^{(1)}$ (independent variables at the last iteration – they stand for the point where NR linearization is performed).

For each nonlinear branch the stored data is used to obtain an extrapolation of variable $\hat{y}^{(1)}$ at the last iteration and to compare it with the true value derived from the last iteration $y(\mathbf{z}^{(1)})$. This comparison enables us to check (and store) whether the local test for convergence is not satisfied at any particular nonlinear branch. Then, we store $y^{(1)}, \mathbf{g}^{(1)}, \mathbf{z}^{(1)}$ as variables $y^{(0)}, \mathbf{g}^{(0)}, \mathbf{z}^{(0)}, \ldots$, i.e. as corresponding to the last but one iteration.

Now, we create a stamp of the branch (of course for nonlinear and linear branches). Linear branches give stamps immediately, while nonlinear ones do so after linearization at the point $\mathbf{z}^{(1)}$. The stamp is added to the partial matrix $\mathbf{Y}^{(1)}$ and the RHS vector $\mathbf{b}^{(1)}$.

Algorithm 4.1 Basic NR procedure for d.c. nonlinear analysis

Data: starting values of independent variables $z^{(0)}$ for each nonlinear branch, which are stored in *struct*;
accuracy indexes ε, δ and *maximum_number_of_iterations.*
for *iter*=1 **to** *maximum_number_of_iterations* **do** {NR iterations loop}
begin
  stop=1;
  {matrix formulation and test for convergence}
  for all branches
  begin
    if (branch is nonlinear)
      if (*iter*>1) derive independent variables $\mathbf{z}^{(1)}$ of the nonlinear branch by means of the last solution vector $\mathbf{x}^{(1)}$ and store them in struct;
      get $y^{(0)}, \mathbf{z}^{(0)}, \mathbf{g}^{(0)} \equiv \partial y^{(0)}/\partial \mathbf{z}$ from *struct* and predict auxiliary $\hat{y}^{(1)} = y^{(0)} + \mathbf{g}^{(0)t}(\mathbf{z}^{(1)} - \mathbf{z}^{(0)})$;
      calculate $y^{(1)}, \mathbf{g}^{(1)} \equiv \partial y^{(1)}/\partial \mathbf{z}$ at point $\mathbf{z}^{(1)}$ from branch constitutive expression $y = y(\mathbf{z})$;
      if ($|\hat{y}^{(1)} - y^{(1)}| > \varepsilon \max\{|\hat{y}^{(1)}|, |y^{(1)}|\} + \delta$ **or** exists such $i=1,\ldots,\dim \mathbf{z}$ as $|z_i^{(1)} - z_i^{(0)}| > \varepsilon \max\{|z_i^{(1)}|, |z_i^{(0)}|\} + \delta_z$)
        stop = 0 {test for convergence at a particular branch not satisfied}
      $y^{(0)} = y^{(1)}$; $\mathbf{z}^{(0)} = \mathbf{z}^{(1)}$; $\mathbf{g}^{(0)} = \mathbf{g}^{(1)}$; {store last results as results of a last but one iteration}
    endif
    write stamp of the branch into the matrix $\mathbf{Y}^{(1)}$ and RHS vector $\mathbf{b}^{(1)}$;
  end
  if (*stop*≠0) **stop** {NR loop termination – convergence reached}
  solve equations $\mathbf{Y}^{(1)}\mathbf{x}^{(1)} = \mathbf{b}^{(1)}$ by means of the LU method;
  if ($\mathbf{Y}^{(1)}$ is singular) **then**
    stop {NR termination - matrix failure}
  else
    {new solution $\mathbf{x}^{(1)}$ obtained}
  endif
end
stop {NR termination – convergence not reached}

When all branches have been taken, the matrix and RHS are completed and also it is clear whether the signals of all nonlinear branches have converged. If they have, then the NR iterations are terminated. Otherwise, we proceed to the solution of the equations just formulated, produce a new iterative solution $\mathbf{x}^{(1)}$, and then return once more to the equation formulating block.

This basic NR algorithm runs without any improvements, with only one exception. It needs protection of the exponential functions of bjt devices against numerical over/underflow. The simplest protection is to use the approximation to exponents presented in Section 4.3.

To test our understanding, we can try to solve the following exercises:

- Choose a scalar equation and using it explain the principle of NR iterations, introduce the iterative equation, demonstrate it graphically and check whether it is convergent linearly or quadratically,
- Choose a set of equations and create iterative equations for it,
- Take an arbitrary nonlinear resistive circuit, write its canonical equations and prove that NR linearization leads to a substitute iterative circuit, explain and demonstrate on a variety of examples (containing resistors, conductances, controlled sources) how iterative models are created,
- For a chosen controlled source create the usual iterative model and an incremental model, explain the difference, and demonstrate both iterations in a plot,
- Using a scalar equation and then a system of two equations explain the principle of secant iterations,
- Explain the three sorts of tests for convergence, their advantages and weak points,
- For a chosen very simple nonlinear circuit write a d.c. analysis program according to Algorithm 4.1, compare results and convergence with PSPICE and OPTIMA, and explain differences.

In what follows we proceed to modified NR-like algorithms, that are used in practice, including tools for speeding up convergence and protection against numerical difficulties.

4.3 Practical quasi-Newton-Raphson algorithms

4.3.1 Numerical problems with the basic NR algorithm

The NR algorithm can be convergent very quickly, when started from a point sufficiently close to the d.c. solution. In bjt and MOS circuits a rough prediction of interterminal bias voltages, required for the selection of their starting values, can be often achieved, due to the nature of their standard (exponential, quadratic) nonlinearities. However, even then, some circuit configurations (e.g. reverse polarization of transistors) may cause unpredictable nonconvergence. For example, a good starting voltage across a bipolar junction lies on the bend of the exponential curve but for reverse driven junctions it is better to start from zero voltage. The need for the proper selection of the starting point is the weak point of the NR algorithm, especially if it is the basic NR algorithm without improvements which will be discussed later on.

In practice, if the starting point is not properly selected then we meet numerical difficulties in the iterative process. These may fall into four categories:

- *divergence*, when some components of an iterative solution tend to infinity in an aperiodic or periodic manner,
- *limit cycles*, when some components of an iterative solution have numerical oscillations of a constant amplitude (a stable limit cycle gives permanent oscillations while an unstable one leads to convergence or divergence after a number of approximately constant oscillations),
- *overflow and underflow run-time errors* in calculations of nonlinear functions $y(\mathbf{z})$ constituting the matrix and RHS vector components, due to large (unphysical) values of signals $\mathbf{z}$ temporarily appearing during the iterative process,
- *extremely slow convergence*, when NR steps become very small, and the final solution is reached only after a very large number of iterations.

Several numerical experiments are demonstrated to aquaint the reader with the convergence problems of the NR method. The simulator OPTIMA is especially useful when we need to probe into the algorithms. To obtain a basic NR iteration the step-limiting mechanism (to be discussed later on) is switched off, by means of the option LIMU set to 0. Quite often, numerical problems emerge from saturation-like characteristics, as e.g. the nonlinear conductance RSAT $i = I_0 \arctan(v/V_0)$ modeled by external model SATU in the task O4-7.CIR. The output current CURR is dependent on controlling variable U assigned to the circuit voltage V(1). This conductance is connected with the current source I. The controlling variable U of SATU is declared as follows:

CONTROL U=VINI

where parameter VINI defines a starting value for U in NR operating point analysis (set to a default in a declaration line and further redefined by means of .ALT commands). Selecting in turn one of a few possible starting values we may observe the different courses taken by the iterations. In the case of saturation-like characteristics such as SATU the course of NR process is extremely sensitive to the starting point. For $U \approx -2.34678066627177...$, we have an unstable limit cycle. Even infinitesimally more negative values lead to oscillative divergence, while less negative ones lead to considerably slow oscillative convergence as can be observed from experiments introduced in the sample O4-7.CIR. Once the starting point for U approaches the exact solution U=0.72654, convergence improves. In Table 4.3 we have voltages V(1) drawn up in successive iterations for two infinitesimally close starting values of U laying in the vicinity of the limit cycle. Divergence and limit cycles (numerical oscillations) appear not only in saturating elements (as e.g. operational amplifiers) but also in multitransistor circuits. To override them, a NR step limiting mechanism has been investigated. Before we proceed to detailed discussion of it, we may experiment with O4-7.CIR restoring default limiting state by writing LIMU=2 in the OPTIONS line. This provides convergence in 9 iterations for both cases of Table 4.3.

Another problem, quite common in bjt circuits, is known as an overflow or underflow error, due to existence of extremely steep $\exp(x)$ functions in element models. The iterative NR process can temporarily enter a region of large arguments x coresponding to physically unreal voltages across junctions and causing floating-point calculations to exceed the range of floating-point representation available. As is known from the IEEE 754 norm, a real number is represented in a floating-point format

$$x = m \cdot 2^{c} \tag{4.42}$$

where the mantissa $m \in [1, 2)$ is stored in a natural binary code with the sign bit, while the exponent c is encoded in a shifted binary code. Due to the limited number of m and c bits, the set of exactly representable real numbers is limited. In particular, the length of the exponent n determines the maximum positive number which can be represented. In shifted binary code the maximum encodable c is 2^{n-1}. Due to $m_{max} \approx 2$, the maximum representable real is $x_{max} = m_{max} 2^{c_{max}} \approx 2^{2^{n-1}+1}$. For example, double precision numbers have $n = 11$, and thus $x_{max} = 3.595\ 10^{308}$. It means that the maximum acceptable argument of the function exp is about 710.4. Greater arguments will cause an overflow error.

Table 4.3 Testing of oscillative convergence of NR method for experimental saturation-like circuit of the task O4-7.CIR for OPTIMA

NR iteration	V(1) for the starting value:	
	-2.34678066627177	-2.34678066627178
1,3,5,...,19	1.1650E+00	1.1650E+00
2,4,6,...,20	-2.3468E+00	-2.3468E+00
21	1.1649E+00	1.1650E+00
22	-2.3465E+00	-2.3468E+00
23	1.1644E+00	1.1650E+00
24	-2.3448E+00	-2.3469E+00
25	1.1599E+00	1.1652E+00
26	-2.3290E+00	-2.3474E+00
27	1.1202E+00	1.1666E+00
28	-2.1953E+00	-2.3523E+00
29	8.0200E-01	1.1791E+00
30	-1.3412E+00	-2.3962E+00
31	-4.9653E-01	1.2924E+00
32	-7.0527E-01	-2.8216E+00
33	-7.2633E-01	2.5721E+00
34	-7.2654E-01	-1.1352E+01
35	-7.2654E-01	9.9637E+01
36	-7.2654E-01	-2.1635E+04
37	-7.2654E-01	4.4110E+08
38	-	-4.2787E+17
39	-	1.7254E+35

It can be easily shown, that the occurrence of excessively large temporary values of arguments of exp is strongly dependent on the starting point. Especially dangerous is the starting point at zero volts across the junction, which always leads to a huge result from the first iteration. We will demonstrate it by means of the circuit of Figure 4.8, whose description for OPTIMA numerical simulation is given in file O4-8.CIR. Let a diode be described by $i = I_s[\exp(v/V_T) - 1]$, with the parameter $I_s = 0.01$ pA, $V_T = 25$ mV, while $I = 0.1$ mA and $R = 1.4$ MΩ. Starting from the point $v^{(0)} = 0$ we get an iterative model of the diode at first iteration $i^{(1)} = (I_s/V_T)v^{(1)}$, which together with I and R gives the solution $v^{(1)} = I/[(I_s/V_T) + (1/R)] = 140$ V. This voltage disables the second iteration, because $\exp[v^{(1)}/V_T] = \exp(5600)$ quite exceeds the possibilities of the double precision. Similar underflow effects appear in the case of the reverse polarization of diodes. To protect against the overflow, OPTIMA provides limiting of arguments of the function exp to $|x| \leq 700$ and hence neither overflow nor underflow are observed. Using the task O4-8.CIR, the effect of extremely slow convergence is observed instead. The long jump in the first iteration results in infinitesimal further steps. This effect is demonstrated in Table 4.4.

Table 4.4 Extremely slow convergence of NR method for diode circuit of Figure 4.8 using the task O4-8.CIR for OPTIMA

iteration	V(1)
1	140.00
2	139.97
3	139.95
4	139.92
5	139.90
6	139.87
30	139.27

It can be seen from the example given, that long jumps leading to excessively large iterative values of circuit variables is a weak point of the basic NR method. This effect is due to the derivation of the step from a local linearization of a function, which, viewed globally, is strongly nonlinear. Therefore, the locally determined step is often too large, when the algorithm is still operating far from the exact solution. To overcome this problem, different methods of step limiting have been investigated. One of the most efficient approaches relies on a logarithmic limitation of the iterative interterminal signals of the elements. It will be discussed in the next section.

Simple d.c. analysis programs (as e.g. student projects) often take the basic NR form, like that of Algorithm 4.1, and hence, do not implement the step-limiting technique. However, they cannot be left without protection against numerical under and overflows. This can be achieved by introducing an appropriate approximation to nonlinear characteristics in steep physically unreal regions by means of more easy slopes. In such an approach approximation has to be continuous and smooth (of a continuous derivative). However, in professional models of nonlinear elements (e.g. in SPICE and OPTIMA), such an approximation is implemented together with the step limiting technique. The approximation is incorporated in the element model equations. It can be obtained by replacing the standard function exp by an approximated one expo, and introducing its derivative dxpo, which is not the same as expo, though function exp has the same exp derivative.

The approximated exponential function $\mathrm{expo}(x)$ has to be exactly equal to $\exp(x)$, only in a quite narrow forward physical range, e.g. $x \in [0, 40]$. In bjt modeling exp is used in formulae like this: $\exp(v/V_T)$. Hence, for V_T=0.025 (average value) the voltage $40\ V_T = 1$ V will be the safe upper limit of the physical range. Above this limit exp is smoothly extrapolated as a straight line, which efficiently limits the rate of increase. In the negative (i.e. reverse) region, exp very quickly approaches zero. However, due to the equation $i = I_s[\exp(v/V_T) - 1]$, the reverse current varies only from zero to a very small value $-I_s$ (e.g. -10^{-15} A). Therefore, this reverse current is quite small, invisible, masked by other parallel components that model junction breakdown or transistor operation. Hence, we can afford to approximate this reverse region quite inaccurately. In SPICE, a smooth linear extrapolation in the reverse region is used. Instead of that, we suggest the less inaccurate smooth hyperbolic extrapolation. The advantage of a hyperbola is that it has the same asymptote as exp, though it decreases considerably more slowly and underflow error does not appear.

The approximated exp takes the form:

$$expo(x) = \begin{cases} \exp(40)(x-39) & \text{for } x>40 \\ \exp(x) & \text{for } 0 \le x \le 40 \\ \begin{cases} x \quad \text{SPICE version} \\ \dfrac{1}{1-x} \quad \text{hyperbolic version} \end{cases} & \text{for } x<0 \end{cases} \tag{4.43}$$

which has to be used in formulae for currents. In formulae for their derivatives, the derivative of *expo* can be written as

$$dxpo(x) = \begin{cases} \exp(40) & \text{for } x>40 \\ \exp(x) & \text{for } 0 \le x \le 40 \\ \begin{cases} 1 \quad \text{SPICE version} \\ \dfrac{1}{(1-x)^2} \quad \text{hyperbolic version} \end{cases} & \text{for } x<0 \end{cases} . \tag{4.44}$$

Similar approximation techniques are needed in transient analysis, where junction capacitances introduce a function $(1-x)^{-m}$ defined only on the left side of the vertical asymptote $x=1$. Meanwhile, in transient analysis discretized capacitances give resistive models. At each time instant, the companion circuit is solved by means of the NR method. The function given above is called, with the argument $x=v/\phi$, where $\phi \approx 0.9$ V. A temporary iterative value of v can easily exceed ϕ (the vertical asymptote) and cause overflow. Due to the indefinite capacitance value above ϕ, it is reasonable to approximate the function $(1-x)^{-m}$, e.g. from $x=0.9$, to a straight line

$$cpxlam(x) = \begin{cases} (1-x)^{-m} & \text{for } x \le 0.9, \\ 10^m\,[m\,(10x-9)+1] & \text{for } x>0.9 \end{cases} . \tag{4.45}$$

This approximation operates only in the forward region and has no influence on the results of circuit simulation.

Exponential functions do not have such nice convergence properties as polynomials, which are available for modeling nonlinear elements in SPICE. Recently [4.17, 4.18] a concept of polynomial approximation to exponential and many other characteristics of nonlinear elements and macromodels of blocks has been proposed. This approximation is chosen to be sufficiently accurate in the whole physical region. It appears to extend considerably a convergence region of the NR algorithm. In O4-HORN.CIR we demonstrate its implementation for OPTIMA.

4.3.2 Technique of NR step limiting on nonlinear elements

4.3.2.1 Current-type and voltage-type iterations

In the preceding consideration it has been shown that iterative branch models use independent branch variables derived from the last iteration. Let us discuss the conductance $i=i(v)=I_s[\exp(v/V_T)-1]$ of the circuit shown in Figure 4.3. In Figure 4.11, its current has been plotted against the terminal voltage v. Due to the current-defined type of branch, the NR linearization at $v^{(p)}$ gives $i^{(p+1)}=I_s[\exp(v^{(p)}/V_T)-1]+(I_s/V_T)\exp(v^{(p)}/V_T)(v^{(p+1)}-v^{(p)})$. It is an iterative model containing coefficients dependent on the voltage $v^{(p)}$, and hence, represented by a tangent at the point A, whose voltage coordinate is $v^{(p)}$. This tangent crosses the the line $i=I-v/R$ at the next iterative solution. This sort of iteration is called the current-type, as it is derived from a current-defined iterative model. As has been shown in the previous section, such a type of iteration can result in overflow and slow convergence, as $v^{(p)}$ falls far from $\hat{v}^{(p-1)}$.

If the same nonlinear conductance is written in the voltage-defined form $v=V_T\ln[1+(i/I_s)]$, then its iterative voltage-defined model takes the following form $v^{(p+1)}=V_T\ln[1+i^{(p)}/I_s]-[V_T/(i^{(p)}+I_s)](i^{(p+1)}-i^{(p)})$, dependent on the current $i^{(p)}$ derived from the last iteration. This iterative model is represented in Figure 4.11 by a tangent set up at the point B, whose current coordinate is $i^{(p)}$. This tangent is quite different from the one obtained from current-type iteration, and yields another iterative solution. Due to the voltage definition of the iterative model, this case of iteration is called the voltage-type. Notice, that now the overflow is easily overcome, because the linearization point B is closer to $\hat{v}^{(p-1)}$ than A, derived from the current-type iteration. Once the pth iterative solution is represented by the point P having coordinates $(v^{(p)}, i^{(p)})$, the current-type iteration utilizes a linearization at a point which is a projection of P

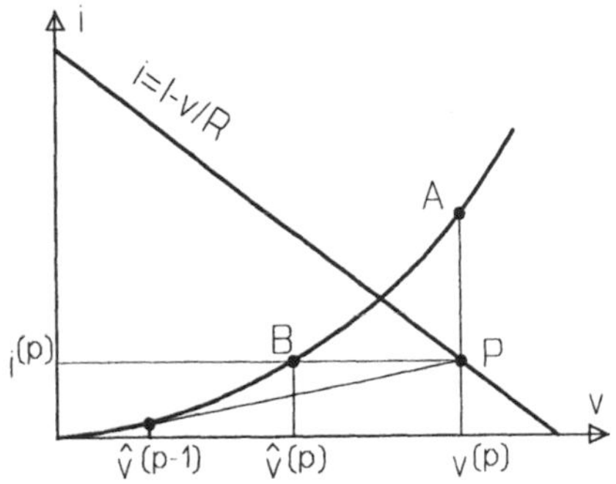

Figure 4.11 Graphical representation of current and voltage iterations from the point P

onto the curve parallel to the i-axis. Similarily, the voltage-type iteration uses a linearization at a point which is the projection of P parallel to the v-axis.

Voltage iteration not only overcomes the overflow but also limits the voltage step, in cases where the step is too large. Thus, the idea of switching the type of iteration provides a good convergence improvement. However, the realization of voltage-iteration by means of a voltage-defined model is not acceptable at all, as it introduces a current variable and changes the structure of circuit equations. To perform the voltage iteration in another way, let us notice that a voltage iteration from P may be viewed as a current iteration from B, that is performed by means of the current-defined model but calculated for a modified voltage $\hat{v}^{(p)}$ instead of $v^{(p)}$. Calculating $i^{(p)}$ by means of a tangent at the point $\hat{v}^{(p-1)}$, and by the increment $v^{(p)}-\hat{v}^{(p-1)}$, we obtain the expression $i^{(p)}=I_s[\exp(\hat{v}^{(p-1)}+(I_s/V_T)\exp(\hat{v}^{(p-1)}/V_T)(v^{(p)}-\hat{v}^{(p-1)})$. Substitution of the latter in $i^{(p)}=I_s[\exp(\hat{v}^{(p)}/V_T)-1]$ yields the modified voltage

$$\hat{v}^{(p)} = \hat{v}^{(p-1)} + V_T \ln\left(1 + \frac{v^{(p)} - \hat{v}^{(p-1)}}{V_T}\right), \qquad (4.46)$$

where the last but one voltage has been assumed to be derived from a similar modification. Modification (4.46) may be viewed as switching from a current-type to voltage-type iteration or equivalently as a nonlinear, continuous NR step limiting formula operating on each nonlinear element separately. For its implementation, the last and last but one junction voltages have to be stored, as was shown in Algorithm 4.1. Switching of the iteration consists of finding a voltage for which the current is equal to the last iterative current $i^{(p)}$. Thus, if $i^{(p)}\leq -I_s$, then there is no such modified voltage, i.e. the voltage iteration does not exist. This results in a negative sublogarithmic term of (4.46).

4.3.2.2 The generalized NR step-limiting technique

The iteration switching technique introduced in Section 4.3.2.1 is only applicable to the exponential characteristics of p-n junctions. However, this idea may be extended to a variety of nonlinearities, not necessarily exponential functions and not necessarily of one variable. First, it has been used in SPICE models of bipolar devices to limit the voltages on diodes and on both junctions of the quasi-Gummel-Poon bjt. Although invented as a method for efficient switching of voltage iterations, the modifying formula (4.46) has become the inspiration for an efficient general-purpose step limitation. This formula provides a logarithmic transformation of the NR step. Due to the properties of logarithms, it weakly limits small steps and strongly limits large steps. If the step limitation is neither exactly voltage nor current type, then we do not have pure NR iterations, but the quasi-Newton-Raphson method. To guarantee its local convergence we need the step transformation to be continuous, defined for all original NR steps and giving a zero limited step for a zero original NR step. Unfortunately, our case (4.46) provides a limitation only if the sublogarithmic term is positive (see subsection 4.3.2.1) and there is no way to override this singularity and obtain a continuous transformation, everywhere defined. Hence, this method cannot guarantee local convergence. However, if the sublogarithmic term is not positive the singularity can be efficiently overcome by substituting for a troublesome voltage another value which causes no problems. This idea can be successfully implemented for bipolar devices due to the exponential character of their nonlinearities.

For diodes, bjts and the parasitic diodes of MOSFETs, in SPICE models the step-limiting procedure is applied only to a range of high junction voltages, where the curve is extremely steep. The value of voltage separating this range from the normal range is called the critical voltage. It has been chosen as the point of maximum curvature of the function exp: $v_{\text{crit}} = V_T \ln[V_T/(I_s\sqrt{2})]$. As this function is RHS-wise steep, the step-limiting procedure dedicated to bipolar devices which has been presented in Algorithm 4.2, is called the RHS-wise limiting. It performs the modification (4.46) only when the last iteration lies in the right range of high voltages, and the last voltage increment is greater then twice V_T (to avoid a purposeless transformation of small increments). The logarithmic modification is used always when a voltage iteration exists ($i^{(p)} \geq -I_s$). Otherwise, the substitution of the critical voltage provides an escape from an inconvenient range to the border of the normal range, where limiting is not necessary. In the forward region the logarithmic modification is of the form (4.46), but in the reverse region, due to a linear approximation, another expression based on the same principle is introduced.

For nonbipolar devices, terms V_T and V_{crit} in Algorithm 4.2 are quite purposeless. For arbitrary nonlinearities, (4.46) may be viewed as a general step-limiting rule. Hence, we replace $1/V_T$ by the general factor k. In that way we obtain the RHS-wise limiting formula applicable not necessarily to voltages but to any independent branch variable x (for example, for the branch $i_a = f(v_a, v_b, i_c)$ limiting can be imposed on v_a, v_b, i_c). As we do not know when to use limiting, we decide to use it only if $v^{(p)}$ exceeds $\hat{v}^{(p-1)}$. Thus, the RHS-wise limiting, valid only for positive original NR steps, takes the form

$$\hat{x}^{(p)} = \hat{x}^{(p-1)} - \frac{1}{k} \ln [1 + k (x^{(p)} - \hat{x}^{(p-1)})]. \tag{4.47}$$

Algorithm 4.2 RHS step-limitting dedicated to bipolar devices

```
{main condition for modification}
if (v^(p) > V_crit & |v^(p) - v̂^(p-1)| > 2V_T) then
begin
  if (v̂^(p-1) ≤ 0) then
    {reverse range limitting}
    v̂^(p) = V_T ln(v^(p)/V_T)
  else
  begin
    arg = 1 + (v^(p) - v̂^(p-1))/V_T;
    if (arg≤0) then {voltage iteration not exists}
      v̂^(p) = V_crit
    else {logarithmic step limitting}
      v̂^(p) = v̂^(p-1) + V_T ln arg;
    end
  end
```

Similarily for the LHS-wise steep nonlinearities (i.e. those whose absolute values of signals increase while an argument decreases) the LHS-wise step-limiting, applicable only for positive original steps, can be expressed as

$$\hat{x}^{(p)} = \hat{x}^{(p-1)} - \frac{1}{k} \ln [1 + k (x^{(p)} - \hat{x}^{(p-1)})]. \tag{4.48}$$

If a shape of function is both-sides-wise steep or if we do not know anything about its shape, then both-sides-wise limiting gives

$$\hat{x}^{(p)} = \hat{x}^{(p-1)} + \frac{1}{k} \operatorname{sgn}(x^{(p)} - \hat{x}^{(p-1)}) \ln [1 + k \, |x^{(p)} - \hat{x}^{(p-1)}|]. \tag{4.49}$$

In practice, the above limiting formulae should be used only if the original NR steps exceed a certain minimum level.

Three options control NR step limiting in the simulator OPTIMA. Option NLIM switches off the RHS-wise limiting in built-in models of bipolar devices. Option LIMU controls step limiting in all external models: value 0 switches off the limiting (to gain pure NR iterations), 1 imposes the RHS limiting, −1 gives the LHS limiting, while default 2 causes both-sides-wise limiting. External model option LIMIT (set to OFF, LHS,RHS,BOTH) controls limiting in a given model.

The parameter k controls the rate of limiting: the greater is k the stronger is the step limiting. In [4.7, 4.15] the step-limiting technique has been tested in detail. For bipolar devices k should be about $1/V_T$=40 (experimentally selected values range from 12 to 20). For MOSFETs and Josephson junctions, an optimal k ranges from 2 to 5. For user-defined models, we need a general purpose choice of k. As strong and adequate limiting is important only at the beginning of the iterative process, a dynamic assignment of k ranging from m=12 to n=2 and decreasing with the number of iterations, according to the formula $k=\max(n, m-iteration)$, appears to be a very good compromise. In OPTIMA we can set up both bounds: the upper bound m by the option or external model option CLUB (ControLled Upper Bound), the lower bound n by CLLB.

Table 4.4 Tests of step-limitting efficiency

benchmark	number of iterations the no limiting case	number of iterations with limiting [n, m]
O4-CHOKE.CIR	48	14
O4-DIF.CIR	singularity	8
O4-D4TTL.CIR	∞ or singularity	40
O4-DTUN1.CIR	26	[1, 2]: 10 [1, 4]: 33 [2, 12]: ∞
O4-DTUN.CIR	3	[1, 2]: 5 [2, 12]: 10

Reference [4.15] proposes an experimental speeding-up factor $\gamma \approx 1.3$ in logarithmic components of the step-limiting transformation.

Step limiting may be easily implemented in Algorithm 4.1. At the beginning of the loop **for** (for all branches) we have to take values $\mathbf{z}^{(0)}$ and $\mathbf{z}^{(1)}$ from *struct,* and apply Algorithm 4.2 to all its components to obtain a new, logarithmically modified $\mathbf{z}^{(1)}$, for further use as a linearization point.

OPTIMA enables us to test the efficiency of the step-limiting method.

To observe how limiting enables or speeds up convergence, the user can execute a few benchmarks. Some of them are collected in Table 4.4. In bjt circuits, e.g. O4-DIF.CIR, O4-D4TTL.CIR we have a specific, RHS-side-steep exponential nonlinearity. Without limiting, usually the solution of linear equations fails to find large enough pivots due to excessively large matrix elements involved at the second iteration. In O4-CHOKE.CIR the matrix is not numerically singular, although number of iterations required is considerably larger. The opposite effect arises in the case of less steep nonlinear circuits, e.g. O4-DTUN1.CIR, or O4-DTUN-.CIR. Such circuits appear to give the best convergence without limiting. The tunnel-diode characteristic is RHS-steep, sensitive only to LIMU 1 or 2. We observe that, the weakest limiting also concludes with the quickest convergence.

4.3.2.3 Other step-limiting techniqes

The techniques of NR step limiting at nonlinear elements so far presented, although heuristic, are quite efficient and have been extensively propagated by virtue of SPICE. Nevertheless, several less heuristic methods have been proposed. Their global convergence (convergence from any starting point) can be proved under certain assumptions which are quite realistic in practice). The simplest globally convergent iterative process for solving the equations $\mathbf{f}(\mathbf{x})=\mathbf{0}$ can be obtained by damping the step of the basic NR algorithm. Let us take the equation $\mathbf{f}(\mathbf{x})=\mathbf{0}$, where $\mathbf{f}$: $R^n \to R^n$ is a 1-1, smooth mapping with both $\mathbf{f}$ and $\mathbf{f}^{-1}$ continuous. At each iteration a damped solution is derived by means of the equation

$$\hat{\mathbf{x}}^{(p+1)} = \hat{\mathbf{x}}^{(p)} + \Delta\hat{\mathbf{x}}^{(p)} \tag{4.50}$$

using from the last solution $\hat{\mathbf{x}}^{(p)}$ and the damped increment $\Delta\hat{\mathbf{x}}^{(p)}$. If the latter equals the Newton step $\Delta\mathbf{x}^{(p)}=\mathbf{x}^{(p+1)}-\hat{\mathbf{x}}^{(p)}=-[\partial\mathbf{f}(\hat{\mathbf{x}}^{(p)})/\partial\mathbf{x}]^{-1}\mathbf{f}(\hat{\mathbf{x}}^{(p)})$, then the nondamped, basic NR iteration arises. In the damped case the increment is slightly limited in comparison with the Newton step. This damping mechanism operates similarly to the step-limiting one of Section 4.3.2.2. It offers protection from numerical overflow, and provides global convergence if the damping factors guarantee a decrease of the residual error to zero: $\|\mathbf{f}(\mathbf{x}^{(p)})\| \to 0$ when $p \to \infty$. The best known are damping techniques in which the work step is a fraction of the Newton-Raphson step, where instead of step limiting on branches, damping of the increments of all circuit variables $\mathbf{x}$ in equation (4.50) is introduced

$$\Delta\hat{\mathbf{x}}^{(p)} = \alpha^{(p)}(\mathbf{x}^{(p+1)} - \hat{\mathbf{x}}^{(p)})\,. \tag{4.51}$$

In fact, the crucial point of (4.51) is a linear search problem, i.e. assignment of the damping factor $\alpha^{(p)}$ to provide the smallest residual error. However, reaching this goal with small computational effort is quite a hard problem. One of the most efficient, globally and quadratically convergent, damped algorithms has been investigated by R.E. Bank and D.J. Rose [4.2] and implemented in the simulator CAzM [4.4]. The damping factor is controlled by the measure of error decrease $M^{(p+1)} = \|\mathbf{f}(\mathbf{x}^{(p+1)})\| / \|\mathbf{f}(\mathbf{x}^{(p)})\|$. It should be at least smaller then unity. As has been proved in [4.2], global, quadratic convergence can be obtained once this measure is slightly smaller, namely

$$M^{(p+1)} \leq 1 - \delta\, \alpha^{(p)} \tag{4.52}$$

where δ is a small number, in practice, taken as the relative machine accuracy ε dependent on the number of floating-point mantissa digits. Once (4.52) is satisfied, error decrease has been proved to be rapid enough to softly weaken the damping towards unity, without loss of quadratic convergence (the smaller the measure $M^{(p+1)}$, the quicker the damping will weaken). In [4.4], this idea has been implemented (slightly modified version of [4.2]) by means of the recursion

$$\alpha^{(p+1)} = \frac{\alpha^{(p)}}{\alpha^{(p)} + (1 - \alpha^{(p)}) M^{(p+1)}/5} \, . \tag{4.53}$$

If (4.52) does not hold, convergence deteriorates and then the damping factor should be lowered by trying the formula with subsequent j's

$$\alpha^{(p+1)(j)} = \alpha^{(p)} \left(\delta \frac{\|\mathbf{x}^{(p)}\|}{\|\Delta \mathbf{x}^{(p)}\|} \right)^{j^2/l^2} \qquad j = 1, \ldots, l \tag{4.54}$$

until condition (4.52) is fulfilled (otherwise the algorithm unconditionally stops). The factor obtained from (4.54) is taken as a subsequent $\alpha^{(p+1)}$.

A family of three efficient global damping techniques has been developed by Agnew *et al* for the simulator SCAMPER [4.1]. The simplest method applies damping of the ratio of step sizes of each two successive iterations, with simultaneous protection from overstepping the Newton increment

$$\Delta\hat{x}^{(p)} = \mathrm{sign}(\Delta x^{(p)})\ \min\left\{|\Delta x^{(p)}|,\ \alpha^{(p)}\ |\Delta\hat{x}^{(p-1)}|\right\}. \tag{4.55}$$

This damping is imposed individually on each scalar signal x (an independent variable of a nonlinear branch, e.g. VGS, VDS in MOSFET), and should be appropriately controlled. The weaker damping $\alpha^{(p)}=1.5$ is applied when the signal does not oscillate: $\mathrm{sign}(\Delta x^{(p)})\,\mathrm{sign}(\Delta\hat{x}^{(p-1)})>0$. The stronger $\alpha^{(p)}=0.5$ when the oscillation $\mathrm{sign}(\Delta x^{(p)})\,\mathrm{sign}(\Delta\hat{x}^{(p-1)})<0$ appears. At the first iteration, when steps are usually too long, it is limited to 0.05.

The NR algorithm fails also for excessively large Jacobi matrix components, when convergence can be slow and standard damping ineffective. To override these effects, simultaneous damping of variables $\mathbf{x}$ and of the Jacobi matrix $\partial\mathbf{f}/\partial\mathbf{x}$ has been proposed [4.1]. As is well known from previous discussions, a damped, and hence inaccurate Jacobi matrix does not introduce error into the solution at all. Calculations are performed in three stages. First, the damped Jacobi matrix is derived from the damped one at the previous iteration and the accurate one at the present iteration. Then, the increment of the solution is calculated from the system of linear equations. Finally, the solution is obtained by damping, namely

$$\begin{aligned}
&\hat{\mathbf{J}}^{(p)} = \hat{\mathbf{J}}^{(p-1)} + \alpha^{(p)}\left[\frac{\mathrm{d}\mathbf{f}(\mathbf{x}^{(p)})}{\mathrm{d}\mathbf{x}} - \hat{\mathbf{J}}^{(p-1)}\right]\\
&\hat{\mathbf{J}}^{(p)}\,\Delta\hat{\mathbf{x}}^{(p)} = -\mathbf{f}(\mathbf{x}^{(p)})\\
&\hat{\mathbf{x}}^{(p+1)} = \hat{\mathbf{x}}^{(p)} + \alpha^{(p)}\,\Delta\hat{\mathbf{x}}^{(p)}.
\end{aligned} \tag{4.56}$$

The algorithm is started from the unit damping factor, and $\hat{J}^{(0)}$ equal to an accurate Jacobian. If current Jacobians are nonsingular, the same happens with the damped ones. To control the damping factor, two cases have to be distinguished. If an error grows up (that is the introduced $M^{(p+1)}>1$), then damping is strengthened, according to $\alpha^{(p+1)}=\alpha^{(p)}/4$. Additionally, a more optimistic error $M^{(p+1)}=M^{(p+1)}/1.5$ is set (to help avoid local minima). Otherwise, damping is weakened, i.e. $\alpha^{(p+1)}=\min\{2\,\alpha^{(p)},\ 0.32+0.7\,\alpha^{(p)},\ 1\}$. This quite complicated formula causes faster undamping of the stronger dampings than of the weaker ones. Unfortunately, in the case of an ill-conditioned matrix, these damping strategies are ineffective.

To improve efficiency for barely convergent circuits, another continuation-like damping method has been proposed in [4.1]. It is dedicated mainly to bjt circuits. At each NR iteration, this method limits the absolute value of the junction voltage step to at most δ. Hence, each iteration shifts the current point by a quantum δ, towards a solution, until steps become smaller than δ. Hence, the algorithm is completed without limitation. The step takes the form $\hat{v}^{(p+1)} = \max\{\hat{v}^{(p)} - \delta,\ \min\{\hat{v}^{(p)} + \delta,\ v^{(p+1)}\}\}$, where $v^{(p+1)}$ is the p+1 Newton solution, while the quantum δ has to be selected (e.g. 0.05 for bjts). Its small value increases the number of initial limited iterations but decreases the number of final unlimited iterations. A large quantum decreases the number of limited iterations but stops far from a solution, and hence the number of final unlimited steps increases or even causes nonconvergence. A variable quantum algorithm is a good idea in this case.

4.3.3 Other improvements and extensions of the NR method

4.3.3.1 Bypassing method

Bypassing is an extension of the convergence test that relies on checking, during the NR procedure, a criterion of signal stabilization, individually for each nonlinear element. If a signal of a particular nonlinear element remains unchanged in subsequent NR iterations, then there is no reason to continue computations of its iterative model, even if other nonlinear elements have yet unstable signals and need more iterations. Hence, the iterative models of branches that have reached convergence are stored, and remain constant in all the following iterations until the whole circuit converges.

Practicing with SPICE has shown [4.9] that bypassing gives savings of a few tens per cent. We can test it using the option NBPS of OPTIMA. The tasks O4-BP733.CIR, O4-BP741.CIR, O4-BPECL.CIR, O4-BPRCA.CIR, O4-BPTTL.CIR and some others can be used to check the influence of switching off the bypassing. This does not reduce the number of iterations, but the computation time is less. To observe this saving we have to turn to the examples involved. The bypassing is especially profitable in circuits containing elements modeled by such complicated devices as advanced model of a MOSFET. A substantial reduction in computation is due to the fact that calculations of nonlinear expressions of device models take a considerable portion (average of 70 per cent in SPICE) of the total simulation time.

Bypassing uses the following parameters of the iterative model: the value of the nonlinear function and a number of its partial derivatives. In the iterative NR process they are permanently updated and stored in the data base representing the circuit. When a bypassing criterion is satisfied, then model calculations are stopped and its stored parameters are recalled. Calculation of the iterative excitation from stored parameters is the only arithmetic operation required. The iterative model $i^{(p+1)} = G^{(p)} v^{(p+1)} + I^{(p)}$ $(I^{(p)} = f(v^{(p)}) - G^{(p)} v^{(p)})$ of the branch $i = f(v)$, usually needs calculation of $G^{(p)}$, $f(v^{(p)})$ at each NR iteration from built-in analytic expressions. If a bypassing criterion indicates a slight change of v between the last two iterations, then the derivation and updating in memory of new $G^{(p)}$, $f(v^{(p)})$ values can be omitted. Stored values can be used instead. This needs two real numbers stored in a structure *struct*, assigned to the element (see Section 4.2.4.2). The source $I^{(p)}$ has to be calculated from the stored numbers, only by means of one multiplication and one subtraction.

In general, bypassing of the branch $y = y(\mathbf{z})$ needs storage of the iterative branch output signal $y(\mathbf{z}^{(1)})$ and its partial derivatives $\mathbf{g}^{(1)} = \partial y(\mathbf{z}^{(1)})/\partial \mathbf{z}$ with respect to the independent variables. If bypassing does not occur, then these quantities are derived at the last iterative point $\mathbf{z}^{(1)}$. They are used in a test for convergence, stored as $y^{(0)}$, $\mathbf{g}^{(0)}$, and, finally, used for formulation of the iterative model stamp. Otherwise, if bypassing is satisfied, then only the recently stored $y^{(0)}$, $\mathbf{g}^{(0)}$ are recalled and used for the stamp formulation. Hence, bypassing, as an *a priori* test for convergence, excludes the already known *a posteriori* test. The two alternatives presented depend on the bypassing criterion which is checked at the beginning of branch processing, as can be seen in Algorithm 4.3 (the test for convergence is located at the end). This criterion has to compare two neighboring iterative solutions, as demonstrated in (4.40), but increments of both independent and dependent variables are taken into account. If the slope of the characteristic is small, then the increment of the independent variables $\mathbf{z}$ should be checked. If it is large, then the increment of the dependent variable $y(\mathbf{z})$ should be verified. That is why both increments are checked in practice, and hence the discussed branch can be bypassed if for all components z_i of $\mathbf{z}$ the criterion $|z_i^{(1)} - z_i^{(0)}| < \varepsilon \max\{|z_i^{(1)}|, |z_i^{(0)}|\} + \delta_z$ and also $|\mathbf{g}^{(0)t}(\mathbf{z}^{(1)} - \mathbf{z}^{(0)})| < \varepsilon \max\{|y^{(0)} + \mathbf{g}^{(0)t}(\mathbf{z}^{(1)} - \mathbf{z}^{(0)})|, |y^{(0)}|\} + \delta_y$ is satisfied.

In Algorithm 4.3 also logarithmic step limiting is implemented. It is active only when the bypassing has not been currently active. If limiting has really occurred, then it returns the flag *check*=1, invoking the test for convergence. Otherwise, it means that signals have changed infinitesimaly, and hence the test for convergence is not necessary. Bypassing is a widely used technique. It is also applicable to circuit blocks.

Algorithm 4.3 Quasi-NR procedure with step limiting and bypassing (continued overleaf)

Data: for each nonlinear branch starting values of its independent variables $\mathbf{z}^{(0)}$ stored in *struct*;
accuracy indexes ε, δ and *maximum_number_of_iterations.*
for *iter*=1 **to** *maximum_number_of_iterations* **do** {NR loop}
begin
  stop=1;
  {matrix formulation and test for convergence}
  for all branches
  begin
    if (branch is nonlinear)
      if ($iter>1$) derive independent variables $\mathbf{z}^{(1)}$ of the nonlinear branch by means of the last solution vector $\mathbf{x}^{(1)}$ and store them in *struct*;
      get $y^{(0)}, \mathbf{z}^{(0)}, \mathbf{g}^{(0)} \equiv \partial y^{(0)}/\partial \mathbf{z}$ from *struct* and predict auxiliary
      $\hat{y}^{(1)}=y^{(0)}+\mathbf{g}^{(0)t}(\mathbf{z}^{(1)}-\mathbf{z}^{(0)})$; *check*=0;
      if ($|z_i^{(1)}-z_i^{(0)}|<\varepsilon \max\{|z_i^{(1)}|, |z_i^{(0)}|\}+\delta_z$ for $i=1, ..., \dim \mathbf{z}$
        and $|\hat{y}^{(1)}-y^{(0)}|<\varepsilon \max\{|\hat{y}^{(1)}|, |y^{(0)}|\}+\delta_y$) **then**
      begin {bypassing}
        $\mathbf{z}^{(1)}=\mathbf{z}^{(0)}$; $\mathbf{g}^{(1)}=\mathbf{g}^{(0)}$; $y^{(1)}=y^{(0)}$;
      end
      else
        logarithmic limiting of $\mathbf{z}^{(1)}$;
        if (NR step has been limited) *check*=1;
        calculate $y^{(1)}, \mathbf{g}^{(1)} \equiv \partial y^{(1)}/\partial \mathbf{z}$ at point $\mathbf{z}^{(1)}$ from branch constitutive expression $y=y(\mathbf{z})$;
        if (*check*=0)**then**
          stop=0;
        else
          if ($|\hat{y}^{(1)}-y^{(1)}|>\varepsilon \max\{|\hat{y}^{(1)}|, |y^{(1)}|\}+\delta_y \|$
            $\bigvee_{i=1}^{\dim \mathbf{z}} |z_i^{(1)}-z_i^{(0)}|>\varepsilon \max\{|z_i^{(1)}|, |z_i^{(0)}|\}+\delta_z$) *stop*=0
          {test for convergence at a particular branch not satisfied}
          $y^{(0)}=y^{(1)}$; $\mathbf{z}^{(0)}=\mathbf{z}^{(1)}$; $\mathbf{g}^{(0)}=\mathbf{g}^{(1)}$; {store last results as results of the last but one iteration}
        endif
      endif
    endif
    write stamp of the branch into the matrix $\boldsymbol{Y}^{(1)}$ and vector $\boldsymbol{b}^{(1)}$;
  end

Algorithm 4.3 Continued

```
    if (stop ≠0) stop {NR loop termination – convergence reached}
    solve equations Y(1) x(1)=b(1) by means of the LU method;
    if (Y(1) is singular) then
        stop {NR termination - matrix failure}
    else
        {new solution x(1) obtained}
    endif
end
stop {NR termination – convergence not reached}
```

4.3.3.2 Overriding ill-conditioned circuits containing floating nodes

It has been shown in Section 4.1.2, that reactances can cause problems in the case of ill-conditioned floating nodes and singular loops. The matrices of these two cases are singular, the former having a zero row. The voltages of floating nodes and currents of singular loops are not defined. In practice, their d.c. state is constituted during the transient phase by charging up capacitors and setting up linked magnetic fluxes. Thus, it can be drawn up in a time-domain simulation and obtained as its limit state. The substitute d.c. circuit does not model these cases, and causes numerical inconvenience. Problems are also caused by MOS devices. First of all, the gate terminal is insulated from the d.c. circuit and connected only through capacitances. Moreover, in the commonly encountered series connection of channels, shown in Figure 4.12, if both transistors are switched off, the node N floats between two open circuits and its voltage is not defined. In practice, there are capacitances connected to N, but their influence on setting up voltages in the transient state cannot be simulated in a d.c. steady state with a substitute resistive circuit. To overcome these hard problems in iterative NR analysis, the singularity of the matrix has to be eliminated by assembling auxiliary

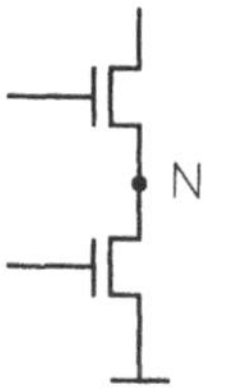

Figure 4.12 A series connection of MOSFETs

elements to floating nodes and singular loops which tend to zero in the d.c. steady state.

The simplest heuristic method [4.7, 4.15] is to replace a capacitor C by a conductance $G^{(p)}$, starting from large values but tending to zero with the number of iterations. If the number of iterations is rather small, then the conductance must decrease very fast. To match the physics of capacitor operation, in the transient state this conductance should be proportional to C. Thus, the rule $G^{(p)}=(C/\Delta^{(p)})$ with $\Delta^{(p)}\rightarrow\infty$ has been proposed. The idea is initially to use $\Delta^{(p)}=1$ until the accuracy of NR iterations reaches 20 times the final required level, and then, $\Delta^{(p)}$ is decreased by a factor of ten in each iteration, up to a certain maximum, e.g. $\Delta_{max}=10^{12}$. This method introduces large conductances at the initial stage of NR iteration, and efficiently stabilizes the calulations.

To protect the matrix against excessively small element values, it is reasonable to introduce a minimum conductance, e.g. $GMIN=10^{12}$, and then to connect it in parallel with all nonlinear conductances. This efficiently covers all effects from excessively small partial derivatives of iterative models and relative singularities. This GMIN in both SPICE and OPTIMA can be set using options. Simulating the example O4-GMIN.CIR with $GMIN=0$, we can observe the failure of a sparse matrix procedure, due to its numerical singularity. To protect MOSFETs against a floating drain or source [4.15] the connection of auxiliary grounded resistors R to drain and source has been proposed. Iterations start from an arbitrary value, e.g. $R=3\,10^3$, and after a rough convergence has been reached it is then arbitrarily adjusted to $R=10^9$.

A less heuristic deflation technique for dealing with numerically singular matrices has been proposed in CAzM [4.4]. Let us consider the equations $\mathbf{f}(\mathbf{x})=\mathbf{0}$. If the norm of the residual vector $\|\mathbf{f}(\mathbf{x}^{(p)})\|$ indicates an ill-conditioned system or a singularity is detected, then instead of $[\partial\mathbf{f}(\mathbf{x}^{(p)})/\partial\mathbf{x}](\mathbf{x}^{(p+1)}-\mathbf{x}^{(p)})=-\mathbf{f}(\mathbf{x}^{(p)})$ a system

$$\left[\frac{\partial\mathbf{f}(\mathbf{x}^{(p)})}{\partial\mathbf{x}}+\mathbf{1}\lambda\,\|\mathbf{f}(\mathbf{x}^{(p)})\|\right](\mathbf{x}^{(p+1)}-\mathbf{x}^{(p)})=-\mathbf{f}(\mathbf{x}^{(p)}) \qquad (4.57)$$

is solved. A specific additional damping factor λ is set. It modifies the Jacobian, and hence does not change the circuit solution. It increases the diagonal matrix elements, and therefore, eliminates singularity. An iterative increase of λ is performed until the ill-conditioned case is overcome. Then this damping begins to be mildly decreased. Since simultaneously convergence is resumed, and the norm of the residual vector tends to zero, the quadratic Newton convergence remains. Reference [4.2] suggests a starting value $\lambda < 1/\|\Delta\mathbf{x}^{(0)}\|$.

To test our understanding of the topics presented in this section, we can try the following exercises:

- List the types of numerical problem arising in the basic NR algorithm.
- For a chosen nonlinear branch, demonstrate current iteration, voltage iteration and general logarithmic step limiting.
- Explain the damping technique and compare it with the step limiting, previously discussed.
- Create an arbitrary nonlinear branch, its iterative model and explain the bypassing technique using this example.
- What types of circuit elements cause singularities and ill-conditioned Jacobi matrices? How can these problems be overriden in d.c. analysis?

4.4 Continuation methods

Continuation methods for solving a nonlinear algebraic problem rely on treating it as an instant of a dynamic problem. As is known from Chapter 1, the d.c. analysis of nonlinear circuits is, in fact, the steady state of time-domain transient analysis. In methods discussed so far the problems have been solved by means of a substitute algebraic approach instead of handling algebraic differential equations. However, the simplified algebraic approach, discussed in Sections 4.2 and 4.3, is sometimes too hard to be solved, due to the nonconvergence of NR iterations. Thus, in this section we are coming back to the beginning, and once more try to solve a dynamic problem, though not the original dynamic circuit but a simpler auxiliary problem, more efficient in analysis.

As an example, let us consider the equation $f(x)=0$, which can be solved by integrating the differential equation $[\mathrm{d}f(x)/\mathrm{d}x]\,(\mathrm{d}x/\mathrm{d}t)=f(x)$. Starting integration from an initial condition $x(0)$, for t $\rightarrow \infty$, a steady state is obtained when $\mathrm{d}x(t)/\mathrm{d}t \rightarrow 0$. Then $f[x(t)] \rightarrow 0$, that is $x(t)$ approaches a solution of $f(x)=0$. This is a differential approach, rather not developed at present.

Another algebraic approach uses homotopies. Let the auxiliary equation $f(x)-(1-\lambda)f(x_0)=0$ be constructed, which for $\lambda=0$ has the known solution x_0. While λ moves from 0 towards 1, the solution of this equation moves from x_0 to an unknown solution of the equation $f(x)=0$. This dynamic, though algebraic, problem can be solved by stepping the continuation parameter λ and solving a sequence of algebraic problems.

The dynamic problem contains all the solutions of the original static problem and theoretically enables us to find all of them, which was not the case in the static problem. As the solution of a dynamic problem can be viewed as a trajectory in a solution space with a continuous parameter (say time) moving along this trajectory, we can subsequently enter all static solutions. Hence, a continuation method can be classified as one of the global d.c. analysis methods that yields all the existing solutions. However, in practice, a continuation problem meets discontinuities of the trajectory or bifurcation points, where the trajectory splits into a few branches. When the algorithm enters a bifurcation it requires a switching mechanism to select an appropriate branch, and to move along this branch. These problems appear to be hard. Moreover, if one does not limit the trajectories searched or assume a specific class of nonlinear problems (e.g. piece-wise linear), then finding all solutions needs infinite computational effort. Methods of finding all the solutions of nonlinear algebraic equations are beyond the scope of our consideration. In this section, continuation is discussed only as a globally convergent method for finding a single solution. Its advantage is that one is able to select a starting point far from the unknown solution.

4.4.1 Homotopies and their applications

Consider the algebraic equation $\mathbf{f}(\mathbf{x}) = \mathbf{0}$, where $\mathbf{f}$: $R^n \rightarrow R^n$ is a continuous mapping. An equation

$$\mathbf{H}(\mathbf{x}, \lambda) = \mathbf{0} \tag{4.58}$$

is drawn up using a continuous function $\mathbf{H}(\mathbf{x}, \lambda)$, such that $\mathbf{H}(\mathbf{x}, 0) = \mathbf{0}$, has the known solution $\mathbf{a}$, and for a continuation parameter $\lambda = 1$ reaches $\mathbf{H}(\mathbf{x}, 1) = \mathbf{f}(\mathbf{x})$. It is called a homotopy corresponding to the function $\mathbf{f}$. Homotopy for a given function $\mathbf{f}$ is not unique. Its solution is a homotopy path $\mathbf{x}(\lambda)$ originating at $\mathbf{a}$ and passing through all solutions of the equation. The solution of homotopies is similar to numerical integration of ODE and will be discussed in the next section. In general, the path may have discontinuities and bifurcations (where the Jacobi matrix vanishes while paths bifurcate). Discontinuities with $\partial\mathbf{H}/\partial\mathbf{x}$ singular are known as limit points. It has been proved [4.20] that if $\partial\mathbf{H}(\mathbf{x}, \lambda)/\partial\mathbf{x}$ is nonsingular at a point $(\mathbf{x}_0, \lambda_0)$ and continuous over its neighborhood, then there exists a neighborhood of this point where the solution $\mathbf{z}(\lambda)$ is continuous. Thus, if the Jacobi matrix is continously differentiable and is nonsingular at the whole path $\mathbf{z}(\lambda)$ then no discontinuities are met during the solution process. Particularily, a continuously differentiable, monotonic character for the function **H** is sufficient to ensure continuity of paths. To omit bifurcations along the currently searched interval of the homotopy path, an appropriate starting point has to be selected. Coercivity conditions for its selection are known [4.9].

In circuit simulation, homotopies first appear in *d.c. sweep analysis*, when a certain circuit parameter λ (e.g. a input voltage source) is selected. Then, circuit equations become a system with a parameter $\mathbf{f}(\mathbf{x}, \lambda) = \mathbf{0}$, and for a sequence of $\lambda_1, \lambda_2, \ldots$, selected by the user, a sequence of solutions, i.e. $\mathbf{x}(\lambda_1), \mathbf{x}(\lambda_2), \ldots$ is found. In other cases, homotopies appear as so called artificial parameter homotopies, where λ is added for scaling some circuit parameters. While the scaling parameter is stepped towards 1, the homotopy solution approaches the d.c. point being calculated. The latter is a continuation method of operating point analysis, very reliable and widely convergent, but requiring more iterations by about one order of magnitude than the methods introduced in Section 4.3. A software package for solving d.c. analysis problems has been recently developed [4.14].

4.4.1.1 Homotopy of variable d.c. sources

The method of variable d.c. sources relies on parameterization of the d.c. excitations and successive solution of the circuit, with d.c. sources starting from zero and tending to given final values. Let the circuit equations be of the following general form

$$\mathbf{f}(\mathbf{x}) = \mathbf{s} \tag{4.59}$$

where $\mathbf{s}$ is the vector of circuit d.c. excitations and $\mathbf{f}(\mathbf{x})$ stands for a passive part of the circuit, i.e. $\mathbf{f}(\mathbf{0}) = \mathbf{0}$. A class of nonlinear circuits with continuous monotonic branch constitutive functions provides continuity of the response $\mathbf{x}(\mathbf{s})$, i.e. any small change in $\mathbf{s}$ causes also a small change of $\mathbf{x}$. Homotopy of a variable d.c. source of the form $\mathbf{f}(\mathbf{x}) = \lambda\mathbf{s}$, and solved for λ increasing from 0 to 1, produces a trajectory that begins at the trivial solution $\mathbf{0}$ and reaches the solution of (4.59). After discretization in the λ domain, a circuit can be solved for $\mathbf{x}_i$, with a scaled excitation $\lambda_i\mathbf{s}$, by means of the NR process starting from $\mathbf{x}_{i-1}$ or from an appropriately predicted starting point.

This method has been implemented in SPICE (automatically executed when a common method fails) and in OPTIMA (when a descriptor CONT in the .OP line arises). This continuation approach needs more NR iterations than a common method with an automatic starting point (presented in Section 4.3). However, its convergence is more reliable. For example the task O4-CONT.CIR for OPTIMA gives a successful convergence of the common (AUTO) NR method in iterations ranging from 34 to 51. Using option CONT the same solution is obtained after 260 to 263 iterations. However, there are such examples as O4-CONT1.CIR which do not converge using the common method, but can be solved successfully using the continuation method.

As discussed, homotopy is solved as a dynamic problem, so the continuation parameter λ can be treated as time. Efficient implementation needs a variable time step, because the trajectory $\mathbf{x}(\lambda)$ is usually strongly nonlinear and the step should vary over a wide range. Efficient step assignment in OPTIMA takes into account local error, i.e. the difference between the solution and the predictor from which the former has been iterated: $|\mathbf{x}(\lambda_i) - \mathbf{x}_i^{(0)}|$. Another method may assign a step according to a the number of NR iterations at each point λ_i. This form of homotopy is not the most efficient technique but is very popular.

4.4.1.2 Variable stimulus and variable gain homotopies

According to references [4.9, 4.12], the most efficient are variable stimulus and variable gain homotopies

$$\mathbf{H}(\mathbf{x},\lambda) = (1-\lambda)G(\mathbf{x}-\boldsymbol{a}) + \mathbf{f}(\lambda\mathbf{x}) - \mathbf{s} \tag{4.59}$$

where the continuation parameter scales the nonlinear branch constitutive equations. As for the variable stimulus, the nodal voltages and currents controlling the nonlinear devices are scaled with the homotopy parameter λ. Hence, at the beginning when $\lambda=0$, a linear circuit arises in which nonlinear devices are open or short-circuited, and, therefore, it can be easily solved to find a starting point for the homotopy path. For example, a diode $i=I_s[\exp(\lambda v/V_T)-1]$ for $\lambda=0$ becomes an open circuit. Moreover, a component containing **a**, scaled by means of G, is added to avoid bifurcations on the path to be passed through. When parameter λ varies from 0 to 1, the homotopy generates a path starting from a solution of the linear circuit and ending at the solution of the original nonlinear circuit. If nodal equations are used, then the added component can be viewed as a connection to the kth node (for $k=1, ..., n$), the series connection of the conductance $G(1-\lambda)$ and the voltage source a_k. During the continuation process these voltages tend to be disconnected.

An alternative approach, also quite efficient, is a variable gain homotopy, where coefficients of controlled sources in transistors models, instead of controlling signals, are scaled with λ. Thus, at the beginning a passive nonlinear circuit is obtained, and it should be solved by means of an efficient quasi-Newton method. Then, along the homotopy path, stepping of the circuit activity arises.

The disadvantages of all homotopy methods are their sensitivity to rapid turning points and strong nonlinearities in nonlinear characteristics. Fortunately, in practice device characteristics are usually regular and smooth.

4.4.2 Efficient algorithms for solving homotopies

A conventional technique for solving the homothopy $\mathbf{H}(\mathbf{x}, \lambda) = \mathbf{0}$ relies on a given discretization $\lambda_n, n = 1, 2, \ldots$ and solving a sequence of equations $\mathbf{H}(\mathbf{x}, \lambda_n) = \mathbf{0}, n = 1, 2, \ldots$ by means of the efficient quasi-Newton-Raphson method starting from a point $\mathbf{x}_n^{(0)}$. To minimize local NR iterations, this point should be as close to the solution $\mathbf{x}_n$ as possible. The only available way of selecting this point is extrapolation from a number of already calculated preceding solutions $\ldots, \mathbf{x}_{n-3}, \mathbf{x}_{n-2}, \mathbf{x}_{n-1}$, known as prediction. It can be shown, from Lagrange interpolation, that a predicting formula takes the form

$$\mathbf{x}_n^{(0)} = \sum_{i=1}^{l+1} \alpha_i \mathbf{x}_{n-i} \tag{4.60}$$

where the coefficients α_i depend on the step size of λ discretization (see Chapter 5 for methods of assigning the coefficients of such, so-called differentiation formulae) and l is the order of the extrapolating polynomial. Quite often first order is sufficient. The step size of the parameter λ should vary over a wide range as the path $\mathbf{x}(\lambda)$ is usually strongly nonlinear. To select the step size a number of NR iterations or a local error, i.e. a difference between a solution and a prediction $\mathbf{x}_n - \mathbf{x}_n^{(0)}$, can be taken into account. The step size can be predicted using, for example, the following rule

$$\Delta\lambda_{n+1} = \max\{\Delta_{\min}, \min\{\Delta_{\max}, \Delta\lambda\}\} \text{ where}$$
$$\Delta\lambda = \min_i \Delta\lambda_n \left(\frac{\varepsilon \max\{|x_i|, |x_i^{(0)}|\} + \delta_i}{|x_i - x_i^{(0)}|} \right)^{1/2}. \tag{4.61}$$

Thus, the final d.c. sweep algorithm includes an internal NR loop and an external loop controlled by the homotopy parameter. This method is implemented in OPTIMA. A weak point of this method lies in fact that the variability of a step does not exactly follow the curvature of a path. This causes convergence problems. More reliable methods use the arc-length concept.

4.4.2.1 Arc-length method

The basic arc-length method of A.Ushida and L.O.Chua [4.11] relies on the transformation of a homotopy into a very convenient ODE problem, whose response is free from sharp turning points. This is achieved by taking the arc-length concept, well known from differential geometry. The arc length is a real coordinate s (length) measured from a given point (where $s=0$) along the curve. Let the curve $\mathbf{x}(\lambda)$ be a unicursal (that is it can be drawn with a pencil permanently touching the paper) path of the homotopy (4.58) in the $n+1$-dimensional space $(\mathbf{x}, \lambda)$. Such a homotopy stands for n circuit equations with n variables $\mathbf{x}$ and the homotopy parameter λ. The solution curve can be drawn up as a parameterized equation $[\mathbf{x}(s), \lambda(s)]$, $s \in R$ with the arc length s as a parameter. The goal of this method is to obtain a solution of the homotopy (4.58) in such a parameterized form. Its advantage lies in fact that functions $\mathbf{x}(s), \lambda(s)$ are more regular and free from sharp turning points, which is not the case in $\mathbf{x}(\lambda)$ form. In fact, from Pythagoras' theorem it is known that the differential of arc length is connected with the differentials of rectangular coordinates by means of an equation of the form $\mathrm{d}s^2 = \mathrm{d}x_1^2 + \ldots + \mathrm{d}x_n^2 + \mathrm{d}\lambda^2$ that gives the relation

$$\left(\frac{\mathrm{d}x_1}{\mathrm{d}s}\right)^2 + \ldots + \left(\frac{\mathrm{d}x_n}{\mathrm{d}s}\right)^2 + \left(\frac{\mathrm{d}\lambda}{\mathrm{d}s}\right)^2 = 1 \tag{4.61}$$

or equivalently

$$\left(\frac{\mathrm{d}\lambda}{\mathrm{d}s}\right)^2 + \left(\frac{\mathrm{d}\mathbf{x}}{\mathrm{d}s}\right)^t \left(\frac{\mathrm{d}\mathbf{x}}{\mathrm{d}s}\right) = 1\,. \tag{4.62}$$

It proves that all $|\mathrm{d}x_i/\mathrm{d}s|$ $(i=1, \ldots, n)$, $|\mathrm{d}\lambda/\mathrm{d}s|$ are less than or equal to 1, and hence, the parameterized $\mathbf{x}(s), \lambda(s)$ do not have sharp turning points as long as their local slopes do not exceed 45°. To obtain a parameterized homotopy path, it is necessary to combine (4.58) and (4.61) together and solve them as a system of $n+1$ algebraic differential equations with unknown functions $\mathbf{x}(s), \lambda(s)$. Numerical methods for ODE, described in Chapter 5, are available for this purpose.

At each discrete point s_n, derivatives are expressed by means of a differentiation formula $\dot{x}_n = \gamma_0 (x_n - d_x)$, which is equivalent to interpolation of $x(s)$ by means of a polynomial of a selected degree (dot means a derivative with respect to s, d_x is a constant, standing for the history of variable x). Right now, it is sufficient to know that such a form of differentiation formula exists; details will be presented in Chapter 5. Substituting in (4.61) derivatives from the differentiation formula, we obtain a set of nonlinear algebraic equations for the instant s_n

$$\begin{aligned} &\mathbf{H}(\mathbf{x}_n, \lambda_n) = \mathbf{0} \\ &(\lambda_n - d_\lambda)^2 + (\mathbf{x}_n - \mathbf{d}_{xn})^t(\mathbf{x}_n + \mathbf{d}_{xn}) = 1/\gamma_0^2 . \end{aligned} \tag{4.63}$$

These equations should be solved by means of NR iterations, starting from predicted values (4.60) (similar to prediction for λ). Linearized equations for a NR iteration, derived from (4.63), are of the form

$$\begin{aligned} &\frac{\partial \mathbf{H}(\mathbf{x}_n^{(p)}, \lambda_n^{(p)})}{\partial \mathbf{x}} \Delta \mathbf{x}_n^{(p)} + \frac{\partial \mathbf{H}(\mathbf{x}_n^{(p)}, \lambda_n^{(p)})}{\partial \lambda} \Delta \lambda_n^{(p)} = -\mathbf{H}(\mathbf{x}_n^{(p)}, \lambda_n^{(p)}) \\ &2\,(\gamma_0 \mathbf{x}_n^{(p)} - \mathbf{d}_x)^t \Delta \mathbf{x}_n^{(p)} + 2\,(\gamma_0 \lambda - d_\lambda)\,\Delta \lambda_n^{(p)} = \\ &\qquad -(\mathbf{x}_n^{(p)} - \mathbf{d}_x)^t(\mathbf{x}_n^{(p)} - \mathbf{d}_x) - (\lambda_n^{(p)} - d_\lambda)^2 + 1/\gamma_0^2 \end{aligned} \tag{4.64}$$

Internal iterations (4.64) over p are continued until convergence is reached, while external iterations over n are continued until a preset maximum λ (typicaly 1) is reached.

4.4.2.2 Pseudo-arc-length method

The arc-length equations (4.64) can be viewed as d.c. circuit equations with an additional nonlinear differential equation. The nonlinearity of the latter is a disadvantage of this method, because it implies considerable senstitivity to the starting point and troubles with convergence. Furthermore, a general integration approach is useless in this case. In [4.4, 4.19] the application of forward Euler discretization is proposed, to treat the quadratic formula (4.61) as a linear interdependence between increments. Namely, a derivative of $\mathbf{x}(s)$ at the point s_n can be expressed by the forward-Euler formula as $\dot{\mathbf{x}}_n = (\mathbf{x}_{n+1} - \mathbf{x}_n)/\Delta s_n$ and similarily for λ. Hence, (4.61) can be approximated at s_n as

$$\dot{\mathbf{x}}_n^t \Delta \mathbf{x}_n + \dot{\lambda}_n \Delta \lambda_n = \Delta s_n \tag{4.65}$$

which is a linear interdependence of increments, where $\mathbf{u}_n = [\dot{\mathbf{x}}_n^t, \dot{\lambda}_n]^t$ is a vector tangent to the trajectory $[\mathbf{x}(s), \lambda(s)]$ at s_n. The method using this form instead of (4.61) is not exactly arc-length, and we call it pseudo-arc-length. Now, formula (4.65) will be imposed on the NR iterative formula for solving (4.58), to connect increments of **x** and λ. Instead of (4.64) we obtain much better converging and more efficient iterative NR pseudo-arc-length equations

$$\frac{\partial \mathbf{H}(\mathbf{x}_n^{(p)}, \lambda_n^{(p)})}{\partial \mathbf{x}} \Delta \mathbf{x}_n^{(p)} + \frac{\partial \mathbf{H}(\mathbf{x}_n^{(p)}, \lambda_n^{(p)})}{\partial \lambda} \Delta \lambda_n^{(p)} = -\mathbf{H}(\mathbf{x}_n^{(p)}, \lambda_n^{(p)}) \tag{4.64}$$

$$\dot{\mathbf{x}}_n^{\,t} \Delta \mathbf{x}_n^{(p)} + \dot{\lambda}_n \Delta \lambda_n^{(p)} = \Delta s_n \,.$$

They will be repeatedly solved within the NR loop to obtain $\Delta \mathbf{x}_n, \Delta \lambda_n$ corresponding to a given Δs_n. It is essential that derivatives $\dot{\mathbf{x}}_n^{t}, \dot{\lambda}_n$ have to be known at each continuation point before we enter the NR loop. Their calculation will be considered later on.

Extracting $\Delta x_n^{(p)}$ from the first equation and substituting the result into the second equation, after two auxiliary vectors **z** and **y** have been introduced, we find that the solution of (4.64) can be obtained from the following four steps of calculation (indexes and arguments have been omitted) that stands for one NR iteration.

$$\text{solve a system } \frac{\partial \mathbf{H}(\mathbf{x}_n^{(p)}, \lambda_n^{(p)})}{\partial x} \mathbf{z} = -\mathbf{H}(\mathbf{x}_n^{(p)}, \lambda_n^{(p)}),$$

$$\text{solve a system } \frac{\partial \mathbf{H}(\mathbf{x}_n^{(p)}, \lambda_n^{(p)})}{\partial x} \mathbf{y} = \frac{\partial \mathbf{H}(\mathbf{x}_n^{(p)}, \lambda_n^{(p)})}{\partial \lambda}, \tag{4.65}$$

$$\text{calculate } \Delta \lambda_n = \frac{\Delta s_n - \dot{\mathbf{x}}_n^{t} \mathbf{z}}{\dot{\lambda}_n - \dot{\mathbf{x}}_n^{t} \mathbf{y}},$$

$$\text{calculate } \Delta \mathbf{x}_n = \mathbf{z} - \Delta \lambda_n \mathbf{y} .$$

Derivatives $\dot{\mathbf{x}}_n^{t}, \dot{\lambda}_n$, needed before NR iterations at the continuation point s_n, can be obtained from a solution at the preceding continuation point s_{n-1}. Let us show how to calculate these derivatives recursively at the next point s_{n+1}, i.e. $\dot{\mathbf{x}}_{n+1}^{t}, \dot{\lambda}_{n+1}$. They can be obtained from an auxiliary interdependence, after convergence of the iterations (4.64) controlled by *p*. As they are derivatives of two functions interconnected with a given equation (4.58), then it is natural to adopt a variational approach, i.e. by differentiating (4.58) with respect to *s* at the point s_{n+1}. The equation thus obtained is combined with the arc-length equation (4.59) at the same point and we obtain

$$\frac{\partial \mathbf{H}}{\partial \mathbf{x}}\dot{\mathbf{x}}_{n+1} + \frac{\partial \mathbf{H}}{\partial \lambda}\dot{\lambda}_{n+1} = \mathbf{0}$$
$$\dot{\mathbf{x}}^{t}_{n+1}\dot{\mathbf{x}}_{n+1} + \dot{\lambda}^{2}_{n+1} = 1\,. \tag{4.66}$$

Due to the availability of $\mathbf{y}$ after convergence at the preceding continuation point, the first equation yields $\dot{\mathbf{x}}_n = \dot{\lambda}_n \mathbf{y}$, and hence the second gives $\dot{\lambda}_n = \pm 1/(1+\mathbf{y}^t\mathbf{y})^{1/2}$. The tangent vector $\mathbf{u}$ obtained is unit. The sign is chosen to select an appropriate direction of movement along the path, positive meaning that λ increases along the path. The user always knows what direction change of continuation parameter occurs at the beginning of his problem (e.g. in d.c. sweep analysis a voltage may change from 0 to 5 V or from 5 V to 0). However, when a path is traced which is not monotonic or even not a function in the mathematical sense (e.g. N or S type negative resistance), the direction of the λ increment may vary, and then the direction of the preceding change of λ is taken to determine the sign.

The NR iterations starts from a very good predicted increment obtained from the forward Euler formula

$$\begin{bmatrix} \mathbf{x}^{(0)}_{n+1} \\ \lambda^{(0)}_{n+1} \end{bmatrix} = \begin{bmatrix} \mathbf{x}_n \\ \lambda_n \end{bmatrix} + \Delta s_n \begin{bmatrix} \dot{\mathbf{x}}_n \\ \dot{\lambda}_n \end{bmatrix}. \tag{4.67}$$

This initial guess satisfies the pseudo-arc-length equation (4.64). In circuits where the path is a function in the mathematical sense this initial guess is sufficient.

The algorithm presented is very efficient and has been implemented in CAzM [4.4]. It very carefully traces the homotopy path with automatic selection of the steps of the circuit variables and homothpy parameter. An additional mechanism for selecting the step of arc length s, adapted from [4.6], is used to improve this tracing. Thus, in sharp regions the step becomes shorter while elsewhere it is enlarged.

Recently, a very efficient global algorithm for homotopies has been developed by L. T. Watson [4.13] and implemented in HOMEPACK [4.14]. Also, the original arc-length algorithm has been slightly improved by C. He and A. Ushida [4.5] by introducing an improved prediction and more efficient step control.

The following exercisesmay serve to test our understanding of the topics:

- Explain two sorts of continuation methods: using ODE and homotopies.
- Write NE of a circuit containing a diode driven by a current source; write a variable source and variable stimulus homotopy.
- Explain a basic, common method of tracing homotopy paths, and list its weak points.
- Explain the idea of the arc length method for tracing paths; how are its disadvantages overcome in the pseudo-arc-length method?

References

4.1 Agnew, D, *Techniques for Robust d.c. Algorithms for Circuit Simulation*, Proc. Int. Symp. Circ. Syst., p. 1198-1201 (1982)

4.2 Bank, R.E., Rose, D.J., *Global Approximate Newton Methods*, Numerische Mathematik, vol. 37, p. 279-295 (1981)

4.3 Chua, L.O., Lin, P.M., *Computer-Aided Analysis of Electronic Circuits: Algorithms and Computational techniques*, Englewood Cliffs, NJ: Prentice-Hall, chapters 5-7, (1975)

4.4 Coughran, W.M., Grosse, E.Jr.,Rose, D.J., *CAzM: A Circuit Analyser with Macromodeling*, IEEE Trans. El.Dev., vol. 30, no. 9, p. 1207-1213 (September 1983)

4.5 He, C., Ushida, A., *An Efficient Algorithm for Solving Nonlinear Resistive Circuits*, Proc. Int. Symp. Circ. Syst., p. 2328-2331 (1991)

4.6 den Heijer, C. Rheinboldt, W.C., *On Steplength Algorithms for a Class of Continuation Methods*, SIAM J. Numerical Analysis, vol. 18, p. 925-948 (1981)

4.7 Ho, C.W., Zein, D.A., Ruehli, A.E., Brennan, P.A., *An Algorithm for d.c. Solution in an Experimental General Purpose Interactive Circiut Design Program*, IEEE Trans. Circ. Syst., vol. 24, no. 8, p. 416-421 (August 1977)

4.8 Melville, R.C., Trajkowic, L., Fang, San-chin, Watson. L.T., *Artificial Homotopy Methods for the d.c. Operating Point Problem*, IEEE Trans. CAD of Int. Circ. Syst., vol. 12, no. 6, p. 861-877 (June 1983)

4.9 Nagel, L.W., *SPICE2: A Computer Program to Simulate Semiconductor circuits*, Memo. no. ERL-M520, Univ. of Calif., Berkeley (9 May 1975)

4.10 Ortega, J.M., Rheinboldt, W.C., *Iterative Solution of Nonlinear Equations in Several Variables*, New York: Academic Press (1970)

4.11 Ushida, A., Chua, L.O., *Tracing Solution Curves of Nonlinear Equations with Sharp Turning Points*, Int. J. Circ. Theory Appl., vol. 12, p. 1-21 (1984)

4.12 Watson, L.T., *Globally Convergent Homotopy Methods: A Tutorial*, Appl. Math. and Comp., vol. 31, p. 369-396 (May 1989)

4.13 Watson, L.T., *Globally Convergent Homotopy Algorithm for Nonlinear Systems of Equations*, Nonlinear Dynamics, vol. 1, p. 143-191 (February 1990)

4.14 Watson, L.T., Billups, S., Morgan, A., *HOMEPACK: A Suite of Codes for Globally Convergent Homotopy Algorithms*, ACM Trans. Math. Software, vol. 13, no. 3, p. 281-310 (September 1987)

4.15 Zein, D.A., *Solution of a Set of Nonlinear Algebraic Equations for Genaral Purpose CAD programs*, In: *Circuit Analysis Simulation and Design. General Aspects of Circuit Analysis and Design.* Ed. A. E. Ruehli, Amsterdam: North Holland (1986)

4.16 Ogrodzki, J., Bukat, D., *Compact Modeling in Circuit Simulation: the General-Purpose Analyzer OPTIMA-3.* Proc. IEEE Int. Symp. Circ. Syst., London (1994)

4.17 Voight, B., Wilkens, H., Horneber, E.H. *Nonlinear Limitations in electronic circuits: A Modelling Approach.* Proc. Europ. Simul. Symp. "Simulation and AI in Computer-Aided Techniques", Dresden, Germany, November 5-8 (1992)

4.18 Voight, B., Thoellmann, K., Horneber, E.H., *Modeling of Limitation Efects in Electronic Circuits.* 35th Midwest Symp. Circ. Syst., Washington, D.C., August 9-12 (1992)

4.19 Bolstad J.H., Keller H.B. *A Multigrid Continuation Method for Elliptic Problems with Folds.* SIAM J. Sci. Stat. Comp., vol. 7, no. 4, p. 1081-1104 (October 1986)

4.20 Rudin W., *Principles of Mathematical Analysis.* McGraw-Hill (1976)

Chapter 5

Time-domain Analysis of Nonlinear Circuits

In Chapter 1 general circuit equations have been introduced. Through an appropriate selection of circuit variables they have been transformed into a set of Ordinary Algebraic Differential Equations (OADE) (1.39) with unknown nodal voltages and currents of voltage-defined branches. In general, such equations will be dependent on time t, if the circuit contains time-varying excitations or other branches of time-dependent coefficients, e.g. parametric resistors. Denoting the unknown circuit variables by $\mathbf{x}$ and state variables (whose time derivatives appear in equations) by $\mathbf{z}$, the basic circuit OADE can be written as

$$\mathbf{F}\,[\mathbf{x},\, \dot{\mathbf{z}}(\mathbf{x}),\, t] = \mathbf{0} \tag{5.1}$$

where $\mathbf{F}$ is a continuous and differentiable vector function. Time-domain circuit analysis can be viewed as integrating such circuit equations to obtain a solution $\mathbf{x}(t)$. This is an initial-value problem [5.9] which may have a unique solution if we assume initial values $\mathbf{z}(0)=\mathbf{z}_0$. In electrical circuits, due to the strongly nonlinear character of Equation (5.1) only efficient numerical integration methods can be utilized. Chapter 5 is dedicated to such methods. It covers basic theoretical aspects of numerical OADE integration and the practical problems involved in creating efficient algorithms.

5.1 Introduction to integrating circuit equations

5.1.1 Basic polynomial methods

5.1.1.1 Introductory example

Universally used for time-domain circuit analysis, numerical methods for integrating OADE exploit the idea of interpolating the next solution with a polynomial of specified degree, spanning a number of points of the solution derived so far. Such methods have been well explored and have been excellently presented in books on numerical analysis and circuit simulation, e.g. [5.4, 5.9, 5.10, 5.15]. Also, they are discussed in works on circuit simulation. We begin their presentation here with a very simple example.

Consider a Hermite interpolation problem consisting of looking for a quadratic polynomial $w(t)$, which satisfies at instant t_1 a constraint $w(t_1)=w_1$, and also a constraint $\dot{w}(t_1)=\dot{w}_1$ for its derivative. Moreover, let another constraint $\dot{w}(t_2)=\dot{w}_2$ for the derivative hold at another point $t_2>t_1$. The number of constraints equals the number of coefficients of a second order polynomial, which equals 3. Hence, the problem has a unique solution as is shown in Figure 5.1. To simplify calculations, the polynomial will be drawn up in an incremental form around the point t_2, i.e. $w(t)=a_0+a_1(t-t_2)+a_2(t-t_2)^2$. Its derivative is $\dot{w}(t)=a_1+2a_2(t-t_2)$. The three constraints can be written now as a system of linear equations

$$\begin{aligned} a_0+a_1(t_1-t_2)+a_2(t_1-t_2)^2 &= w_1^2 \\ a_1+2a_2(t_1-t_2) &= \dot{w}_1 \\ a_1+2a_2(t_2-t_2) &= \dot{w}_2 \end{aligned} \tag{5.2}$$

suitable for deriving the three unknown coefficients. After some algebra we obtain $a_1=\dot{w}_2$, $a_2=0.5(\dot{w}_2-\dot{w}_1)/(t_2-t_1)$, $a_0=w_1+0.5(\dot{w}_1+\dot{w}_2)(t_2-t_1)$.

Notice, that in this problem, there was no constraint for the polynomial value at instant t_2. This value can be derived from the polynomial once it is known. For fixed t_1, t_2 it is a function of $w_1, \dot{w}_1, \dot{w}_2$. Since, an appropriate polynomial form has been chosen, w_2 is equal to a_0. Hence

$$w_2 \equiv w(t_2) = w_1 + \frac{1}{2}(\dot{w}_1+\dot{w}_2)(t_2-t_1). \tag{5.3}$$

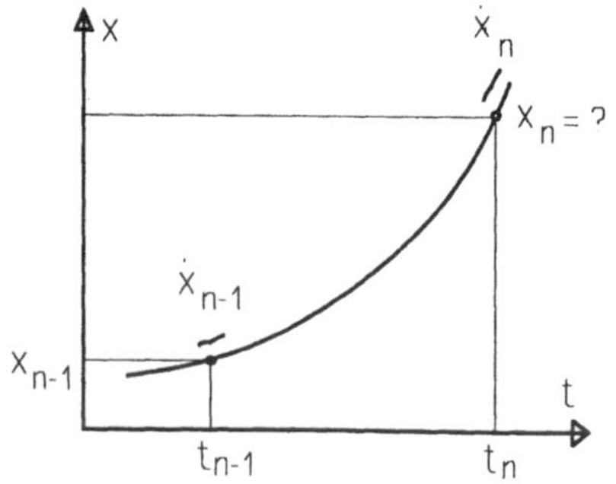

Figure 5.1 A quadratic polynomial spanning over 3 constraints

This expression is known as the differentiation formula (DF). Thanks to interpolation, the DF enables us to join the instantaneous value of response with the instantaneous derivative, using a number of values at previous points. The inaccuracy of such a relation is caused by the interpolation, and hence can be well controlled.

To demonstrate the purpose of introducing DF let us consider the ODE $\dot{x}=\lambda x$. We assume that a solution $x(t)$ and its derivative $\dot{x}(t)$ is known at instant t_1, namely $x(t_1)=x_1$, $\dot{x}(t_1)=\dot{x}_1$. At the next instant $t_2>t_1$ we calculate a solution $x_2=x(t_2)$ as if it lay on a quadratic polynomial spanned over points t_1 and t_2. First, the derivative at t_2 is calculated from the DF (5.3), written as follows:

$$\dot{x}_2 = -\dot{x}_1 + 2(x_2-x_1)/(t_2-t_1). \tag{5.4}$$

Then, the differential equation is represented as an algebraic relation between the function and its derivative at instant t_2

$$\dot{x}_2 = \lambda x_2. \tag{5.5}$$

Finally, after the DF (5.4) has been substituted into (5.5), the purely algebraic equation $-\dot{x}_1+2(x_2-x_1)/(t_2-t_1)=\lambda x_2$ arises. Its solution takes the form $x_2=-[\dot{x}_1(t_2-t_1)+2x_1]/[\lambda(t_2-t_1)-2]$. It has been expressed as a recursion, joining the solutions at t_2 and at t_1 with a second order polynomial approximation embedded. Now, due to the availability of x_2, a derivative at t_2 can also be calculated from DF (5.4), namely $\dot{x}_2=2(x_2-x_1)/(t_2-t_1)-\dot{x}_1$. Hence, there is complete information to repeat these calculations with x_2, $\dot{x}_2$ as the initial state, in order to obtain a solution at t_3, etc.

5.1.1.2 Differentiation formulae

A more general approach to differentiation formulae (DF) can be introduced by extending the above example to an arbitrary l-degree polynomial. Let t_n be the current instant where a solution is being calculated, while t_{n-k} $(k=p, ..., 1)$ are chronologically ordered previous instants. We span over these points an l-degree polynomial (i.e. of $l+1$ unknown coefficients) fitting $l+1$ constraints. They will be selected from the set $\{w_{n-k}, \dot{w}_{n-k}\ (k=p, ..., 1),\ \dot{w}_n\}$. However, these $2p+1$ constraints form too large a set in comparison with the $l+1$ constraints needed. To obtain a unique solution, the inequality $2p\geq l$ has to hold and an arbitrary $2p-l$ redundant constraints have to be omitted. For the example of Section 5.1.1.1, $p=1$ and $l=2$. Thus, $2p+1=l+1=3$, and all three constraints $\dot{w}_n$, w_{n-1}, $\dot{w}_{n-1}$ are necessary. However, if one spans a quadratic polynomial over two points of the past ($p=2$), then, $2p-l=2$ constraints among $\dot{w}_n$, w_{n-1}, w_{n-2}, $\dot{w}_{n-1}$, $\dot{w}_{n-2}$ are redundant and can be omitted. For example, by omitting $\dot{w}_{n-1}$, $\dot{w}_{n-2}$ the so-called backward differentiation formula would arise.

Thus, we formulate $l+1$ equations using constraints for the function

$$a_0 + a_1(t_{n-k}-t_n) + a_2(t_{n-k}-t_n)^2 + \dots + a_l(t_{n-k}-t_n)^l = w_{n-k} \tag{5.6}$$

and for its derivative

$$a_1 + 2a_2(t_{n-k}-t_n) + \dots + la_l(t_{n-k}-t_n)^{l-1} = \dot{w}_{n-k}\,. \tag{5.7}$$

These can be written all together in a matrix form

$$\mathbf{V}\,\mathbf{a} = \mathbf{w}\,, \tag{5.8}$$

where **a** contains the polynomial coefficients a_k, **w** comprises all active constraints from the set $\dot{w}_n$, w_{n-k}, $\dot{w}_{n-k}$ ($k=1, \dots, p$), **V** is a nonsingular matrix dependent on increments $t_n - t_{n-k}$.

It is essential that **w** does not include the constraint w_n. The solution of (5.8) will have the form $\mathbf{a} = \mathbf{V}^{-1}\mathbf{w}$. Thus, each a_i will take the form of a linear combination of constraints, collected in the vector **w**. The required value of polynomial at t_n equals a_0, and hence, it is also a linear combination of **w** components

$$w_n = \beta_0 \dot{w}_n + \sum_{k=1}^{p} (\alpha_k w_{n-k} + \beta_k \dot{w}_{n-k})\,. \tag{5.9}$$

Among the above $2p+1$ coefficients: β_0, α_k, β_k ($k=1, \dots, p$), there will be only $l+1$ independent ones. The rest are zero or dependent on the former. Expression (5.9) is known as the Differentiation Formula (DF). The interpolating polynomial degree l is called the order of the DF. As has been said, the degree should satisfy the inequality $2p > l$. Let us stress the meaning of DF. It describes a function x_n at a current point t_n as being proportional to the derivative at this point: $\dot{x}_n$, with an additive past storing term d_x. Hence, (5.9) can be written briefly as $x_n = \beta_0 \dot{x}_n + d_x$. In mathematical analysis a function and its derivative are not algebraically connected. That connection can be written only for a particular class of functions, e.g. for polynomials, and therefore it has an inaccuracy embedded. If the condition $\beta_0 \neq 0$ is satisfied it means that the DF uses a constraint for the derivative at the current point. Hence, we can solve the DF for that derivative to obtain

$$\dot{x}_n = \gamma_0 (x_n - d_x) \tag{5.10}$$

where $\gamma_0 = 1/\beta_0$ exists. This compact formula is very important for OADE discretization. A differentiation formula satisfying $\beta_0 \neq 0$ can be written in the form (5.10), and is called implicit. Otherwise, the DF has an explicit form $x_n = d_x$, and contains only the past storing term. Such a DF is predictive, i.e. the next point is extrapolated from previous ones.

To calculate the coefficients of DF one has to preselect the polynomial order l and substitute a general form $x(t) = a_0 + a_1 t + \ldots + a_l t^l$ to (5.9). Then, DF coefficients can be chosen obtain identity for any set of coefficients of the polynomial involved. The lth degree polynomial is a linear combination of basic elementary polynomials $x_i(t) = t^i$ $(i=0,1, \ldots, l)$. Due to the linear form of (5.9), DF holds as identity, iff it holds for all these basic polynomials. Thus, the easiest method of DF identification is to substitute basic polynomials and create $l+1$ equations, each one for a particular degree $i = 1, \ldots, l$. These equations will have unknown coefficients of DF. To simplify calculations, it is better to write these basic polynomials in the form $(t-t_n)^i$.

For example, let $x_n = \beta_0 \dot{x}_n + \alpha_1 x_{n-1} + \beta_1 \dot{x}_{n-1} + \alpha_2 x_{n-2} + \beta_2 \dot{x}_{n-2}$ (so $p=2$) be of order $l=2$. Taking basic polynomials $(t-t_n)^0, (t-t_n)^1, (t-t_n)^2$, their corresponding derivatives, and calculating all their values required in the DF we can construct Table 5.1.

Table 5.1 A data set for identification of a second order DF

i	$x_i(t)$	x_n	x_{n-1}	x_{n-2}	dx^i/dt	dx_n/dt	dx_{n-1}/dt	dx_{n-2}/dt
0	1	1	1	1	0	0	0	0
1	$t-t_n$	0	$t_{n-1}-t_n$	$t_{n-2}-t_n$	1	1	1	1
2	$(t-t_n)^2$	0	$(t_{n-1}-t_n)^2$	$(t_{n-2}-t_n)^2$	$2(t-t_n)$	0	$2(t_{n-1}-t_n)$	$2(t_{n-2}-t_n)$

It is convenient to define the steps between time-points $h_n = t_n - t_{n-1}$, $h_{n-1} = t_{n-1} - t_{n-2}$, and also the partial sums of steps $s_2 = t_n - t_{n-2} = h_n + h_{n-1}$, $s_1 = t_n - t_{n-1} = h_n$. Substituting the quantities of Table 5.1 into the DF, the following system of three linear equations with five unknowns arises

$$\begin{gathered} \alpha_1 + \alpha_2 = 1 \\ -\alpha_1 s_1 - \alpha_2 s_2 + \beta_0 + \beta_1 + \beta_2 = 0 \\ \alpha_1 s_1^2 + \alpha_2 s_2^2 - 2\beta_1 s_1 - 2\beta_2 s_2 = 0\,. \end{gathered} \tag{5.11}$$

Two coefficients will be chosen arbitrarily. The assumption $\beta_1=\beta_2=0$ yields the 2nd order Gear's DF, known also as the 2nd order Backward Differentiation Formula (BDF)

$$\beta_0=\frac{h_n(h_n+h_{n-1})}{2h_n+h_{n-1}}, \quad \alpha_1=\frac{(h_n+h_{n-1})^2}{h_{n-1}(2h_n+h_{n-1})}, \quad \alpha_2=\frac{-h_n^2}{h_{n-1}(2h_n+h_{n-1})}. \tag{5.12}$$

Another preselection of two coefficients will give another DF. In practice, existing degrees are exploited, to provide such advantageous DF properties as accuracy and stability. They will be discussed later on.

Equations (5.11) which are satisfied by a DF can be easily generalized. Using the same basic polynomials technique and creating an appropriate Table 5.1, after some algebra, we obtain $l+1$ conditions

$$\begin{gathered}\sum_{k=1}^{p}\alpha_k=1\\ \sum_{k=1}^{p}(\alpha_k s_k-\beta_k)-\beta_0=0\\ \sum_{k=1}^{p}[(\alpha_k s_k-i\beta_k)(s_k)^{i-1}]=0 \quad (i=2,\ldots,l)\end{gathered} \tag{5.13}$$

where $h_{n-k}=t_{n-k}-t_{n-k-1}$ are time steps, while $s_k=t_n-t_{n-k}=h_n+\ldots+h_{n-k+1}$ are partial sums of steps. Assuming arbitrary values for $2p-l$ coefficients, according to such additional criteria as accuracy and stability, we can derive $l+1$ coefficients uniquely.

So far, we have taken into account the general case of nonuniform time points. A uniform time grid is a specific case with a fixed numerical integration step. Fixed step methods are very well studied, but their usefulness is limited to extremely simple dynamic problems only. Due to their considerable simplicity, fixed step methods are often used in theoretical considerations as the first attempt (e.g. in stability theory). Fixed step DF have simple coefficients. For example, recalling a variable step 2nd order BDF (5.11) and setting $h_{n-1}=h_n=h$, we obtain the fixed step 2nd order BDF

$$x_n=\frac{2}{3}h\dot{x}_n+\frac{4}{3}x_{n-1}-\frac{1}{3}x_{n-2}. \tag{5.14}$$

Conditions (5.13) can be easily simplified to the fixed step case by substituting h for all steps, and assuming that the partial sums of steps are $s_k = kh$ instead of $t_n - t_{n-k}$. It can be shown that a fixed step DF is of the form (5.9) with rational α_k and β_k of the form: a rational number mutiplied by h.

5.1.1.3 The most important practical differentiation formulae

In practice, an important role is played by one-step DF ($p=1$) because they use only one preceding solution x_{n-1} at point t_{n-1}, thus minimizing the ammount of stored data. Unfortunately, the family of one-step DF is small. As $2p > l$, then one-step DF can be at most of second order. Formulating three conditions, according to (5.13), we obtain the only one possible 2nd order DF, the Trapezoidal DF (TDF)

$$x_n = \frac{1}{2} h_n \dot{x}_n + \frac{1}{2} h_n \dot{x}_{n-1} + x_{n-1} . \tag{5.15}$$

By virtue of the fact that it has very good accuracy and stability, TDF is widely used to integrate OADE of circuits, e.g. in SPICE.

Other one-step DFs are first order. The Implicit Euler DF (IEDF) is very popular among them. Due to bad accuracy it is used mainly for rough integration, e.g. in timing simulation.

$$x_n = h_n \dot{x}_n + x_{n-1} . \tag{5.16}$$

Another useful Explicit Euler DF (EEDF) has the form

$$x_n = h_n \dot{x}_{n-1} + x_{n-1} . \tag{5.17}$$

It is a convenient way of predicting a solution from one preceding point. EEDF has been used e.g. to integrate psudo-arc-length equations in Section 4.4.2.2.

Apart from TDF, the other DFs having excellent numerical properties are Backward Differentiation Formulae (BDF). These implicit formulae make use of only function values (not derivatives) at several past points to construct an interpolative polynomial. The first BDFs were introduced by Gear [5.8, 5.9]. Now, we rather prefer the BDFs based on Lagrange interpolation [5.2], or even newer approaches using Newton interpolation [5.1, 5.14] which will be discussed later on. Each implicit DF is used in OADE to substitute a derivative, such that it is convenient to solve the DF for a derivative. Thus the lth order BDF is usually expressed in the form

$$\dot{x}_n = \sum_{k=0}^{l} \gamma_k^{(l)} x_{n-k} . \tag{5.18}$$

It can be derived by spanning the l-degree polynomial over $l+1$ constraints x_{n-k} $(k=0, ..., l)$. This Lagrange interpolation problem, formulated as a system of conditions (5.13) because of the Vandermonde form of its matrix, has been solved analytically as [5.2]

$$\gamma_0^{(l)} = \sum_{i=1}^{l} \frac{1}{s_i}, \quad \gamma_k^{(l)} = -\frac{1}{s_k} \prod_{i=1}^{l} e_k^{(i)} \quad (k=1, ..., l), \tag{5.19}$$

where

$$e_k^{(i)} = \begin{cases} \dfrac{s_i}{s_i - s_k} = \dfrac{t_n - t_{n-i}}{t_{n-k} - t_{n-i}} & \text{for } i \neq k \\ 1 & \text{for } i = k \end{cases} \tag{5.20}$$

are Lagrange factors expressed through the already specified partial sums of steps $s_i = t_n - t_{n-i}$. The first order BDF is simply the implicit Euler DF.

The above DF is usually accompanied by an lth order forward formula obtained from the lth degree polynomial spanned over $l+1$ past solutions x_{n-k} $(k=1, ..., l+1)$ (compared to BDF of the same order one more past point is used)

$$x_n = \sum_{k=1}^{l+1} \alpha_k^{(l)} x_{n-k} . \tag{5.21}$$

Coefficients derived from Lagrange interpolation

$$\alpha_k^{(l)} = \prod_{i=1}^{l+1} e_k^{(i)} \quad (k=1, ..., l) \tag{5.22}$$

include the same Lagrange factors (5.20). Forward formulae are applied to predict a rough solution at the current instant.

The advantage of BDFs and forward formulae is that the coefficients of the latter can be recursively derived, starting from the order 0 and increasing, i.e. $\alpha_k^{(l)} = \alpha_k^{(l-1)} e_k^{(l+1)}$. Hence, the forward formula arises from

$$
\begin{aligned}
&\text{for } k=1, \ldots, l+1 \\
&\quad \alpha_k^{(-1)} = 1 \\
&\quad \text{for } j=0, \ldots, l \\
&\qquad \alpha_k^{(j)} = \alpha_k^{(j-1)} e_k^{(j+1)},
\end{aligned}
\tag{5.23}
$$

while the corresponding BDF can be obtained from the above as

$$\gamma_k^{(l)} = \alpha_k^{(l)} \left(\frac{1}{s_{l+1}} - \frac{1}{s_k} \right) \quad (k=1, \ldots, l), \quad \gamma_0^{(l)} = -\sum_{i=1}^{l} \gamma_i^{(l)}. \tag{5.24}$$

The above recursions can be proved if we compare (5.19) and (5.22). A weak point of this BDF and forward formula is that once we want to increase the order, even if only by one, then all the DF coefficients have to be recalculated.

Formulae (5.18), (5.19) and (5.21), (5.22) can be easily reduced to a fixed step. For example, fixed step 1st, 2nd and 3rd order forward formulae are

$$
\begin{aligned}
x_n &= 2x_{n-1} - x_{n-2} \\
x_n &= 3x_{n-1} - 3x_{n-2} + x_{n-3} \\
x_n &= 4x_{n-1} - 6x_{n-2} + 4x_{n-3} - x_{n-4}
\end{aligned}
\tag{5.25}
$$

while 2nd and 3rd order BDFs take the form

$$
\begin{aligned}
\dot{x}_n &= \frac{1}{h}\left(\frac{3}{2} x_n - 2x_{n-1} + \frac{1}{2} x_{n-2} \right) \\
\dot{x}_n &= \frac{1}{h}\left(\frac{11}{6} x_n - 3\, x_{n-1} + \frac{3}{2} x_{n-2} - \frac{1}{3} x_{n-3} \right).
\end{aligned}
\tag{5.26}
$$

5.1.2 Realization of an algorithm for integrating OADE

In circuit simulation, equations are created automatically from the netlist. Their creation can be viewed as a separate problem, to be discussed later on. In this section we discuss the main blocks of time-domain transient analysis algorithms, temporarily assuming that circuit OADEs are already known (e.g. analyticaly).

Algorithm 5.1 General method for integrating OADEs

data: initial conditions z_0, initial time step h, range of integration from 0 to *timemax*, maximum number of NR iterations *nitermax* at each time point, accuracy indexes for a test for convergence of NR iterations, accuracy indexes for controlling the time-step, minumum and maximum acceptable step;

initialization of the past storing tables by means of initial conditions;

$time = 0$;

for $n = 1$ **to** *ntimemax* **do** {main loop over time}
begin
    calculation of coefficients of DF and forward formula for prediction;
    calculation of $\mathbf{x}_n^0$ from the forward formula (prediction);
b: $time = time + h$;
    for $k = 1$ **to** *nitermax* **do** {NR loop}
    begin
        creation of equations discretized with DF $\dot{\mathbf{z}}_n = \gamma_0(\mathbf{z}_n - \mathbf{d}_z)$ and linearized in a vicinity of $\mathbf{x}_n^0$;
        solving the above equations for a kth iterative solution $\mathbf{x}_n^1$;
        if (test for NR loop convergence satisfied) **then goto** a;
    end
    if (step after being shortened would remain within the acceptable range) **then**
    begin
        $time = time - h$;
        $h = new\_shortened\_step$;
        goto b;
    end
    stop {no NR loop convergence with the shortest step}
a: storage of $\mathbf{x}_n^1$ in the solution buffer or on disk;
    if ($time \geq timemax$) **then stop** {successful termination}
    calculation of $\mathbf{z}_n = \mathbf{z}(\mathbf{x}_n^1)$ and of $\dot{\mathbf{z}}_n = \gamma_0(\mathbf{z}_n - \mathbf{d}_z)$ from DF;
    updating the past storing tables;
    selection of a new step h;
    updating of the table storing preceding steps;
end

This enables us to better understand the discretization and explains how to handle dynamic problems, not necessarily circuit simulation. Thus, let the problem be generally described by implicit equations (5.1). We will discuss Algorithm 5.1 in which an implicit DF (5.10) has been used.

As the first stage, the OADEs (5.1) are discretized and derivatives $\dot{\mathbf{z}}$ are substituted from the DF. In this way those substitute algebraic equations are solved by means of the Newton-Raphson (NR) method (see Chapter 4), starting from a rough solution obtained from a prediction block. Due to the nature of DF, which includes the past storing component, and because of the step-wise advancing in time, updating of the past storing tables (or structures) is necessary.

5.1.2.1 Discretization of OADEs

The first approach to the concept of discretization of OADEs has been presented in Section 5.1.1.1, where the connection between a function and its derivative, by means of polynomial interpolation, has been established. This connection enabled us to eliminate derivatives from equations. We call this manipulation the discretization, because it is based on entering a domain of discrete instants. Hence, differential equations are replaced by difference equations. Let us consider a system of n OADEs (5.1) with n unknowns $\mathbf{x}$ and m state variables $\mathbf{z}$, which are functions $\mathbf{z}(\mathbf{x})$.

Let us consider a simple nonlinear dynamic circuit shown in Figure 5.2, where $I_1(t)$ is a current source, $\varphi_3(i_3)=L_3 i_3$ is a flux linked with a linear coil, and $q_4(v_4)$ is the charge of a nonlinear capacitance. Modified nodal equations (see Section 1.3.4) can be written as follows

$$\begin{aligned} \dot{q}_4(v_{n1}) + i_3 + G_2 v_{n1} &= I_1(t) \\ \dot{\varphi}_3(i_3) - v_{n1} &= 0 \end{aligned} \tag{5.27}$$

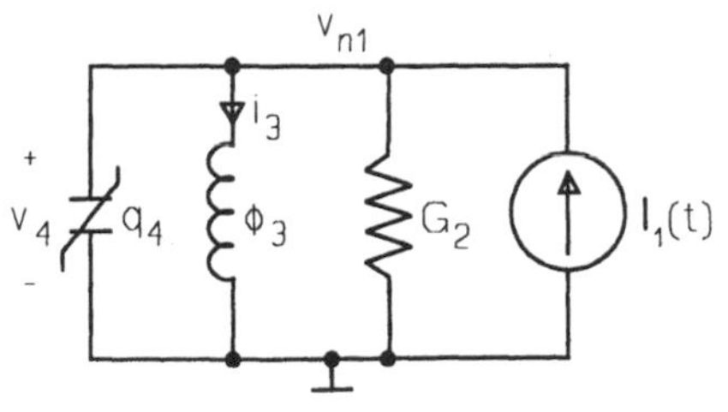

Figure 5.2 A simple dynamic nonlinear circuit

where: $\mathbf{x}=[v_{n1}, i_3]^t$ are circuit variables, while $\mathbf{z}(\mathbf{x})=[q_4(v_{n1}), \varphi_3(i_3)]^t$ are state variables (we use this term without any regard to state equations, which are beyond our consideration). Notice, that selection of state variables is arbitrary to some degree, i.e. we can transform $\dot{q}_4=(\mathrm{d}q_4/\mathrm{d}v_4)\dot{v}_4$, $\dot{\varphi}_3=L_3\dot{i}_3$ and obtain

$$\frac{\mathrm{d}q_4(v_{n1})}{\mathrm{d}v_{n1}}\dot{v}_{n1}+i_3+G_2v_{n1}=I_1(t) \qquad (5.28)$$
$$L_3\dot{i}_3-v_{n1}=0$$

where state variables $\mathbf{z}=[v_1, i_3]^t$ are now identical with $\mathbf{x}$.

To prepare an OADE for numerical solution we start by writing it at an instant t_n. This can be expressed in the following form:

$$\mathbf{F}[\mathbf{x}_n, \dot{\mathbf{z}}(\mathbf{x}_n), t_n]=\mathbf{0}. \qquad (5.29)$$

It is an algebraic relation between functions $\mathbf{x}_n$ and derivatives $\dot{\mathbf{z}}_n$. Combining this relation with the DF $\dot{\mathbf{z}}(\mathbf{x}_n)=\gamma_0[\mathbf{z}(\mathbf{x}_n)-\mathbf{d}_z]$, (a counterpart of (5.10) for the state variables $\mathbf{z}$) derivatives are eliminated and the following purely algebraic equations appear

$$\mathbf{F}(\mathbf{x}_n, \gamma_0[\mathbf{z}(\mathbf{x}_n)-\mathbf{d}_z], t_n)=\mathbf{0}. \qquad (5.30)$$

Differentiation formulae introduce into discretized equations (5.30) the past storing vector $\boldsymbol{d}_z$, corresponding to the state variables $\mathbf{z}$

$$\mathbf{d}_z=\sum_{k=1}^{p}(\alpha_k\mathbf{z}_{n-k}+\beta_k\dot{\mathbf{z}}_{n-k}), \qquad (5.31)$$

which is a linear combination of some preceding samples of solution. As has been already discussed, different DF will need the storage of different sets of samples. Each state variable from $\mathbf{z}$ will have its own past storing variable from $\mathbf{d}_z$.

For example, circuit equations (5.27) need two such variables: $d_{q4}=\sum_{k=1}^{p}(\alpha_k q_{4,n-k}+\beta_k\dot{q}_{4,n-k})$ corresponding to the capacitor charge q_4, and $d_{\varphi3}=\sum_{k=1}^{p}(\alpha_k\varphi_{3,n-k}+\beta_k\dot{\varphi}_{3,n-k})$ corresponding to the coil flux φ_3. Writing equations (5.27) for the instant t_n and substituting derivatives from DF, the following algebraic formula is obtained

$$\gamma_0[q_4(v_{n1,n})-d_{q4}]+i_{3,n}+G_2v_{n1,n}=I_1(t_n) \qquad (5.32)$$
$$\gamma_0[\varphi_3(i_{3,n})-d_{\varphi3}]-v_{n1,n}=0$$

where in $v_{n1,n}$ the first n stands for the nodal voltage v_{n1} while the second one denotes the instant t_n.

Notice, that the discretization involved means that the relation between a function and its derivative is not exactly differential but rather calculated from interpolation. This introduces an error of interpolation of an exact solution with a polynomial, known as a Local Truncation Error. Truncation here means ending an infinite Taylor expansion of a solution after a finite number of terms.

5.1.2.2 Linearization of discretized equations for NR iterations

Discretization produces nonlinear algebraic equations (5.30). They can be solved using the Newton-Raphson (NR) method (see Chapter 4). It is quite efficient in this application, because at each instant extrapolation provides a good starting point for iterations. Denoting the last, i.e. the pth solution by $\mathbf{x}_n^{(p)}$ and linearizing (5.30) in the vicinity of this value we obtain iterative equations

$$\left[\frac{\partial \mathbf{F}(\mathbf{x}_n^{(p)}, \gamma_0[\mathbf{z}(\mathbf{x}_n^{(p)})+\mathbf{d}_z], t_n)}{\partial \mathbf{x}_n} + \gamma_0 \frac{\partial \mathbf{F}(\mathbf{x}_n^{(p)}, \gamma_0[\mathbf{z}(\mathbf{x}_n^{(p)})-\mathbf{d}_z], t_n)}{\partial \dot{\mathbf{z}}_n} \frac{\mathrm{d}\mathbf{z}(\mathbf{x}_n^{(p)})}{\mathrm{d}\mathbf{x}_n}\right] \times$$
$$\times (\mathbf{x}_n^{(p+1)} - \mathbf{x}_n^{(p)}) = -\mathbf{F}(\mathbf{x}_n^{(p)}, \gamma_0[\mathbf{z}(\mathbf{x}_n^{(p)})+\mathbf{d}_z], t_n). \quad (5.33)$$

Considering the example discussed in Section 5.1.2.1, discretized equations (5.32) give, after linearization (5.33), the following equation

$$\begin{bmatrix} \gamma_0 \dfrac{\mathrm{d}q_4(v_{n1,n}^{(p)})}{\mathrm{d}v_{n1,n}} + G_2 & 1 \\ -1 & \gamma_0 \dfrac{\mathrm{d}\varphi_3(i_{3,n}^{(p)})}{\mathrm{d}i_{3,n}} \end{bmatrix} \begin{bmatrix} v_{n1,n}^{(p+1)} - v_{n1,n}^{(p)} \\ i_{3,n}^{(p+1)} - i_{3,n}^{(p)} \end{bmatrix} = \quad (5.34)$$
$$= \begin{bmatrix} I_1(t_n) - \gamma_0[q_4(v_{n1,n}^{(p)}) - d_{q4}] - i_{3,n}^{(p)} - G_2 v_{n1,n}^{(p)} \\ -\gamma_0[\varphi_3(i_{3,n}^{(p)}) - d_{\varphi 3}] + v_{n1,n}^{(p)} \end{bmatrix}.$$

Newton-Raphson iterations over p=0, 1, ..., according to (5.33), will continue until a test for convergence is satisfied or a maximum number of iterations reached, as has been presented in Chapter 4. If convergence is not reached, the current time is reset back to the preceding point, the time step is shortened (e.g. 10 times) and new NR iterations at this back-spaced point are performed. This is possible, if the shortened step does not exceed the minimum acceptable step size. Otherwise, a fatal execution error arises. Theoretically, if a circuit response is continuous, then a solution after shortening the step approaches the extrapolated starting solution. Hence, this manipulation should guarantee convergence. In practice, when the slopes of signals are too rapid (e.g. in very small time constant cases) the minimum available step size may not be sufficient. On the other hand, analysis with very short steps becomes extremely slow.

Choice of a good starting point $\mathbf{x}_n^{(0)}$ for NR iterations (5.33) is essential for efficient time-domain analysis. The starting point should be as close to the true solution as possible. This, so-called prediction can use explicit DFs, which provide an extrapolation of the solution at t_n by means of a polynomial spanned over a number of preceding samples $\mathbf{x}_{n-1}, \mathbf{x}_{n-2}, \ldots$. The degree of the polynomial is known as the order of prediction. Forward formulae (5.21) are ubiquitous ways of prediction. In simple programs the linear prediction $\mathbf{x}_n^{(0)} = \alpha_1 \mathbf{x}_{n-1} + \alpha_2 \mathbf{x}_{n-2}$ is used. In advanced algorithms with local truncation error control, prediction should be of the same order as the DF used for integration.

5.1.2.3 Updating of the past storing tables

The DF so far introduced, before calculations at point t_n can be started, need a set of state variable values and sometimes their derivatives at a number of preceding points. Which quantities and at what time points have to be stored, depends on the specific DF. The trapezoidal DF (5.15) needs state variables and their derivatives at one point t_{n-1}, implicit Euler (5.16) uses only state variables at this point, BDF of order l (5.18) needs state variables at l recent points, while the lth order forward formula (5.21) uses the most recent $l+1$ points. If derivatives have to be stored, then they are calculated from DF after termination of the solution at each time-point, as it has been given in Algorithm 5.1.

Let m preceding n-dimensional state vectors and their derivatives be stored in tables $ZM[n, m]$, $DM[n, m]$. During calculations at point t_n, they are filled by solutions at points $t_{n-1}, t_{n-2}, \ldots$. After the solution $\mathbf{z}_n$, $\dot{\mathbf{z}}_n$ has been accomplished, the contents of the tables have to be shifted by one column and new solutions are added, as shown in Algorithm 5.2. Once the past storing tables have been updated the past storing variables $\mathbf{d}_z$ are recalculated, according to equation (5.31).

Algorithm 5.2 Updating the past storing tables

```
for i=1 to n do
begin
   for j=m downto 2 do
   begin
      ZM[i,j]=ZM[i,j-1]; DM[i,j]=DM[i,j-1];
   end
   ZM[i,1]=z_{n,i}; DM[i, 1]=ż_{n,i};
end
```

In variable-step analysis, the table of $m-1$ partial sums of steps is used for calculating the coefficients of DF. Before discretization at t_n this table contains: $HS(1)=t_{n-1}-t_{n-2}=h_{n-1}$, $HS(2)=t_{n-1}-t_{n-3}=h_{n-1}+h_{n-2}$, ..., $HS(m-1)=t_{n-1}-t_{n-m}=h_{n-1}+...+h_{n-m+1}$. It is used together with the current step h_n for calculation of coefficients of a DF at instant t_n. At instant t_{n+1} table HS should contain: $HS(1)=t_n-t_{n-1}=h_n$, $HS(2)=t_n-t_{n-2}=h_n+h_{n-1}$, ..., $HS(m-1)=t_n-t_{n-m+1}=h_n+...+h_{n-m+2}$. Hence, we can write Algorithm 5.3 for its updating.

Algorithm 5.3 Updating of partial sums of steps

```
for i=m-1 downto 2 do HS[i]=HS[i-1]+h_n;
HS[1] = h_n;
```

5.1.3 BDF based on Newton's interpolation

5.1.3.1 Newton's interpolation

Newton's interpolation is an alternative to Lagrange interpolation where an lth order polynomial is fitted to $l+1$ constraints $x_n, ..., x_{n-l}$ at points $t_n, ..., t_{n-l}$. The polynomial is calculated in a Newton form. To explain the Newton form of polynomial, let us consider an lth degree polynomial $x(t)$. To look for its Newton form we divide it $l$times successively by $t-t_n$, $t-t_{n-1}$, ... up to $t-t_{n-l+1}$. We denote residua from these divisions by r_n^k $(k=0, ..., l)$.

For example, let $x(t)=7+t+2t^2-3t^3$, $t_n=1$, $t_{n-1}=2$, $t_{n-2}=3$. From the first division by $t-1$ we obtain $x(t)=7+(t-1)(-t-3t^2)$. Then, dividing by $t-2$ we find $x(t)=7+(t-1)[-14+(t-2)(-7-3t)]$. Finally, dividing by $t-3$ we obtain $x(t)=7+(t-1)\{-14+(t-2)[-16+(t-3)(-3)]\}$, or equivalently $x(t)=7-14(t-1)-16(t-1)(t-2)-3(t-1)(t-2)(t-3)$. This is the Newton form of polynomial, and in general it can be written as

$$x(t) = r_n^0 + \sum_{k=1}^{l} r_n^k (t-t_n)(t-t_{n-1})\dots(t-t_{n-(k-1)}), \tag{5.35}$$

where r_n^k are the residua of subsequent divisions. Newton's interpolation problem is to obtain residua directly from a set of $l+1$ constraints.

To solve this problem a triangular system of quotients can be created. Let the given constraints $x(t_{n-k})=x[n-k]$ $(k=0, 1, \dots)$ be of zero order quotients. First order quotients will be

$$x[n, n-1] = \frac{x[n]-x[n-1]}{t_n - t_{n-1}},\; x[n-1, n-2] = \frac{x[n-1]-x[n-2]}{t_{n-1} - t_{n-2}},\dots \tag{5.36}$$

Then, we create the second order quotients

$$\begin{aligned} x[n, n-1, n-2] &= \frac{x[n, n-1]-x[n-1, n-2]}{t_n - t_{n-2}}, \\ x[n-1, n-2, n-3] &= \frac{x[n-1, n-2]-x[n-2, n-3]}{t_{n-1} - t_{n-3}}, \;\dots \end{aligned} \tag{5.37}$$

and so on. In general, the development in this manner can be expressed as the recursion

$$\begin{aligned} &\text{let } x[i]=x(t_i) \;\; (i=n, \dots, n-l) \\ &\text{for } k=1, 2, \dots, l \\ &\text{for } i=n, n-1, \dots, n-k \\ &x[i, \dots, i-k] = \frac{w[i, \dots, i-k+1] - w[i-1, \dots, i-k]}{t_i - t_{i-k}}. \end{aligned} \tag{5.38}$$

It generates quotients of orders from 0 to l spanned on points from t_n to t_{n-l}. Each kth order quotient is obtained from the difference between two quotients of $(k-1)$th order, divided by the difference between two time points. The first of these quotients is from the same instant as the quotient being calculated, while the second one is from the preceding instant. In the denominator the first instant is the current one while the second is the oldest one influencing the quotient involved. After the whole triangular structure of quotients has been set up, a sequence of quotients at the point t_n gives the residual coefficients of the Newton form of an interpolating polynomial, i.e.

$$r_n^k = x[n, ..., n-k] \quad (k = 0, ..., l). \tag{5.39}$$

This result can be proved after some algebra, based on manipulating the Newton form (5.35).

As introduced quotients are convenient and easy to understand but they need too large a computational effort. To optimize its efficiency, the Newton interpolation has been reformulated using the concept of Scaled Differences (SD). Let $g_n^k = (t_n - t_{n-1})(t_n - t_{n-2}) \cdot ... \cdot (t_n - t_{n-k})$ be the *k*th order coefficients at instant t_n. Then, the term $\Delta_n^k = x[n, ..., n-k]\, g_n^k$ is called the *k*th order SD at point t_n. Multiplying by g_n^k both sides of the recursive relation (5.38), due to SD definition, we obtain the recursive expression

$$x[n, ..., n-k]\, g_n^k = x[n, ..., n-k+1]\, g_n^{k-1} - x[n-1, ..., n-k]\, g_n^{k-1}, \tag{5.40}$$

where on the LHS we have the *k*th order SD at t_n expressed as the $(k-1)$th order SD at t_n minus $x[n-1, ..., n-k]\, g_{n-1}^{k-1} (g_n^{k-1} / g_{n-1}^{k-1})$, the latter being the $(k-1)$th order SD at t_{n-1} multiplied by a scaling factor denoted by α_{k-1}. In general, this factor is defined as

$$\alpha_0 = 1, \quad \alpha_i = \frac{g_n^i}{g_{n-1}^i} = \frac{(t_n - t_{n-1}) \dots (t_n - t_{n-i})}{(t_{n-1} - t_{n-2}) \dots (t_{n-1} - t_{n-i-1})}. \tag{5.41}$$

Finally, we can use the recursive relationship between SD, obtained from (5.40)

$$\Delta_n^k = \Delta_n^{k-1} - \alpha_{k-1} \Delta_{n-1}^{k-1} \tag{5.42}$$

instead of (5.38), to generate a triangular system of SDs. Hence, residual coefficients of the Newton polynomial (5.35), instead of using the form (5.39), can be easily obtained from SD as

$$r_k = \frac{\Delta_n^k}{g_n^k} \quad (k=0, ..., l). \tag{5.43}$$

5.1.3.2 Differentiation formulae based on scaled differences

Newton's formulation of polynomial interpolation gives a past storing system for BDF quite different from the Lagrange formulation. Let us start from the extrapolation problem involved in prediction of starting NR iterations. Let the solution $x_{n-1}, \ldots, x_{n-1-l}$ be given at $l+1$ points $t_{n-1}, \ldots, t_{n-l-1}$. To match a Newton polynomial to these solutions we have to write the similar expression as (5.35), but with coefficients corresponding to instant t_{n-1} (not t_n)

$$x(t) = r_{n-1}^{0} + \sum_{k=1}^{l} r_{n-1}^{k} (t-t_{n-1}) \ldots (t-t_{n-k}) \tag{5.44}$$

where, due to (5.43), $r_{n-1}^{k} = \Delta_{n-1}^{k} / g_{n-1}^{k}$. Thus, calculating $x(t_n)$ we obtain the lth order forward predicting formula

$$x_n = \sum_{k=0}^{l} \alpha_k \Delta_{n-1}^{k} \tag{5.45}$$

using the coefficients (5.41), and $l+1$ SD of order from 0 to l, which store the past solutions. This forward formula is an alternative to formula (5.21), where the past was stored in the form of solutions at $l+1$ preceding points.

As can be seen from this, the SDs form the past storing table. Hence, they will have to be updated at subsequent time points. Recalling relation (5.42) and applying it recursively to Δ_n^{k-1}, Δ_n^{k-2}, etc. on the RHS we obtain the updating rule

$$\Delta_n^{k} = \Delta_n^{0} - \sum_{j=0}^{k-1} c_n^{j} \Delta_{n-1}^{j} . \tag{5.46}$$

This rule includes a system of SDs corresponding to preceding points and scaling factors. Moreover, let us notice that $\Delta_n^{0} = x_n$.

To obtain the lth order BDF we span the Newton polynomial over points $x_n, \ldots, x_{n-l}$. We obtain the formula (5.35), and then calculate its derivative at instant t_n: $\dot{x}(t_n) = \sum_{k=1}^{l} r_n^{k} (t_n - t_{n-1}) \ldots (t_n - t_{n-k+1})$. Substituting (5.43) and (5.41) for the above we obtain the expression $\dot{x}(t_n) = \sum_{k=1}^{l} \Delta_n^{k} / (t_n - t_{n-k})$. To obtain a DF, the SDs have to be expressed in terms of their preceding generation using (5.46)

$$\dot{x}(t_n) = \sum_{k=1}^{l} \frac{\Delta_n^{0}}{t_n - t_{n-k}} - \sum_{k=1}^{l} \sum_{j=0}^{k-1} \frac{\alpha_j}{t_n - t_{n-k}} \Delta_{n-1}^{j} . \tag{5.47}$$

Interchanging the order of summation we obtain

$$\dot{x}(t_n) = x_n \sum_{k=1}^{l} \frac{1}{t_n - t_{n-k}} - \sum_{j=0}^{l-1} \Delta_{n-1}^{j} \, \alpha_j \sum_{k=j+1}^{l} \frac{1}{t_n - t_{n-k}} . \tag{5.48}$$

Then, after new coefficients have been introduced

$$\gamma_j = \sum_{k=j+1}^{l} \frac{1}{t_n - t_{n-k}} \tag{5.49}$$

the following BDF appears

$$\dot{x}_n = \gamma_0 x_n - \sum_{j=0}^{l-1} \alpha_j \gamma_j \Delta_{n-1}^{j} . \tag{5.50}$$

The introduction of BDF and its respective predicting forward formula is very advantageous. For different orders of BDF, the coefficients of the BDF and the past storing SD are always the same, since they depend only on discretization steps. Only their number need be adjusted to suit the current BDF order. The organization of a practical algorithm will be as follows. At each time point calculations start from updated SD. We calculate α_j, γ_j, perform prediction (5.45) and enter the discretization procedure using BDF (5.50). After NR iterations are accomplished, the solution obtained is used to update SD, according to (5.46), and the procedure is repeated for the next time point. Such an algorithm has been implemented in NAP2 [5.14] and OPTIMA.

To start the procedure we initialize SD with $\Delta_0^0 = x_0$, where x_0 is the assumed initial condition at instant $t_0 = 0$. Calculations at $t_1 = h_1$ are started from zero-order prediction $x_1^{(0)} = \alpha_0 \Delta_0^0 = 1\, x_0$, and then equations are discretized with the first order BDF $\dot{x}_1 = \gamma_0 x_1 - \alpha_0 \gamma_0 \Delta_0^0$ which is the implicit Euler DF $\dot{x}_1 = (1/h_1)x_1 - (1/h_1)x_0$. After x_1 has been calculated, we update (5.46): $\Delta_1^0 = x_1$, $\Delta_1^1 = x_1 - \alpha_0 \Delta_0^0 = x_1 - x_0$. Now, the next step h_2 is chosen. The first order prediction $x_2^{(0)} = \alpha_0 \Delta_1^0 + \alpha_1 \Delta_1^1$ (that is $x_2^{(0)} = x_1 + (h_1/h_2)(x_1 - x_0)$) is executed and the first order discretization is once more applied $\dot{x}_2 = \gamma_0 x_2 - \alpha_0 \gamma_0 \Delta_1^0$ (that is $\dot{x}_2 = (1/h_2)x_2 - (1/h_2)x_1$). After x_2 has been calculated, SD can be updated, and the process is continued with order 1 or 2. At each time step the possible order increases by 1.

5.1.3.3 Backward differentiation formulae based on predictors

Backward DF can be expressed in terms of different past storing quantities. In Gear's works [5.4, 5.8] the Nordsieck vector was used. In the preceding section the SD vector has been introduced. In this section another equivalent approach of W.M.G. Van Bokhoven is recomended [8.1], where the past solutions are stored in a vector of predictors. Let us recall equation (5.47), where the internal sum in the right-hand component (comparing with forward formula (5.45)) appears to contain a predictor of order $k-1$. In general, a vector of l predictors of orders from 0 to $l-1$ can be defined as

$$\bar{x}_n^k = \sum_{j=0}^{k-1} \alpha_j \Delta_{n-1}^j \tag{5.51}$$

(to have error estimation, the $l+1$ predictor will be useful also). Hence, (5.47) gives the predictors-based BDF

$$\dot{x}_n = \gamma_0 x_n - \sum_{k=1}^{l} \frac{1}{t_n - t_{n-k}} \bar{x}_n^k . \tag{5.52}$$

To efficiently manipulate predictors, we should update them recursively. From (5.51) the jth predictor can be expressed in terms of the $j-1$ one, namely $\bar{x}_n^j = \bar{x}_n^{j-1} + \alpha_{j-1}\Delta_{n-1}^{j-1}$, where SD is equal to $x_{n-1} - \bar{x}_{n-1}^{j-1}$, due to (5.46) which has the form $\Delta_{n-1}^k = \Delta_{n-1}^0 - \bar{x}_{n-1}^k$. Hence, for given $\bar{x}_{n-1}^1, \ldots, \bar{x}_{n-1}^l$ and calculated x_n we obtain predictors

$$\begin{aligned} &\bar{x}_n^1 = x_n, \\ &\text{for } j = 2, \ldots, l \\ &\quad \bar{x}_n^j = \bar{x}_n^{j-1} + \alpha_{j-1}(x_{n-1} - \bar{x}_{n-1}^{j-1}) \end{aligned} \tag{5.53}$$

updated at time point t_n. Initialization of the algorithm needs one predictor $\bar{x}_1^1 = x_0$. After the first step we already have two predictors $\bar{x}_2^1 = x_1$, $\bar{x}_2^2 = \bar{x}_2^1 + \alpha_1(x_1 - \bar{x}_1^1)$, etc.

Backward DFs introduced in Section 5.1.3, can be implemented in Algorithm 5.1. All the BDFs and forward formulae (Lagrangian, SD and predictor based) differ only in the principles used to store the past, and then in its updating. All of them are of the general form (5.10). The only method-specific features of their implementation will be the past storing matrices, and procedures for their updating and calculating the past storing variable d_x of (5.10). The rest will remain the same.

To test your understanding, try to solve the following problems:

- Consider the general three-step DF. Explain how its coefficients can

be obtained from matching an interpolating polynomial to certain previous points. Using the basic polynomials technique calculate the coefficients of: 2nd order predictive forward formula, 3rd order BDF, any 4th order DF. Show how many degrees of freedom remain in each case. Reduce the DF obtained to the fixed step case.
- For each of the above derived DFs explain what should be stored and how the storing tables can be updated at each time step.
- Take any first order nonlinear ODE and using it explain discretization and linearization for NR iterations. Write discretized and linearized equations for the implicit Euler DF, the trapezoidal DF and for a general notation of DF involved.
- Explain prediction and its purpose.
- Compare construction of the past storing variables d_x and their updating in three of the BDFs introduced here.

5.2 Formulation of circuit equations for time-domain transient analysis

We have assumed that ordinary algebraic differential circuit equations for time domain transient analysis are given analytically. Then, we have demonstrated discretization and linearization using analytic circuit description. However, such an approach is quite inefficient in computer simulation. In this section, we are going to present general methods of creating discretized and linearized equations directly from circuit net-list definitions.

5.2.1 The companion circuit method

5.2.1.1 The concept of a companion circuit

Discretization and linearization dedicated to NR iterations were operations on circuit equations. If we are not concerned with the particular circuit, then these operations can be applied to general, canonical circuit description (see Chapter 1). Hence, let us recall the general canonical description (1.23) of a lumped circuit

$$\begin{aligned} &\mathbf{A\,i} = \mathbf{0} \\ &\mathbf{B\,v} = \mathbf{0} \\ &\mathbf{f}\left(\mathbf{i}, \mathbf{v}, \frac{d\mathbf{q}(\mathbf{v})}{dt}, \frac{d\boldsymbol{\psi}(\mathbf{i})}{dt}, \mathbf{s}(t)\right) = \mathbf{0} \end{aligned} \tag{5.54}$$

with branch currents $\mathbf{i}$, voltages $\mathbf{v}$, and state variables $\mathbf{q}(\mathbf{v})$, $\boldsymbol{\psi}(\mathbf{i})$. They can be charges and flux linkages, but also in the simplest case, they can be voltages and currents themselves. The problem of choosing state variables will be studied in detail in this section.

If we discretize (5.54), by writing it at instant t_n and substituting for the derivatives of state variables by means of differentiation formulae (5.10), then we obtain algebraic equations

$$\begin{aligned}
&\mathbf{A}\,\mathbf{i}_n = \mathbf{0}\\
&\mathbf{B}\,\mathbf{v}_n = \mathbf{0}\\
&\mathbf{f}\{\mathbf{i}_n, \mathbf{v}_n, \gamma_0[\mathbf{q}(\mathbf{v}_n)-\mathbf{d}_q], \gamma_0[\boldsymbol{\psi}(\mathbf{i}_n)-\mathbf{d}_\psi], \mathbf{s}(t_n)\} = \mathbf{0}.
\end{aligned} \tag{5.55}$$

They can be viewed as canonical equations of a circuit structurally identical with the original (5.54) but resistive, and in general nonlinear. It contains branches obtained from original branches by discretizing their equations with DFs. Branch variables $\mathbf{i}_n$, $\mathbf{v}_n$ of this circuit physically are instantaneous responses (at instant t_n) of the original circuit. However, they are not accurate due to interpolation error. This substitute circuit is called a nonlinear companion circuit.

Creation of the nonlinear companion circuit is quite easy. We take the original circuit topology as the outline, and put into it discretized branches in place of the original ones. Of course, resistive branches need no changes. Their branch variables have the subscript n denoting that signals are instantaneous, e.g. a nonlinear conductance $i_1=f_1(v_1)$ after discretization takes the form $i_{1,n}=f_1(v_{1,n})$. In the case of reactive branches a DF has to be applied. After discretization, a nonlinear capacitance $i=\mathrm{d}q(v)/\mathrm{d}t$ becomes a nonlinear conductance $i_n=\gamma_0[q(v_n)-d_q]$, where γ_0 is the DF coefficient, while d_q is the known past storing variable. In general it is equal to $d_q=\sum_{k=1}^{p}(\gamma_k\, q_{n-k}+\delta_k\,\dot{q}_{n-k})$. In the case of the trapezoidal DF we have: $\gamma_0=2/h_n$, $d_q=-2\,q_{n-1}/h_n-\dot{q}_{n-1}$. We should emphasize that companion models need appropriate past storing variables, which at a given point of analysis are constant.

To solve a nonlinear companion circuit we use the Newton-Raphson method that consists of linearization at a pth point $(\mathbf{v}_n^{(p)}, \mathbf{i}_n^{(p)})$. This results in iterative equations, which are canonical equations of a linear circuit structurally identical with the original one. They describe a linear companion circuit, called for short a companion circuit, whose response stands for an iterative solution at instant t_n. These equations arise from linearization of each nonlinear resistive branch (including discretized one) at its recent NR solution, namely

$$\mathbf{A}\,\mathbf{i}_n^{(p+1)} = \mathbf{0}$$

$$\mathbf{B}\,\mathbf{v}_n^{(p+1)} = \mathbf{0}$$

$$\begin{aligned}&\mathbf{f}\{\mathbf{i}_n^{(p+1)}, \mathbf{v}_n^{(p+1)}, \gamma_0[\mathbf{q}(\mathbf{v}_n^{(p+1)}) - \mathbf{d}_q], \gamma_0[\psi(\mathbf{i}_n^{(p+1)}) - \mathbf{d}_\psi], \mathbf{s}(t_n)\} + \\ &+ \left(\frac{\partial \mathbf{f}^{(p)}}{\partial \mathbf{v}_n} + \gamma_0 \frac{\partial \mathbf{f}^{(p)}}{\partial q}\frac{\partial \mathbf{q}^{(p)}}{\partial \mathbf{v}_n}\right)(\mathbf{v}_n^{(p+1)} - \mathbf{v}_n^{(p)}) + \\ &+ \left(\frac{\partial \mathbf{f}^{(p)}}{\partial \mathbf{i}_n} + \gamma_0 \frac{\partial \mathbf{f}^{(p)}}{\partial \psi^{(p)}}\frac{\partial \psi^{(p)}}{\partial \mathbf{i}_n}\right)(\mathbf{i}_n^{(p+1)} - \mathbf{i}_n^{(p)}) = \mathbf{0}.\end{aligned} \tag{5.56}$$

Hence, the companion circuit is built of companion branch models that are linear resistive two-poles. It can be characterized e.g. using Modified Nodal Equations (MNE) (see Chapter 2) leading to an efficient linear algebraic system (see Chapter 3).

The equations of a companion circuit can be created utilizing the stamps technique (see Section 2.1.3.4). Branches of a nonlinear dynamic circuit are usually given analytically. In such simulators as SPICE or CAzM [5.5], they can be piecewise linear. After discretization and linearization, the formulae for coefficients of the companion model arise and a corresponding stamp can be formulated. The process of automatic formulation of modified nodal companion equations can be easily computerized in this way. In Algorithm 5.1 the block of discretization and linearization of analytically given equations should be replaced by a procedure which calculates stamp components for all branches and puts them on appropriate matrix positions. Thanks to this approach, circuit equations no longer have to be known analytically.

Proceeding to companion models, we observe that time-varying sources transform into constant sources of the instantaneous value at t_n, e.g. an independent voltage source $V(t)$ becomes a respective independent source $V(t_n)$. Hence, it remains constant during NR iterations at t_n and it only changes when we proceed to the next time instant.

Resistive branches do not need DF, and so produce the same linearized iterative models as in d.c. analysis, but operating on instantaneous signals at t_n. For example, a controlled current source $i_a = \varphi(v_a, v_b)$ after discretization produces $i_{a,n} = \varphi(v_{a,n}, v_{b,n})$, while after linearization it gives

$$i_{a,n}^{(p+1)} = \varphi^{(p)} + (\partial\varphi^{(p)}/\partial v_{a,n})(v_{a,n}^{(p+1)} - v_{a,n}^{(p)}) + (\partial\varphi^{(p)}/\partial v_{b,n})(v_{b,n}^{(p+1)} - v_{b,n}^{(p)})$$

where $\varphi^{(p)} = \varphi(v_{a,n}^{(p)}, v_{b,n}^{(p)})$.

There is an essential difference between iterative d.c. and linearized companion models only in the case of reactive branches. In the next two sections we will show companion models for typical reactive branches. Their form depends on the selection of state variables. In practice, we have to choose between charge-flux and voltage-current state variables. The base of selection will be explained later on.

5.2.1.2 Companion models of branches in terms of charge-flux state variables

As will be shown, due to stability and conservation of charge, the best choice of state variables will be to select charges and fluxes. Thus, we start our consideration from such a case. Let us consider the nonlinear capacitance shown in Figure 5.3(a), which is the most common element of MOS ICs. It stores a charge $q=q(v)$, imposed by a terminal voltage, while its current $i=\dot{q}$ is the time derivative of the charge. Discretizing q at instant t_n and substituting the DF (5.10) for the charge derivative, we obtain the nonlinear current-defined branch $i_n(v_n)$ shown in Figure 5.3(b)

$$i_n = \gamma_0 [q(v_n) - d_q] \tag{5.57}$$

This branch includes a term d_q storing the past values of the charge q according to the specific DF involved. For common DF (5.9) it takes the form $d_q = \sum_{k=1}^{p} (\gamma_k q_{n-k} + \delta_k \dot{q}_{n-k})$. Due to the adoption of charge-flux state variables, the charges (and their derivatives if necessary) are those quantities which have to be stored in the tables of the algorithm.

After linearization in the vicinity of a given pth iterative voltage $v_n^{(p)}$, the final companion model arises

$$i_n^{(p+1)} = \gamma_0 [q(v_n^{(p)}) - d_q] + \gamma_0 \frac{dq(v_n^{(p)})}{dv_n} (v_n^{(p+1)} - v_n^{(p)}) . \tag{5.58}$$

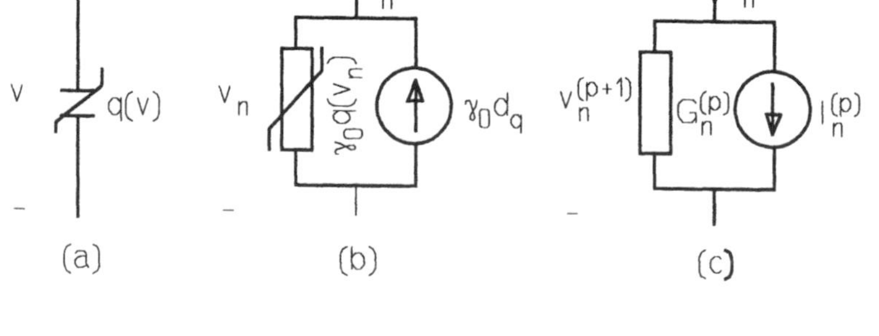

Figure 5.3 Modeling of a nonlinear capacitance for time-domain analysis: (a) the original, (b) nonlinear companion model, (c) linear companion model

This model is shown in Fig. 5.3(c), and it comprises a conductance $G_n^{(p)}=\gamma_0 C(v_n^{(p)})$, and a parallel current source $I_n^{(p)}=\gamma_0[q(v_n^{(p)})-d_q]-G_n^{(p)}v_n^{(p)}$. Both components depend on dynamic capacitance, which is a derivative of charge with respect to voltage: $C(v_n^{(p)})=\mathrm{d}q(v_n^{(p)})/\mathrm{d}v_n$.

Let us consider an example of capacitance appearing in a model of a bipolar junction transistor $i=\dot{q}$, $q=q_0\exp(v/V_T)$. Assuming Implicit Euler DF $\dot{q}_n=(q_n-q_{n-1})/h_n$, where $q_n=q(v_n)$, and substituting it for the charge derivative, we obtain: $i_n=[q_0\exp(v_n/V_T)-q_{n-1}]/h_n$. After Taylor linearization, the following companion model arises:

$$i_n=[q_0\exp(v_n^{(p)}/V_T)-q_{n-1}]/h_n+[(1/h_n)(q_0/V_T)\exp(v_n^{(p)}/V_T)](v_B^{(p+1)}-v_n^{(p)}).$$

This is a current-defined branch as shown in Figure 5.3(c).

Well-defined capacitance is expressed by means of an analytical formula for charge, which appears only on the equations' RHS. After convergence is reached, the current of the capacitance is equal to $\gamma_0[q(v_n)-d_q]$. Thus, its accuracy obviously depends on DF (due to interpolation) and on the accuracy of formulae for charge, which should be exact and obey Gauss's Law. Concerning the dynamic capacitance, which can be easily derived from the charge, it appears only in the circuit matrix. Hence, some slight inaccuracy in it is acceptable, since it does not change the solution, although it can influence convergence.

Due to the NR method, there is a danger of temporarily getting out of the range where the function $q(v)$ is defined and finite. For instance, a junction capacitance $q(v)=[C\phi/(1-m)]\,[1-(1-v/\phi)^{1-m}]$ is defined only for $v<\phi$. Due to $\phi\approx 0.9$ V, and since physical voltages are about 0.7 V the region above ϕ can be easily entered. The common way of guarding against this trouble is to continuously and smoothly approximate $q(v)$ above 0.9ϕ, with a quadratic function that gives a linear approximation to the dynamic capacitance. This approximation produces a function defined everywhere and introduces error only in the forward-biased region, where junction charge is well masked by the much greater charge of carriers in the base.

These results can be easily extended to charge-type multipoles $\mathbf{q}(\mathbf{v})$ controlled by a number of independent voltages $\mathbf{v}$, which are common elements of MOSFET models. We obtain a current-defined linear resistive multipole $\mathbf{i}_n^{(p+1)}=\mathbf{G}_n^{(p)}\mathbf{v}_n^{(p+1)}+\mathbf{I}_n^{(p)}$ (extension of (5.58)), with $\mathbf{G}_n^{(p)}=\gamma_0\mathbf{C}(\mathbf{v}_n^{(p)})$ and $\mathbf{I}_n^{(p)}=\gamma_0[\mathbf{q}(\mathbf{v}_n^{(p)})-\mathbf{d}_q]-\mathbf{G}_n^{(p)}\mathbf{v}_n^{(p)}$, where $\mathbf{C}(\mathbf{v}_n^{(p)})$ is a matrix of derivatives of $\mathbf{q}$ with respect to $\mathbf{v}$. The vector $\mathbf{d}_q$ contains the past storing variables for every multipole terminal.

Omitting the analogous mathematical development we write a companion model for a nonlinear inductor $v=\varphi(i)$

$$v_n^{(p+1)}=\gamma_0[\varphi(i_n^{(p)})-d_\varphi]+\gamma_0\frac{d\varphi(i_n^{(p)})}{di_n}(i_n^{(p+1)}-i_n^{(p)}), \qquad (5.59)$$

which is a combination of series resistance $R_n^{(p)}=\gamma_0 L(i_n^{(p)})$, and a voltage source $V_n^{(p)}=\gamma_0[\varphi(i_n^{(p)})-d_\varphi]-R_n^{(p)}i_n^{(p)}$. The past storing variable d_φ is now adequate for the DF combination of fluxes and their derivatives (voltages) at preceding instants. The inductor introduces its flux $\varphi(i)$ into the vector of state variables. Extension of the above inductance to an inductive multipole (whose mutual inductance is a specific case) is a quite simple task.

5.2.1.3 Companion models of branches in terms of voltage-current state variables

In this section let us demonstrate how a companion model of a nonlinear capacitance looks if we select its voltage as the state variable. Assuming that the capacitance is written as $i=C(v)\dot{v}$, where $C(v)$ is the directly given dynamic capacitance, we can apply the DF to voltage, which is now a state variable, namely $\dot{v}_n=\gamma_0(v_n-d_v)$. The past storing variable d_v is of voltage type, and for common DF (5.9) it can be expressed as a linear combination of voltages and their derivatives at the past instants $d_v=\sum_{k=1}^{p}(\gamma_k v_{n-k}+\delta_k\dot{v}_{n-k})$. Substituting the DF for the instantaneous derivative, the nonlinear companion model arises as $i_n=\gamma_0 C(v_n)(v_n-d_v)$. It is quite different from that shown in Figure 5.3(b), obtained for a charge-type state variable. The essential difference between these two models lies in the fact that if voltage is a state variable, then charge is not determined at all, which causes numerical difficulties.

After linearization of this nonlinear companion model at the point $v_n^{(p)}$ we obtain the following expression (5.60)

$$i_n^{(p+1)}=\gamma_0 C(v_n^{(p)})(v_n^{(p)}-d_v)+\gamma_0\left[\frac{dC(v_n^{(p)})}{dv_n}(v_n^{(p)}-d_v)+C(v_n^{(p)})\right](v_n^{(p+1)}-v_n^{(p)})$$

This equation stands for a parallel combination shown in Figure 5.3(c), but with the conductance $G_n^{(p)}=\gamma_0\{(v_n^{(p)}-d_v)[dC(v_n^{(p)})/dv_n]+C(u_n^{(p)})\}$ and the source $I_n^{(p)}=\gamma_0 C(v_n^{(p)})(v_n^{(p)}-d_v)-G_n^{(p)}v_n^{(p)}$. After convergence is reached, this equation reduces to $i_n=\gamma_0 C(v_n)(v_n-d_v)$. It means that the accuracy of the current depends on the capacitance, while the charge is not considered.

A corresponding companion model for the capacitive multipole given by $\mathbf{i}=\mathbf{C}(\mathbf{v})\dot{\mathbf{v}}$ can be described by $\mathbf{i}_n^{(p+1)}=\mathbf{G}_n^{(p)}\mathbf{v}_n^{(p+1)}+\mathbf{I}_n^{(p)}$, where the conductance matrix is $\mathbf{G}_n^{(p)}=\gamma_0\{[d\mathbf{C}(\mathbf{v}_n^{(p)})/d\mathbf{v}_n](\mathbf{v}_n^{(p)}-\mathbf{d}_v)+\mathbf{C}(\mathbf{v}_n^{(p)})\}$, while the source vector $\mathbf{I}_n^{(p)}=\gamma_0\mathbf{C}(\mathbf{v}_n^{(p)})(\mathbf{v}_n^{(p)}-\mathbf{d}_v)-\mathbf{G}_n^{(p)}\mathbf{v}_n^{(p)}$. After convergence, the currents of the multipole are of the form $\mathbf{i}_n=\gamma_0\mathbf{C}(\mathbf{v}_n)(\mathbf{v}_n-\mathbf{d}_v)$. Hence, the *j*th current is a sum $i_{n,j}=\gamma_0[C_{j,1}(v_{n,1}-d_{v,1})+\ldots C_{j,m}(v_{n,m}-d_{v,m})]$, including all partial capacitances. Therefore, the accuracy of the current is determined by the accuracy of these capacitances. On the other hand, these partial capacitances are partial derivatives of the *j*th terminal charge with respect to all controlling voltages. Hence, if a multipole is defined in a capacitive form $\mathbf{i}=\mathbf{C}(\mathbf{v})\dot{\mathbf{v}}$, then we should provide that all partial capacitances $\mathbf{C}(\mathbf{v})$ originated from a certain charge $\mathbf{q}(\mathbf{v})$. Otherwise, simulation is inaccurate and beyond the physical sense.

Recalling the single capacitance we notice that the expression (5.60) for the conductance and source are complicated and they need a derivative of the capacitance (second derivative of charge). Since inaccurately calculated derivatives are acceptable in the NR method, we can introduce a simplified companion model $G_n^{(p)}=\gamma_0 C(v_n^{(p)})$, in which the derivative of capacitance has been neglected. Experiments show that this considerably improves efficiency without any loss of convergence. This approach has been tested in OPTIMA.

5.2.1.4 General example of the companion equations formulation

To demonstrate how linear companion equations for time-domain transient analysis can be formulated, let us consider the circuit shown in Figure 5.4(a). Its two nonlinear elements are taken to have the form $q_3(v_3)=q_0\exp(v_3/V_T)$, and $i_4=i_s[\exp(v_4/V_T)-1]$. We create a companion circuit shown in Figure 5.4(b), with a structure identical with the original. The voltage source gives $v_1=V_1(t_n)$; linearized conductance gives $G_{4n}^{(p)}=i_s\exp(v_{4n}^{(p)}/V_T)/V_T$, $I_{4n}^{(p)}=i_s[\exp(v_{4n}^{(p)}/V_T)-1]-G_{4n}^{(p)}v_{4n}^{(p)}$. The inductor gives $R_{2n}^{(p)}=\gamma_0 L_2$, $V_{2n}^{(p)}=-\gamma_0 L_2 d_{i2}$ (where d_{i2} stands for a variable storing the past values of the current flowing through this inductor). The nonlinear capacitance, if its charge is a state variable (Section 5.2.1.2), gives $G_{3n}^{(p)}=\gamma_0 q_0\exp(v_{3n}^{(p)}/V_T)/V_T)$, and $I_{3n}^{(p)}=\gamma_0[q_0\exp(v_{3n}^{(p)}/V_T)-d_{q3}]-G_{3n}^{(p)}v_{3n}^{(p)}$, where d_{q3} is a variable storing the past values of the charge on this capacitor. All listed elements are drawn in Figure 5.4(b). The modified nodal equations of the companion circuit thus obtained are

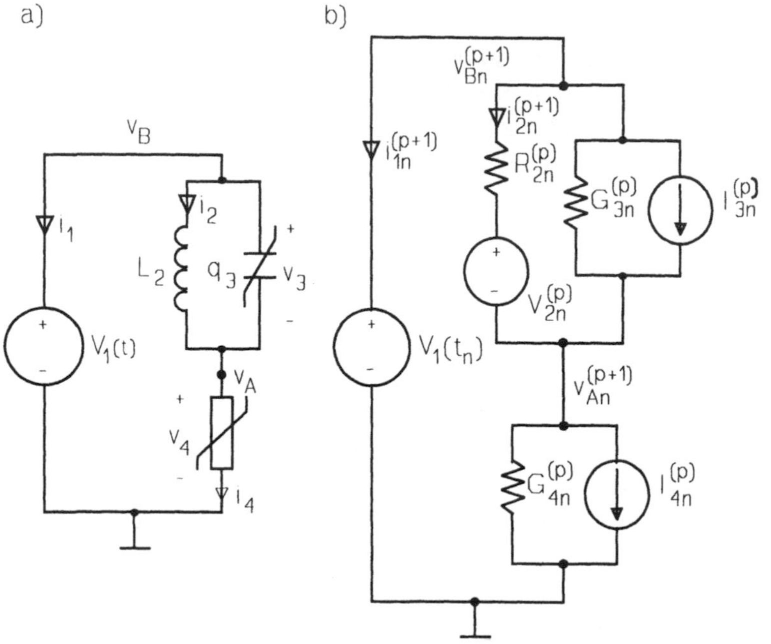

Figure 5.4 (a) A dynamic nonlinear circuit and (b) its companion counterpart

$$\begin{bmatrix} G_{4n}^{(p)}+G_{3n}^{(p)} & -G_{3n}^{(p)} & 0 & -1 \\ -G_{3n}^{(p)} & G_{3n}^{(p)} & 1 & 1 \\ 0 & -1 & 0 & 0 \\ 1 & -1 & 0 & R_{2n}^{(p)} \end{bmatrix} \begin{bmatrix} v_{An}^{(p+1)} \\ v_{Bn}^{(p+1)} \\ i_{1n}^{(p+1)} \\ i_{2n}^{(p+1)} \end{bmatrix} = \begin{bmatrix} -I_{4n}^{(p)}+I_{3n}^{(p)} \\ -I_{3n}^{(p)} \\ -V_1(t_n) \\ -V_{2n}^{(p)} \end{bmatrix}.$$

To perform time-domain analysis of the circuit given in Figure 5.4(a) these equations are repetitively solved within the NR and time loop.

5.2.1.5 Implementation in extended MNE

In the OPTIMA simulator an extension of MNE has been proposed [5.13] to adapt state variables. In simulators where models can be user-defined, and elements like $i(v, \dot{v}) = av\dot{v}$ are available, it is reasonable to treat all circuit variables as state variables. All the charges and fluxes of the nonlinear elements are also necessary and they have been added to the circuit variables.

A general nonlinear charge-defined branch $i_a = \dot{q}_a$, where $q_a = q_a(v_a, v_b)$, after discretization and linearization gives the following linear equation $i_a^{(p+1)} = \gamma_0 q_a^{(p+1)} - \gamma_0 d_{qa}$ accompanied by $q_a^{p+1)} = C_{aa} v_a^{(p+1)} + C_{ab} v_b^{(p+1)} + s_a^{(p)}$, where the auxiliary parameter is $s_a^{(p)} = q_a(v_a^{(p)}, v_b^{(p)}) - C_{aa} v_a^{(p)} - C_{ab} v_b^{(p)}$. This leads to the stamp

	m_a	n_a	m_b	n_b	q_a	
m_a					γ_0	$\gamma_0 d_{qa}$
n_a					$-\gamma_0$	$-\gamma_0 d_{qa}$
q_a	C_{aa}	$-C_{aa}$	C_{ab}	$-C_{ab}$	-1	$-s_a$

Similarily, a general nonlinear flux branch $v_a = \dot{\varphi}_a$, where $\varphi_a = \varphi_a(i_a, i_b)$, after discretization and linearization gives the following linear equation $v_a^{(p+1)} = \gamma_0 \varphi_a^{(p+1)} - \gamma_0 d_{\varphi a}$ accompanied by $\varphi_a^{(p+1)} = L_{aa} i_a^{(p+1)} + L_{ab} i_b^{(p+1)} + w_a^{(p)}$, where the auxiliary parameter is $w_a^{(p)} = \varphi_a(i_a^{(p)}, i_b^{(p)}) - L_{aa} i_a^{(p)} - L_{ac} i_c^{(p)}$. This leads to the stamp

	m_a	n_a	m_b	n_b	i_a	i_c	ϕ_a	
m_a					1			
n_a					-1			
i_a	-1	1					γ_0	$\gamma_0 d_{\phi a}$
ϕ_a					L_{aa}	L_{ac}	-1	$-w_a$

5.2.2 Numerical problems due to selection of state variables

The choice between charge-flux and voltage-current state variables is essential for the numerical properties of the integrating algorithm. It has been proved that the selection of charge-flux variables is more advantageous and it is based on the following two facts:

(i) its accuracy and stability (that is cumulative inaccuracy) is better, as has been proved by D. Calahan [5.3].

(ii) the charge of capacitive elements, which is constant in cases drawn up according to Gauss's Law (see Chapter 1), should remain conserved in time also during the simulation. However, in general this is achieved only with charge-flux state variables.

We will discuss these two problems in detail.

5.2.2.1 The problem of stability

In this section we will explain a problem pointed out for the first time by D. Calahan [5.3]. Let us consider the simple circuit shown in Figure 5.5, where a nonlinear capacitance has been loaded with a conductance. If v is a state variable, then the equation of this circuit is $C(v)\dot{v} = -Gv$, where $C(v)$, G are positive. We rewrite it in the form $\dot{v} = -Gv/C(v)$.

Let $v(t)$ be the solution of this equation under a given positive initial condition $v(0)$. We look for a perturbed solution $v(t)+\Delta v(t)$, obtained from a slightly perturbed initial condition $v(0)+\Delta v(0)$ (perturbation is sufficiently small, and then it can be handled with a linear approximation). Substituting a perturbed solution to the equation $\dot{v} = -Gv/C(v)$ we obtain a perturbed solution $\dot{v}+\Delta\dot{v} = -(Gv+G\Delta v)/C(v+\Delta v)$. Linearizing its RHS in a vicinity of $v(t)$ and subtracting the nonperturbed equation we obtain

$$\Delta\dot{v} = -\frac{G}{C(v)}\left(1 - \frac{v}{C(v)}\frac{dC(v)}{dv}\right)\Delta v . \tag{5.61}$$

Let us set up the similar equation for a charge state variable. The circuit equation now is $\dot{q}(v) = -Gv$. By similar linearization, with respect to a perturbed solution, instead of (5.61) we obtain the equation

$$\Delta\dot{v} = -\frac{G}{C(v)}\Delta v . \tag{5.62}$$

Both (5.61), and (5.62) are of the form $\Delta\dot{v} = p(v)\Delta v$ but they differ in the characteristic function p. To examine the time-profile of Δv we solve it for this variable by integration

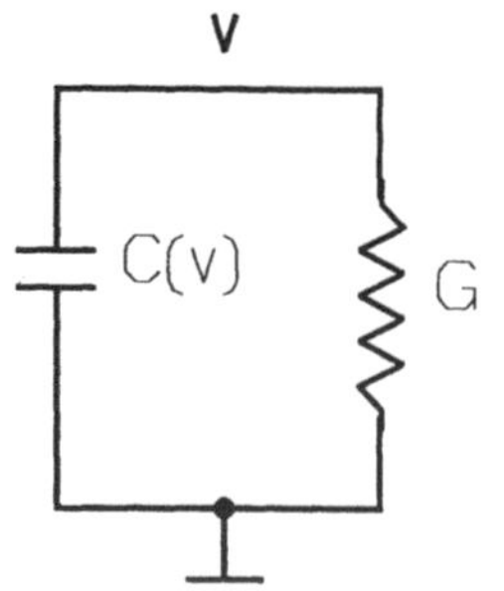

Figure 5.5 A sample for testing how selection of state variables implies stability problems

$$\Delta v(t) = \Delta v_0 \exp \int_0^t p[v(\tau)]\,d\tau . \qquad (5.63)$$

From (5.63) it can be readily shown [5.3] that, if for the response $v(t)$ the subintegral function satisfies $p(v) \leq 0$, then $\Delta v(t)$ does not increase. We call this stability. It means that any initial perturbation is not amplified during the integration process.

For a charge-type state variable it can be seen from (5.62) that p is always negative, and so provides this stability. However, if voltage is the state variable, then the sign of p in (5.61) can be positive, and variation of the capacitance with voltage can be so rapid that $[dC(v)/dv]\,v/C(v) > 1$.

5.2.2.2 Introductory charge conservation problems

Let us consider the capacitor shown in Figure 5.6(a). We surround a plate of this capacitor with a closed surface (so-called Gauss's surface). A current i flows from the outside to the inside of the surface. Due to Gauss's Law as introduced in Chapter 1 the total charge q inside the surface is constant iff the total current flowing through the surface is zero (i.e. interior of the surface is insulated from the exterior). The charge $q(v)$ is essential for the capacitor's operation. It is a function of the voltage v across the capacitor. Hence, if q is constant, then also the voltage v remains fixed. If the charge q varies in time, then a nonzero current flow i arises. This current can be expressed as the time derivative of the charge: $i = \dot{q}$. Equivalently, a difference between values of charge at the endpoints of the interval $[t_1, t_2]$ is equal to the integral of the current over this interval

$$q(v_1) - q(v_2) = \int_{t_1}^{t_2} i(t)\,dt . \qquad (5.64)$$

This relation shows that, if a physical system provides such a time-profile of current that its integral over a certain interval is zero, then the charge and voltage at the endpoints of this interval remain the same. This property is called conservation of charge.

Unfortunately, this property does not hold in circuit simulation, due to inaccurate integration. Hence, even when the integral of current is zero, in circuit simulation we may observe that $q(v_1) \neq q(v_2)$ and $v_1 \neq v_2$. Such inaccurate charge calculation can cause considerable error in the voltage balance. Inaccuracy of voltages can be considerable. This happens espe-

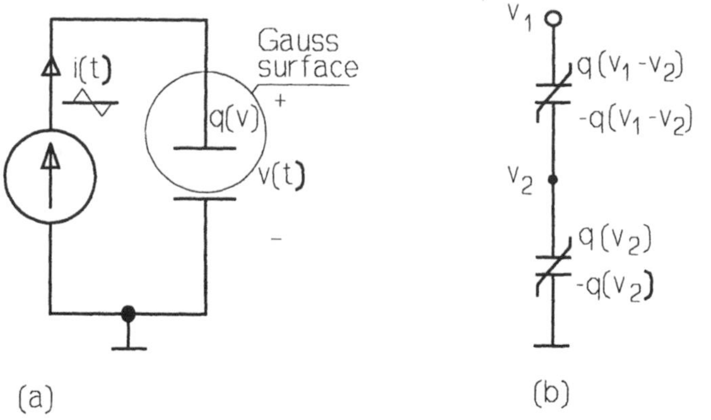

Figure 5.6 Illustration of the charge conservation problem: (a) on a common capacitor, (b) in a series connection of capacitors

cially in circuits incorporating nodes connected only with capacitors and/or having general capacitive elements whose charge is controlled by several voltages. Practical problems with conservation of charge have been observed in such MOS circuits as dynamic RAMs and switched-capacitor circuits.

As the first example, let us consider the OPTIMA task O5-1.CIR. The circuit shown in Figure 5.6(a) contains a periodical current excitation, whose waveform is built of alternate positive and negative symmetric triangle pulses. The integral of this waveform over one cycle is zero. However from simulation we can observe that the level of voltage v after each period increases slightly, due to inaccurate numerical integration by the simulator. The charge q and so the voltage v are not conserved. Decreasing the relative accuracy index ERTR, which controls the integration stepsize improves the accuracy of integration, and so we can reduce this lack of charge conservation. This so-called first type of charge nonconservation is due to the accumulation of interpolation error during analysis. As the experiment with O5-1.CIR shows this problem can appear even in linear circuits and the only way of overcoming it is to improve the accuracy of integration.

The second type of charge nonconservation arises in nonlinear circuits due to NR iteration error. As an instructive example we consider the series connection of two identical nonlinear capacitors $q(v)$ given in OPTIMA task O5-2.CIR and shown in Figure 5.6(b). Its node 1 is driven by a sinusoidal voltage source. Due to the series connection of the capacitors, their charges should be the same. This can be provided by an appropriately established voltage $v_2 = v_1/2$ for all t. Hence, satisfaction of this relation in simulation will indicate conservation of the charge. We integrate this circuit over a few cycles, and observe an error V(2)-V(1)/V(2). If both capacitors are linear, then this sum is very accurately zero. Otherwise, this error becomes nonzero at some instants, and its magnitude is dependent on the accuracy index RELTOL. This indicates the numerical nonconservation of charge q on the capacitors. For sufficiently small RELTOL charge is approximately conserved.

These two types of nonconservation are secondary in comparison with the third type, which will be discussed below. Let us consider the capacitive multipole shown in Figure 5.7. Its kth terminal charge $q_k(\mathbf{v})$, is, in general, dependent on all terminal voltages $\mathbf{v}$. The relation $i_k(t) = \dot{q}_k(\mathbf{v})$ between this charge and its current should satisfy the same conservation properties as were discussed before for the individual capacitor shown in Figure 5.6(a). However, in this case we have a multidimensional $\mathbf{v}$-space. Introducing a row vector of partial dynamic capacitances $\mathbf{C}_k(\mathbf{v}) = dq_k/d\mathbf{v}$ we can express the difference between two charges $q_k(\mathbf{v}_1)$ and $q_k(\mathbf{v}_2)$ as an integral of the capacitances vector over $\mathbf{v}$ ranging from $\mathbf{v}_1$ to $\mathbf{v}_2$. This difference can be nonzero even if $i_k(t)$ satisfies the relation (5.64), since the integral of capacitance depends on the trajectory of $\mathbf{v}(t)$ between $\mathbf{v}_1$ and $\mathbf{v}_2$. In circuit simulation a variety of different trajectories may appear, but only one of them is correct, and consistent with physical phenomena, especially conservation of charge. Hence, we usually observe nonconservation of charge, causing inaccurate voltage calculations. It is a consequence of the involvement of dynamic capacitances, and the representation of the capacitive element by means of the formula $i_k(t) = C_k(\mathbf{v})\dot{\mathbf{v}}$, where voltage is the state variable. Two main reasons for this effect can be emphasized: (1) charge not being defined when dynamic capacitances are involved, and (2) bad properties of numerical integration (i.e. of DF) when voltage state variables are chosen. Item (1) obliges us to describe capacitive elements by means of charge, since a description using dynamic capacitance can be very imprecise, as will be shown in Section 5.2.2.4. However, even if a model is charge-defined, while state variables are voltages and currents, the results of simulation can be wrong due to inaccurate integration. Thus, item (2) obliges us to use charge state-variables, which we present in Section 5.2.2.3.

5.2.2.3 Charge conservation properties of DF

Differentiation formulae provide the relation between the present solution at t_n and past solutions. By virtue of their application we obtain

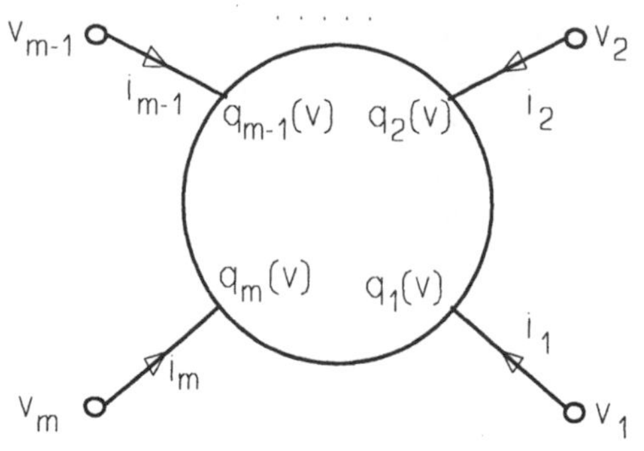

Figure 5.7 A charge-type multipole

iterative circuit solutions $\mathbf{v}_n$ moving in time along a certain trajectory in the solution space. Along this trajectory a charge $q(\mathbf{v}_n)$, which should remain constant according to physical laws, may appear to vary due to numerical nonconservation of charge. This effect can cause spectacular simulation inaccuracy, especially when voltage solutions are determined by the balance of charges in capacitive elements, as e.g. in a capacitive voltage divider. We will show that DFs based on voltage-type state variables do not conserve charge. If a charge, which should be theoretically constant, actually remains constant during numerical integration, then a DF used for this integration provides charge conservation.

To examine charge conservation let us consider the multipole shown in Figure 5.7. It can be drawn up with the equation $\mathbf{C}(\mathbf{v})\,\dot{\mathbf{v}} = \mathbf{i}$, where according to KCL the sum of its currents is zero, and hence, the sum of its terminal charges (called the total charge Q) remains constant, although voltages $\mathbf{v}$ vary in time. The DF provides charge conservation, iff the total charge Q remains constant at each point of numerical integration.

Let us take the kth row of the equations involved. After discretization at instant t_n we obtain

$$\gamma_0\, \mathbf{c}_k^t(\mathbf{v}_n)\cdot(\mathbf{v}_n - \mathbf{d}_v) = i_k, \tag{5.65}$$

where a derivative has been substituted for DF, while $\mathbf{c}_k^t(\mathbf{v})$ is the kth row of $\mathbf{C}(\mathbf{v})$. We want to examine whether for all possible solutions, $\mathbf{v}$, the sum of charges $q_k(t_n)$ over $k=1, ..., m$ at instant t_n is equal to the sum of charges at t_{n-1}. First let us expand $q_k(\mathbf{v})$ in a Taylor series in the vicinity of point $\mathbf{v}_n$

$$q_k(\mathbf{v}_{n-1}) = q_k(\mathbf{v}_n) + \mathbf{c}_k^t(\mathbf{v}_n)\cdot(\mathbf{v}_{n-1} - \mathbf{v}_n) + r_k, \tag{5.66}$$

where r_k is the second order residuum, nonzero if the charge is not a linear function of voltage. Next, the equation (5.65) can be combined with the the linear term of (5.66), and we obtain

$$q_k(\mathbf{v}_{n-1}) = q_k(\mathbf{v}_n) - \frac{1}{\gamma_0} i_k + \mathbf{c}_k^t(\mathbf{v}_n) \cdot (\mathbf{v}_{n-1} - \mathbf{d}_v) + r_k. \qquad (5.67)$$

Now, we sum up both sides of equation (5.67) over $k=1, ..., m$ (that is over all terminals). On the LHS we obtain a total charge at t_{n-1}. From the first terms of the RHS we obtain the total charge at t_n. From the second terms we obtain zero, since the sum of currents flowing into the multipole is zero (KCL). However, the sums of the third and fourth components do not become zero, and hence charge conservation does not occur.

Only in the specific case of Implicit Euler DF are the sums of the third terms zero. If charge functions $\mathbf{q}(\mathbf{v})$ are linear, then the fourth residual terms become zero. Hence, in a voltage state-variables case, charge is not conserved, except for linear capacitors and implicit Euler DF.

This is not the case, when charge state-variables are selected. To show this, let us write circuit equations in the form $\dot{\mathbf{q}}(\mathbf{v})=\mathbf{i}$. The equation corresponding to the kth terminal is now $q_k(\mathbf{v})=i_k$. After discretization with DF (5.9) we obtain

$$\gamma_0 q_{k,n} + \sum_{i=1}^{p} \gamma_i q_{k,n-i} + \sum_{i=1}^{p} \delta_i \dot{q}_{k,n-i} = i_k. \qquad (5.68)$$

Summing up over $k=1, ..., m$, on the RHS we obtain zero, due to KCL. Since $\dot{q}_{k,n-i}=i_{k,n-i}$ is the kth terminal current at t_{n-i}, then the sum of the third terms of the LHS will be zero as well. The remaining two terms of the LHS give

$$\gamma_0 \sum_{k=1}^{m} q_{k,n} + \sum_{i=1}^{p} \gamma_i \sum_{k=1}^{m} q_{k,n-i} = 0. \qquad (5.69)$$

Since DF satisfies the zero order interpolative condition $\gamma_0 + \sum_{i=1}^{p} \gamma_i = 0$, (it is analogous to the first equation (5.13) and can be proved using the coefficients identification method described in Section 5.1.1.2), then, assuming that at all past points $t_{n-1}, t_{n-2}, ...$ charge is constant and equal Q, it will be Q also at t_n. Hence, selection of charge state-variables provides that if the exact solution $\mathbf{v}$ traces the trajectory of constant charge, then the numerical solution $\mathbf{v}_n$ also traces that trajectory where the charge remains constant.

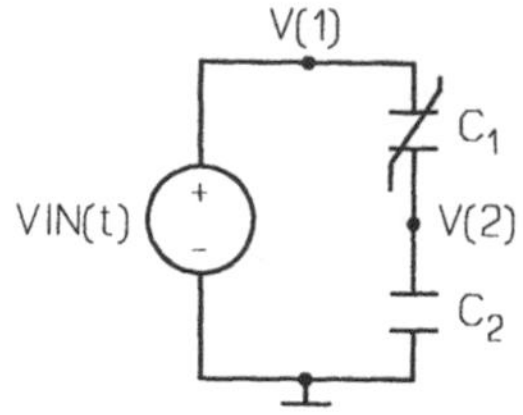

Figure 5.8 A simple charge pumping circuit

A series connection of a linear and nonlinear capacitor has been demonstrated in Figure 5.8. This circuit is driven by a periodical voltage source, which pumps up the capacitors with charge. This example is the most spectacular test for charge conservation. Since both capacitors are identically charged up, the voltage across the linear capacitor is a sensitive indicator of the instantaneous charge. In the real system it has to be zero at all instants when the exciting voltage is zero. Let us consider the task O5-3.CIR, where the nonlinear capacitor is described by means of an OPTIMA external, user-defined model. If its parameter IFL is set to 0, then the model stands for a capacity-defined capacitor. If IFL=1, then the capacitor is charge-defined. We perform the time-domain analysis in these two cases, and observe the voltage V(2) across the linear capacitor C2. Figure 5.9 shows, that for the voltage state variable the mean value of the charge unphysically increases with the slope, depending on the capacitor nonlinearity, in a wholly unphysical way. This effect is consistent with the discussion presented here.

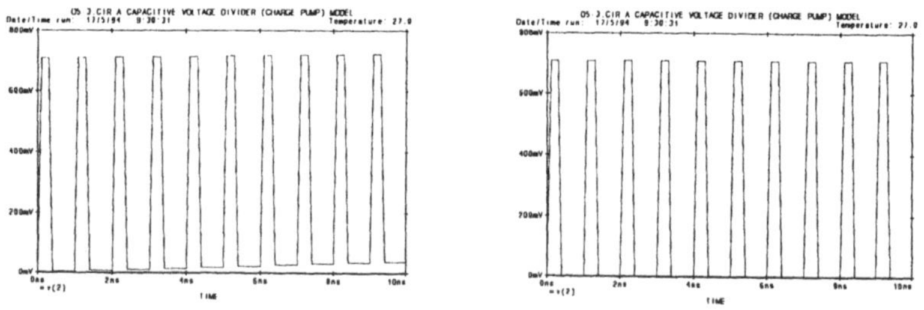

Figure 5.9 A response V(2) of the charge pump shown in Figure 5.8 in the case of voltage (the left plot) and charge (the right plot) type of the state variable

A more complex charge pump circuit can be built for a MOS transistor, including nonlinear capacitances, especially between the gate and channel. Since the gate is insulated from the channel, the capacitive divider effect occurs. Examples O5-4.CIR and O5-5.CIR show that using the Meyer model of MOSFET [5.12] we cannot provide charge conservation, while for Ward-Dutton [5.16] and other models [5.7] charge conservation occurs.

Therefore, to provide charge conservation we have to select charges as state-variables. However, in this case we require charge-defined capacitive elements. In the next section we will explain how to create such models.

5.2.2.4 Creating of proper charge models

It has been shown in the preceding section, that to provide accurate circuit response maintaining charge conservation, charges should be selected as state-variables. A problem of charge conservation in MOSFET models has been investigated by P. Yang *et al* [5.17]. To use charge as state-variables, models have to be expressed in terms of charges instead of dynamic capacitances. This necessity has been often misunderstood. For example, in the Meyer model of MOSFETs [5.12], charge storage elements are characterized in terms of three nonlinear capacitances C_{GS}, C_{GD}, C_{GB} in a way that does not guarantee conservation of charge (the full model albeit for the a.c. case, is shown in Figure 2.21). To discuss whether a model is charge conservant, we will consider the example shown in Figure 5.10(a), where a capacitor is charged up from a constant voltage source through a resistor. However, the value of capacitance is controlled by an external voltage v_2 stimulated from the source $V_2(t)$. If we have a charge controlled by more than one voltage, then we often start thinking about it in the same terms as a capacitance dependent on some external signal. We will show where this approach leads to. In this example the capacitor was modeled as $i_1 = C(v_2)\dot{v}_1$. The capacitance $C(v_2)$ does not depend on the voltage v_1 across it. Thus, its charge can be written in the form $q = C(v_2)v_1$. However, this charge is dependent on v_2, and hence the current produced should rather be $i_1 = \dot{q} = C(v_2)\dot{v}_1 + [\mathrm{d}C(v_2)/\mathrm{d}v_2]v_1\dot{v}_2$, which contradicts the capacitive model already introduced. We have been introduced to this problem also in Section 1.2.1. The involved charge of terminal 1 is stored not only on a plate of C but also on a mutual capacitor connected between nodes 1 and 2. In Figure 5.10(a) there is no mutual capacitor, and hence, the element cannot be a true charge controlled by both v_1 and v_2. The way of modeling by expressing a dynamic capacitance as a function of a controlling

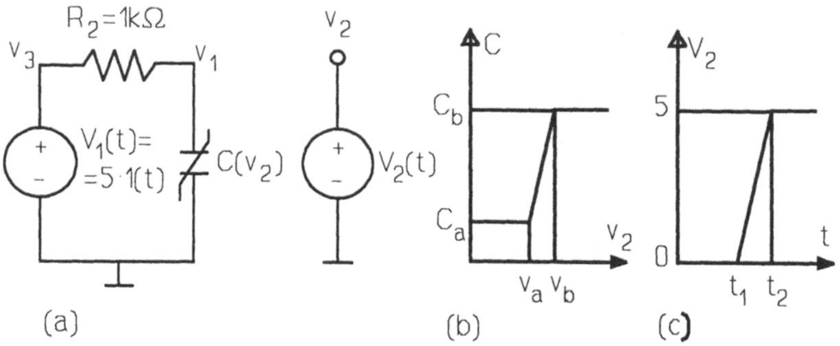

Figure 5.10 Instructive example of modelling controlled capacitances

variable is not physical. We are unable to decide what is the true charge in this example and there is no way to resolve the problem.

Moreover, if we try to simulate this example, the results are even more confusing. Let the capacitance $C(v_2)$ change from C_a to C_b, as demonstrated in Figure 5.10(b), while the controlling voltage shown in Figure 5.10(c) ranges from 0 to 5 V. In the sample O5-6.CIR for OPTIMA we can distinguish three cases: (A), (B), (C).

In case (A), the capacitance is modeled exactly as in Figure 5.10(a), with a voltage state-variable, i.e. there is no mutual capacitance. The response $v_1(t)$ increases almost smoothly as if there were no change in capacitance. This is a very inaccurate result.

However, in case (B), where a mutual capacitance is considered (taking into account the derivative of $C(v_2)v_1$ with respect to v_2), the response gives a characteristic spike at a switching instant. Unfortunately, if we increase the time grid in such a way that the point of analysis does not fall into the narrow sloped segment of the $C(v_2)$ characteristic (see Figure 5.10(b)) or if we make the slope of $V_2(t)$ in Figure 5.10(c) infinite by setting $t_1 = t_2$, then the spike disappears and so the response becomes the same as in case (A).

This problem arises from the charge modeling. Hence, in case (C), if the capacitor is modeled using charge and with the charge-type state-variable, then a spike is obtained once more. This result is the most accurate.

The results of O5-6.CIR above discussed can be better understood by demonstrating a small mathematical development. In cases (A) and (B), the equation of the circuit of Figure 5.10(a) is $v_1 + R_2C(v_2)\dot{v}_1 = V_1(t)$. Thus, after implicit Euler discretization we obtain the following equation $v_{1,n} + R_2C(v_{2,n})v_{1,n}/h = V_1(t_n) + R_2C(v_{2,n})v_{1,n-1}/h$ giving the final solution

$$v_{1,n} = \frac{V_1(t_n) + R_2 C[V_2(t_n)]\, v_{1,n-1}/h}{1 + R_2 C[V_2(t_n)]/h} \tag{5.70}$$

which is slightly greater then the preceding $v_{1,n-1}$. Even if the capacitance at t_n jumps up considerably, the voltage cannot jump down as it appears in both numerator and denominator. Thus modeling of the capacitor is wrong.

In case (B), we introduce the mutual capacitance, giving the equation $\dot{v}_1 + R_2 C(v_2)\dot{v}_1 + R_2 v_1 C'(v_2)\dot{v}_2 = V_2(t)$, where C' denotes the derivative of C with respect to v_2. This derivative (see Figure 5.10(b)) is zero everywhere outside the narrow middle range, where it is a nonzero constant. The greater the slope of C in this range, the greater will be this derivative. After disretization we solve the equation, obtaining

$$v_{1,n} = \frac{E + R_2 C[V_2(t_n)]\, v_{1,n-1}}{1 + R_2 C[V_2(t_n)]/h + R_2 C'[V_2(t_n)]\,(v_{2,n} - v_{2,n-1})/h} \tag{5.71}$$

where, if C' is zero, then the solution slightly increases in comparison with the preceding one, i.e. the algorithm behaves as in the case (A). If the point of analysis falls into the range of nonzero C', then the denominator rapidly decreases. We obtain the spike, but unfortunately this narrow range can be easily overstepped by too large a time-step.

In case (C) we have the equation $v_1 + R_2\dot{q} = V_2(t)$. After discretization it gives $v_{1,n} + R_2 C(v_{2,n}) v_{1,n}/h = R_2 C(v_{2,n-1}) v_{1,n-1}/h + V_2(t_n)$. Thus,

$$v_{1,n} = \frac{R_2 C[V_2(t_{n-1})]\, v_{1,n-1} + V_2(t_2)}{1 + R_2 C[V_2(t_n)]/h}. \tag{5.72}$$

If the capacitance is constant between two subsequent points, then the response increases as in cases (A) and (B). However, if the capacitance increases between these points, then the response immediately jumps down.

Summing up the above discussion, we can say that:

- the best way of modeling reactive multiterminal devices is to formulate expressions for the charges and fluxes of each terminal; they should be physical and satisfy Gauss's Law (i.e. their sum should be zero); they should be selected as state-variables;
- however, if we want to model capacitive elements in terms of dynamic capacitances, then all their mutual components should be involved (none neglected) to be sure that a corresponding charge exists; if all dynamic

capacitors are continuous, then a sufficiently small step of transient analysis improves the accuracy of analysis, though it does not guarantee charge conservation.

Transformation charge to dynamic capacitances requires differentiation, and then is unique. The opposite transformation is ambiguous, since it takes the form of the integral

$$\mathbf{q}(\mathbf{v}) = \mathbf{q}(\mathbf{v}_0) + \int_{\mathbf{v}_0}^{\mathbf{v}} \mathbf{C}(\tilde{\mathbf{v}})\, d\tilde{\mathbf{v}} \tag{5.73}$$

along a trajectory $\mathbf{v}(t)$ originating from $\mathbf{v}_0$. Thus, the charge of the multipole depends on the specific trajectory of the solution $\mathbf{v}(t)$. That is why writing models in terms of dynamic capacitances is not a good idea.

Thus the solution is to express them directly by charges originating from physical consideration, as has been done in the newest MOSFET models [5.7]. Some of them are provided in a bank of external models provided for OPTIMA (see directory B:\EMDL) and tested using the example O5-MOS.CIR enclosed. Charge-flux state-variables have become standard in circuit simulators.

5.3 Accuracy of differentiation formulae

5.3.1 Theory of the local truncation error

5.3.1.1 Local and global error concepts

Methods for integrating circuit equations introduced in this chapter use polynomial interpolation to obtain a solution at instant t_n, based on solutions at several preceding instants. In this recursive method local inaccuracies at each step are propagated, and hence, accumulated. At each step three components of the local error can be distinguished:

- A systematic interpolation error, due to the fact that the true solution is not a polynomial. This is the Local Truncation Error (LTE).
- A random arithmetic error caused by the finite accuracy of NR iterations, and also by inaccurate arithmetic operations. We call it Local Arithmetical Error.
- A Local Propagated Error originating in the past as an effect of interpolating over inaccurate constraints.

These three errors are combined in a manner, which is difficult to formalize, to give the cumulative Global Error (GE) of the solution. This

GE depends mostly on the LTE and a numerical property of the simulation problem, known as stability. A general characterization of global accuracy of time-domain analysis is impossible.

A practical simplified way of GE consideration involves selecting DFs to provide a limit to the variability of the numerical solution within the range $[0, \infty]$, known as the stability of DF. It implies a weaker than linear accumulation of local errors (LE). If accumulation were linear in the worst case, i.e. $GE=\sum_i |LE_i|$, for $i \rightarrow \infty$, then the GE would continue to increase infinitely. Thus, providing the response is limited, we obtain compensation of local errors. In such cases, having reduced local errors (e.g. by shortening the step) we will decrease the GE, as well.

Hence, application of the stable DF to the stable OADE theoretically can provide global accuracy by only a local error control. Among local errors the LTE is dominant and will be discussed below. Problems of stability will be treated later on.

Local Truncation Error (LTE) is that portion of the error introduced at one step of time-domain transient analysis by the DF, due to its interpolative foundations. Let us assume that for a certain OADE its exact solution $x(t)$ and derivative $\dot{x}(t)$ are known. The DF (5.9) spans over an instant t_n and p preceding instants $t_{n-1}, ..., t_{n-p}$. Now, we perform one time-step with the DF to determine x_n from an exact present derivative $\dot{x}(t_n)$ and the exact past solutions $x(t_{n-k}), \dot{x}(t_{n-k})$ $(k=1, ..., p)$. Hence, x_n is the value of an interpolating polynomial of degree l, spanned over exact constraints. Thus it differs from $x(t_n)$ only due to interpolation error. A difference between an interpolated solution x_n and the exact $x(t_n)$ under these assumptions is called the **Local Truncation Error** (LTE) at point t_n corresponding to the variable x. We denote this error ξ_n:

$$\xi_n = \beta_0 \dot{x}(t_n) + \sum_{k=1}^{p} [\alpha_k x(t_{n-k}) + \beta_k \dot{x}(t_{n-k})] - x(t_n). \qquad (5.74)$$

Similarily, if $\beta_0 \neq 0$, then a polynomial may be spanned over an exact value $x(t_n)$ and the same past solutions $x(t_{n-k})$, $\dot{x}(t_{n-k})$ $(k=0, ..., p)$ to obtain $\dot{x}_n$. A difference between $\dot{x}_n$ calculated from DF and an accurate solution $\dot{x}(t)$ is known as Local Truncation Error at t_n corresponding to the derivative. We denote this type of LTE as η_n. Now,

$$x(t_n) = \beta_0 [x(t_n) + \eta_n] + \sum_{k=1}^{p} [\alpha_k x(t_{n-k}) + \beta_k \dot{x}(t_{n-k})]. \qquad (5.75)$$

Both quantities are in fact measures of the same phenomena. For implicit DFs they both exist and are related by the formula

$$\xi_n = -\beta_0 \eta_n. \tag{5.76}$$

Explicit DFs have only the error ξ_n.

LTE is a theoretical concept that does not express the true local errors of the current solution. Since at the current time-point both variable and derivative are inaccurate, the LTE is somehow partitioned into errors of the variable and the derivative in such a way that

$$[x_n - x(t_n)] - \beta_0 [\dot{x}_n - \dot{x}(t_n)] = \xi_n. \tag{5.77}$$

This partition depends on the particular OADEs that have to be solved. Let us take discretized equations $\mathbf{f}(\mathbf{x}_n, \dot{\mathbf{x}}_n, t_n) = \mathbf{0}$. An increment of the derivative, calculated by linearization and substituted into (5.77), gives

$$\left[\beta_0 \left(\frac{\partial \mathbf{f}}{\partial \dot{\mathbf{x}}_n}\right)^{-1} \frac{\partial \mathbf{f}}{\partial \mathbf{x}_n} + 1\right] [\mathbf{x}_n - \mathbf{x}(t_n)] = \xi_n. \tag{5.78}$$

The first component in brackets determines that part of LTE connected with derivatives. This component has the coefficient β_0 whose value is close to an integration step. The term $(\partial\mathbf{f}/\partial\dot{\mathbf{x}})^{-1}\partial\mathbf{f}/\partial\mathbf{x}$ introduces elements not greater then the smallest time-constant of the circuit. To obtain accurate analysis the step is kept small in comparison with this time-constant, and so $\|\beta_0 (\partial\mathbf{f}/\partial\dot{\mathbf{x}}_n)^{-1} (\partial\mathbf{f}/\partial\mathbf{x}_n)\| < 1$. This proves that the inaccuracy of derivatives is negligible in comparison with the inaccuracy of variables. Hence, in practical time-domain analysis the inaccuracy of variables represents the whole LTE, i.e. $\mathbf{x}_n - \mathbf{x}(t_n) = \xi_n$.

5.3.1.2 The l+1-order LTE approximation

Local truncation error is a discrepancy between a simulated signal and its true counterpart, and it is difficult to calculate it since we do not know the true solution. That is why a polynomial approximation to LTE has been developed. Differentiation formulae are satisfied by polynomials up to the lth order. Thus, LTE can be approximated assuming the $l+1$-degree polynomial model for an exact reference signal $x(t)$. This gives the $l+1$-order expression for LTE. We can obtain the LTE by expanding $x(t)$ into Taylor series up to its $l+1$-order component.

These can be demonstrated by an example of a fixed-step 2nd order BDF $x_n = 2h\dot{x}_n/3 + 4x_{n-1}/3 - x_{n-2}/3$. From the definition (5.74) LTE can be written as

$$\xi_n = \frac{2}{3} h\, x(t_n) + \frac{4}{3}\, x(t_{n-1}) - \frac{1}{3}\, x(t_{n-2}) - x(t_n). \tag{5.79}$$

The accurate signal $x(t)$ will be approximated by the 3rd degree polynomial using the 3rd order Taylor expansion in the vicinity of t_n.

$$x(t) \approx x(t_n) + \dot{x}(t_n)(t-t_n) + \frac{1}{2} x^{(2)}(t_n)(t-t_n)^2 + \frac{1}{6} x^{(3)}(t_n)(t-t_n)^3, \tag{5.80}$$

where the 2nd degree component satisfies the DF, while the 3rd degree residuum will form the LTE. Differentiating (5.80) we obtain the corresponding expression for the derivative

$$\dot{x}(t) = \dot{x}(t_n) + x^{(2)}(t_n)(t-t_n) + \frac{1}{2} x^{(3)}(t_n)(t-t_n)^2. \tag{5.81}$$

From the equations (5.80), (5.81) the values of $x(t_{n-k}), \dot{x}(t_{n-k})$ can be evaluated and then substituted in (5.79). We obtain

$$\begin{aligned}
\xi_n = {} & \frac{2}{3} h\, [\dot{x}(t_n)] + \\
& + \frac{4}{3}\left[x(t_n) + \dot{x}(t_n)(t_{n-1}-t_n) + \frac{1}{2} x^{(2)}(t_n)(t_{n-1}-t_n)^2 + \frac{1}{6} x^{(3)}(t_n)(t_{n-1}-t_n)^3 \right] + \\
& - \frac{1}{3}\left[x(t_n) + \dot{x}(t_{n-2}-t_n) + \frac{1}{2} x^{(2)}(t_n)(t_{n-2}-t_n)^2 + \frac{1}{6} x^{(3)}(t_n)(t_{n-2}-t_n)^3 \right] + \\
& - \left[x(t_n) \right].
\end{aligned}$$

After all time increments have been expressed by means of the step h, the components corresponding to polynomials up to the 2nd degree can be cancelled, since the DF is satisfied by the polynomials of zero, first and 2nd degree. The rest of the components incorporate the 3rd order derivative. They give finally $\xi_n = 2h^3 x^{(3)}(t_n)/9$.

The above development can be easily applied to a general DF. Performing the $l+1$-order Taylor's expansion of $x(t)$ in the vicinity of t_n we obtain the sum of the lth degree polynomial $x_l(t)$ and the $l+1$-degree residuum, namely

$$\begin{aligned} x(t) &\approx x_l(t) + \frac{1}{(l+1)!}(t-t_n)^{(l+1)}\, x^{(l+1)}(t_n), \\ \dot{x}(t) &\approx \dot{x}_l(t) + \frac{1}{l!}(t-t_n)^l\, x^{(l+1)}(t_n). \end{aligned} \tag{5.83}$$

Calculating from the above values at discrete points and substituting them into the LTE definition (5.74), after cancellation of terms relating to $x_l(t)$, a general formula arises

$$\xi_n = (-1)^{l+1} \frac{x^{l+1}(t_n)}{(l+1)!} \sum_{k=1}^{p} \left[(t_n - t_{n-k})^{l+1} \left(\alpha_k - \frac{\beta_k (l+1)}{(t_n - t_{n-k})} \right) \right], \tag{5.84}$$

In the fixed step case this formula takes the form

$$\xi_n = K\, h^{l+1}\, x^{(l+1)}(t_n) \tag{5.85}$$

including the factor K dependent on DF coefficients

$$K = \frac{(-1)^{l+1}}{(l+1)!} \sum_{k=1}^{p} k^{l+1} \left(\alpha_k - \frac{l+1}{k} \beta_k' \right). \tag{5.86}$$

For one-step DF, equation (5.86) holds also in a variable-step case, e.g. for Explicit Euler DF: $l=1$, $K=1/2$, while for Trapezoidal DF: $l=2$, $K=1/12$.

For a family of BDF from a Taylor series based development applied to (5.18), we can evaluate the LTE of the derivative, namely

$$\eta_n = -\frac{1}{(l+1)!} (t_n - t_{n-1})(t_n - t_{n-2}) \ldots (t - t_{n-l})\, x^{(l+1)}(t_n), \tag{5.87}$$

Similarily, for a respective explicit forward formula (5.21) we have a very similar formula for LTE ξ_n

$$\xi_n = -\frac{1}{(l+1)!} (t_n - t_{n-1})(t_n - t_{n-2}) \ldots (t_n - t_{n-l-1})\, x^{(l+1)}(t_n). \tag{5.88}$$

which is exactly $t_n - t_{n-l-1}$ times (5.87). The above formulae hold either for Lagrangian or for Newtonian formulation of BDF (see sections 5.1.1.3 and 5.1.3). For fixed-step cases, values of the factor K for several DFs have been collected in Table 5.4.

Table 5.4 Values of LTE factors K for a set of fixed-step DFs

DF	l	K
Explicit Euler	1	1/2
Trapezoidal	2	1/12
BDF 2	2	2/9
BDF 3	3	3/22
BDF 4	4	12/125
BDF 5	5	10/137
BDF 6	6	20/343

5.3.1.3 General properties of DF accuracy

The $l+1$-order approximation of LTE is not quite exact, and therefore conclusions deriving from such formulae as (5.84), (5.85) are only qualitative:

(1) ξ_n, η_n are proportional to the $l+1$ the derivative of a solution. Hence, the smoother the solution the smaller is LTE. By means of OPTIMA, where the user is able to view the maximum LTE of all variables on the screen, we can execute the task S5-LTE.CIR. We use a fixed step and observe the LTE. In ranges where the response is flat and smooth the LTE is small. Once the response varies rapidly, the LTE increases considerably. Thus the LTE traces the rate of variability of the response.

(2) If the step is fixed, then ξ_n is proportional to its $l+1$ power (and η_n to its lth power). Hence, for any steps less then 1 the higher order DFs are of a smaller factor h^{l+1}. This cannot provide lower LTEs, due to different factors K in different DFs. In general, it is not true that higher order DFs are more accurate.

(3) The rate at which the LTE decreases as the step is shortened increases as the order of the DF is raised. Thus theoretically a halving of step length makes LTE ξ_n decrease four times, when the order is 1, 8 times, when the order is 2, 16 times, when the order is 3 (η_n decreases correspondingly more slowly). The tasks O5-LTE1.CIR, O5-LTE2.CIR enables us to perform an experiment with OPTIMA, to demonstrate how this effect can be observed in practical simulations. Hence, for a given DF, to obtain a given ratio of decreasing LTE, we require smaller ratios of step decrease for higher DF orders than for lower orders.

(4) In practical time domain analysis there is the need to select a step

short enough not to overstep the maximum acceptable LTE level. At the same time it should not be so short as to extend the solution time needlessly. Methods of optimal step assignment will be considered later on.

In the above item (2) there was the problem of comparing the accuracy of two DFs. We will discuss this problem by means of the introduced $l+1$-order approximation (5.84), (5.85) to LTE. If the DFs are of the same order, then for a given step the more accurate DF is that whose factor $|K|$ in (5.85) is smaller. For example, from Table 5.4 it can be concluded that the Trapezoidal DF is always more accurate then BDF 2.

It is interesting to compare LTEs of two DFs of different orders, say l_1 and l_2. Let $l_2>l_1$. In LTE analysis, for a given signal $x(t)$ a factor $A_l(h)=|K_l|h^{l+1}$ of (5.85) is essential. Its plots against h for two orders involved, are shown in Figure 5.11. They are crossed at point $h_{opt}=|K_1/K_2|^{1/(l_2-l_1)}$. The higher order DF is more accurate for steps shorter then h_{opt}. For steps exceeding h_{opt} the lower order DF is more accurate. The factor K_2 of the DF of the higher order becomes the greater, as h_{opt} approaches zero. Hence, the range where the higher order DF is more accurate diminishes. There is no way to compare the accuracies of different order DFs without taking into account step-size.

However, it can be stated that for a sufficiently small $|K_2/K_1|$ the point h_{opt} lies above the range of practically useful steps, and thus a DF of higher order is in practice more accurate. For example, comparison of Explicit Euler and Trapezoidal DF gives $h_{opt}=6$, which is an extremely large value. Thus, the Trapezoidal DF is in practice always more accurate than the Explicit Euler DF.

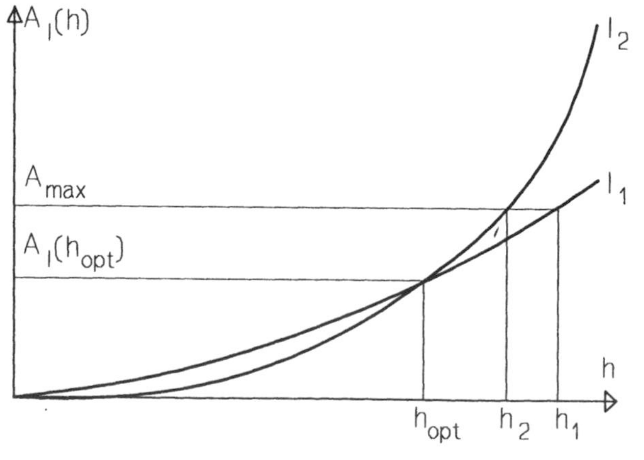

Figure 5.11 A comparison of accuracy for two DFs of different orders

If, however, h_{opt} lies in a useful region, then a lower order DF is more accurate with steps $h > h_{opt}$. If the user imposes a certain level of accuracy $A_l(h) \leq A_{max}$, and if $A_{max} > A_l(h_{opt})$, where after some algebra $A_l(h_{opt}) = (|K_1|^{l_2+1} / |K_2|^{l_1+1})^{1/(l_2-l_1)}$, then to provide this accuracy a lower order DF needs a step h_1 larger than the value h_2, which was calculated from the lower order, as shown in Figure 5.11. Of course, due to the efficiency of analysis, we prefer a step length as large as possible and surprisingly a lower order would be more advantageous than a higher one. However, if the accuracy level A_{max} lies below $A_l(h_{opt})$ the higher order will be better.

For example, subsequent BDFs presented in Table 5.4 provide a sequence of $A_l(h_{opt})$ decreasing with l. On the level of the most inaccurate analysis, order 1 is the optimal. If more accurate analyses are required, the optimal order increases. Of course, in this discussion accuracy is measured in terms of $A_l(h)$, while the LTE, which is controlled by the user in reality, is yet multiplied by a derivative $x^{(l+1)}(t_n)$. Thus the optimality of DF orders will depend not only on the user's LTE requirements but on the shape of the response as well. This discussion cannot be directly applied for optimal order selection, but it shows that during analysis such a selection can be advantageous.

5.3.2 A LTE controlled variable step time-domain analysis

5.3.2.1 Estimation of LTE during analysis

In Section 5.3.1.1 we have shown that during a sufficiently accurate analysis practically all LTEs can be assigned to circuit variables rather than to their derivatives. In this section we concentrate on methods of calculating LTEs during analysis, and this fact, that the LTE is the same as the current error of the variables, will be essential to us.

The estimated LTE ξ_n can be approximated according to (5.84):

$$\xi_n = A(t_n)\, x^{(l+1)}(t_n), \tag{5.89}$$

where the symbol $A(t_n)$ for a factor accompanying the $l+1$-order derivative has been introduced. We are considering a general variable step DF. Let an lth-order prediction be performed at the current point t_n to obtain a predicted state variable $\mathbf{x}_n^0$ from a chosen forward formula (5.21), e.g. explicit DF (5.9) with $\beta_0 = 0$. The approximated LTE $\xi_n^{pred} = \mathbf{x}_n^0 - \mathbf{x}(t_n)$ of this predictive formula contains the same $l+1$-order derivative, and hence, is of similar form to (5.89), i.e. $\xi_n^{pred}(t_n) = A^{pred}(t_n)\mathbf{x}^{(l+1)}(t_n)$. Combining the latter with (5.89) we can write it in the form

$$\xi_n^{\text{pred}} = A^{\text{pred}}(t_n)\,\mathbf{x}^{(l+1)}(t_n) = \frac{A^{\text{pred}}(t_n)}{A(t_n)}\,\xi_n. \tag{5.90}$$

The LTE of the prediction can be rewritten as $\xi_n^{\text{pred}} = (\mathbf{x}_n^0 - \mathbf{x}_n) + [\mathbf{x}_n - \mathbf{x}_n(t_n)]$. Substituting ξ_n from (5.89) we obtain $\xi_n^{\text{pred}} = \mathbf{x}_n^0 - \mathbf{x}_n + \xi_n$. Then, eliminating ξ_n^{pred} by (5.90), a practical result arises

$$\xi_n = \frac{A(t_n)}{A^{\text{pred}}(t_n) - A(t_n)}\,(\mathbf{x}_n^0 - \mathbf{x}_n). \tag{5.91}$$

Therefore, while prediction and DF are of the same order and $A^{\text{pred}}(t_n) \neq A(t_n)$, the LTE can be evaluated from the distance between the calculated and predicted solutions. This rule is efficient in estimating the LTE only if the denominator is not too small and prediction error not too large.

For example, when one uses the Trapezoidal DF with the 2nd order predictive forward formula (5.19), then the LTE can be estimated as follows

$$\xi_n = -\frac{h_n^2}{2(h_n + h_{n-1})(h_n + h_{n-1} + h_{n-2}) + h_n^2}\,(\mathbf{x}_n^0 - \mathbf{x}_n). \tag{5.92}$$

Similarily, in algorithms including the lth order BDF (5.18) with predictive FF (5.21), using (5.86), (5.87), the formula $\xi_n = -\eta_n/\gamma_0$, and (5.91) we obtain

$$\xi_n = -\frac{1}{\gamma_0 s_{l+1} + 1}\,(\mathbf{x}_n^0 - \mathbf{x}_n) = -\frac{\dfrac{1}{s_{l+1}}}{\dfrac{1}{s_1} + \ldots + \dfrac{1}{s_{l+1}}}\,(\mathbf{x}_n^0 - \mathbf{x}_n). \tag{5.93}$$

Such a way of estimating LTE has been implemented in many simulators. In OPTIMA we have an empirically modified version of (5.93).

5.3.2.2 The need for step control during analysis

In time-domain analysis we impose a certain maximum error index and intend that in the whole range of analysis the global error should remain no greater than the required error level introduced. If the analyzed

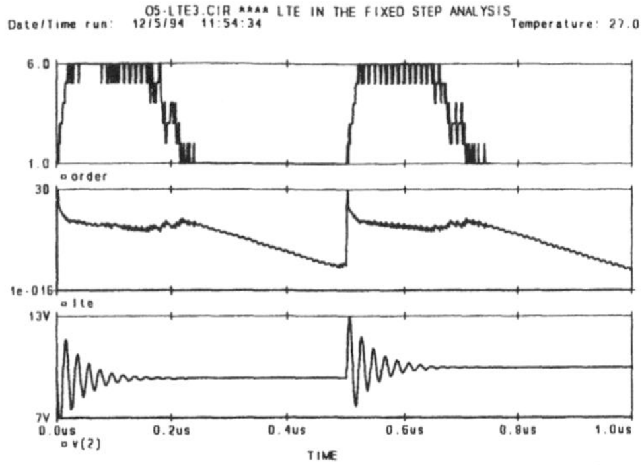

Figure 5.12 Relation between the time-domain response V(2) (bottom), LTE in logarithmic scale (middle) and order of DF (top) obtained from the example O5-LTE3.CIR

task is stable, then this global error is bounded on the whole positive time axis. Thus LTE values at all instants do not accumulate. A sufficient method of controlling the global accuracy is to control the course of LTE in time and to maintain it at the sufficiently small level. As we know from previous consideration, the LTE is dependent on step length and, what is essential, on participation of higher-order Taylor series terms in the time-profile of the response. Performing a fixed-step analysis we can plot the LTEs in time and observe how large they are for sharp circuit responses, and how small they remain when the response is smooth. We can observe the relations between responses and their relative LTE (a maximum over all components of the response of LTEs divided by a current maximum of the component) by running the example O5-LTE3.CIR and displaying the voltage V(1) and LTE using the command ".PROBE TRAN V(1) LTE", and then a postprocessor. Results are demonstrated in Figure 5.12. Another example O2-LTE4.CIR shows how the LTE oscillates with the oscillating response of the circuit.

Due to the coexistence of smooth and sharp sectors of transient responses, the LTE can vary within a wide range. Hence, to maintain a uniform accuracy level the step-size should be precisely in accordance with the shape of the response. At the same time, we want to integrate using as large a step-size as possible, since efficiency of analysis within the presumed interval [0, *T*max] requires it to be covered in the smallest number of time steps. More precisely, as the local curvature of the response becomes greater, so a smaller step-size is required. Hence, the step can vary over a wide range and an efficient procedure is required to follow changes in LTE and ensure that it remains on approximately a constant level throughout the whole response.

To show how the step-size should be varied in circuit integration let us consider the resonant network of Figure 5.13. Its equations take the form

$$\begin{bmatrix} \dot{v} \\ \dot{i} \end{bmatrix} = \begin{bmatrix} 0 & -\dfrac{1}{C} \\ \dfrac{1}{L} & -\dfrac{R}{L} \end{bmatrix} \begin{bmatrix} v \\ i \end{bmatrix} + \begin{bmatrix} \dfrac{I}{C} \\ 0 \end{bmatrix} \tag{5.93}$$

Their matrix has two eigenvalues:

$$\lambda_{1,2} = -\frac{R}{2L} \pm \left(\frac{R^2}{4L^2} - \frac{1}{LC} \right)^{1/2} . \tag{5.94}$$

To observe the responses of this circuit we can run the SPICE example S5-STIFF.CIR or OPTIMA example O5-STIFF.CIR. Two important cases (a) and (b) have been distinguished in these simulations.

In case (a), $R = 1\ k\Omega$ and the network is strongly damped, eigenvalues become real negative and introduce two time constants. The small time constant $\tau_1 = -1/\lambda_1 \approx L/R = 0{,}001$ ns stands for the magnetic inertia of L, while the large constant $\tau_2 = -1/\lambda_2 = RC = 1$ ns corresponds to the inertia of the charge of capacitor C. The response to the unit jump in the current I has a fast component, visible on the front slope of the pulse of the voltege $V(2) = v_L$ accross the inductor. Exact analysis of this slope requires a step of the order of the smaller time constant, e.g. $0.1\,\tau_1 = 0.1$ ps.

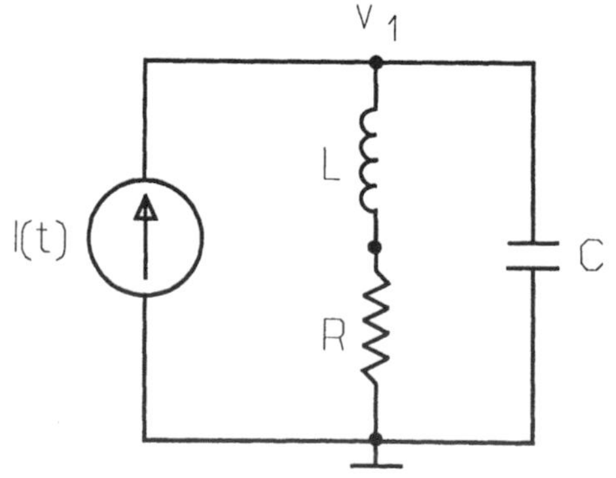

Figure 5.13 Example of a resonant circuit, whose equations require a step-size variable over a wide range

However, if this short step is maintained for analysis of the back slope of the pulse, it unnecessarily prolongs the whole analysis, since that back slope of the pulse of voltage V(2) across the inductor is determined by the greater time constant τ_2, e.g. 0.1 τ_2, which is 1000 times longer. We call this case the first type of stiffness.

In case (b), R=0.01 Ω and hence, the network is weakly damped. High frequency oscillations appear. Eigenvalues are complex conjugated $\lambda_{1,2}=-1/\tau \pm j \cdot 2\pi/T$, the time constant of damping is equal to $\tau=2L/R=0.2\,\mu s$, while the period of oscillation is equal to $T=2\pi\sqrt{LC}=0.2\,ns$. This is 1000 times shorter than the time constant of damping. To represent a profile of oscillation exactly, the step should be of the same order as the period order, e.g. 0.1 T=0.02 ns. However, after the oscillations have decayed, e.g. for $t>3\tau=0.6\mu s$, such a step will be too short. In this case a very long step can applied. We call this case the second type of stiffness.

Both cases demonstrate how flexible the step assignment procedure should be to follow the time constants of responses that can vary over an extremely wide range.

5.3.2.3 Analysis with a priori and a posteriori step control

To control step-size by means of the LTE we introduce accuracy indexes E_a, E_r of time-domain analysis. They should not be confused with the indexes of accuracy of the NR iterations, introduced in Chapter 4. Let E_a be the index of maximum absolute error, while E_r is the index of maximum relative error of the LTE $|\xi_n|$. Since the LTE should be controlled for all components of the state variables vector, we introduce the maximum absolute error of the ith component of the state vector at instant t_n: $e_{ni}=E_a+E_r|x_{ni}^{max}|$, where x_{ni}^{max} is the maximum value of the ith component of state variables vector $\mathbf{x}_n$ appearing during analysis up to the instant t_n. Integration satisfies the user's requirements, if for all components of the state variables the inequality $|\xi_{ni}|\le e_{ni}$ holds. In OPTIMA, E_a is definable for charges and fluxes, by CHGTOL and FXTOL options, while E_r is definable by the option ERTR. In SPICE, we have the option CHGTOL, but E_r is fixed, while the user can increase accuracy by increasing the option TRTOL, which scales the LTE, i.e. TRTOL $|\xi_{ni}|\le e_{ni}$ (in OPTIMA we have a similar scaling option LTEA).

The basic mechanism for step assignment is its prediction for the next time iteration by means of the information available on LTE at the current instant. Let us try to integrate the circuit equations at instant t_{n+1}, and hence, try to select the step $h_{n+1}=t_{n+1}-t_n$, having the circuit response only at instant t_n and before. Theoretically, we have no credible information for this purpose, as we do not know what will be the profile of the response (theoretically, future variations in the signal are not determined by its profile so far). Hence, prediction is realized by extrapolation for the instant $t_n + h_{n+1}$ of the previous course of the LTE of the response. To provide such an extrapolation, an efficient formula for LTE $\xi_{n+1}(h_{n+1})$ as a function of current step is required, which is not simply and exactly available.

In such a case, a quasi-fixed-step approach can be applied, since the fixed-step formula $\xi_{n+1} = K h_{n+1}^{l+1}$ (recalled from (5.85)) can be viewed as an approximation to (5.84) in the vicinity of zero, i.e. for small steps. Thus, the LTE can be extrapolated as

$$\xi_{n+1} = \left(\frac{h_{n+1}}{h_n} \right)^{(l+1)} \xi_n . \tag{5.95}$$

Introducing the e_{ni} as the required LTE level $|\xi_{n+1}|$ at instant t_{n+1} we obtain a step predictor for each state vector component separately. To provide the required accuracy for all state variables, the minimum step should be chosen from the step predictions obtained. Finally, we have the following formula for the step:

$$h_{n+1} = h_n \min_i \left(\frac{E_a + E_r |x_{ni}^{\max}|}{|\xi_{ni}|} \right)^{1/(l+1)} . \tag{5.96}$$

where the current LTE in the denominator has to be estimated by means of methods described in Section 5.3.2.1. This we call *a priori* step control.

In practice, the step (5.96) is additionally bounded to be greater than a minimum step $h_{\min}$, smaller than a maximum step $h_{\max}$ and smaller than $2h_n$ (the latter due to stability). Furthermore, a step cannot be longer than the current distance to the endpoint of integration and to the nearest breakpoint. Breakpoints are special instants, which the integration procedure should necessarily find. They arise from characteristic turning points of time-domain excitations (automatically generated when sources are modeled by built-in models), or (in the case of option ALLP) from the display grid imposed by the command .TRAN. In OPTIMA, breakpoints can be optionally defined by the command .SBRP. User-defined breakpoints are especially useful if user-defined time-domain sources are introduced that do not generate breakpoints automatically. If ALLP is omitted then the solution at the grid is obtained from the interpolation (5.21).

These rules of step control react only after some delay, and hence sharp turning points in the response do not shorten the step immediately but cause its shortening according to (5.96) after at least one step of successful analysis and LTE estimation. In practice, the delay while the algorithm adapts to new step requirements takes a few steps. Hence, though theoretically the variable step method should provide a constant course for the LTE during analysis, in practice we observe regions of adaptation to sharp changes in signals. Running OPTIMA examples OP5-ADAP.CIR and O5-ADAP1.CIR we can observe the adaptation mechanism of the algorithm discussed. We can find that simulation can be inaccurate at the very beginning (if an initial step was inadequate) and on steep slopes of waveforms. In some simulators (also in OPTIMA) a default initial step is equal to a minimum step. This is more accurate approach, though in some cases it consumes slightly more computation time. In OPTIMA we can change the initial step by the option STEP.

An auxiliary tool for the reduction of the above observed inaccuracy in *a priori* step control is the so-called *a posteriori* step correction. It consists of estimating of the current LTE ξ_n at t_n (using methods introduced in Section 5.3.2.1) and comparing it with the maximum acceptable error $\mathbf{e}_n$. If the estimated error exceeds that which is acceptable, then a shortened step h_n' is calculated, by means of the relation $\xi_n(h_n)$. Then, we come back to the preceding point $t_{n-1}=t_n-h_n$, cancel the step h_n, introduce a new shortened one h_n' $(h_n'<h_n)$, and repeat the analysis once more at $t_n'=t_{n-1}+h_n'$. Now, we are able to reestimate the LTE once more, to verify our satisfaction with its accuracy. If it is not satisfactory we should repeat the step rejection. Otherwise, we proceed to the next time iteration. In practice, an *a posteriori* control is performed only if it gives an essential adjustment of accuracy, for example if $h_n'<0.9h_n$ (to avoid wasting time for small accuracy corrections).

To shorten the step we can use the fixed-step approximation of the relation $\xi_n(h_n)$, in a manner similar to that of the *a priori* approach. Hence, we assume the relation $(\xi_n'/\xi_n)=(h_n'/h_n)^{(l+1)}$. Therefore, an adjusted step h_n' is calculated for each state variable $\mathbf{x}_n$ separately from the equation $(h_n'/h_n)^{(l+1)}=e_{ni}/|\xi_{ni}|$, and the shortest step is chosen:

$$h_n'=h_n\,\min_i\left(\frac{E_a+E_r\,|x_{ni}^{\max}|}{|\xi_{ni}|}\right)^{1/(l+1)}. \tag{5.97}$$

Components of ξ_n are estimated by simulation. Rejecting the step, we test whether it does not fall below the minimum level $h_{\min}$.

This *a posteriori* approach increases simulation time. In OPTIMA, the option POST switches it on. Handling examples O5-PST.CIR and O5-PST1.CIR we can observe quite good maintenance of the LTE at a constant level, but unfortunately there is a considerable time increase in transient analysis time, as well.

5.3.2.4 Analysis with the variable order

It has been proved in Section 5.3.1.3, that there are theoretical foundations for choosing the order of DF in such a way that for a given level of accuracy the step length is maximal, and hence, the number of time points within a given range of analysis is minimal. In practice, a variable order can be used only for BDF, where, due to stability, we can select from orders ranging from 1 to 6.

The selection of DF order at the point t_{n+1} can be performed after termination of analysis at the point t_n. Using the predictive formula (5.45) of orders $r=1, ..., l+1$ we can calculate predictors $\mathbf{x}_n^{(r)0}$. Then, combining the available solution $\mathbf{x}_n$ with these predictors we can estimate the LTE $\xi_n^{(r)}$, which will occur at the point t_n, if the order r were to be used. From the calculated LTE $\xi_n^{(r)}$, the steps $h_{n+1}^{(r)}$ required for a given accuracy are predicted from (5.96) (where l should be replaced by r). Now, we choose that order r for which an available step is the longest. If the resulting order is l or if the step provided by this optimal order is not significantly (e.g. 1.1-times) longer than the step $h_{n+1}^{(l)}$ provided by the order l, then we retain this order unchanged. Otherwise, we select the new order. Notice, that due to stability the order should not be increased by more than one at a single step.

Simulating the example O5-ORD.CIR for OPTIMA, we observe that an OPTIMA default of variable order BDF integrates more rapidly than any fixed order method, set by option ORDE. The orders from 1 to 6 require 246, 113, 85, 94, 134 and 274 time iterations respectively. The most efficient is the variable order method with its 71 iterations.

The original version of SPICE contained both variable-order BDF and trapezoidal (TDF) integration methods. In PSPICE the BDF has been deleted, since it has been considered too time-consuming. Practice in using PSPICE proves that the stability and accuracy of TDF are quite sufficient, while the numerical complication is less than that of BDF. However, for many examples TDF fails to draw exactly the correct details of a waveform. In OPTIMA, we prefer to use the BDF. The version of it that we use is very similar to the one implemented in NAP2 [5.14]. This simulator has been checked for the best numerical properties for complicated stiff equations. It implements many tricks that makes it unlike SPICE resistant to numerical difficulties. We can test these problems by handling examples S5-DTL1.CIR, O5-DTL1.CIR with PSPICE and OPTIMA respectively.

To improve the numerical properties of TDF while maintaining the low complexity of this method, the creators of CAzM [5.5] have developed a new DF, which is a combination of TDF with BDF2. Its stability is similar to that of BDF2, while the "ringing" property (see Section 5.4) of TDF is eliminated. The integration algorithm of CAzM appears to be more advantageous than TDF implemented in SPICE. We will present this DF in the next section.

5.4 Global properties of differentiation formulae

5.4.1 Stability of differentiation formulae

5.4.1.1 Fundamental test for differentiation formulae

The numerical features of the solution of an OADE are dependent on the numerical properties of differentiation formula and on the equation itself. To start discussing the global properties of the solution of an OADE, i.e. its global accuracy problem, we assume that exact solution $\mathbf{x}(t)$ of the OADE is bounded, i.e. such $M<\infty$ exists that $\|\mathbf{x}(t)\|\leq M$ holds within the set $[0, \infty)$. This case is usually met in practice. Moreover, if, for $t\in[0,\infty)$ a numerical solution remains bounded, we call this numerical problem stable. However, we want to know whether this stability is a property of the DF, and hence disregard the influence of equation features. In general, this is quite impossible, but fortunately in physical problems we deal with typical tasks which can be linear (and have exponential-sinusoidal responses) or nonlinear (but for small signals locally behave closely to the linear case). Therefore, the linear elementary task can be viewed as a basic ODE for testing the stability of numerical problems. We call it a basic standard test. Several standard tests have been developed, though in this treatment we use the most important one, the linear first order ODE with one time constant:

$$\dot{x}=\lambda x, \quad x(0)=1, \quad x,\lambda\in C, \ \mathrm{Re}\ \lambda\leq 0. \tag{5.98}$$

Solution of this ODE takes the following form:

$$x(t)=\mathrm{e}^{\lambda t}=\mathrm{e}^{\mathrm{Re}\,\lambda\, t}\,[\cos(\mathrm{Im}\,\lambda\, t)+\mathrm{j}\,\sin(\mathrm{Im}\,\lambda\, t)]. \tag{5.99}$$

Hence, due to the assumption Re $\lambda \leq 0$, the solution has a bounded response, i.e. $|x(t)| \leq 1$.

Several special cases can be distinguished:
Once Im $\lambda = 0$, the response is aperiodical, namely:

(i) if Re $\lambda = 0$, then it is constant: $x(t) = 1$,

(ii) if Re $\lambda < 0$, then it is damped (time constant is $\tau = -1/\text{Re}\,\lambda$).

Otherwise, the response is oscillating with the period Im λ, namely:

(iii) if Re $\lambda = 0$, then the amplitude of oscillations is eqal to 1,

(iv) otherwise, oscillations are damped with time constant τ.

This equation repesents all basic cases of electrical circuit operation and appears to be an excellent test for the global behavior of integration methods.

5.4.1.2 Stability of differentiation formula

Basic stability tests for DF are performed for the fixed-step discretization, and hence, we deal with the formula (5.9) in the form

$$x_n = \beta_0' h \dot{x}_n + \sum_{k=1}^{p} (\alpha_k x_{n-k} + \beta_k' h \dot{x}_{n-k}) \tag{5.100}$$

where, coefficients β_0', α_k, β_k' ($k=1, ..., p$) are real numbers (α_k, $\beta_k = h\,\beta_k'$ satisfy constraints (5.13)). This DF is said to be stable when the complex sequence $\{x_n\}$ arising from the solution of the test (5.98) by means of the DF (5.100) is limited for $n \to \infty$.

We discretize the ODE (5.98) at instants t_{n-k}, and obtain $\dot{x}_{n-k} = \lambda x_{n-k}$. After $\dot{x}_{n-k}$ have been substituted in the DF (5.100), we obtain a complex algebraic equation that joins the current solution x_n with previous ones x_{n-k} ($k=1, ..., p$):

$$\sum_{k=0}^{p} (\alpha_k + \beta_k' h\,\lambda)\, x_{n-k} = 0, \quad \alpha_0 = -1. \tag{5.101}$$

This linear difference equation, when started from p initial conditions, generates the sequence $\{x_n\}$, which is the numerical solution of (5.98). Hence, it is an approximation to the exact solution $\{x(t_n)\} = \{(e^{\lambda h})^n\}$. Hence we predict a numerical solution in a form $x_n = u^n$, substitute it to (5.101) and look for u. This substitution gives the so-called characteristic equation:

$$a(u) + q\,b(u) = 0, \tag{5.102}$$

where:

$$a(u)=\sum_{k=0}^{p}\alpha_{p-k}u^k,\quad b(u)=\sum_{k=0}^{p}\beta_{p-k}u^k,\quad q=h\lambda. \tag{5.103}$$

In general, the equation (5.102) has p complex roots, where m of them are different, denoted as u_i. Each one can be multiple, with the multiplicity factor σ_i ($\sigma_i \geq 1$, $i=1, ..., m$). Hence, $\sum \sigma_i = p$. Due to the linearity of the equation (5.102), its general solution is a linear space stretched over the base $\{n^{j-1}u_i^n\}$, $j=1, ..., \sigma_i$, $i=1, ..., m$, namely [5.10]:

$$x_n=\sum_{i=1}^{m}\left[c_{i0}+nc_{i1}+n^2c_{i2}+...+n^{\sigma_i-1}c_{i\sigma_i}\right]u_i^n. \tag{5.104}$$

We can state the stability criterion. The sequence (5.104) is bounded for $n\to\infty$, i.e. the discussed DF (5.98) is stable for q, iff all single complex roots u_i ($\sigma_i=1$) of the characteristic equation (5.102) obey the criterion $|u_i|\leq 1$, while the multiple roots u_i ($\sigma_i>1$) obey $|u_i|<1$.

As can be seen, this solution is composed of m components. In the stable case all of them decay, but only that one of the maximal magnitude $|u_i|$ is dominant, useful and stands for the solution of the test ODE (5.98). All the others are parasites giving the quick decay of the parasitic response on the background of the useful response component.

Hence, the examination of stability of DF consists of finding the relation between the maximum modulus of roots ($\max|u_i|$ of (5.102)) and the variable $q=h\lambda$, which is a synthetic parameter describing together a type of response variability (through λ) and step size $h>0$. A convenient means of examination uses plots on a complex plane $q\in Q$. A set of those q, which provide DF stability is known as the stability region. From (5.102), it can be seen that

$$\phi(u)=-\frac{a(u)}{b(u)},\quad |u|<1 \tag{5.105}$$

maps the interior of the sphere $|u|<1$ onto the interior of the stability region, whose outline is drawn up with the complex function $\phi[\exp(j\zeta)]$, $\zeta\in[0,2\pi)$.

For a given λ ($\mathrm{Re}\,\lambda\leq 0$), when $h\to\infty$, point q moves from the origin $0+j0$ along a radial line located in the left closed halfplane. The common part of this radial line and the stability region define a range of $q=\lambda h$, where stability holds. The DF stable for all λ, provided by (5.98), and all $h\geq 0$ should have the left halfplane as their stability region. However, most DF are stable only in some subsets of this region.

This is a basis for distinguishing certain types of stability. In the next section we demonstrate a few practically useful types of stability.

5.4.1.3 Types of stability

A minimum amount of stability is the A_0-stability, i.e. stability for the step $h=0$. All useful DFs should be stable for $h \to 0$, hence, A_0-stability is a limit of stability for small steps. In this case, roots u_i tend to roots of the equation $a(u)=0$. These roots should satisfy the stability criterion given in Section 5.4.1.2. It can be proved, that if the DF obeys constraints (5.13) of degree 0 and 1, and the condition $\sum_{k=0}^{p} \beta_k' \neq 0$, then $a(u)$ has a single root $u=1$, responsible for the useful, i.e. nonparasitic) component of a response. Hence, it is A_0-stable, iff its other roots satisfy $|u|<1$. This feature is required for all DF used in practice.

At the opposite extreme we have the maximum amount of stability, i.e. stability for all $h>0$, i.e. in the whole left closed halfplane, which is called A-stability. It happens iff $\phi(u)$ (5.105) maps the sphere $|u| \leq 1$ onto the left closed halfplane, or in other words, iff the inequality $\mathrm{Re}\,[-a(u)/b(u)] \leq 0$ holds. From this criterion we can prove, following Dahlquist [5.6], that A-stability can be only for DF of order $l \leq 2$. This essentially limits the accuracy of A-stable DF. Examples of A-stable DF are: trapezoidal DF (TDF) and BDF1,2.

Stability itself appears to be insufficient to test the numerical properties of DF. For example, we examine the TDF: $x_n = h\dot{x}_n/2 + h\dot{x}_{n-1}/2 + x_{n-1}$. Substituting into TDF the discretized test $\dot{x}_{n-k} = \lambda x_{n-k}$ $(k=0,\,1)$ we obtain $x_n(2-h\lambda) - x_{n-1}(2+h\lambda) = 0$. It gives the characteristic equation of the form $u(2-h\lambda)-(2+h\lambda)=0$, whose single root $u=(2+h\lambda)/(2-h\lambda)$, after some algebra, appears to be lower than or equal to unity, for $h\lambda$ located in the left closed halfplane. This shows that TDF is A-stable. The numerical solution of the test ODE will take the form $\{u^n\}$. Unfortunately, though the A-stability occurs, a parasitic "ringing" of u^n appears, since the real part of the root takes the form

$$\mathrm{Re}\,u = \frac{4-h^2|\lambda|^2}{|2-h\lambda|^2}. \tag{5.106}$$

It becomes negative for $h>2/|\lambda|$. This "ringing" can be easily observed, if the test ODE is real. Then, a real negative root u causes alternately changing sign of solutions u^n, though its modulus $|u|^n$ might seem to be a quite good response. In general, we find out that "ringing" does not occur in DFs, where the root-loci of $u(h)$ remain in the right half-plane while $h \to \infty$.

Among DFs that have this feature there are A_0-stable DFs, i.e. those providing a dominating root $u=1$ and other roots $0<u<1$ for $h=0$, whose roots all tend to 0 for $h\to\infty$. To explain this requirement let us notice that for $h\to\infty$ the decay of the useful component of the response should correspond to the exponential test ODE solution, while decay of all parasites should be even quicker. Since $t_n=nh$, the exact solution takes the form: $x(t_n)=\exp(\lambda t_n)=[\exp(\lambda h)]^n=\hat{u}^n$, where $\hat{u}=\exp q$. For $\mathrm{Re}\,\lambda<0$, if $q\to\infty$, then the exact solution has the property $\hat{u}\to 0$. Concerning a numerical solution, we want $x_n=u_i^n$ (where u_i is the dominating root) to have the similar property $u_i\to 0$, when $h\to\infty$. Furthermore, we want the same for all parasitic roots $u_i\to 0$. Finally, since for $q\to\infty$ roots u_i tend to roots of $b(u)=0$, then all roots tend to zero iff $b(u)=\beta_0 u^p$. It means that the DF has to be implicit ($\beta_0\neq 0$) and obey $\beta_1=\ldots=\beta_p=0$. This condition produce the family of BDFs. A-stable DFs providing all roots of the characteristic equation tending to zero with $h\to\infty$ are known as L-stable. Hence, all BDF1-6 are L-stable.

In Reference [5.5] the composite TDF-BDF2 has been introduced, to obtain the 2nd order DF. It is one-step as TDF, but free of any "ringing" effect and L-stable as BDF2. The DF obtained consists of dividing each step $h_n=t_n-t_{n-1}$ into two pieces by introducing an intermediate point $t_{n-1+\gamma}$. The first piece is $\gamma h_n=t_{n-1+\gamma}-t_{n-1}$, while the second is $(1-\gamma)h_n=t_n-t_{n-1+\gamma}$. At the intermediate point, discretization by means of the TDF: $x_{n-1+\gamma}=\gamma h_n\dot{x}_{n-1+\gamma}/2+\gamma h_n\dot{x}_n/2+x_n$ is performed to calculate an intermediate solution. Then, using x_{n-1} and $x_{n-1+\gamma}$, the BDF2 is stretched to calculate a solution x_n. The TDF for discretization at the intermediate point, after introduction of the step γh_n instead of h_n, takes the form

$$x_{n-1+\gamma}=\frac{\gamma h_n}{2}\dot{x}_{n-1+\gamma}+\frac{\gamma h_n}{2}\dot{x}_{n-1}+x_{n-1}\,. \tag{5.107}$$

After the intermediate solution has been calculated, we can use the BDF2 for solution at the point t_n. Expressing steps in equations (5.12) by h_n and γ we obtain the BDF2 of the form

$$x_n=\frac{1-\gamma}{2-\gamma}h_n\dot{x}_n+\frac{1}{\gamma(2-\gamma)}x_{n-1+\gamma}-\frac{(1-\gamma)^2}{\gamma(2-\gamma)}x_{n-1}\,. \tag{5.108}$$

In this method, each time-iteration is performed in two stages: TDF as the preliminary stage, and BDF2 as the main stage. To provide identical β_0 for both DF and to minimize LTE the parameter $\gamma=2-\sqrt{2}\approx 0.586$ is selected. This combined method is L-stable, identically to BDF2.

Ch. W. Gear has introduced a stronger stability requirement useful in the analysis of stiff ODEs, and proved that his BDF1-6 satisfy this so called stiff-stability [5.8, 5.9]. Studying step-size requirements for analysis of stiff-equations Ch. W. Gear has noticed that if the solution is not oscillating ($\mathrm{Im}\,\lambda = 0$, $\mathrm{Re}\,\lambda \leq 0$), then analysis utilizes the whole range $h > 0$, and therefore the whole left closed Re q-axis.

If the solution is oscillating, then first of all the analysis may need short steps, corresponding to a period of oscillation. Then, such $m \geq 1$ (e.g. $m=10$) should exist, that $h \leq T/m = 2\pi/(m\,\mathrm{Im}\,\lambda)$. This gives a requirement for the existence of $\zeta \geq 0$, such that the condition $\mathrm{Im}\,q \leq \zeta = 2\pi/m$ holds.

Moreover, a step should correspond to the time constant of oscillation damping, i.e. the value of $\exp(-h/\tau)$ should be distinguished from zero, so as not to jump over oscillations. Hence, such $r \geq 0$ should exist (e.g. $r=5$), that $h \leq r\tau = -r/\mathrm{Re}\,\lambda$ holds. Therefore, the following condition arises: $\mathrm{Re}\,q \geq -r$, $r \geq 0$. Furthermore, oscillations require even longer steps, exceeding the range of quick responses (i.e. $h \geq -r/(\mathrm{Re}\,\lambda)$). Hence, a range of q obeying $\mathrm{Re}\,q \leq -r$, $r \geq 0$ is necessary.

Concluding the above discussion, we can say that analysis of stiff equations requires such steps that q is located in a subset of the q plane described by two numbers $\zeta \geq 0$ i $r \geq 0$, and the following three conditions: $\mathrm{Im}\,q = 0$ i $\mathrm{Re}\,q \leq 0$, $|\mathrm{Im}\,q| \leq \zeta$ i $0 \geq \mathrm{Re}\,q > -r$ and $\mathrm{Re}\,q \leq -r$, as is shown in Figure 5.14. If, for a DF which is L-stable, we can find these two numbers describing the rectangular region where the DF is stable, then stability in this region we call stiff-stability. BDF1-6 obey this type of stability, though, for increasing orders, the size of this stability region reduces. The stiff-stability region always contains infinity, corresponding to the infinite step-size. The stiffly-stable DF is always L-stable, and hence it is stable in the vicinity of an infinite step.

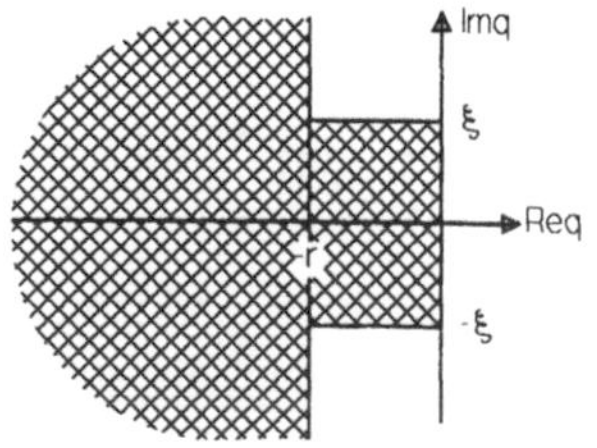

Figure 5.14 A region of stiff-stability

5.4.2 Convergence of differentiation formulae

To introduce the concept of convergence of DF let us consider an instant t. To obtain a solution at this instant we have to integrate the test ODE (5.98) from 0 to t. Let us divide this range into n intervals and integrate by means of a fixed-step DF, whose step-size is $h=t/n$. Theoretically, the solution should be $e^{-\lambda t}$. Due to the inaccuracy of numerical analysis, we obtain a solution x_n different from the exact one. While n is increased towards infinity, the step h decreases to zero. If for any t solution x_n tends to $e^{-\lambda t}$, then DF is called convergent.

Convergence depends on properties of the DF for $h \to 0$. One of its necessary conditions is a bounded response for $h \to 0$, i.e. A_0-stability. This is not sufficient, since an additional requirement for accuracy occurs, i.e. for $h \to 0$ an asymptotic convergence towards an appropriate response value must occur. The latter is possible, iff for $h \to 0$ a dominant (useful) root of the characteristic equation (5.102) obeys $u \to 1$. As can be shown, this occurs when the DF obeys constraints (5.13) of orders 0 and 1. Hence, the DF is convergent iff it is A_0-stable, and at least of the 1st order.

For example, TDF applied to the test ODE gives the solution:

$$x_n = \left(\frac{1+h\lambda/2}{1-h\lambda/2} \right)^n . \qquad (5.109)$$

Introducing $t=nh$ and the complex variable z such that $n=(z+0.5)\lambda t$, we can calculate $\lim_{n\to\infty} x_n$:

$$\lim_{n\to\infty} \frac{1+\dfrac{t\lambda}{2n}}{1-\dfrac{t\lambda}{2n}} = \lim_{z\to\infty} \left(1+\frac{1}{z}\right)^{(z+0.5)\lambda t} = e^{\lambda t} . \qquad (5.110)$$

The limit of the solution thus obtained is the same as the exact solution. Hence, the TDF is convergent.

5.4.3 Contractive differentiation formulae

5.4.3.1 The concept of contractive differentiation formula

The differentiation formula (5.9) manipulates p preceding solutions. They can be in general written as the vector $\mathbf{z}_{n-1} = [x_{n-1}, x_{n-2}, ..., x_{n-p}]^t$. After one time-step, a solution at instant t_n is obtained. At this instant, we can write the following p-dimensional vector of recent solutions:

$\mathbf{z}_n = [x_n, x_{n-1}, \ldots, x_{n-p+1}]^t$. This vector is of course available only for the particular equation being solved. Let us examine the test ODE (5.98), which has been also considered for stability purposes. Stability of DF has provided that the sequence of numerical solutions of (5.98) is bounded. In this section we need a stronger requirement, that the mapping of the vector $\mathbf{z}_{n-1}$ on $\mathbf{z}_n$ should be contractive. Contractivity means that for an arbitrary norm the inequality $\|\mathbf{z}_n\| \leq \|\mathbf{z}_{n-1}\|$ holds for any n. A differentiation formula satisfying this condition for the test ODE (5.98) is called the contractive DF.

This is a sensible requirement, as the exact solution of the test ODE is nonincreasing. In practice, a maximum-norm $\|\ldots\|_{\max}$ can be applied, which is defined as a maximum modulus of components. In the case of the test ODE, all components are positive, and hence, the condition of contractivity transforms into the inequality $x_n \leq \max\{x_{n-1}, x_{n-2}, \ldots, x_{n-p}\}$, which should hold for all n. It can be seen that a solution obtained by means of the contractive DF is upper-bounded by initial values, and hence a contractive DF is also stable.

The numerical solution of the test ODE considered in the previous section obeys the difference equation (5.101), introduced for stability examination purposes. It can be written as follows:

$$(\alpha_0 + \beta_0' q)\, x_n = \sum_{k=1}^{p} (\alpha_k + \beta_k' q)\, x_{n-k}. \tag{5.111}$$

Taking absolute values of both sides of this equation and estimating its RHS by means of the triangle-rule, we obtain the inequality:

$$|\alpha_0 + \beta_0' q|\, |x_n| \leq \|\mathbf{z}_{n-1}\|_{\max} \sum_{k=1}^{p} |\alpha_k + \beta_k' q|. \tag{5.112}$$

It proves that $|x_n| \leq \|\mathbf{z}_n\|_{\max}$, i.e. the DF is contractive iff its coefficients obey the inequality:

$$|\alpha_0 + \beta_0' q| \geq \sum_{k=1}^{p} |\alpha_k + \beta_k' q| \tag{5.113}$$

Let us consider two important specific cases. The DF is 0-contractive, if it is contractive for $q=0$. From (5.113) arises the condition $|\alpha_0| \geq \Sigma\, |\alpha_k|$, which, due to $\alpha_0 = -1$ and the constraint (5.13) of 0 order, leads to the requirement, that all the α_k should be positive.

Moreover, DF is called ∞-contractive, if it is contractive for $q=\infty$. From (5.113) the condition $\beta_0' \geq \Sigma\, |\beta_k'|$ is implied. If both these specific cases occur, then DF is contractive on the whole negative real axis. It is then called A_0-contractive.

For practical purposes, contractivity in the whole left closed halfplane is useful, and we call this the A-contractivity. It happens, if DF is A_0-contractive and contractive on the whole imaginary axis. Assuming that in (5.113) $q=\pm i y^{1/2}$, we obtain, after some algebra, a criterion for A-contractivity:

A DF is A-contractive iff:

it is A_0-contractive, i.e. all α_k are positive and also $\beta_0' \geq \sum |\beta_k'|$ holds,

and for positive y the equation

$$\sum_{k=1}^{p} \sqrt{(\alpha_k^2+\beta_i^2\, y)} \leq \sqrt{(\alpha_0^2+\beta_0^2\, y)} \quad \text{foreach } y \geq 0 \qquad (5.114)$$

holds.

Contractive DFs are a subset of stable DFs, and hence, the A-contractive DFs are a subset of the A-stable. Since only DFs of at most 2nd order can be A-stable, then also A-contractive DFs can be at most of the 2nd order. Their number of steps can be any number $p \geq 1$.

5.4.3.2 Examples of A-contractive differential formukae

Among one-step DFs, all satisfying $0.5 \leq \beta_0' \leq 1$ are A-contractive, particularly the implicit Euler DFs and TDFs. The latter is the only one-step 2nd order A-contractive DFs. To create other A-contractive DF of the 2nd order we have to assume at least $p=2$.

Let us demonstrate an important family of A-contractive DFs: $x_n = h\,\beta_0'\,\dot{x}_n + \alpha_1 x_{n-1} + h\,\beta_1'\,\dot{x}_{n-1} + \alpha_2 x_{n-2} + h\,\beta_2'\,\dot{x}_{n-2}$ of the 2nd order, introduced in reference [5.11]. Its coefficients have to obey constraints (5.13) of orders 0, 1 and 2, the introduced conditions of A_0-contractivity, and a condition (5.114). It can be shown, that their simultaneous satisfaction is possible only if, $F = \sum_{k=0}^{2} \beta_k' / \alpha_k = 0$ holds. This results from the fact that the 2nd order constraint (5.13) implies $F \leq 0$, while a necessary condition of (5.114) is $F \geq 0$. Finally, we choose such coefficients of DF, which maximize the LHS of (5.114). This maximum reaches $F=0$. In that way, the problem of selecting five coefficients is solved for four given conditions. A one-parameter family of fixed step, A-contractive DFs of the 2nd order arises. We introduce the parameter $0 \leq c \leq 1$, describing the existing one degree of freedom. For both extreme values of *c*, i.e. 0 and 1, the TDF arises, while within this range we have a continuum of practical A-contractive DFs. After some algebra, their coefficients take the form:

$$\beta_0' = \frac{1}{2}\,\frac{2+c-c^2}{1+c},\quad \beta_1' = \frac{c^2}{1+c^2},\quad \beta_2' = \frac{1}{2}\,\frac{2-c-c^2}{1+c},$$
$$\alpha_1 = \frac{2c}{1+c},\quad \alpha_2 = \frac{1-c}{1+c}. \tag{5.115}$$

From further experiments we find that one of the most useful in practice is the instance when $c=2/3$:

$$x_n = \frac{2}{3}h\dot{x}_n + \frac{4}{5}x_{n-1} + \frac{4}{15}h\dot{x}_{n-1} + \frac{1}{5}x_{n-2} + \frac{4}{15}h\dot{x}_{n-2}. \tag{5.116}$$

The variable-step A-contractive DF has a more complicated form [5.11]. Implementation of the A-contractive DF provides a considerably better global accuracy than is the case with the A-stable DF. The A-contractive DFs represent the details of responses better, and provide a better numerical damping, than implicit Euler DFs and BDF1-6. To examine how TDFs can mislead a user we can use the sample tasks S5-MIS.CIR, S5-MIS1.CIR of SPICE. Running OPTIMA with sample circuits O5-MIS.CIR, and O5-MIS1.CIR shows how BDFs can be faulty due to stability and damping problems.

5.4.4 Numerical damping of differentiation formulae

Numerical damping is a feature of DFs, introduced for testing how exactly they can analyze oscillating responses. For this purpose, the complex test ODE has been introduced:

$$\dot{x} = \mathrm{j}\,\omega x,\quad \omega \in R^{+}, \tag{5.117}$$

Its solution has a complex oscillating nondamped profile, described by the formula $x(t) = \cos(\omega t) + \mathrm{j}\sin(\omega t)$. The amplitudes of its real and imaginary parts are constant, and equal 1. Hence, the instantaneous value of the modulus of the solution is constant and equals unity.

Trials, solving this test equation by means of DF, lead to the complex sequence x_n of moduli representing the amplitude of numerical oscillation. Due to errors, this amplitude is usually not constant, but is either damped or amplified. We call this phenomenon the numerical damping of oscillations.

Let us create a difference equation describing the solution of the test ODE (5.117). After substituting derivatives in the DF we obtain a difference equation:

$$(-1+\mathrm{j}\,\beta_0'\delta)\,x_n+\sum_{k=1}^{p}(\alpha_k+\mathrm{j}\,\beta_k'\delta)\,x_{p-k}=0, \qquad (5.118)$$

where $\delta=h\,\omega$. Its solution is described by means of the roots of a characteristic equation, similar to the one discussed in our consideration of stability. The same problem is treated in damping and stability cases. Substituting in (5.118) $x_n=u^n$, we obtain the characteristic equation:

$$(-1+\mathrm{j}\,\beta_0'\,\delta)\,u^p+\sum_{k=1}^{p}(\alpha_k+\mathrm{j}\,\beta_k'\,\delta)\,u^{p-k}=0, \qquad (5.119)$$

having probably many roots, though only one of them is a main root u_i, responsible for a useful component $x_n=u_i^n$. The modulus of this component takes the form $|x_n|=|u_i|^n$, which should be constant, equal to unity, if the DF does not introduce any error. In practice, it will increase or more often decrease with *n*, the rate being dependent on the modulus of the main root involved. This modulus should be as close to unity as possible. As can be seen, a plot of this modulus is an objective measure of the damping introduced by the DF. The modulus of the root itself expresses the ratio of damping (or amplifying) amplitude at one step of analysis. We call it the damping factor.

In the case of one-step DFs, the characteristic equation takes the form:

$$(-1+\mathrm{j}\,\beta_0'\,\delta)\,u+[1+\mathrm{j}\,(1-\beta_0')\,\delta]=0, \qquad (5.120)$$

giving the modulus of the solution $|x_n|=|u|^n$, where:

$$|u| = \left[\frac{1+(1-\beta_0')^2\delta^2}{1+\beta_0'^2\delta^2}\right]^{\frac{1}{2}}. \qquad (5.121)$$

In the explicit Euler DF case ($\beta_0'=0$), the amplitude increases geometrically with the quotient $1+\delta^2/2$. Similarly, in the implicit Euler DF case, the amplitude is geometrically damped with the quotient $1-\delta^2/2$. These dampings are extremely strong, not acceptable in practice. In the case of TDFs, numerical damping does not appear at all; the modulus of the root is equal to unity.

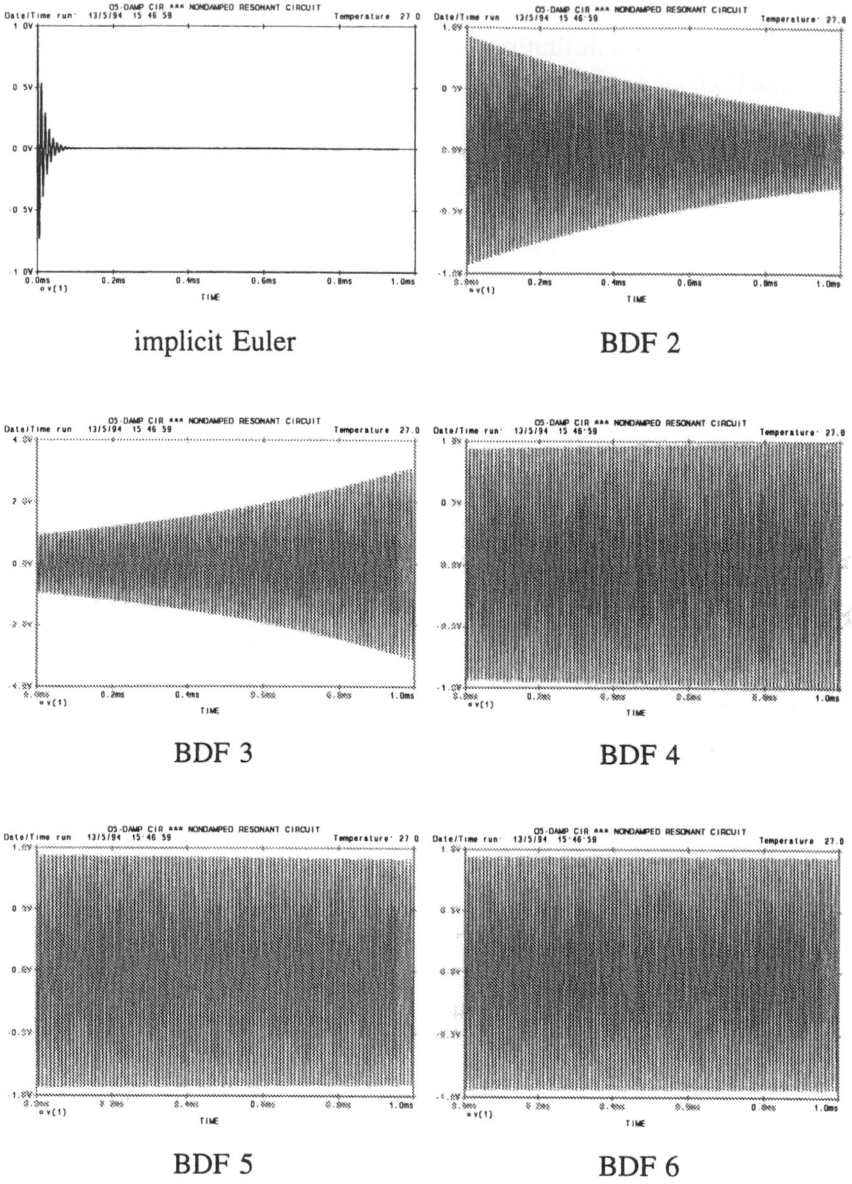

Figure 5.15 Damping of the ideal sinusoidal waveform by means of a family of BDFs

For comparison with those DFs, let us examine the damping of the BDF2. Its characteristic equation $(-1+\mathrm{j}\,2\delta/3)\,u^2+4u/3-1/3=0$ gives the dominating root $u=[2+\exp(\mathrm{j}\,\phi/2]/(3-\mathrm{j}2\delta)$. Its square of modulus $|u|^2=[4+4\cos(\phi/2)+1]/(9+4\delta^2)$ can be approximated for small steps by means of a linear part of the Maclaurin series $|u|^2 = 1 - 4\delta^2/9$. It implies the modulus, i.e. damping factor, in the following form: $|u| = 1 - 2\delta^2/9$. We can find that this quantity is closer to unity than in Euler DF cases, though also not sufficiently close. After similar development, for the A-contractive DF (5.116) we can find that the DF (5.116) is, among those discussed so far, one of the closest to unity damping factor.

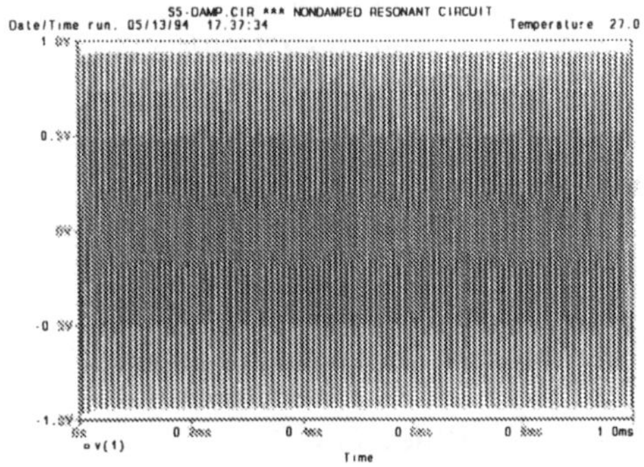

Figure 5.16 Damping of the ideal sinusoidal waveform obtained frm PSPICE using the trapezoidal DF

The task S5-DAMP.CIR enables us to test numerical damping of TDF implemented in SPICE. A non-damped result of this task is shown in Figure 5.16. A similar task O5-DAMP.CIR can be used to test implicit Euler and higher order BDF in OPTIMA. Damping obtained from this test is shown in Figure 5.15. We notice that the implicit Euler DF has very bad damping properties, BDF2 and BDF3 are slightly better, though the latter amplifies oscillations. The formulae BDF4÷6 are the best.

References

5.1 Van Bokhoven, W.M.G. *Linear Implicit Differentiation Formulas of Variable Step and Order,* IEEE Trans. Circ. Syst., vol. 22, no 2, p. 109-115 (1975)

5.2 Brayton, R.K., Gustavson, F.G., Hachtel, G.D. *A New Efficient Algorithm for Solving Differential-Algebraic Systems Using Implicit Backward Differentiation Formulas.* Proc. IEEE, vol 60, no. 1, p. 98-108 (1972)

5.3 Calahan, D.A., *Computer-Aided Network Design,* New York: McGraw-Hill (1968)

5.4 Chua, L.O., Lin, P.M., *Computer-Aided Analysis of Electronic Circuits: Algorithms and Computational Techniques,* Englewood Cliff, New Jersey: Prentice-Hall (1975)

5.5 Coughran, W.M. Jr, Grosse, E., Rose, D.J., *CAzM: A Circuit Analyzer with Macromodelling*, IEEE Trans. Electron. Dev., vol. 30, no. 9, p. 1207-1213 (Sept. 1983)

5.6 Dahlquist, G.G. *A Special Stability Problems for Linear Multistep Methods*, BIT, vol. 3, p. 27-43 (1963)

5.7 Divekar, D.A., *FET Modeling for Circuit Simulation*, Boston, Dordrecht, Lancaster: Kluwer Academic Publishers (1988)

5.8 Gear, Ch.W., *Simultaneous Numerical Solution of Differential-Algebraic Equations*, IEEE Trans. Circ. Theor., vol. 18, no. 1, p. 89-95, (1971)

5.9 Gear Ch.W., *Numerical Initial Value Problems in Ordinary Differential Equations*, Englewood Cliffs: Prentice-Hall (1971)

5.10 Henrici, P., *Discrete Variable Methods in Ordinary Differential Equations,* New York: Wiley (1962)

5.11 Liniger, W., Odeh F., Ruehli A. *Integration Methods for the Solution of Circuit Equations.* In: *Circuit Analysis Simulation and Design.* Amsterdam: North Holland (1986)

5.12 Meyer, J.E. *MOS Models and Circuit Simulation.* RCA Rev., vol. 32, p. 42-63 (1971)

5.13 Ogrodzki, J., Bukat, D., *Compact Modelling in Circuit Simulation: The General purpose Analyser OPTIMA 3*, IEEE Int. Symp. Circ. and Syst., London, U.K. (1994)

5.14 Rübner–Petersen, T. *NAP 2–Mathematical Background.* Rep. Inst. Circ. Theor. Telecom., Tech. Univ. of Denmark, Lyngby, no. 22/2 (1973)

5.15 Vlach J., Singhal, K., *Computer Methods for Circuit Analysis and Design,* (Chapter 13), New York: Van Nostrand (1983)

5.16 Ward, D.E., Dutton R.W., *A Charge-Oriented Model for MOS Transistor Capacitances*, IEEE J. Sol. State Circ., vol 13, no. 5, p. 703-708 (Oct. 1978)

5.17 Yang, P., Epler, B.D., Chatterjee, P.K., *An Investigation of the Charge Conservation Problem for MOSFET Circuit Simulation.* IEEE J. Solid State Circ., vol. 18, no. 1, p. 128-138 (Feb. 1983)

Chapter 6

Periodic Steady-state Time-domain Analysis

6.1 Introductory topics

6.1.1 General overview of the problem

The time-domain transient analysis presented in Chapter 5 consisted of the numerical integration of Ordinary Algebraic Differential Equations (OADEs) of nonlinear circuits. Hence, the methods discussed so far have grasped the problem in a general way. Theoretically, we can use these methods for solving all specific cases. However, in practice there is a category of stiff OADEs, where the problem arises of a large discrepancy between the time constants of different components. In Section 5.3.2.2 we introduced the 1st and the 2nd types of stiffness. The 2nd type gives rise to an especially hard problem, since the high-frequency oscillating response can be modulated by a much slower signal. Then, to accurately model these oscillations, the step size has to be chosen according to their period. However, if we want to observe the signal as a whole, including the modulation, the simulation has to cover a wide range. Thus, integration over a great number of periods is required. Such computations require an extremely large number of integration steps. For example, to see how time-consuming is such an analysis, we can turn back to the resonant circuit file O5-STIFF.CIR for OPTIMA, where oscillations were damped, or to O6-STFF1.CIR where oscillations were modulated.

In many cases, a periodic signal is neither modulated, nor damped but rather tends to a periodic steady state, which is reached very slowly, since the time constant of approaching the steady state is very large in comparison with the period. In file S6-SLOW.CIR for PSPICE, we have an example of such a task, where we obtain a steady state after integration over a wide range with a small step size. In practice, such problems always appear in dynamic periodically excited circuits if the period is short in comparison with the time constants of the circuit, as for example is shown in sample tasks S6-SLOW1.CIR and O6-SLOW1.CIR. Similar problems appear in generators, where after triggering (or self-triggering) oscillations increase slowly and reach the periodic steady state after a great number of periods of oscillation. The latter can be observed in S6-OSC.CIR and O6-OSC.CIR.

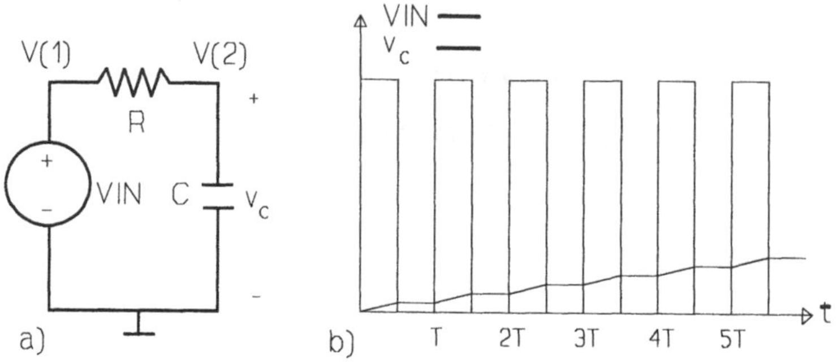

Figure 6.1 Instructive example of a signal tending to a periodic steady state: (a) RC circuit, (b) plot of responses V(1) and V(2)

To better understand how a slowly increasing oscillatory response can be obtained, let us consider an instructive example shown in Figure 6.1(a), where an RC integrating circuit ($R = 1\ \mathrm{M\Omega}$, $C = 1\ \mu\mathrm{F}$) is driven with a rectangular waveform *VIN* of amplitude *E*, period *T*, and PDM 0.5. A few initial cycles of the voltage v_c accross the capacitor *C* are shown in Figure 6.1(b) against the background of input waveform *VIN*. At each period, the capacitor is slightly charged-up, and then it slightly discharges. The mean voltage monotonically, though very slowly, increases. The small, almost trangular, oscillation slowly tends to a steady state. At the $k+1$ cycle the voltage v_c can be expressed as

$$y_{k+1}(t) = \begin{cases} y_k + (E - y_k)\left(1 - e^{-\frac{t-t_k}{\tau}}\right) & (t_k \le t \le t_k + 0.5T) \\ \left(y_k + (E - y_k)\left(1 - e^{-\frac{0.5T}{\tau}}\right)\right) e^{\frac{-(t-t_k-0.5T)}{\tau}} & (t_k + 0.5T \le t \le t_k + T). \end{cases} \tag{6.1}$$

Letting there be $y_{k+1}(t_k + T) = y_k \equiv g$, we obtain its value at the endpoints of a period, corresponding to a steady state:

$$g = E\,\frac{e^{-\frac{0.5T}{\tau}}}{1 + e^{-\frac{0.5T}{\tau}}} \underset{T \ll \tau}{\approx} E\,\frac{1 - \frac{0.5T}{\tau}}{2 - \frac{0.5T}{\tau}} \underset{T \ll \tau}{\approx} \frac{E}{2}. \tag{6.2}$$

We can simulate this example using the PSPICE task S6-RC.CIR or the OPTIMA task O6-RC.CIR and observe how time-consuming even such a simple case can be, if we use a usual .TRAN analysis only. After uncommenting in file S6-RC.CIR the second .TRAN line instead of the first one, we can observe in detail a few cycles of the response and compare it with Figure 6.1(b). The operation of the steady-state analysis algorithm in OPTIMA (command .STDS) will be discussed later on.

As can be seen from these examples, simulation of stiff circuits can take an extremely long time. However, we are often interested only in the steady state, and hence, brute force methods of integrating from the very beginning are not satisfactory. We require faster methods. In periodic steady-state analysis it is essential to find proper initial conditions for numerical integration. Considering Figure 6.2, it can be seen that the response at the end of a period depends on the value from which it has started at the beginning of the period. If a steady state is reached, then the beginning value $f(t+T)$ is the same as the end value $f(t+2T)$, as obtained from integration over one period. Hence, it will be the same also after any number of following periods. The steady state is a limit cycle, where the path of the response repeatedly passes through the same points. Coming back to the task O6-RC.CIR, we can use the postprocessor NV to observe a limit cycle assigning the voltage V(1) to the X-axis and V(2) to the Y-axis. A limit cycle has a useful property: if we know any point on it, we can immediately obtain the whole cycle by integrating over one period from this point. If we integrate from a point slightly perturbed, then we observe that after a few cycles the path comes back to a limit cycle. This can be simulated by uncommenting the .IC command and the .TRAN command following it in the task O6-RC.CIR. The results of this experiment are shown in Figure 6.3(a,b).

The process of simply integrating from the very beginning can be viewed as a sequence of integrations over one period, where at each integration the initial conditions are taken from the endpoint of the preceding integration. Hence, even a brute force method can be interpreted as a process of selecting initial conditions. In dedicated algorithms for

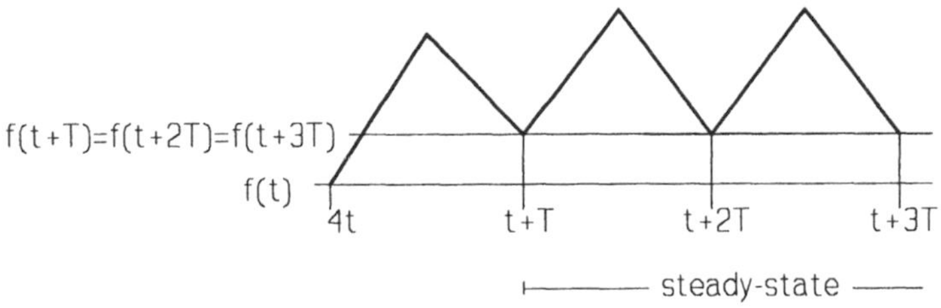

Figure 6.2 Explanation how the periodic steady-state is reached

periodic steady-state analysis the essence is to find the initial conditions in a more efficient way than by taking the result of the preceding integration. Recalling methods of predicting starting values for NR iterations in time-domain analysis, we can say that our brute force method is like a zero-order prediction, where the initial condition was the result of the preceding iteration.

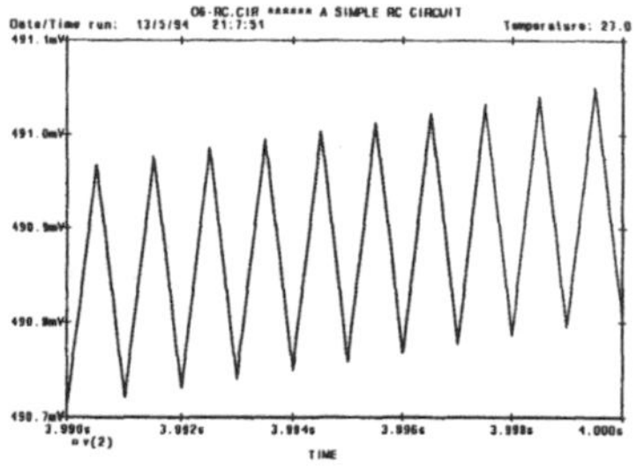

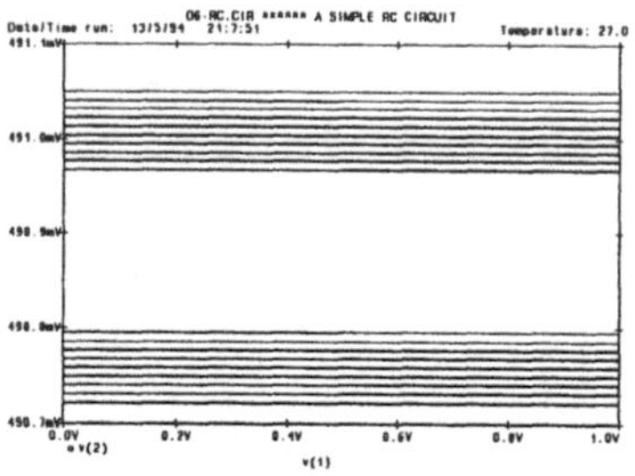

Figure 6.3 (a) A time-profile of V(2) approaching a steady state (b) cycles in [V(1),V(2)]-plane approaching a rectangular limit cycle

Efficient methods of periodic steady-state analysis have to determine the initial conditions. There are frequency and time-domain methods for calculating periodic steady-state responses. Frequency methods are not within the scope of our consideration. Among time-domain methods, we present two algorithms representing the class of shooting methods, where we try to shoot towards a good initial point and integrate from it. The Newton-Raphson approach to the problem of finding initial conditions (IC) has been applied in the algorithm of T.J. Aprille, F.R. Colon, T.N. Trick [6.1, 6.3]. The secant approach is utilized by M. Bukowski's method [6.2]. In the Newton method, derivatives with respect to ICs are calculated by sensitivity analysis. In the secant method, derivatives are approximated. These two algorithms are the most efficient. The Bukowski method has been implemented in OPTIMA. Thus, we can test Bukowski's method practically, while the Aprille *et al* method can only be described.

6.1.2 The periodic steady-state calculation problem

6.1.2.1 Formulation of the problem

Let us consider a dynamic, nonlinear circuit described by means of Ordinary Algebraic Differential Equations (OADE) $\mathbf{F}[\mathbf{x}, \dot{\mathbf{z}}(\mathbf{x})] = \mathbf{s}(t)$ and having a solution under initial conditions $\mathbf{z}(0) = \mathbf{z}_0$. The variables $\mathbf{z}$ are known as state variables. We assume that this circuit contains periodic excitations of a frequency f or commensurate frequencies satisfying $m_1 f_1 = m_2 f_2 = \ldots = m_n f_n$. (the latter provide a periodical character to their superposition). The response $\mathbf{x}(t)$ of this circuit is not periodic in general, though it can have a periodic steady state with predictable period T. This period is equal to $1/f$ or $1/(m_1 f_1)$. A periodic steady state need not always exist, but it can be proved that if the equations can be transformed into the state form $\dot{\mathbf{z}} = \varphi(\mathbf{z}, t)$, where φ is periodic with respect to time t and has a period T, i.e. on the RHS we have additive periodic excitations, then the response has a periodic steady state whose period is T. For example, coming back to the circuit given in Figure 6.1(a) we have a state equation of the form $\dot{v}_c = [VIN(t) - v_c]/(RC)$, where due to a periodic character of $VIN(t)$ the RHS is also periodic.

For each initial condition $\mathbf{z}_0$ at instant 0, a certain solution $\mathbf{x}(t)$ is obtained which, after one period, gives the values $\mathbf{z}[\mathbf{x}(T)]$ of the state variables. These values can be viewed as a picture of the initial conditions $\mathbf{z}_0$ in a mapping. This mapping characterizes the equations involved. It is called the **basic mapping**, and denoted as $\mathbf{\Phi}$:

$$\mathbf{z}[\mathbf{x}(T)] = \mathbf{\Phi}(\mathbf{z}_0). \tag{6.3}$$

A periodic steady state exists, iff the basic mapping has a fixed point $\mathbf{z}_0^\star$, where $\mathbf{z}_0^\star = \mathbf{\Phi}(\mathbf{z}_0^\star)$. In the example shown in Figure 6.1(a) we have the response $y_{k+1}(t)$ (6.1) starting from the value y_k. Hence, taking $y_{k+1} = y_{k+1}(t_k + T)$ we obtain the basic mapping of y_k onto y_{k+1}:

$$y_{k+1} = \mathrm{e}^{-\frac{0.5T}{\tau}} \left[y_k + (E - y_k)\left(1 - \mathrm{e}^{-\frac{0.5T}{\tau}}\right)\right]. \tag{6.4}$$

As we can see, this mapping is linear.

In general, the basic mapping is linear iff the analyzed circuit is linear. Hence, it can be expressed by means of the linear function:

$$\mathbf{z}[\mathbf{x}(T)] = \mathbf{A}(\mathbf{z}_0 - \mathbf{z}_0^\star) + \mathbf{z}_0^\star, \tag{6.5}$$

where $\mathbf{A}$ is a state transit matrix. It transforms the difference between two initial conditions onto the difference between their respective solutions, after one period, namely: $\mathbf{z}[\mathbf{x}^{(1)}(T)] - \mathbf{z}[\mathbf{x}^{(2)}(T)] = \mathbf{A}(\mathbf{z}_0^{(1)} - \mathbf{z}_0^{(2)})$.

Provided that the derivative matrix of the basic mapping (6.3) is continuous and differentiable, we can construct its linear approximation by means of a tangent set in the vicinity of the fixed point. Hence, we obtain a linear formula identical with (6.5):

$$\mathbf{z}[\mathbf{x}^{(k)}(T)] = \mathbf{z}_0^\star + \frac{\partial \boldsymbol{\Phi}(\mathbf{z}_0^\star)}{\partial \mathbf{z}_0}(\mathbf{z}_0^{(k)} - \mathbf{z}_0^\star). \tag{6.6}$$

The matrix of derivatives in (6.6) determines a way of approaching the steady state. Subtracting side-wise the vector $\mathbf{z}_0^{(k)}$ from (6.6) we obtain the relation $\boldsymbol{\Phi}(\mathbf{z}_0^{(k)}) - \mathbf{z}_0^{(k)} = [(\partial\boldsymbol{\Phi}(\mathbf{z}_0^{(k)})/\partial\mathbf{z}) - \mathbf{1}](\mathbf{z}_0^{(k)} - \mathbf{z}_0^\star)$. Using the spectral norm we can prove that if the matrix of derivatives has a maximum eigenvalue close to unity, then the inequality $\|\boldsymbol{\Phi}(\mathbf{z}_0^{(k)}) - \mathbf{z}_0^{(k)}\| \ll \|\mathbf{z}_0^{(k)} - \mathbf{z}_0^\star\|$ holds. Hence, the difference between the values of the response at the boundaries of the period is small in comparison with the distance of the current initial condition from the steady state. It means that the process of approaching the steady state will take a lot of periods. This observation explains the main inconvenience of the brute force method, which is too time-consuming. We try to overcome this problem by developing more efficient algorithms for steady-state analysis.

6.1.2.2 First and second order iterative methods

Initial conditions corresponding to a steady state are zeros of the function:

$$\eta(\mathbf{z}_0) = \boldsymbol{\Phi}(\mathbf{z}_0) - \mathbf{z}_0. \tag{6.7}$$

The simplest numerical method of working out the problem is a simple first order iteration method. It involves the construction of a sequence of approximate solutions, according to the rule $\mathbf{z}_0^{(k+1)} = \boldsymbol{\Phi}(\mathbf{z}_0^{(k)})$. This means, that a solution obtained after one period at the $k-1$ iteration is taken as an initial condition at the kth iteration of periodic steady-state analysis. This is a brute force method.

We prefer the 2nd order method, i.e. the Newton-Raphson (NR) method with respect to the initial conditions $\mathbf{z}_0$. In Chapter 4, the NR approach with respect to circuit variables $\mathbf{x}$ has been described. In this section, the NR approach consists of the linearization of the equation (6.7) in the vicinity of the approximate initial conditions $\mathbf{z}_0^{(k)}$. Then, we obtain $\mathbf{z}_0^{(k+1)}$ from the formula:

$$\left[\frac{d\boldsymbol{\Phi}(\mathbf{z}_0^{(k)})}{d\mathbf{z}_0} - \mathbf{1}\right](\mathbf{z}_0^{(k+1)} - \mathbf{z}_0^{(k)}) = \mathbf{z}_0^{(k)} - \boldsymbol{\Phi}(\mathbf{z}_0^{(k)}). \tag{6.8}$$

If the circuit is linear, then its basic mapping is also linear, and hence,

(6.8) after substituting (6.5) gives $(\mathbf{A}-\mathbf{1})\mathbf{z}_0^{(k+1)}=(\mathbf{A}-\mathbf{1})\mathbf{z}_0^\star$, which proves that the Newton step gives steady-state initial conditions at one step. It can be concluded that the Newton step (6.8) in nonlinear circuits is equivalent to the construction of a local linear approximation to a circuit, i.e. its basic mapping (6.3), and to directly calculating a steady state as a fixed point of (6.3). In practice, if linearization is not performed exactly at the current point but utilizes preceding points, then we have the quasi-Newton step.

Two types of method are available:

(i) exact identification of a matrix comprising the derivatives and the RHS vector of equation (6.8) from nonlinear circuit analysis. This requires the utilization of time-domain sensitivity analysis introduced in Chapter 7,

(ii) use of several previous initial conditions for identification of the linearized basic mapping (6.8).

To summarize and consolidate comprehension of the topics discussed, we can work through the following problems:

- Explain the concept of a periodic steady state, and a basic mapping; show on appropriate diagrams, that the problem of finding a steady state is equivalent to finding appropriate initial conditions;
- Describe a brute-force steady-state analysis method, state the computational effort it requires and explain why we view it as inefficient.

6.2 The method of Aprille, Colon and Trick

6.2.1 A general principle for realization of NR iterations

The NR method with respect to initial conditions has been applied to the periodic steady-state problem by T.J. Aprille, F.R. Colon and T.N. Trick [6.1, 6.3]. It consists of calculating the zeros of the equation (6.7) by means of the NR method, where the matrix of derivatives has been exactly calculated. Linearization of (6.7) with respect to the initial conditions $\mathbf{z}_0$ leads to the iterative equation (6.8), which describes the NR iteration. One NR iteration, according to (6.8), requires:

(i) calculation of a matrix of the derivatives of the circuit response $\mathbf{\Phi}(\mathbf{z}_0)$ at the end of period T with respect to the initial conditions $\mathbf{z}_0$ (i.e. values at the beginning of the period) - this requires time-domain sensitivity analysis over one period,

(ii) calculation of the basic mapping $\mathbf{\Phi}(\mathbf{z}_0)$, which requires integration of the circuit equations (time domain-analysis) over one period.

To formulate a set of n independent equations (6.8) requires that n independent state variables be selected as unknowns. Once MNEs are used, we can choose all independent currents of inductors (or in general, reactive voltage-defined branches) and all independent voltages across capacitors (or in general, reactive current-defined branches). If a cut-set of inductors or a loop of capacitors appears, then one of its components should be omitted.

The main problem is to calculate the matrix $\mathrm{d}\boldsymbol{\Phi}(\mathbf{z}_0)/\mathrm{d}\mathbf{z}_0$. Its columns can be obtained by differentiation of the basic mapping with respect to all components of the initial conditions vector. An algorithm should be constructed, which calculates in the range of one period $t_k \le t \le t_k + T$, originating from initial conditions $\mathbf{z}_0^{(k)}$, time-profiles of the derivatives of time-domain responses with respect to initial conditions. These derivatives at instant $t_n + T$, for each component of the response calculated for each component of the initial conditions, give the matrix required. Hence, we have a time-domain sensitivity analysis problem as discussed in Chapter 7. Since the number of the response components is equal to the number of scalar initial conditions, then we can use either incremental or adjoint methods of sensitivity analysis. The incremental method is recommended due to its simplicity (mainly because of the availability of forward integration).

The incremental method of sensitivity analysis uses an auxiliary differential equation, obtained by differentiating the circuit equations with respect to a variable parameter at a point, which is a current circuit response. This is a differentiation along the path of the solution, leading to the variational equations of a substitute circuit. In the case discussed, we differentiate with respect to the initial conditions. To create an efficient method of formulating variational equations, let us start from a general canonical representation of the circuit:

$$\begin{gathered} \mathbf{A}\,\mathbf{i} = \mathbf{0}, \\ \mathbf{B}\,\mathbf{u} = \mathbf{0}, \\ \mathbf{f}\,[\mathbf{i}, \mathbf{v}, \dot{\mathbf{q}}(\mathbf{v}), \dot{\varphi}(i)] = \mathbf{s}(t) \end{gathered} \tag{6.9}$$

with initial conditions $\mathbf{q}(0) = \mathbf{q}_0$, $\varphi(0) = \varphi_0$, and denote the solution of this initial values problem by $\mathbf{i}(t), \mathbf{v}(t)$. This solution depends on initial conditions, though they do not appear in the equation. We can artificially introduce them into the equations by virtue of the fact that they are additive constants connected with the state variables. Hence, e.g. instead of equation $f(\dot{q}) = 0$ we can write $f(\dot{q} - \dot{q}_0) = 0$, while instead of (6.9) we can write:

$$\mathbf{A\,i} = \mathbf{0},\qquad \mathbf{B\,v} = \mathbf{0},\qquad \mathbf{f}\,[\mathbf{i},\mathbf{v},\dot{\mathbf{q}}(\mathbf{v})-\dot{\mathbf{q}}_0,\dot{\boldsymbol{\varphi}}(i)-\dot{\boldsymbol{\varphi}}_0] = \mathbf{s}(t). \tag{6.10}$$

Differentiating the canonical equations (6.10) along their solution with respect to a certain initial condition c, which can be any component of the initial conditions vectors $\mathbf{q}_0$, $\boldsymbol{\varphi}_0$, we obtain the incremental equations:

$$\mathbf{A}\frac{\partial \mathbf{i}}{\partial c} = \mathbf{0}$$

$$\mathbf{B}\frac{\partial \mathbf{v}}{\partial c} = \mathbf{0} \tag{6.11}$$

$$\frac{\partial \mathbf{f}}{\partial \mathbf{i}}\frac{\partial \mathbf{i}}{\partial c} + \frac{\partial \mathbf{f}}{\partial \mathbf{v}}\frac{\partial \mathbf{v}}{\partial c} + \frac{\partial \mathbf{f}}{\partial \dot{\mathbf{q}}}\frac{\mathrm{d}}{\mathrm{d}t}\left[\frac{\partial \mathbf{q}}{\partial \mathbf{v}}\frac{\partial \mathbf{v}}{\partial c} - \frac{\partial \mathbf{q}_0}{\partial c}\right] + \frac{\partial \mathbf{f}}{\partial \dot{\boldsymbol{\varphi}}}\frac{\mathrm{d}}{\mathrm{d}t}\left[\frac{\partial \boldsymbol{\varphi}}{\partial \mathbf{i}}\frac{\partial \mathbf{i}}{\partial c} - \frac{\partial \boldsymbol{\varphi}_0}{\partial c}\right] = \mathbf{0}$$

where the sensitivities $\partial\mathbf{i}(t)/\partial c, \partial\mathbf{v}(t)/\partial c$ of circuit responses with respect to an initial condition c are unknown variables, while $\partial\mathbf{q}_0/\partial c, \partial\boldsymbol{\varphi}_0/\partial c$ are their initial conditions. Equations (6.11) describe a linear, nonstationary dynamic substitute circuit, structurally identical with (6.10), whose branch signals at any instant are derivatives of responses of the original circuit with respect to a selected initial condition. Branches of this substitute circuit are obtained from original circuit branches by differentiation with respect to c. Hence, so-called incremental (or variational) models of branches arise. Coefficients of these branches are time variant, obtained from a time variant solution to the original equations of the nonlinear circuit.

To calculate sensitivities we have to solve numerically the incremental circuit represented by (6.11), using integration methods described in Chapter 5. Let the differentiation formula (DF) for any state variable x: $\dot{x}_n = \gamma_0(x_n - d_x)$, be applied for discretization of the incremental circuit (6.11). We obtain:

$$\mathbf{A}\frac{\partial \mathbf{i}}{\partial c} = \mathbf{0}$$

$$\mathbf{B}\frac{\partial \mathbf{v}}{\partial c} = \mathbf{0} \tag{6.12}$$

$$\left(\frac{\partial \mathbf{f}}{\partial \mathbf{v}} + \gamma_0\frac{\partial \mathbf{f}}{\partial \dot{\mathbf{q}}}\frac{\partial \mathbf{q}}{\partial \mathbf{v}}\right)\frac{\partial \mathbf{v}}{\partial c} + \left(\frac{\partial \mathbf{f}}{\partial \mathbf{i}} + \gamma_0\frac{\partial \mathbf{f}}{\partial \dot{\boldsymbol{\varphi}}}\frac{\partial \boldsymbol{\varphi}}{\partial \mathbf{i}}\right)\frac{\partial \mathbf{i}}{\partial c} =$$

$$= \gamma_0\left(\frac{\partial \mathbf{f}}{\partial \dot{\mathbf{q}}}\mathbf{d}_{\partial q/\partial c} + \frac{\partial \mathbf{f}}{\partial \dot{\boldsymbol{\varphi}}}\mathbf{d}_{\partial \varphi/\partial c}\right)$$

which is the variational companion circuit. Comparing this circuit with the companion circuit for time-domain analysis (5.56), it can be seen that at any instant the matrix of the incremental companion circuit (6.12) is identical with the matrix of the discretized and linearized original circuit (5.56). In fact, they differ only in their excitations and meaning of variables. The incremental companion circuit is built of incremental companion models of branches. The way in which it is created will be described in the next section.

6.2.2 Incremental companion models for calculation of derivatives

6.2.2.1 General principles

In this section we discuss incremental models for calculating time-domain sensitivities with respect to a chosen initial condition *c*. It is essential to notice that branches of the incremental circuit are linear nonstationary, i.e. their coefficients follow in time the companion models of time-domain analysis. Its original and substitute incremental excitations vanish, and hence, the RHS of (6.11) is zero. However, after discretization a nonzero RHS arises due to introduction of the past storing components. Since the incremental circuit is created using instantaneous values of the time-solution of the original circuit, both circuits have to be integrated simultaneously. Integration of the variational circuit is initialized from its own initial conditions $\partial q_0/\partial c$, $\partial \varphi_0/\partial c$, which appear within brackets in the branch equations of (6.11). Since these initial conditions are derivatives of original initial conditions with respect to a chosen variable *c*, they are usually numbers 0 or ±1.

As is known from the discussion so far, an incremental model of a branch is obtained by differentiation of the branch equation with respect to the initial condition *c* along the path of the original solution. We will consider representative types of branches and show their variational models. Variational models of linear resistive branches are almost the same as the original branches. The conductance $i = Gv$ transforms into the conductance $\partial i/\partial c = G\,\partial v/\partial c$. A nonlinear resistive element produces a linear nonstationary counterpart, which varies according to the path of the original solution. For example, the conductance $i = f(v)$, whose voltage $v(t)$ is presumed, gives the current-defined model, whose conductance varies in time: $\partial i/\partial c = \{\partial f[v(t)]/\partial v\}\,\partial v/\partial c$. At the present instant of integration this model is the same as the companion model for time-domain analysis after the convergence of NR iterations.

Concerning voltage and current sources, as a result of differentiation with respect to c we obtain short and open circuits respectively. For example, $v = V$ transforms into the incremental model $\partial v/\partial c = \partial V/\partial c \equiv 0$, which is a short-circuit.

Now, let us proceed to reactive elements. The linear capacitance, charged up to the initial voltage v_0, can be represented in the integral form $v = v_0 + (1/C)\int_0^t i(\tau)\mathrm{d}\tau$ or in the differential form:

$$i = C\frac{\mathrm{d}}{\mathrm{d}t}(v - v_0). \tag{6.13}$$

Differentiating (6.13) with respect to an initial condition c, and interchanging the order of differentiation we obtain:

$$\frac{\partial i}{\partial c} = C\frac{\mathrm{d}}{\mathrm{d}t}\left(\frac{\partial v}{\partial c} - \frac{\partial v_0}{\partial c}\right), \tag{6.14}$$

i.e. the same capacitance C, though with another initial voltage condition. This condition is $\partial v_0/\partial c$, which is unity if c is the initial voltage of that actual capacitance, -1 if c is that initial voltage taken with the minus sign, and 0 if c corresponds to other elements. The initial condition obtained, $\partial v_0/\partial c$, has to be considered in the time-domain analysis of the incremental circuit.

This time-domain analysis requires discretization by substituting the time derivatives of state variables for DF $\dot{x}_n = \gamma_0(x_n - d_x)$, where state variables x are here derivatives with respect to c, unlike in time-domain analysis. After discretization (6.14) gives the companion incremental model for a linear capacitor:

$$\frac{\partial i}{\partial c} = C\gamma_0\left(\frac{\partial v}{\partial c} - d_{\partial v/\partial c}\right), \tag{6.15}$$

where the past storing variable $d_{\partial v/\partial c}$ stores previous sensitivities $\partial v/\partial c$ and hence propagates the initial condition $\partial v_0/\partial c$.

As an exercise, we can create a similar model for a linear inductor. In the case of the inductor the initial condition concerns the inductor's current. In the incremental circuit this initial condition is equal to 1 if c corresponds to the current of an inductor discussed, and 0 otherwise.

In the case of a charge-wise described nonlinear capacitance:

$$i = \frac{\mathrm{d}}{\mathrm{d}t}\left[q(v) - q(v_0)\right]. \tag{6.16}$$

After differentiation with respect to an initial condition c, we obtain a charge-wise described incremental model:

$$\frac{\partial i}{\partial c} = \frac{\mathrm{d}}{\mathrm{d}t}\left[\frac{\partial q[v(t)]}{\partial v}\frac{\partial v}{\partial c} - \frac{\partial q(v_0)}{\partial c}\right], \tag{6.17}$$

which expresses a capacitance whose current is $\partial i/\partial c$, and voltage is $\partial v/\partial c$. It stores a charge $\partial q/\partial c = \{\mathrm{d}q[v(t)]/\mathrm{d}v\}\,\partial v/\partial c$, and hence is linear nonstationary. Its capacitance $\mathrm{d}q[v(t)]/\mathrm{d}v$ does not depend on the incremental circuit variables but follows the original solution $v(t)$. Provided that the charge of this incremental capacitor, as given above, is selected as a state variable, then also its initial condition $\partial q(v_0)/\partial c$ should be the charge. It will be 1, if c is the charge initial condition on the capacitor discussed. If c correspond to other elements, then this initial condition will be 0. Presuming the charge to be state variable, after discretization we have the companion incremental model:

$$\frac{\partial i}{\partial c} = \gamma_0\left[\frac{\partial q[v(t)]}{\partial v}\frac{\partial v}{\partial c} - d_{\partial q/\partial c}\right] \tag{6.18}$$

where the past storing variable $d_{\partial q/\partial c}$ stores $\partial q/\partial c$ at previous instants according to the DF requirements. It propagates the charge-wise initial condition $\partial q(v_0)/\partial c$.

If the capacitor is described in terms of a voltage state variable, i.e. $i = C(v)(\dot{v} - \dot{v}_0)$, then after differentiating with respect to c we obtain a parallel connection of a capacitance and conductance, namely

$$\frac{\partial i}{\partial c} = \frac{\partial C[v(t)]}{\partial v}\dot{v}\frac{\partial v}{\partial c} + C[v(t)]\frac{\mathrm{d}}{\mathrm{d}t}\left[\frac{\partial v}{\partial c} - \frac{\partial v_0}{\partial c}\right] \tag{6.19}$$

where the nonstationary conductance equal to $\partial C/\partial v\,\dot{v}$ has to follow $v(t)$ (its time derivative is also required), while the nonstationary capacitance is equal to $C(v)$. Moreover, this model requires a voltage initial condition $\partial v_0/\partial c$ for incremental circuit analysis. In this case the initial condition is of voltage-type, and hence, $\partial v_0/\partial c$ is 1 or -1 if c corresponds to the capacitance discussed at the moment, while it is 0 for c corresponding to other state variables.

Similarly, we can calculate incremental models for a nonlinear inductor assuming either linked flux or current as state variables. This calculation could be a good exercise.

To calculate the whole matrix of derivatives we must create incremental circuits for c covering all initial conditions of elements, i.e.

charges (or voltages) of capacitors and fluxes (or currents) of inductors. Hence, in each of these incremental circuits the displacement of the initial conditions of elements will be different.

Now, we briefly explain how the algorithm discussed can be implemented within a standard procedure of time-domain analysis. After the convergence of the NR loop, additional calculations should be executed at each integration step. Since variational and original companion models differ only in the RHS, and due to the availability of the current matrix in a LU decomposed form (see Chapter 3), we only have to formulate separate RHS vectors for each incremental circuit excitation (it will contain only the past storing terms, see (6.12), (6.15)), and then, perform the forward and backward substitution. Since companion variational circuits have their specific past storing terms, then in comparison with time-domain analysis additional past storing tables are necessary. Thus, analysis of the whole matrix of derivatives is quite memory-consuming, while its time-consumption is equivalent to the time required by one more NR iteration, at each time-point.

6.2.2.2 Example of the calculation of derivatives with respect to initial conditions

Let us consider a parallel nonlinear resonant network driven by a periodic current source $I_1(t)$, as shown in Figure 6.4(a). This circuit contains two state variables q_2, φ_3. To perform its time-domain analysis we have to create a companion circuit, according to Chapter 5. This circuit is shown in Figure 6.4(b). Its elements take the form:

$$G_{2n}^{(p)} = \gamma_0 \frac{q_0}{V_T} \exp\left(v_n^{(p)}/V_T\right), \quad I_{2n}^{(p)} = -\gamma_0 d_{q2}, \tag{6.20}$$
$$R_{3n}^{(p)} = \gamma_0 L_3, \quad V_{3n}^{(p)} = -\gamma_0 d_{\varphi 3},$$

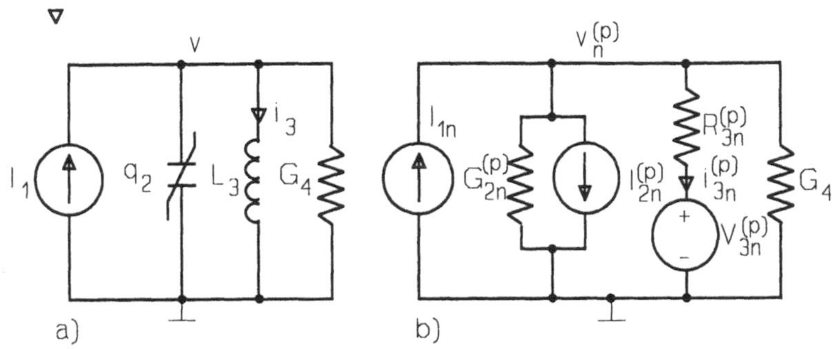

Figure 6.4 (a) The resonant network, and (b) its companion counterpart for time-domain analysis

where the trapezoidal DF has been assumed, i.e. $\gamma_0 = 2/h$ and the past storing variable $d_x = 0.5h\dot{x}_{n-1} + x_{n-1}$. The variable x is of course formal, and in our consideration it will be replaced by the state variables required. In (6.19) we have two variables storing the past of the state variables q_2, φ_3. At each NR iteration of periodic steady-state analysis, we integrate the circuit given in Figure 6.4(a) over one period. This means a repetitive solution (within time and NR loops) of the companion circuit in Figure 6.4(b).

At each integration point t_n, if calculation of the solution v_n, i_{3n} is completed, then we can proceed to creating incremental circuits for calculating the derivatives with respect to initial conditions q_{20}, $\varphi_{30} = L_3 i_{30}$. Differentiating branch equations, according to the rules presented in the preceding section, we obtain for $i_1 = I_1(t)$ the open-circuit $\partial i_1/\partial c = 0$, and for $i_4 = G_4 v_4$ the conductance $\partial i_4/\partial c = G_4 \partial v_4/\partial c$.

The linear inductor $v_3 = \dot{\varphi}(i_3)$, where $\varphi(i_3) = L_3 i_3$, after differentiation with respect to c and then discretization by means of DF gives the incremental companion model of the form $\partial v_3/\partial c = \gamma_0(L_3 \partial i_3/\partial c - d_{\partial\varphi_3/\partial c})$. It is equivalent to the voltage-defined branch $\partial v_3/\partial c = R_{3n} \partial i_3/\partial c + V_{3n}$, where R_{3n} is the same as $R_{3n}^{(p)}$ after convergence of the NR loop at t_n, while $V_{3n} = -\gamma_0 d_{\partial\varphi_3/\partial c}$. This model requires the current-type initial condition $\partial\varphi_{30}/\partial c$, which will be unity if $c = \varphi_{30}$, and 0 if $c = q_{20}$.

Finally, we proceed to the nonlinear capacitor $i_2 = \dot{q}_2(v_2) - \dot{q}_{20}$, which after differentiation with respect to c gives the incremental model

$$\frac{\partial i_2}{\partial c} = \frac{\mathrm{d}}{\mathrm{d}t}\left[\frac{q_0}{V_T}\exp\left[v(t)/V_T\right]\frac{\partial v}{\partial c} - \frac{\partial q_{20}}{\partial c}\right]. \tag{6.21}$$

It indicates that an initial condition for this model is $\partial q_{20}/\partial c$, and hence it will be 1 if $c = q_{20}$, and 0 if $c = \varphi_{30}$. Discretizing this model at t_n we obtain the companion incremental model

$$\frac{\partial i_2}{\partial c} = \gamma_0\left[\frac{q_0}{V_T}\exp(v_{2n}/V_T)\frac{\partial v_2}{\partial c} - d_{\partial q_2/\partial c}\right]. \tag{6.22}$$

Taking $c = q_{20}$ and φ_{30} respectively, we obtain two incremental companion circuits of the same form, shown in Figure 6.5, and hence having the same companion matrix. This matrix is identical with the matrix of the companion original circuit of Figure 6.4(b) after convergence of the NR loop. These two incremental companion circuits should be integrated with different sets of initial conditions: the first one with $\partial q_{20}/\partial q_{20} = 1$ and $\partial\varphi_{30}/\partial q_{20} = 0$ on a capacitor and inductor respectively, and the second

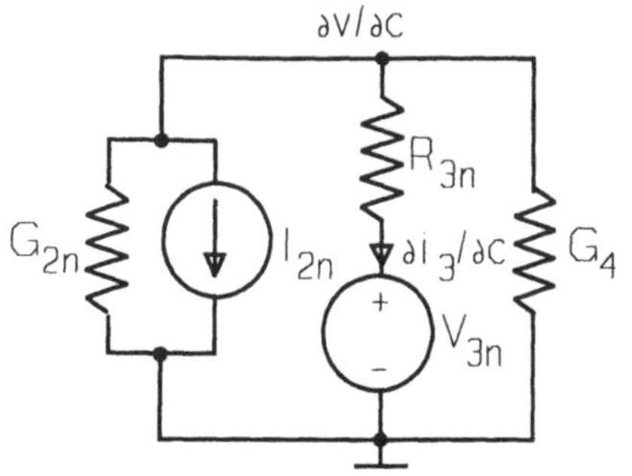

Figure 6.5 Companion incremental counterpart of the circuit given in Figure 6.4(a) for calculation of partial derivatives of time-response with respect to the initial condition *c*.

one with $\partial q_{20}/\partial \varphi_{30}=0$ and $\partial \varphi_{30}/\partial \varphi_{30}=1$ on a capacitor and inductor respectively. Hence, both circuits differ in their past storing excitations, i.e. RHS vectors. From these circuits we calculate two sets of MNE variables at instant t_n: $[\partial v_n/\partial q_{20}, \partial i_{3n}/\partial q_{20}]$ and $[\partial v_{2n}/\partial \varphi_{30}, \partial i_{3n}/\partial \varphi_{30}]$ respectively.

6.2.3 Application of the NR method to autonomous circuits

Autonomous circuits (oscillators) cannot be analyzed using the approach discussed so far, because the matrix of derivatives $\partial \mathbf{\Phi}(\mathbf{z}_0)/\partial \mathbf{z}_0$ of the basic mapping (6.4), calculated for $\mathbf{z}_0$ lying on the solution path, has one eigenvalue equal to unity. Therefore, the matrix of the NR iterative system (6.8) is singular and cannot be solved for the required initial conditions. Moreover, in this case the period length is not known *a priori*, and hence, we do not know an interval of integration required for calculating a basic mapping value.

We have to introduce their period T as an unknown among the set of initial conditions to analyze oscillators. Authors of the methods discussed propose to introduce the period instead of the pth component of the initial conditions vector $\mathbf{z}_0$. Thus, we solve the problem with the following unknowns

$$\hat{\mathbf{z}}_0 = [z_{01}, \ldots, z_{0,p-1}, T, z_{0,p+1}, \ldots, z_n]^t. \tag{6.23}$$

We can do this, since the basic mapping may not use all state variables. There is only one requirement to ensure the independence of the selected variables. In the integration process we have to select, for the initial condition z_{0p} which has been omitted in the vector (6.23), a value within the range taken during oscillations.

Modification (6.23), slightly changes the form of the iterative formula (6.8), where the period becomes the pth unknown. Therefore, the linearization of $\mathbf{\Phi}(\mathbf{z}_0)-\mathbf{z}_0$ with respect to $\hat{\mathbf{z}}_0$, in the vicinity of $\hat{\mathbf{z}}_0^{(k)}$, gives the following iterative NR formula:

$$\left[\frac{\partial\hat{\mathbf{\Phi}}(\mathbf{z}_0^{(k)})}{\partial\hat{\mathbf{z}}_0}-\hat{\mathbf{I}}\right](\hat{\mathbf{z}}_0^{(k+1)}-\hat{\mathbf{z}}_0^{(k)}) = \mathbf{z}_0^{(k)}-\mathbf{\Phi}(\mathbf{z}_0^{(k)}), \tag{6.24}$$

where $\hat{\mathbf{I}}=\partial\mathbf{z}_0/\partial\hat{\mathbf{z}}_0$ is the unity matrix except for position (p, p), where 0 appears, while matrix $\partial\hat{\mathbf{\Phi}}(\hat{\mathbf{z}}_0^{(k)})/\partial\hat{\mathbf{z}}_0$ is identical with $\partial\mathbf{\Phi}(\mathbf{z}_0^{(k)})/\partial\mathbf{z}_0$ at all positions except for the pth column, which is equal to the derivative $\partial\mathbf{\Phi}(\mathbf{z}_0^{(k)})/\partial T$. This derivative of the circuit response with respect to period, calculated after the period, can be obtained by virtue of the fact that the value of this response is the value of a state variable (charge of a capacitor and flux of an inductor). For example, the voltage across a capacitor after one period can be written as:

$$v(T)-v(0) = \frac{1}{C}\int_0^T i_C(t)\,\mathrm{d}t, \tag{6.25}$$

Its derivative with respect to period is $\partial v(T)/\partial T = i_C(T)/C$, while the derivative of charge is $\partial q/\partial T=[\partial q(T)/\partial v]\, i_C/C$. Identically, for the flux of an inductor: $\partial\varphi(T)/\partial T = [\partial\varphi(T)/\partial i]\, v_L(T)/L$. Hence, the pth column of the matrix, equal to $\partial\mathbf{\Phi}(\mathbf{z}_0^{(k)})/\partial T$, will contain terms dependent on the currents of those capacitors and voltages of those inductors, which define state variables collected in $\mathbf{z}$. These terms are nonzero, and hence the diagonal position (p, p) will be nonzero. Thus, the matrix of derivatives is nonsingular. To provide the best numerical properties for the system (6.24), the derivative introduced at position (p, p) should be the maximum one, i.e. among n components of $\partial\mathbf{\Phi}(\mathbf{z}_0^{(k)})/\partial T$ one of the largest magnitude should be selected.

6.2.4 Realization of algorithm

6.2.4.1 Basic algorithm

The basic implementation of the Aprille-Colon-Trick method is demonstrated in Algorithm 6.1. Its essential feature is to simultaneously integrate three circuits: a companion nonlinear circuit corresponding to the original circuit analyzed at the present instant t, and two companion incremental circuits required for calculating derivatives with respect to initial conditions. They have the same matrix $\mathbf{Y}$. The first circuit is integrated from initial conditions $\mathbf{z}_0^{(0)}$, the next two from initial conditions $\partial\mathbf{z}_0/\partial\mathbf{z}_0$, obtained from the companion incremental models discussed in Section 6.2.3.2.

Algorithm 6.1 Basic Newton-Raphson algorithm for periodic steady-state analysis

```
    initialization of z0(0);
c:  initialization of past storing tables for original state variables z
    and incremental ∂z/∂z0; t=0; h=initial_step;
b:  t=t+h; prediction of MNE variables x(0);
a:  setting up the circuit matrix Y(0), and RHS vector b(0);
    LU decomposition of Y(0);
    if (Y(0) is singular) then stop
    solution of Y(0) x(1) = b(0);
    if (x(0), x(1) do not obey the test for NR loop convergence) then
    begin
       x(0) = x(1); goto a;
    end
    calculation and storing of z(x(1));
    for i=1 to n do
    begin
       creation of the RHS vector b_i for ith incremental circuit;
       solution of Y(0) (∂x/∂z_i0) = b_i; calculation of ∂z/∂z_i0 using
       ∂x/∂z_i0 and storage of it as the ith column of ∂z/∂z0;
    end
    updating of the past storing tables for variables z and ∂z/∂z0;
    if (t<T) goto b;
    Solution of (6.8) or (6.24) for new initial conditions z0(1);
    if (a distance between z0(0), z0(1) does not satisfy the steady-state
    analysis test for convergence) then
    begin
       z0(0) = z0(1); goto c;
    end
    stop
```

Let us focus on a certain instant t. First, NR iterations are completed. Then, for n state variables $\mathbf{z}$, n RHS vectors $\mathbf{b}_i$ are subsequently created, according to MNE, using models given in Section 6.2.3.2. Next, incremental companion equations are solved. From their solutions subsequent columns $\partial \mathbf{z}/\partial z_{i0}$ of the matrix of derivatives $\partial \mathbf{z}/\partial \mathbf{z}_0$ are obtained. All past storing variables of the original and incremental circuits are updated and we proceed to the next point. The process in continued until the instant T (end of period) is reached. Now, the matrix $\partial \mathbf{z}/\partial \mathbf{z}_0$ and variables $\mathbf{z}$ correspond exactly to the endpoint of the period, and NR formula (6.8) or (6.24) can be used to calculate the next initial conditions $\mathbf{z}_0^{(1)}$. The test for convergence of steady-state analysis compares the distance between $\mathbf{z}_0^{(0)}$ and $\mathbf{z}_0^{(1)}$. If it is not satisfied, then $\mathbf{z}_0^{(1)}$ are stored as $z_0^{(0)}$ and the integration procedure is repeated from new initial conditions.

This basic method can cause several numerical problems: no convergence; in oscillators the algorithm can converge to a local solution, which is an unstable limit cycle; moreover, considerable numerical complexity in this algorithm can a large amount of time consume. Convergence problems appear when we do not have good starting values approximating the initial conditions. These can be found using the brute-force method, i.e. performing an initial time-domain analysis over few periods. As in the initial phase of analysis, we notice the influence of some fast components of the response. They can be omitted by starting integration over a few periods. The range of this initial brute-force analysis can be detected using the following error index:

$$E^{(k)} = \frac{\sum_{i=1}^{n} w_i^2 \, [\phi_i(z_0^{(k)}) - z_{0i}^{(k)}]^2}{\sum_{i=1}^{n} w_i^2 \, (z_{0i}^{(k)})^2}, \tag{6.26}$$

where ϕ_i is the ith component of state variables after a period, while x_{0i} is the same component before a period, k is the NR iteration count, and the weight w_i is equal to the capacitance or inductance, where this initial condition is defined. This error index is a measure of the relative variability of state variables over one period. The initial analysis is continued for so long as this index remains greater than unity. If it becomes less than unity, then we proceed to the steady-state NR iteration. However, if during these iterations the index once more exceeds unity, then we return to brute-force iterations.

Reference [6.3] proposes damping the iterations to improve the convergence of the steady-state NR iterations. The damping factor $\alpha^{(k)} = 1 - (E^{(k)})^6$ is controlled by the error index. Therefore, in autonomous circuits, instead of (6.24) the following damped system is solved:

$$\left[\alpha^{(k)}\frac{\partial \Phi_p^\star(\mathbf{z}_0^{(k)})}{\partial \hat{\mathbf{z}}_0} - \mathbf{I}_p\right](\hat{\mathbf{z}}_0^{(k+1)} - \hat{\mathbf{z}}_0^{(k)}) = \mathbf{z}_0^{(k)} - \Phi(\mathbf{z}_0^{(k)}), \tag{6.27}$$

where the matrix $\partial \Phi_p^\star / \partial \hat{\mathbf{z}}_0$ is the same as $\partial \hat{\Phi} / \partial \hat{\mathbf{z}}_0$, with the exeption that element (p, p) is divided by $(\alpha^{(k)})^2$.

For nonautonomous (nonoscillating) circuits, the damping iterative NR formula takes the form:

$$\left[\alpha^{(k)}\frac{\partial \Phi(\mathbf{z}_0^{(k)})}{\partial \mathbf{z}_0} - \mathbf{1}\right](\mathbf{z}_0^{(k+1)} - \mathbf{z}_0^{(k)}) = \mathbf{z}_0^{(k)} - \Phi(\mathbf{z}_0^{(k)}). \tag{6.28}$$

To decrease the numerical complexity of the algorithm, an auxiliary bypassing of the matrix updating is applied. Bypassing is performed when initial conditions in subsequent steady-state NR iterations are changed infinitesimally from those for which the matrix has been recently calculated. An index of relative change of these initial conditions (similar to (9.25)) is defined as:

$$EP^{(k)} = \frac{\sum_{i=1}^{n} w_i^2 (z_{0i}^{(k+1)} - z_{0i}^{(k)})^2}{\sum_{i=1}^{n} w_i^2 (z_{0i}^{(k)})^2}. \tag{6.29}$$

and can be used as the bypassing criterion. The matrix of derivatives for the next steady-state NR iteration is updated iff the index *EP* is greater than 0.2 or the index *E* has increased in comparison with recent iteration.

For a further improvement in efficiency, secondary state variables (e.g. corresponding to internal capacitances of transistors) can be neglected. The modified version of the steady-state analysis, as described above, is presented in Algorithm 6.2. Another approach is introduced in [6.6].

To ensure asimilation of the foregoing topics, the reader may work through the following main problems and exercises:

- Explain a NR method of steady-state analysis; what sort of circuit analyses and computational effort is required;
- Take a simple RC circuit and explain the concept of incremental circuit and incremental equations for sensitivity analysis with respect to initial conditions. Demonstrate that this circuit is parametric and that its elements trace in time a solution of the original circuit;

Algorithm 6.2 Periodical steady-state algorithm according to the Colon and Trick method

$k=0$; set initial conditions $\mathbf{z}_0^{(0)}$ and initial period $T^{(0)}$; {in non-autonomous circuits it will be constant}

a: $k=k+1$; {initial brute-force iterations}
integration of a circuit over one period $T^{(k)}$ starting from initial conditions $\mathbf{z}_0^{(k)}$; calculation of $E^{(k)}$ from (6.25);
$\mathbf{z}_0^{(k+1)} = \Phi(\mathbf{z}_0^{(k)})$; $T^{(k+1)} = T^{(k)}$;
if ($E^{(k)} \geq 1$) **then goto** a;

b: $k=k+1$; $\mathbf{z}_0^{\star} = \mathbf{z}_0^{(k)}$; {calculation of derivatives matrix}
integration of circuit equations over one period $T^{(k)}$ starting from initial conditions $\mathbf{z}_0^{(k)}$ with simultaneous calculation of a derivatives matrix; for generating circuits also the characteristic position p is selected and derivatives with respect to the period are calculated;
calculation of $E^{(k)}$ from (6.25);

c: **if** ($E^{(k)} \leq \varepsilon$) **then stop** {algorithm termination}
if ($E^{(k)} \geq 1$) **then**
begin
  $\mathbf{z}_0^{(k+1)} = \Phi(\mathbf{z}_0^{(k)})$; **goto** a;
end
$\alpha^{(k)} = 1 - (E^{(k)})^6$;
for generating circuits the NR step is performed according to (6.27), for nongenerating, according to (6.28);
calculate $EP^{(k)}$ from (6.29);
if ($EP^{(k)} \geq 0{,}2$ || $E^{(k)} > E^{(k-1)}$) **goto** b;
$k=k+1$; {integration of equations without calculating a new derivatives matrix}
integration of the circuit over one period $T^{(k)}$ from initial conditions $\mathbf{z}_0^{(k)}$;
goto c;

- Choose a charge-described nonlinear capacitance and calculate its incremental companion model; show which part of it is identical with its companion model, and which part is different; repeat the exercise for a nonlinear flux-described inductance; will similar models occur for voltage-described capacitors and current-described inductors?
- Take a simple linear capacitor C; explain what initial conditions should be assigned, in a variational companion circuit, for sensitivity analysis; are they somehow connected with the initial conditions of the original circuit ?

6.3 Method of Bukowski

6.3.1 Theoretical formulation of the method

In the method introduced by M. Bukowski [6.2] the simple idea of exploiting secant iterations instead of NR ones has been introduced. In fact, steady-state iterations are performed according to the formula (6.8). However, the derivatives matrix is estimated from previous solutions in a manner corresponding to the multivariable secant method [6.4]. Let us assume that, at the *k*th iteration of the steady-state analysis, we know the sequence of solutions, at the end of period $\mathbf{\Phi}(\mathbf{z}_0^{(i)})$ $(i=k-n, k)$, of analyses originated from initial conditions $\mathbf{z}_0^{(i)}$ at the start of that period. These solutions have been stored during the $n+1$ preceding iterations of steady-state analysis. In the steady-state NR method only one preceding solution has been used, obtained from the last iteration. Using linearization at this point, a linear circuit was created and a Newton step was performed. This approach requires at the last iteration a solution and (very inconveniently) its derivative with respect to the initial conditions.

To override this awkward calculation of derivatives M. Bukowski has proposed the use of a number of preceding iterative solutions. If they obey a linear basic mapping (6.5), then the identification of this mapping is possible. However, since they are in general the result of a nonlinear analysis, we only can stretch on them an interpolating linear mapping and then treat it as a linearization at a recent point. This will be a secant linearization, different from a tangent-based NR linearization.

To identify the linear mapping (6.5) at the *k*th iteration of the secant steady-state analysis let us presume a sequence of $n+1$ vectors of initial conditions, enumerated with index *i* ranging from $k-n$ to *k*. Solutions, at the end of the period obtained from analyses started from these vectors of initial conditions, obey a linear basic mapping iff, the vectors of differences (see a comment connected with formula (6.5)) are connected by means of a constant state transit matrix $\mathbf{A}^{(k)}$. Let these differences at the begining of the period be denoted as $\mathbf{y}^{(i)}=\mathbf{z}_0^{(i)}-\mathbf{z}_0^{(i-1)}$, while at the end of the period they are $\mathbf{r}^{(i)}=\mathbf{\Phi}(\mathbf{z}_0^{(i)})-\mathbf{\Phi}(\mathbf{z}_0^{(i-1)})$, where $i=k-n+1, ..., k$. All these differences satisfy the linear basic mapping, and hence obey $\mathbf{A}^{(k)}\mathbf{y}^{(i)}=\mathbf{r}^{(i)}$. Since we have *n* such relations, they can be collected to define matrices of differences $\mathbf{Y}^{(k)}$, $\mathbf{R}^{(k)}$ and to formulate *n* equations:

$$\underbrace{\left[\mathbf{r}^{(k-n+1)}|\dots|\mathbf{r}^{(k)}\right]}_{\mathbf{R}^{(k)}} = \mathbf{A}^{(k)}\underbrace{\left[\mathbf{y}^{(k-n+1)}|\dots|\mathbf{y}^{(k)}\right]}_{\mathbf{Y}^{(k)}}. \tag{6.30}$$

where | denotes location of columns side by side).

If the $n+1$ points $[\mathbf{z}^{(i)}, \Phi(\mathbf{z}^{(i)})]$ obey the linear mapping, then the matrix $\mathbf{Y}^{(k)}$ is nonsingular and the state transit matrix $\mathbf{A}^{(k)} = \mathbf{R}^{(k)}(\mathbf{Y}^{(k)})^{-1}$ of the linear mapping, interpolating the nonlinear basic mapping, can be identified. Hence, to obtain a local linear approximation passing through the points involved, it is sufficient to substitute the matrix in (6.8) for the matrix of derivatives. After some algebra we obtain a set of two equations, which gives the iterative initial conditions for a linear mapping

$$\begin{aligned} (\mathbf{Y}^{(k)} - \mathbf{R}^{(k)})\,\mathbf{v} &= \Phi(\mathbf{z}_0^{(k)}) - \mathbf{z}_0^{(k)} \\ \mathbf{z}_0^{(k+1)} &= \Phi(\mathbf{z}_0^{(k)}) + \mathbf{R}^{(k)}\mathbf{v}. \end{aligned} \tag{6.31}$$

This system constitutes the kth secant iteration. To perform it, the first equation should be solved for $\mathbf{v}$, and then a new initial condition $\mathbf{z}_0^{(k+1)}$ is calculated from the second equation.

Considering equation (6.8) or (6.31) we see that, if the described method is convergent, then its iterative solution tends to a fixed point of the basic mapping $\mathbf{z}_0^{\star} = \Phi(\mathbf{z}_0^{\star})$ for any nonsingular matrix $\mathbf{A}^{(k)}$. It can be proved that in the vicinity of this point the method has a local convergence exponent greater than 1 but lower than 2. We omit a wider discussion of the local and global convergence properties of this algorithm [6.2, 6.4]. Let us observe, that as the algorithm converges, the vector $\mathbf{v}$ tends to $\mathbf{0}$, and the component of the second equation (6.31) added to the value of the state variable at the period endpoint, also gradually vanishes. Thus, in the vicinity of the solution the algorithm discussed approaches the brute force method.

Concerning numerical complexity, we can state that Bukowski's method requires $n^3/3 + 2n^2 - n/3$ multiplications and divisions per time-domain analysis over one period. This is a negligible computational effort in comparison with circuit analysis over this interval.

6.3.2 Extension of the secant method to autonomous circuits

In autonomous circuits we do not know the period of the steady-state solution, but as we know from Section 6.2.3 the period can be incorporated into the unknown vector of initial conditions. The transit matrix $\mathbf{A}^{(k)} = \mathbf{R}^{(k)}(\mathbf{Y}^{(k)})^{-1}$ of the basic mapping (6.5), calculated for $\mathbf{z}_0$ lying on the solution path, has one eigenvalue equal to unity, similar to the matrix of derivatives in Aprille *et al*'s method. Therefore, if $\mathbf{A}^{(k)}$ is substituted in the iterative equation (6.8), its matrix becomes singular along the solution path. Thus, the steady-state of an autonomous circuit cannot be found by solving (6.8) for initial conditions. Incorporating the period into the initial conditions vector instead, of one component, eliminates this singularity.

Bukowski proposes another approach. His treatment of autonomous circuits consists of a small perturbation of a regular structure of $\mathbf{R}^{(k)}$ to avoid singularity of the first equation (6.31)

$$(\mathbf{Y}^{(k)} - \hat{\mathbf{R}}^{(k)})\nu = -\mathbf{z}_0^{(k)} + \Phi(\mathbf{z}_0^{(k)}) \tag{6.32}$$

We will discuss this below in details.

Combining both equations (6.31) it can be readily shown that the second equation (6.31) (which expresses the secant step) can be written in the following equivalent form

$$\mathbf{z}_0^{(k+1)} = \mathbf{z}_0^{(k)} + \mathbf{Y}^{(k)}\nu . \tag{6.33}$$

This can be easily applied to the period T, since we know its kth approximation and we can create a row vector $\mathbf{Y}_T^{(k)}$ which is a counterpart of $\mathbf{Y}^{(k)}$ corresponding to the period T:

$$\mathbf{Y}_T^{(k)} = [T^{(k)} - T^{(k-1)}, T^{(k-1)} - T^{(k-2)}, \ldots, T^{(k-n+1)} - T^{(k-n)}]$$

Thus, combining the second equation (6.31) with (6.33) we obtain the augmented the secant step for autonomous circuits

$$\begin{bmatrix} \mathbf{z}_0^{(k+1)} \\ T^{(k+1)} \end{bmatrix} = \begin{bmatrix} \Phi(\mathbf{z}_0^{(k)}, T^{(k)}) \\ T^{(k)} \end{bmatrix} + \begin{bmatrix} \mathbf{R}^{(k)} \\ \mathbf{Y}_T^{(k)} \end{bmatrix} \nu \tag{6.34}$$

Iterative solution of the period by means of the secant step stands for linearization of the relationship joining a time interval with initial conditions. Due to its highly nonlinear character, our algorithm requires an accurate initial guess at the period (not worse than 10 per cent).

This algorithm uses integration of the analyzed circuit over the interval $T^{(k)}$, starting from initial conditions $\mathbf{z}^{(k)}$, to calculate their new value $\Phi(\mathbf{z}^{(k)}, T^{(k)})$ after this interval. Then, in both autonomous and nonautonomous circuits, we calculate auxiliary variables ν using the first equation (6.31) or equivalently (6.32) (the perturbation will be explained later on).

In practice, we use a set of more than n circuit variables instead of n state variables $\mathbf{z}$, and hence (6.32) is realized by the formulation of a system with a rectangular matrix $\mathbf{Y} - \hat{\mathbf{R}}$. Then, we use the full pivoting procedure, consisting of rows and columns interchanging (see Section 3.2.3), to solve this system for n circuit unknowns, maximizing the magitudes of the pivots. In this way, an automatic selection of n independent rows and columns (unknowns) is performed. Due to such an approach, we can use the circuit variables of MNE (Section 2.1.3) or extended MNE (Section 5.2.1.5) instead of state variables $\mathbf{z}$. Hence, we automatically solve for n chosen components circuit variables. Obtaining these variables, we then calculate, using (6.34), new values of initial conditions for the state variables $\mathbf{z}^{(k+1)}$ or, which is equivalent, for the circuit variables. Then a new interval of integration $T^{(k+1)}$ is predicted using the second equation (6.34).

The length of this interval tends to the period as the initial conditions converge. Convergence of this algorithm, composed of equations (6.32) and (6.34), is almost the same as convergence by the Aprille *et al* method, though it is incomparably less time-consuming, since it does not require sensitivity analysis at all. This algorithm has been implemented by Bukowski in a version of the NAP2 program and in OPTIMA 3. Examination of its properties and an explanation of some practical aspects of its use are given in the next section.

6.3.3 Practical implementation of the method

The method just introduced can be easily implemented as a general-purpose algorithm, which for nonautonomous circuits treats the period, which has been inputted, as a constant, while in the autonomous circuits case it treats it as an initial guess and performs iterations according to (6.34). OPTIMA provides the command:

.STDS *tincr tmax* [*tmin* [*stepmax*]] PER=*period* [AUT] [UIC] .

If the keyword AUT does not appear, then the algorithm will operate as if *period* is the given period of a nonautonomous circuit. Otherwise, *period* is used as an initial guess at the period of an autonomous circuit. Surprisingly, a nonautonomous circuit can be analyzed using a version of the algorithm dedicated to autonomous circuits. However, the algorithm for nonautonomous circuits does not converge, unless the initial guess at the period is successful. Unfortunately our algorithm is considerably sensitive to the initial guess at the period. The command .STDS provides a calculation of steady-state initial conditions, the period, and then executes time-domain analysis from instant 0 to *tmax* with display increment *tincr* and a maximum step-size *stepmax* (if it has been introduced). Integration starts from the initial conditions thus found. If the calculated initial conditions are not accurate, then the profiles of responses obtained will not be the steady-state. This may happen if the accuracy of this analysis is slightly different from the accuracy of analyses performed during the course of Bukowski's algorithm.

Now we proceed to practical experiments with an OPTIMA implementation of this algorithm. The test O6-RC.CIR deals with the example linear circuit of Figure 6.1(a). Because the circuit is linear, the initial conditions are obtained in only one secant iteration. We can compare the results of this example with an analytical description (6.2) and Figure 6.1(b), where the signal very slowly approaches its steady state.

As a good test for the algorithm, we can select three error-levels of time-domain analysis, and observe their influence on finding a steady state. We execute the task O6-ERTR.CIR, with three alternative settings of an option ERTR, controlling the relative local truncation error (see Section 5.3.2.3). For decreasing accuracy, the steady state obtained becomes more and more incorrect, which can be seen by comparing it with the accurate transient analysis. This stems from the fact, that if the transient analysis used in the steady-state algorithm is inaccurate, then those inaccurate initial conditions are satisfactory from the point of view of a test for convergence of the steady-state algorithm.

In general, it can be stated that the algorithm of periodic steady-state analysis gives the steady state of a numerical response calculated by the simulator, not of a response of a physical system. It means that, if the transient analysis gives a considerably inaccurate response, then a steady state of this inaccurate response is found, instead of a steady state of an unknown exact time-domain response. Steady-state analysis appears to be extremely sensitive to inaccuracy of time-domain analysis which gives rise to additional secant iterations.

A crucial point in the efficiency of the algorithm discussed is a steady-state accuracy index ERST which is used for breaking secant iterations. In linear circuits theoretically it is of no importance, since iterations should converge in one steps. In practice several steps may be required when a time-domain analysis is insufficiently accurate. In such a case, but especially in highly nonlinear circuits, as for example O6-NL1.CIR, ERST is of great importance. If its value is too big (see O6-NL1.CIR) this will cause great inaccuracy in a steady state.

Table 6.1 shows how initial conditions approach the steady state in the case of a nonautonomous circuit O6-NL1.CIR. A similar problem in an autonomous circuit is tested in the example O6-GEN1.CIR. In Table 6.1 absolute inaccuracy $z_{0i}^{(k)} - z_{0i}^{\star}$ for each component is calculated at each iteration with respect to a final value $z_{0i}^{\star}$ obtained after convergence. We observe an irregular character of initial iterations. Then the rate of convergence ranges from linear to superlinear in final iterations.

Table 6.1 Secant algorithm: convergence of O6-NL1.CIR (absolute errors in parentheses)

it.	V(5)	V(4)	V(1)	V(3)	I(L8)	I(L10)
1	41.70 (13.32)	4.170 (4.459)	-9.394e-1 (9.033e-1)	1.397 (-0.02)	9.402e-3 (9.26e-3)	-0.5342e-1 (0.4538e-1)
2	39.99 (11.61)	4.545 (4.834)	-8.432e-1 (-8.071e-1)	1.400 (-0.017)	6.797e-3 (6.655e-3)	-0.5459e-1 (0.4421e-1)
3	29.61 (1.23)	4.271 (4.560)	-6.604e-1 (-6.243e-1)	1.394 (-0.023)	4.181e-3 (4.039e-3)	-0.6074e-1 (0.3806e-1)
4	21.81 (-6.57)	3.383 (3.672)	-2.890e-1 (-2.529e-1)	1.409 (-0.008)	-7.106e-3 (-7.248e-3)	-1.089e-1 (0.0101e-1)
5	21.56 (-6.82)	3.091 (3.380)	-.2218e-1 (-0.1397e-1)	1.482 (0.065)	-10.59e-3 (-10.73e-3)	-1.207e-1 (0.219e-1)
6	23.59 (-4.79)	2.799 (3.088)	0.1882e-1 (0.5497e-1)	1.483 (0.066)	-10.46e-3 (-10.60e-3)	-1.236e-1 (-0.248e-1)
7	29.00 (0.62)	-0.9152 (-0.6262)	-0.1141e-1 (-0.2474e-1)	1.380 (-0.037)	3.550e-3 (3.408e-3)	-0.9848e-1 (0.0032e-1
8	28.33 (-0.05)	-0.2413 (0.0477)	-0.4947e-1 (-0.1332e-1)	1.419 (0.002)	0.2154e-3 (0.0736e-3)	-1.018e-1 (-0.030e-1)
9	28.34 (0.04)	-0.2486 (0.0404)	-0.4771e-1 (-0.1156e-1)	1.418 (0.001)	0.2124e-3 (0.0706e-3)	-1.006e-1 (-0.0180e-1)
10	28.38 (0.0)	-0.2888 (0.0002)	-0.3762e-1 (-0.0147e-1)	1.417 (0.0)	0.1679e-3 (0.0261e-3)	-0.9858e-1 (-0.0022e-1)
11	-	-0.2891 (-0.001)	-0.3638e-1 (-0.0023e-1)	1.417 (0.0)	0.1478e-3 (0.006e-3)	-0.9890e-1 (0.0010e-1)
12	-	-0.2890 (0.0)	-0.3619e-1 (-0.0004e-1)	-	0.1428e-3 (0.001e-3)	-0.9882e-1 (-0.0002e-1)
13	-	-	-0.3615e-1 (0.0)	-	0.1418e-3 (0.0)	-0.9880e-1 (0.0)

Diskette A contains some additional tests for steady-state analysis, namely: O6-NL2.CIR, O6-NL3.CIR, O6-GEN2.CIR, *etc*. We can observe that the algorithm for autonomous circuits requires that a discrepancy between an initial guess at the period and a true period value be not greater than about 10 per cent. This is a serious weak point in this algorithm.

Now, we proceed to more details of the implementation of this algorithm. Its initialization is a hard problem, in practice. To obtain a steady-state in only a few iterations requires a good initial estimation of the state transit matrix. Moreover, in the autonomous case the matrix of the first equation (6.31) has to be perturbed to avoid singularity of (6.32). To initalize the matrices $\mathbf{Y}^{(0)}, \mathbf{R}^{(0)}, \mathbf{Y}_T^{(0)}$ we integrate the circuit over three periods with N breakpoints $t_{ij}=T^{(0)}(i-j/N)$, $i=1,2,3$, $j=N-1,N-2,...,0$, uniformly distributed within each period. The number of breakpoints N should be greater than the number of state variables n. However, we use only the last n points of each of the three periods. We store the values of circuit variables at these points and calculate the following differences

$$\begin{aligned}
\mathbf{y}^{(j)} &= \mathbf{z}(t_{2j}) - \mathbf{z}(t_{1j}) \quad (j=n-1, ..., 0)\\
\mathbf{r}^{(j)} &= \mathbf{z}(t_{3j}) - \mathbf{z}(t_{2j}) \quad (j=n-1, ..., 1)\\
\mathbf{r}^{(0)} &= \mathbf{z}(t_{30}) - \mathbf{z}(t_{21})\\
y_T^{(j)} &= T^{(0)} - T^{(0)} = 0 \quad (j=n-1, ..., 1)\\
y_T^{(0)} &= T^{(0)} - T^{(0)}(1-1/N) = \frac{T^{(0)}}{N}
\end{aligned} \tag{6.35}$$

to construct matrices

$$\begin{aligned}
\mathbf{Y}^{(0)} &= [\,\mathbf{y}^{(n-1)} | \ldots | \mathbf{y}^{(0)}]\\
\mathbf{R}^{(0)} &= [\,\mathbf{r}^{(n-1)} | \ldots | \mathbf{r}^{(0)}]\\
\mathbf{Y}_T^{(0)} &= \left[0, ..., 0, \frac{T^{(0)}}{N}\right].
\end{aligned} \tag{6.36}$$

To avoid the singularity of these initial matrices, the column corresponding to $j=0$ has been perturbed by one point. This nonsingularity will be next propagated through all subsequent secant iterations.

To start calculations we must properly select starting values for the period. Using the test O6-GEN2.CIR we can examine how sensitive the algorithm is to this data. Perturbing the starting period we observe the acceptable maximum distance from the final solution.

Now, we proceed to a description of the algorithm structure. In its implementation it is essential that it cooperates with a conventional time-domain analyzer, operating with interrupts on breakpoints generated by the steady-state procedure. Thus, we start by initialization, and then integrate over three periods, gradually reaching breakpoints t_{ij} and storing the state vectors at these breakpoints. When the instant $3T^{(0)}$ is reached, then one secant iteration is performed. It gives a new vector of initial conditions and a new period. Now, the analyzer is started from the calculated initial conditions to integrate over the calculated period. After termination of this analysis, state variables at the endpoint are stored, working tables are updated, and the next secant iteration is executed. It gives a new initial conditions vector and so a new vector of differences. Updating of the working matrices $\mathbf{Y}, \mathbf{R}, \mathbf{Y}_T$ consists of shifting them one column left and filling the right column with the last vector of differences. As a test for convergence, the distance (a norm) between two successive initial conditions vectors $[\mathbf{z}_0^{(k)}, T^{(k)}]$, $[\mathbf{z}_0^{(k+1)}, T^{(k+1)}]$ is tested. If the test is not satisfied and the maximum number of secant iterations is not exceeded, then the process is repeated by proceeding to the next transient analysis from calculated initial conditions and over the calculated period. This procedure is given in Algorithm 6.3.

In conclusion, we state that the algorithm requires very little computational effort in comparison with the time needed for circuit integration. The secant iteration is approximately equivalent to circuit analysis at one more time point. For most examples tested this algorithm has appeared to be convergent in a number of one period integrations which is not greater, and sometimes even smaller, than in the case of the NR algorithm (see Section 6.2). To observe this, we can proceed to literature benchmarks: O6-NL1.CIR, O6-AT_DC.CIR, O6-AT_CLC.CIR, O6-NAKH1.CIR, O6-NAKH2.CIR, *etc.* Bukowski's algorithm also appears to be more efficient and easier in implementation than Skelboe's method [6.5], which has not been discussed in this book.

The following exercise can be helpful in understanding the content of this section:

- Take a function $\mathbf{y} = \mathbf{f}(\mathbf{x})$, $\mathbf{x} \in R^n$ and a set of $n+1$ points $\mathbf{x}^{(i)}$ and explain how a secant plane stretched over these points can be created; explain how the above result can be applied to the basic mapping.

Algorithm 6.3 Periodic steady-state analysis according to Bukowski's method

data: starting period $T^{(0)}$, starting initial conditions $\mathbf{z}_0^{(0)}$, steady-state analysis convergence indexes ε, δ, maximum number of secant iterations *kmax*;

{initialization}
for $i=1$ **to** 3 **do**
for $j=N-1$ **downto** 0 **do**
begin
  $t_{ij}=T^{(0)}(i-j/N)$;
  if $(j\leq n-1)$ **then**
  begin
    time-domain analysis until the instant t_{ij};
    storage of the solution vector $\mathbf{x}(t_{ij})$;
  end
end
creation of $\mathbf{Y}^{(0)}$, $\mathbf{R}^{(0)}$, $\mathbf{Y}_T^{(0)}$ from the above results according to equations (6.35), (6.36);
for $k=1$ **to** *kmax* **do**
begin
  solution of the rectangular system (6.32) for ν using pivoting;
  calculation of circuit variables $\mathbf{z}_0^{(1)}$ from the first equation (6.34);
  $T^{(1)}=T^{(0)}$;
  if $(T^{(0)}<0)$ calculation of $T^{(1)}$ from the second equation (6.34);
  time-domain analysis with initial conditions $\mathbf{z}_0^{(1)}$ from 0 to $T^{(1)}$;
  storage of $\mathbf{x}(T^{(1)})$ as $\mathbf{x}^{(1)}$;
  if (test of a distance between $\mathbf{x}^{(0)}$, $\mathbf{x}^{(1)}$ using indexes ε, δ is satisfied) **then**
  begin
    time-domain analysis with initial conditions $\mathbf{z}^{(1)}$ within a range defined by user in command line;
    stop; {successful termination}
  end
  updating matrices $\mathbf{Y}^{(0)}$, $\mathbf{R}^{(0)}$, $\mathbf{Y}_T^{(0)}$;
end
stop {no convergence of secant iterations}

References

6.1 Aprille, T.J. Jr., Trick, T.N., *Steady-State Analysis of Nonlinear Circuits with Periodic Inputs*, Proc. IEEE, vol. 60, no. 1, p. 108-114 (1972)

6.2 Bukowski, M., *Fast Algorithm for Numerical Calculation of Steady State in Nonlinear Dynamical Circuits of Periodic Excitations*, Proc. XIII Nat. Conf. Circ. Theory and Electronic Netw., Bielsko-Biała, Poland, p. 681-686 (1990)

6.3 Colon, F.R., Trick, T.N. *Fast Periodic Steady-State Analysis for Large Signal Electronic Circuits,* IEEE J. Solid State Circ., vol.8, no. 4, p. 260-269 (1973)

6.4 Ortega, J.M., Rheinboldt, W.C., *Iterative Solution of Nonlinear Equations in Several Variables*, New York: Academic Press (1970)

6.5 Skelboe, S., *Computation of the Periodic Steady-State Response of Nonlinear Networks by Extrapolation Methods,* IEEE Trans. Circ. Syst., vol. 27, no. 3, p. 161-175 (1977)

6.6 Strohband, P.H., Laur, R., Engl, W.L., *TNPT-An Efficient Method to Simulate Forced Nonlinear RF Networks,* IEEE J. Solid State Circ., vol. 12, p. 243-246 (1977)

Chapter 7

Sensitivity Analysis

7.1 Foundations of sensitivity analysis

7.1.1 Definitions of sensitivities

A nonlinear electrical circuit is represented by means of a system of nonlinear OADEs, which can be written in a general form:

$$\mathbf{F}(\mathbf{x},\, d\mathbf{x}/dt,\, \mathbf{p}) = \mathbf{s}(\mathbf{p}), \tag{7.1}$$

where $\mathbf{p}$ is a vector of circuit parameters, e.g. the parameters R, L, C of linear elements, parameters of nonlinear elements I_s, n, and of excitations I. The symbol $\mathbf{x}$ denotes a vector of circuit variables (e.g. according to MNE) describing the circuit response to excitations $\mathbf{s}$. For each $\mathbf{p}$ let there be only one response $\mathbf{x}$ satisfying Equation (7.1). Hence, that response can be viewed as a function of the circuit parameters $\mathbf{p}$. We will call it the **network function** and denote it as $\mathbf{x}(\mathbf{p})$. In nonlinear circuits there are cases when network functions are multivalued, but we shall not take into account such cases in our sensitivity analysis consideration. In practice, we usually consider scalar network functions, e.g. $x_j(\mathbf{p})$ (the jth component of $\mathbf{x}$), or simply $x(\mathbf{p})$, in general.

We define the **small-change sensitivity** of the network function $x(\mathbf{p})$ with respect to a parameter p_i, which is the ith component of $\mathbf{p}$ at a nominal point $\mathbf{p}_0$, as its partial derivative with respect to p_i calculated at this nominal point:

$$D_{p_i}^{f(\mathbf{p}_0)} = \frac{\partial f(\mathbf{p}_0)}{\partial p_i}. \tag{7.2}$$

This sensitivity is a measure of the absolute change in the network function per unit absolute change in a parameter. So, given a differential approximation by increments we can estimate the network function increment $\Delta x = D_{p_i}^{x}\,\Delta p_i$ caused by a parameter change.

If, in a particular problem, more significance attaches to the relative change in the network function due to a parameter change, then we can use a relative small-change sensitivity:

$$S_{p_i}^{f(\mathbf{p}_0)} = \frac{p_{i0}}{f(\mathbf{p}_0)} \frac{\partial f(\mathbf{p}_0)}{\partial p_i}, \tag{7.3}$$

which expresses a differential approximation to the relative change in the network function for a 100 per cent relative change in a parameter.

If the relative change in a parameter is less significant than the absolute change, then we prefer to use a **semi-relative small change sensitivity** with the network function normalized, namely:

$$Q_{p_i}^{f(\mathbf{p}_0)} = \frac{1}{f(\mathbf{p}_0)} \frac{\partial f(\mathbf{p}_0)}{\partial p_i}. \tag{7.4}$$

This sensitivity expresses a differential approximation to the relative change of network function per unit absolute change in the parameter.

Any parameter p_i, with respect to which we calculate a sensitivity, is called a variable parameter, since sensitivities express the influence of changes in this parameter on a change in network function, i.e. a network response. Therefore, they play an important role in circuit design.

As a simple example, let us consider the voltage divider shown in Figure 7.1. We have its three parameters R_1, R_2, V. Selecting the signal v as the response of interest we can write the network function:

$$v(R_1, R_2, V) = V\frac{R_2}{R_1 + R_2}. \tag{7.5}$$

Now, we can calculate three relative sensitivities (7.3) of this function with respect to the parameters involved:

$$\begin{aligned} S_V^v &= \frac{R_2}{R_1 + R_2} = k, \\ S_{R_1}^v &= -\frac{R_1}{R_1 + R_2} = -1 + k, \\ S_{R_2}^v &= \frac{R_1}{R_1 + R_2} = 1 - k. \end{aligned} \tag{7.6}$$

It can be seen that none of them depend on V, and in fact are functions only of the ratio of division $v/V = k$. When this ratio tends to unity, sen-

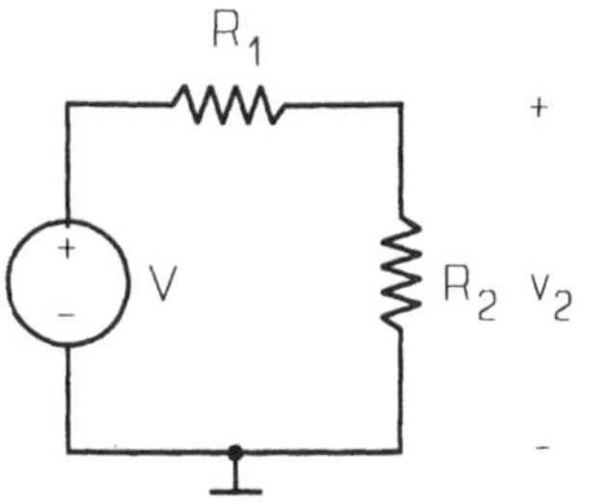

Figure 7.1 A voltage divider

sitivites with respect to resistors decrease towards zero. Hence, the circuit becomes insensitive only if it does not work. Sensitivity is its immanent property. Furthermore, in this case we have no degree of freedom to reduce the sensitivities. In circuits containing more elements there are often many such degrees.

So far the sensitivities introduced are of first order, since they are defined by means of first order derivatives. In some design and optimization problems **second** (or even higher) **order sensitivities** can also be useful. Let us demonstrate for example the 2nd order absolute sensitivity with respect to a parameter p_i:

$$DD_{p_i, p_i}^{f(\mathbf{p}_0)} = \frac{1}{2} \frac{\partial^2 f(\mathbf{p}_0)}{\partial p_i^2} \tag{7.7}$$

and the mixed 2nd order absolute sensitivity with respect to p_i and p_j:

$$DD_{p_i, p_j}^{x(\mathbf{p}_0)} = \frac{\partial^2 x(\mathbf{p}_0)}{\partial p_i \partial p_j}. \tag{7.8}$$

Such sensitivites can also be expressed in a relative form, e.g.:

$$SS_{p_i, p_j}^{x(\mathbf{p}_0)} = \frac{p_i p_j}{x(\mathbf{p}_0)} \frac{\partial^2 x(\mathbf{p}_0)}{\partial p_i \partial p_j}. \tag{7.9}$$

but their practical meaning is small. Therefore, higher-order sensitivity consideration has been neglected in this book.

7.1.2 Sensitivities as quality measures of circuit response spreads

An important goal in circuit design is to control and keep to a sufficiently low level the sensitivities of circuit responses with respect to the parameters of circuit elements. As we know, circuits always have non-zero sensitivities, since, if they were not sensitive to parameters, they could not be sensitive to signals and therefore could not operate. So, the designing process must take into account sensitivities and aim at their minimization. To provide the engineer with sensitivity analysis it is usually implemented in general-purpose circuit simulators. PSPICE 5.02 enables the user to calculate the sensitivities of selected d.c. responses with respect to all circuit parameters. SPICE3 also offers sensitivity analysis in the time and frequency domains. From examples S7-SENS1.CIR and S7-SENS2.CIR we obtain large tables of sensitivities, and can determine the critical parameters of these circuits. In practical silicon implementations, such parameters should have small tolerances assigned.

Usually, knowing a network function change due to changes in parameters, is more important than knowledge of particular sensitivities. Thus, assuming absolute increments in parameters **p**, we can obtain an increment of the network function $x(\mathbf{p})$ using a differential evaluation:

$$|x(\mathbf{p})-x(\mathbf{p}_0)| \approx \sum_{i=1}^{\dim \mathbf{p}} |D_{p_i}^{x(\mathbf{p}_0)}| \, |p_i - p_{i0}| . \tag{7.10}$$

We call this formula for a response increment the **small-change worst-case evaluation**. If relative changes in parameters are given, then we can evaluate the relative change in a network function using relative sensitivities:

$$\left|\frac{x(\mathbf{p})-x(\mathbf{p}_0)}{x(\mathbf{p}_0)}\right| \approx \sum_{i=1}^{\dim \mathbf{p}} S_{p_i}^{x(\mathbf{p}_0)} \left|\frac{p_i - p_{i0}}{p_{i0}}\right| . \tag{7.11}$$

For example, in the example S7-WCASE.CIR we have assigned tolerances (worst case spreads) to parameters of certain elements, and calculate small-change worst-case increments in output signals using sensitivity analysis and tools for worst-case analysis implemented in PSPICE.

7.1.3 Large-change sensitivities and sensitivity nonlinearities

The large-change sensitivity is an extension of the small-change sensitivity to arbitrary increments in a variable parameter. The idea behind such an extension is the same as the consideration of difference quotients instead of derivatives. Thus, as the **large-change sensitivity** of a network function $x(\mathbf{p})$ with respect to a parameter p_i at a point $\mathbf{p}_0$ we take the difference quotient:

$$DL_{p_i}^{x} = \frac{x(\mathbf{p})-x(\mathbf{p}_0)}{p_i-p_{i0}}, \tag{7.12}$$

where $\mathbf{p}=\mathbf{p}_0+\mathbf{e}_i(p_i-p_{i0})$ is a point, which differs from $\mathbf{p}_0$ only in the ith coordinate ($\mathbf{e}_i$ is a vector having zeros at all positions except the ith, which is unity). The large-change sensitivity describes the network function with a single variable parameter.

We can also define a **relative large-change sensititvity**, similarly to (7.3), namely:

$$SL_{p_i}^{x(\mathbf{p}_0)} = \frac{p_{i0}}{x(\mathbf{p}_0)} \, \frac{x(\mathbf{p})-x(\mathbf{p}_0)}{p_i-p_{i0}}. \tag{7.13}$$

If the network function is linear with respect to a variable parameter p_i, then the large-change sensitivity is equal to the corresponding small-change sensitivity. Nonlinearity in a network function causes them to diverge. As a measure of this divergence a **nonlinearity index** has been introduced [7.7]:

$$NI_{p_i}^{x} = \frac{x(\mathbf{p})-x(\mathbf{p}_0) - D_{p_i}^{x}(p_i-p_{i0})}{x(\mathbf{p})-x(\mathbf{p}_0)} = \frac{SL_{p_i}^{x} - S_{p_i}^{x}}{SL_{p_i}^{x}}. \tag{7.14}$$

It expresses the relative error of network function increments calculated by means of small-change sensitivities.

Recalling the voltage divider shown in Figure 7.1, and assuming the increment ΔR_1, we can calculate the relative large-change sensitivity with respect to R_1:

$$SL_{R_1}^{v} = \frac{R_1}{\dfrac{R_2}{R_1+R_2}} \, \frac{\dfrac{R_2}{R_1+\Delta R_1+R_2} - \dfrac{R_2}{R_1+R_2}}{\Delta R_1} = \\ = -\frac{R_1}{R_1+\Delta R_1+R_2} = -\frac{1-k}{1+(1-k)\dfrac{\Delta R_1}{R_1}} \tag{7.15}$$

where the nonlinearity index takes the form

$$NI_{R_1}^{v} = (1-k)\frac{\Delta R_1}{R_1} \tag{7.16}$$

and hence it is proportional to the change in the parameter.

Summing up this section we underline the main problem which has been discussed:

- Different definitions of sensitivities and their importance in circuit and system design.

7.2 Small-change sensitivity analysis of static circuits

In this section we consider static circuits, that is those described by algebraic equations. This case comprises linear and nonlinear circuits in steady state under d.c. excitations, or small-signal a.c. excitations. In the d.c. case, a circuit is described by means of real nonlinear algebraic equations. In the a.c. case, that circuit is linearized and symbolically transformed into the $j\omega$ domain (see Chapter 2) where it can be described by complex linear equations. Thus, the canonical equations (see Chapter 1) of a static circuit can take real or complex algebraic form. Introducing variable parameters $\mathbf{p}$ we can write these equations in the form:

$$\begin{aligned} \mathbf{A}\,\mathbf{i} &= \mathbf{0} \\ \mathbf{B}\,\mathbf{v} &= \mathbf{0} \\ \mathbf{f}(\mathbf{i}, \mathbf{v}, \mathbf{p}) &= \mathbf{s}(\mathbf{p}) \end{aligned} \quad . \tag{7.17}$$

where $\mathbf{p}$ can be branch parameters, e.g. R, I_s, ω appearing on the LHS, or parameters of excitations, e.g. d.c. value, a.c. amplitude or phase appearing on the RHS. In a specific case the circuit may be linear, and then its branches have the form:

$$\mathbf{K}(\mathbf{p})\,\mathbf{i} + \mathbf{L}(\mathbf{p})\,\mathbf{v} = \mathbf{s}(\mathbf{p}) \tag{7.18}$$

where the matrices of branch coefficients $\mathbf{K}$, $\mathbf{L}$, and excitations $\mathbf{s}$ can be dependent on the parameters $\mathbf{p}$.

7.2.1 The incremental equations method

Let us consider a system of nonlinear algebraic equations with an embedded parameter p. This system has a vector of unknowns $\mathbf{x}$. In this section we deal with the following problem. Let p be perturbed by an increment Δp. Then a resultant change $\Delta\mathbf{x}$ arises. To calculate this increment exactly we have to solve the equations twice: first for the nominal p to obtain $\mathbf{x}$, and then for $p+\Delta p$ to obtain $\mathbf{x}+\Delta\mathbf{x}$. However, if the parameter perturbation is small, we can approximate increments in a differential manner. This consists of looking for a differential increment $\Delta\mathbf{x}_d = [\mathrm{d}\mathbf{x}(p)/\mathrm{d}p]\,\Delta p$ instead of looking for $\Delta\mathbf{x}$. In the former case the derivative is calculated at the nominal solution $\mathbf{x}(p)$, at the point p. Since the particular value of perturbation Δp is not important, the problem can be viewed as the calculation of a solution derivative with respect to a parameter. This calculation is performed at a nominal solution of the equation concerned. To solve this problem efficiently we formulate substitute equations, where differential increments are the unknowns. Such an approach is called incremental. It is equivalent to the problem of calculating the derivatives of the solution with respect to a parameter. Such an approach is called variational. To simplify our terminology we will treat both approaches as the same and call them incremental. Equations with unknown solution derivatives will be called the incremental equations. Now we proceed to the incremental appproach to circuit sensitivity analysis.

7.2.1.1 The substitute incremental circuit

To discuss the incremental method of sensitivity analysis we consider a general static circuit represented by means of canonical equations (7.17) (in linear case with (7.18)). Since the incremental approach is suitable for the case of only one variable parameter, we substitute in these equations p for $\mathbf{p}$.

Differentiating equations (7.17) with respect to p at the nominal solution $(\mathbf{i}, \mathbf{v}, p)$ we have to remember that $\mathbf{f}$ is a direct function of p, and moreover that it depends on p through the solutions $\mathbf{i}$, $\mathbf{v}$, which are affected by the change in p. Since differentiation is performed at the solution point we must treat $\mathbf{i}$, $\mathbf{v}$ as if they vary with p in such a way that the point $(\mathbf{i}, \mathbf{v}, p)$ remains a solution of (7.17). Hence, we obtain a system of equations:

$$\begin{aligned} &\mathbf{A}\frac{\partial \mathbf{i}}{\partial p} = \mathbf{0} \\ &\mathbf{B}\frac{\partial \mathbf{v}}{\partial p} = \mathbf{0} \\ &\frac{\partial \mathbf{f}}{\partial \mathbf{i}}\frac{\partial \mathbf{i}}{\partial p} + \frac{\partial \mathbf{f}}{\partial \mathbf{v}}\frac{\partial \mathbf{v}}{\partial p} = -\frac{\partial \mathbf{f}}{\partial p} + \frac{d\mathbf{s}}{dp} \end{aligned} \quad , \qquad (7.19)$$

where partial derivatives are calculated at the nominal point satisfying Equation (7.17). The system (7.19) of canonical equations stands for a linear circuit structurally identical with the original one (7.17). Its variables are equal to the sensitivities of the corresponding original circuit variables with respect to p, while its branches are obtained by differentiation of the original branches. As an effect of this differentiation, substitute branches arise, which are called incremental models of the original branches. If (7.17) is an immitance circuit (2.51) for a.c. analysis, then comparing the LHS of the branch equations of (7.19) with (7.17) we observe that the incremental and immitance model coefficients are identical, except for excitations. Similarly, comparing with (4.20) we find identity between incremental models and iterative d.c. models after convergence is reached, except for excitations. The whole circuit described by (7.19) is called the incremental (or variational) circuit. Its matrix will be identical with the original matrix at the original solution, only RHS vectors will be different. To create incremental models of branches a solution of the original circuit is necessary.

7.2.1.2 The incremental models of branches

In this section we deal with linear and nonlinear branches of circuits. Their incremental models can be obtained by differentiating the branch equations with respect to a variable parameter p. As an introductory example let us consider the conductance $i = G\,v$. Differentiating this equation with respect to p, and remembering that both G and v can be dependent on p (the latter due to the influence of p on the circuit solution) we obtain $\partial i/\partial p = G\,(\partial v/\partial p) + (\partial G/\partial p)\,v$. Since the quantities $\partial i/\partial p$, $\partial v/\partial p$ are branch variables in the incremental circuit, the formula obtained describes a conductance G connected in parallel with a current source $v\,(\partial G/\partial p)$. This so-called incremental source is dependent on the voltage v across the corresponding original conductance, but it is nonzero only if G is dependent on p, otherwise it is zero. Thus, for example, if we are interested in sensitivites with respect to G, then the incremental source is equal exactly to v. This surprising result with the incremental current source being equal to a voltage is not at all strange in fact. The physical unit of current in the incremental circuit (i.e. the unit of the derivative of current with respect to p) is ampere per unit of p. The unit of the substitute incremental source is volt times ampere per volt per unit of p, hence the same. The unit of voltage in the incremental circuit is volt per unit of p.

Let us consider another example of a nonlinear conductance: $i = I_s\,[\exp(v/V_T) - 1]$. Differentiating both sides with respect to p we obtain:

$$\frac{\partial i}{\partial p} = \frac{\partial I_s}{\partial p}\left(e^{\frac{v}{V_T}} - 1\right) + I_s e^{\frac{v}{V_T}}\left(\frac{1}{V_T}\frac{\partial v}{\partial p} - \frac{v}{V_T^2}\frac{\partial V_T}{\partial p}\right) =$$
$$= \left(\frac{I_s}{V_T}\, e^{\frac{v}{V_T}}\right)\frac{\partial v}{\partial p} + \frac{\partial I_s}{\partial p}\left(e^{\frac{v}{V_T}} - 1\right) - \frac{\partial V_T}{\partial p}\, v\frac{I_s}{V_T^2}\, e^{\frac{v}{V_T}} \tag{7.20}$$

where $(I_s/V_T)\exp(v/V_T)$ is a conductance dependent on the voltage v across the nonlinear conductance, while the additive component is an incremental current source which vanishes if neither i_s nor V_T depend on p.

Now, we proceed to a more general approach to creating incremental models. We begin from linear branches, which have been classified in Section 1.2.1 as belonging to current-defined (CD) and voltage-defined (VD) types (2.13) and (2.14) respectively.

After the CD branch (2.13) has been differentiated with respect to p, we obtain the CD incremental model:

$$\frac{\partial i_a}{\partial p} = G_{aa}\frac{\partial v_a}{\partial p} + G_{ab}\frac{\partial v_b}{\partial p} + K_{ac}\frac{\partial i_c}{\partial p} + I_{as}, \tag{7.21}$$

which, except for the physical sense of the variables, differs from (2.13) only in excitation. Namely, the original current source I_a vanishes, and the incremental current excitation appears:

$$I_{as} = \frac{\partial I_a}{\partial p} + v_a\frac{\partial G_{aa}}{\partial p} + v_b\frac{\partial G_{ab}}{\partial p} + i_c\frac{\partial K_{ac}}{\partial p}. \tag{7.22}$$

It is nonzero only if the branch under consideration is dependent on p. Branches independent of p have incremental models identical to the original ones but free of excitations.

For example, let us recall the linear conductance demonstrated at the beginning of this section. Furthermore, let us consider the current source $i=I$, which transforms into $\partial i/\partial p=\partial I/\partial p$, i.e. an open circuit $\partial i/\partial p=0$ in all cases except when calculating sensitivities with respect either to I or parameters influencing I.

As for the voltage-defined branch (2.14), after differentiation, we obtain the voltage-defined incremental model:

$$\frac{\partial v_a}{\partial p}=R_{aa}\frac{\partial i_a}{\partial p}+T_{ab}\frac{\partial v_b}{\partial p}+R_{ac}\frac{\partial i_c}{\partial p}+V_{as}, \tag{7.23}$$

where the incremental voltage excitation takes the form:

$$V_{as}=\frac{\partial V_a}{\partial p}+i_a\frac{\partial R_{aa}}{\partial p}+v_b\frac{\partial T_{ab}}{\partial p}+i_c\frac{\partial R_{ac}}{\partial p}. \tag{7.24}$$

For example, for CCVS $v_a=pi_c$ we have $\partial v_a/\partial p=p\,(\partial i_c/\partial p)+V_{as}$, where $V_{as}=i_c$. This model requires knowledge of the original controlling current i_c. Similarly, for the voltage source $v=V$ we obtain the branch $\partial v/\partial p=\partial V/\partial p$, which is a short-circuit in all cases except for calculating sensitivities with respect to the value of this source or parameters influencing it.

Now, we proceed to nonlinear branches, which in general can be VD or CD controlled by a voltage v_b, current i_c, and dependent on p:

$$\begin{aligned} i_a&=\phi_{ia}(v_a,v_b,i_c,p) \quad (CD)\\ v_a&=\phi_{va}(i_a,v_b,i_c,p) \quad (VD). \end{aligned} \tag{7.25}$$

Differentiating them with respect to p we obtain their incremental models, which are linear of the same form as the standard linear branches (2.13) and (2.14):

$$\begin{aligned} \frac{\partial i_a}{\partial p}&=\frac{\partial \phi_{ia}}{\partial v_a}\frac{\partial v_a}{\partial p}+\frac{\partial \phi_{ia}}{\partial v_b}\frac{\partial v_b}{\partial p}+\frac{\partial \phi_{ia}}{\partial i_c}\frac{\partial i_c}{\partial p}+I_{as},\\ \frac{\partial v_a}{\partial p}&=\frac{\partial \phi_{va}}{\partial i_a}\frac{\partial i_a}{\partial p}+\frac{\partial \phi_{va}}{\partial v_b}\frac{\partial v_b}{\partial p}+\frac{\partial \phi_{va}}{\partial i_c}\frac{\partial i_c}{\partial p}+V_{as}, \end{aligned} \tag{7.26}$$

but they include incremental sources, which are direct derivatives of branch functions with respect to p: $I_{as}=\partial\phi_{ia}/\partial p$, $V_{as}=\partial\phi_{va}/\partial p$. All partial derivatives are calculated at a d.c. solution point. LHS coefficients of these obtained incremental branches can be compared with coefficients of the iterative models (4.25), (4.26) of the same branches. They appear to be identical when we are exactly at the solution, i.e. after convergence of NR iterations. Thus, as the matrix of the incremental circuit we can take the Jacobi matrix of the original circuit after convergence of this is reached. To check this observation we can recall the example of a diode, discussed at the beginning of this section.

As another example let us take the CCVS $v_a=pi_c^3$. After differentiating, we obtain $\partial v_a/\partial p=(\partial p/\partial p)i_c^3+(3pi_c^2)(\partial i_c/\partial p)$, which is the CCVS of the coefficient $3pi_c^2$ connected in series with the incremental voltage source having the value i_c^3. To compare this with an iterative model for NR iterations, we perform a linearization and obtain the iterative model in the form $v_a^{(k+1)}=[3p(i_c^{(k)})^2]\,i_c^{(k+1)}+p(i_c^{(k)})^3-[3p(i_c^{(k)})^2]i_c^{(k)}$, which is composed of a linear CCVS with the coefficient $3p(i_c^{(k)})^2$, and a voltage source added in series. The iterative model coefficient tends to the formerly calculated coefficient of the incremental model, when $i_c^{(k)}$ tends to the solution i_c.

7.2.1.3 Practical realization of the incremental equations method

The incremental equations method can be used for a.c. sensitivity analysis of linear circuits and d.c. sensitivity analysis of linear and nonlinear circuits. Analysis in the small-signal a.c. case is more complicated and will be discussed later on.

The incremental equations method requires the previous performance of a full analysis, a.c. or d.c. respectively. After this we have a vector of circuit variables, containing the a.c. or d.c. solution, and a circuit matrix (linear or linearized) calculated at that solution. This matrix is decomposed into ***L***, ***U*** factors, and, as we have proved, it is the matrix of the incremental circuit.

For a given parameter p we have to formulate a vector of incremental excitations $\mathbf{b}_s$, e.g. according to MNEs. We can use for this purpose the same stamps given in Section 2.1.3.4 as for a circuit matrix formulation, though only their RHS's are useful. According to these stamps, we fill the vector $\mathbf{b}_s$ with incremental sources (7.23), (7.24) or (7.25), (7.26). The solution of the incremental circuit

$$\mathbf{L}\,\mathbf{U}\frac{\partial\mathbf{x}}{\partial p}=\mathbf{b}_s \tag{7.27}$$

requires only one forward and backward substitution (see Section 3.2.2.4). We obtain the vector of sensitivities of circuit variables $\mathbf{x}$ with regard to the parameter p. If we are interested in the sensitivity of a function $f(\mathbf{x})$, then it can be calculated as $df/dp=\Sigma_i(\partial f/\partial x_i)(\partial x_i/\partial p)$ using components of the sensitivity vector obtained from the incremental equations method.

7.2.1.4 Example of the incremental analysis

Let us consider for a.c. analysis the linear immitance circuit shown in Figure 7.2(a). Its complex MNEs take the form

$$\underbrace{\begin{bmatrix} p+j\omega C_3+G_4 & -G_4 & 0 \\ -G_4 & G_4 & 1 \\ 1/p & -1 & 0 \end{bmatrix}}_{\mathbf{Y}} \underbrace{\begin{bmatrix} v_{n1} \\ v_{n2} \\ i_5 \end{bmatrix}}_{\mathbf{x}} = \underbrace{\begin{bmatrix} J_1 \\ 0 \\ 0 \end{bmatrix}}_{\mathbf{b}}. \tag{7.28}$$

To create an incremental circuit corresponding to the parameter p we consider successively all branches. The current source I_1 transforms into the open-circuit $\partial i_1/\partial p=0$. The 2nd and 5th branches, which are dependent on p, transform into incremental models:

$$\frac{\partial i_2}{\partial p} = G_2 \frac{\partial v_2}{\partial p} + I_{2s}, \quad \text{where } I_{2s} = v_2, \tag{7.29}$$

$$\frac{\partial v_5}{\partial p} = \frac{1}{p}\frac{\partial v_3}{\partial p} + V_{5s}, \quad \text{where } V_{5s} = -\frac{1}{p^2} v_3. \tag{7.30}$$

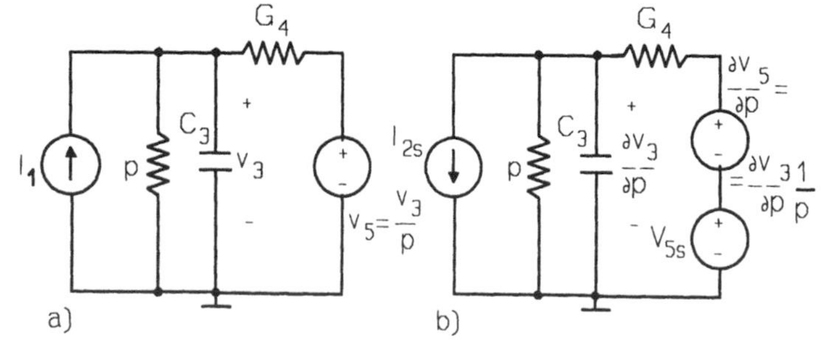

Figure 7.2 (a) A simple linear a.c. circuit and (b) its incremental counterpart for a.c. sensitivity analysis

Other branches, after differentiation with respect to p, remain unchanged, i.e. their incremental sources are zero. The resulting incremental circuit is drawn in Figure 7.2(b). Its MNEs are as follows:

$$\underbrace{\begin{bmatrix} p+j\omega C_3+G_4 & -G_4 & 0 \\ -G_4 & G_4 & 1 \\ 1/p & -1 & 0 \end{bmatrix}}_{\boldsymbol{Y}} \underbrace{\begin{bmatrix} \partial v_{n1}/\partial p \\ \partial v_{n2}/\partial p \\ \partial i_5/\partial p \end{bmatrix}}_{\partial \boldsymbol{x}/\partial p} = \underbrace{\begin{bmatrix} -I_{2s} \\ 0 \\ -V_{5s} \end{bmatrix}}_{\boldsymbol{b}_s} \tag{7.31}$$

From its solution we can obtain the sensitivites of three components of the circuit variables vector.

7.2.2 The adjoint equations method

The incremental equations method introduced in Section 7.2.1 was directed towards calculating the sensitivity of the whole vector of circuit variables with respect to a single variable parameter, by means of forward/backward substitutions in incremental equations. However, if we want to calculate sensitivities with respect to a number (say m) of parameters $\mathbf{p}$, then we have to perform m such substitutions with m different RHS vectors standing for incremental excitations. When m is large, this can be a quite time-consuming process. Unfortunately, calculating sensitivities with respect to many parameters is much more important in practice than their calculation with respect to just one. On the other hand, we often do not require the sensitivities of many circuit variables simultaneously. The number of network functions may be at most a few, while the number of parameters may be hundreds. To solve such a multiparameter sensitivity problem a very efficient adjoint equations method has been developed. This method enables us to obtain the sensitivities of one output signal with respect to a number of variable parameters by means of a single forward/backward substitution process only. We start its presentation from the simplest formulation which does not require arduous adjoint circuit concepts. Then, we will introduce Tellegen's theorem, which is useful in a more general problem formulation.

7.2.2.1 The adjoint equations and efficient multiparameter sensitivities calculation

In Section 7.2.1 we introduced the concept of an incremental circuit. Efficient equations describing this circuit can be formulated, e.g. in a

MNE form, as shown in (7.27). Of course other available types of circuit description (e.g. tableau) are also suitable for a discussion of the adjoint method. Thus, let a general system of incremental equations corresponding to the variable parameter p_i be of the form:

$$\mathbf{Y}\frac{\partial \mathbf{x}}{\partial p_i} = \mathbf{b}_{si}. \tag{7.32}$$

Provided that we are interested only in the sensitivity of the *j*th component of the circuit response vector with respect to the *i*th circuit parameter: $\partial x_j / \partial p_i$, we can introduce the column vector $\mathbf{e}_j$ (of a dimension $n = \dim \mathbf{x}$), containing zeros at all positions for the *j*th which is unity. Then, the sensitivity can be written as:

$$\frac{\partial x_j}{\partial p_i} = \mathbf{e}_j^t \frac{\partial \mathbf{x}}{\partial p_i} = \mathbf{e}_j^t \mathbf{Y}^{-1} \mathbf{b}_{si}. \tag{7.33}$$

This formula takes the form of a product of matrices giving a scalar (i.e. 1×1) result. Hence, even if we transpose this expression the result remains unchanged. However, the transposition of a product of matrices is equal to a reversely-ordered product of their transposes. Moreover, transposition of an inverse matrix is the inverse of its transposition. Thus, after transposition we obtain:

$$\frac{\partial x_j}{\partial p_i} = \mathbf{b}_{si}^t (\mathbf{Y}^t)^{-1} \mathbf{e}_j. \tag{7.34}$$

The second and third factor forms a product of the $n \times n$ matrix and the column vector $n \times 1$, and hence it is a column vector. This vector is a solution of the following system:

$$\mathbf{Y}^t \mathring{\mathbf{x}} = \mathbf{e}_j, \tag{7.35}$$

which is nonsingular iff the matrix $\mathbf{Y}$ of the original circuit is nonsingular. This system corresponds to the *j*th component of the circuit response, i.e. to the particular network function, not to the particular variable parameter. We call it a system of adjoint equations. We solve it only once for each network function whose sensitivites are being calculated. The solution $\mathring{\mathbf{x}}$ is called the vector of adjoint variables.

From Equation (7.34) a formula for sensitivities arises which is based upon the adjoint variables:

$$\frac{\partial x_j}{\partial p_i} = \mathbf{b}^{t}_{si}\,\mathring{\mathbf{x}}\,. \tag{7.36}$$

Thus, the calculation of the sensitivities of the network function x_j with respect to a number of parameters does not require multiple solution of a system of equations. We can solve once the adjoint equations (7.35), and then use the formula for sensitivities as many times as necessary. In this formula we use the adjoint variables and an incremental excitations vector, created by means of the circuit topology and its corresponding stamps.

To show how simple are the operations behind the product of the two vectors (7.36) and to explain how they can be computerized, we cite several examples. Let the current of a conductance $i_a = G\,v_a$ flow from node 1 to node 2. To obtain an incremental model for G being the variable parameter we differentiate the branch function with respect to G. The resulting model $\partial i_a/\partial G = G\,(\partial v_a/\partial G) + v_a$ has a substitute current source with the value v_a. Provided the conductance current is not a circuit variable, the incremental excitations vector formulated according to the MNE stamp takes the form

$$\mathbf{b}_s = [-(v_{n1} - v_{n2}),\ (v_{n1} - v_{n2}),\ 0,\ \ldots,\ 0]$$

and hence, the formula $\partial x_j/\partial G = \mathbf{b}^{t}_{s}\,\mathring{\mathbf{x}}$ transforms into $-(v_{n1} - v_{n2})\,(\mathring{x}_1 - \mathring{x}_2)$, where $\mathring{x}_1 \equiv \mathring{v}_{n1}$, $\mathring{x}_2 \equiv \mathring{v}_{n2}$.

Alternatively, when the conductance current is chosen as a circuit variable, from an adequate MNE stamp we obtain the excitation vector $\mathbf{b}_s = [0, \ldots, 0, -(v_{n1} - v_{n2}), 0, \ldots]$ whose nonzero entry corresponds to the variable i_a. Thus, $\partial x_j/\partial G = -\mathring{i}_a(v_{n1} - v_{n2})$.

As the third example let us recall the example introduced in Section 7.2.1.4. We calculate the sensitivites of the output variable x_3, and hence the adjoint system $\mathbf{Y}^{t}\,\mathring{\mathbf{x}} = \mathbf{e}_3$ has to be solved. To obtain the sensitivity with respect to p we recall the incremental excitations vector from (7.31), and obtain the formula:

$$\frac{\partial x_j}{\partial p} = \mathring{\mathbf{x}}_{\mathbf{j}}\,\mathbf{b}^{t}_{s} = -\mathring{x}_1\,I_{2s} - \mathring{x}_3\,V_{5s}. \tag{7.37}$$

As can be seen from these examples, if we are able to predict the nonzero entries of an incremental excitations vector, then the sensitivities can be calculated with very small computational effort. This prediction is obtained from the RHS of the branch stamps. Thus we proceed to general rules of sensitivity calculation for linear and nonlinear circuits composed of generalized standard branches, which can easily be reduced to all specific cases met in practice.

Let a circuit be composed of generalized linear branches: voltage-defined (VD) (2.14) and current-defined (CD) (2.13). The latter can have the current selected or not selected as part to the variables vector (notation CDX or CD respectively), as was shown in Section 2.1.3. We want to find formulae for network function sensitivities with respect to a parameter p, which can influence all the coefficients of a single branch. Taking into account the RHS of MNE stamps given in Section 2.1.3.4, the formula (7.36) can be transformed into the form:

$$\frac{\partial x_j}{\partial p} = -(\mathring{v}_{nma} - \mathring{v}_{nna})\left(\frac{\partial G_{aa}}{\partial p}v_a + \frac{\partial G_{ab}}{\partial p}v_b + \frac{\partial K_{ac}}{\partial p}i_c + \frac{\partial I_a}{\partial p}\right) \quad \text{for CD}$$

$$\frac{\partial x_j}{\partial p} = -\mathring{i}_a\left(\frac{\partial G_{aa}}{\partial p}v_a + \frac{\partial G_{ab}}{\partial p}v_b + \frac{\partial K_{ac}}{\partial p}i_c + \frac{\partial I_a}{\partial p}\right) \quad \text{for CDX}$$

$$\frac{\partial x_j}{\partial p} = -\mathring{i}_a\left(\frac{\partial R_{aa}}{\partial p}i_a + \frac{\partial T_{ab}}{\partial p}v_b + \frac{\partial R_{ac}}{\partial p}i_c + \frac{\partial V_a}{\partial p}\right) \quad \text{for VD} \tag{7.38}$$

where $v_a = v_{nma} - v_{nna}$, $v_b = v_{nmb} - v_{nnb}$, and i_c are signals obtained from the original circuit solution, while, for example, v_{nna} denotes the voltage of the node n_a stored in the vector $\boldsymbol{v}_n$. Similarily, $\mathring{v}_{nma}$ stands for an entry of the adjoint vector $\mathring{\boldsymbol{x}}$ corresponding to the m_a node voltage numbered as in the vector $\mathbf{x}$. A circuit interpretation of adjoint variables will be introduced later on.

Now, we proceed to the more general case of nonlinear CD and VD branches (7.25). Our example will be based on a situation in which the variable parameter p belongs to the description of only one branch. Consider a CCVS of the form $v_a = p \arctan(i_c/I_0)$. Its current is circuit variable number 1. We differentiate it with respect to p, and hence obtain the incremental equation $\partial v_a/\partial p = [p i_0/(I_0^2 + i_c^2)]\,(\partial i_c/\partial p) + \arctan(i_c/I_0)$. The incremental excitation vector comes from the RHS of the type v stamp of Section 2.1.3.4 and takes the form $\mathbf{b}_s = [-\arctan(i_c/I_0), 0, \ldots]$, and so (7.36) gives the sensitivity $\partial v_a/\partial p = -\mathring{x}_1 \arctan(i_c/I_0)$, where, due to the numbering of the variable i_a as the first one, the first component of the adjoint vector appears.

Proceeding to the general discussion we distinguish three cases: p belongs to the CD branch whose current is not a circuit variable (i.e. CD case), p belongs to the CD branch whose current is a circuit variable (CDX case), and p belongs to the VD branch. We obtain three distinct formulae for sensitivities:

$$\begin{aligned} \frac{\partial x_j}{\partial p} &= -(\overset{\circ}{v}_{nma} - \overset{\circ}{v}_{nna})\frac{\partial \phi_{ia}}{\partial p}, \quad \text{(CD)} \\ \frac{\partial x_j}{\partial p} &= -\overset{\circ}{i}_a \frac{\partial \phi_{ia}}{\partial p}, \quad \text{(CDX)} \\ \frac{\partial x_j}{\partial p} &= -\overset{\circ}{i}_a \frac{\partial \phi_{va}}{\partial p}. \quad \text{(VD)} \end{aligned} \tag{7.39}$$

In general, when p influences many branches simultaneously, we have to create a vector $\mathbf{b}_s$ and directly use Equation (7.36). It will be equivalent to the superposition of the appropriate formulae (7.38), (7.39) applied to all circuit branches dependent on p.

7.2.2.2 Circuit representation of adjoint equations.

The introduction of adjoint equations (7.35) leads to a problem whether they can be treated as equations of a particular linear circuit. To answer this question we have to discuss the stamps of circuit branches. Let our circuit be composed of CD and VD nonlinear branches (7.25), or linear (2.13), (2.14) shown in Figure 2.5. Of course, even in the nonlinear case, branches after linearization at a solution take the linear form (2.13), (2.14). Their MNEs can be formulated by means of stamps given in Section 2.1.3.4. If the transposed stamp of each branch can be interpreted as a certain substitute set of branches, then the adjoint matrix $\mathbf{Y}^t$ is the MNE matrix of a circuit, called the adjoint circuit. The transposition of the CD and VD stamps mentioned gives MNE stamps which stand also for CD and VD adjoint branches shown in Figure 7.3(a) and (b) respectively. If the current of CD branch is selected as a circuit variable (CDX case), then there is no MNE adjoint counterpart.

Analysis of these adjoint models has shown that, taking linear (or linearized at a solution) original branches, the following rules exist for the creation of adjoint models:

(i) Original resistive and conductive branches remain unchanged.

(ii) Original excitations become zero.

(iii) If the original branch contains a voltage-controlled source, then the adjoint model contains a controlled current source connected in parallel with a controlling branch. If the original branch is current-controlled, then the adjoint model contains a controlled voltage source connected in series with a controlling branch. They are added to previously existing contents of controlling branches, drawn in Figure 7.3 by means of elipses.

(iv) If the original branch is CD, then the sources mentioned in item (iii) are voltage controlled, while if the original branch is VD, then these sources are current controlled.

(v) If the types of the original branch and of the controlling variable are consistent, then the sign of the control coefficient is changed, otherwise it remains the same.

The adjoint circuit exhibits a reciprocal direction of signal transmission in comparison with the original one. Due to this, the original circuit transmittance from a particular input to some output is equivalent to a transmittance from this output to the input in the adjoint circuit. Thus, instead of calculating transmittances from many inputs to one output by means of many circuit analyses, we are able to calculate numerically the same adjoint transmittances from one output to many inputs using a single analysis of the adjoint circuit. In this way, the need to stimulate the circuit many times avoided. Such a possibility we call *inter-reciprocity*, as an extension of reciprocity, which means that one circuit has the same transmittances in opposite directions. Here, two circuits have the same transmittances when measured in opposite directions.

Besides the elements shown in Figure 7.3, the adjoint models contain adjoint excitations. They depend only on the way of defining the output signal whose sensitivity has to be calculated. If the output signal is a voltage between nodes (m, n), then the unitary adjoint current source

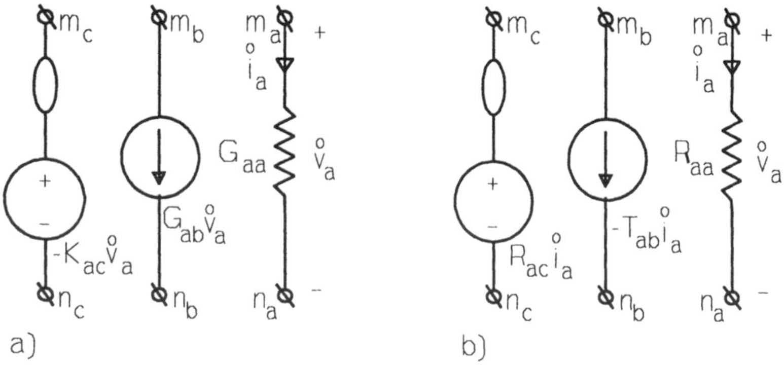

Figure 7.3 Adjoint couterparts of: (a) CD and (b) VD branches

flows from m to n. If the output signal is the current of a CD branch, included in the circuit variables, then the unitary adjoint excitation means a current connected in parallel with this branch and directed consistently with its current. If an output signal is the current of a VD branch, then the unitary adjoint excitation is a voltage added in series into this branch and consistent with its terminal voltage.

As we have shown, not all branches have adjoint models. It depends on the chosen technique of equation formulation, branch models and circuit variables. However, this interpretation has no influence on the general adjoint equations, which can always be applied directly by means of matrix transposition.

7.2.2.3 How to implement the method of adjoint equations in practice

The adjoint equations method requires availability of the original circuit solution. Hence, we implement it after completion of a nonlinear d.c. (if d.c. sensitivities are to be calculated) or linear a.c. (if a.c. frequency domain sensitivities are wanted) analysis. After those analyses a circuit matrix at the solution point is available. It is moreover, split up into LU factors. Its transposition gives the matrix of adjoint equations (7.25). For each signal x_j, whose sensitivities are to be calculated, we create the adjoint excitations $\mathbf{e}_j$, and solve the adjoint equations:

$$\mathbf{U}^t\,\mathbf{L}^t\,\mathring{\mathbf{x}} = \mathbf{e}_j. \tag{7.40}$$

The computational effort required is not greater than the effort in the incremental equations case (7.27), since the calculations include only one pass of forward and backward substitutions. Forward substitutions $\mathbf{U}^t\,\mathbf{b}' = \mathbf{b}$ incorporate $\mathbf{U}^t$ instead of the previously used $\mathbf{L}$, while backward substitutions are of the form $\mathbf{L}^t\,\mathring{\mathbf{x}} = \mathbf{e}_j$. Therefore, if we neglect reordering, these algorithms differ from Algorithms 3.11 and 3.12 in taking a_{ki} instead of a_{ik} (transposition), and in moving divisions through pivots from backward to forward substitutions, due to normalization of the diagonal of $\mathbf{L}$. Thus, we obtain Algorithm 7.1. For simplicity of description, the RHS vector has been denoted by $\mathbf{b}$ instead of $\mathbf{e}_j$. If we have many output signals, then we have to execute Algorithm 7.1 as many times as the number of outputs, with adequate RHS vectors.

Algorithm 7.1 becomes much more complicated if we introduce row and/or column reordering. Then, the $\mathbf{L}$ and $\mathbf{U}$ factors used above correspond to the reordered matrix. For example, if we employ physical interchanging of the rows (Algorithm 3.13), then the RHS vector of adjoint equations need not be reordered. However after the end of the solution process we obtain reordered unknowns. Hence, in the forward substitutions no RHS interchanges are used, while after backward substitutions we must reorder solutions according to the trace *itr* in such a manner that for $i=n-1,\ldots,1$, if $itr_i \neq i$, we interchange x_i with $x_{itr[i]}$.

Algorithm 7.1 Forward and backward substitutions with transposed matrices $\boldsymbol{U}$ and $\boldsymbol{L}$

```
{forward substitutions}
for i=2 to n do
begin
   s=0;
   for k=1 to i-1 do
      s=s - a_ki b_k;
   b_i=(b_i+s)/a_ii;
end
{backward substitutions}
for i=n downto 1 do
begin
   s=0;
   for k=i+1 to n do
      s=s-a_ki b_k;
   b_i=b_i+s;
end
```

After adjoint variables have been calculated, we apply the formulae for sensitivities (7.38), (7.39) for those p which are required in our particular problem. If we have p influencing many branches simultaneously, then the general rule (7.36) is relevant, where $\mathbf{b}_s$ has to be formulated using MNE stamps, to fill the RHS vector properly with incremental sources (7.22), (7.24).

The adjoint equations method is suitable for sensitivity analysis of one output signal with respect to many parameters. Hence, it is very efficient e.g. for calculating the gradients of objective functions in optimization tasks. However, if there are a number of output signals, we have a number of adjoint equations to solve. Neglecting a small computational effort for sensitivity calculation by means of adjoint variables, we can state, that the adjoint method is more efficient than the incremental one iff the number of outputs is less then the number of variable parameters.

The adjoint equations method was introduced in Reference [7.1, 7.2] using Tellegen's theorem. Since then, it has been developed and generalized. Now, it is extensively used for small and large-change sensitivity and noise analysis.

7.2.3 Generalized adjoint circuit method based on Tellegen's theorem

The adjoint equations method has been primarily developed by means of Tellegen's theorem [7.9]. As we have shown, if the sensitivity of a network function x_j, which is a circuit variable, is calculated, then we can describe it simply without using Tellegen's theorem. However, to introduce the more general case of any network function, as well as to demonstrate the classical approach known for many years, Tellegen's theorem appears to be very convenient.

7.2.3.1 Tellegen's theorem and sensitivity analysis

Tellegen's theorem [7.9] is a simple consequence of the fact that branch currents and voltages obey KCL and KVL at any instant. To introduce this theorem we discuss two (in a specific case the same) structurally identical circuits (see Section 1.2.3): the original circuit having branch variables $\mathbf{i}$, $\mathbf{v}$ and the auxiliary circuit having branch variables $\mathring{\mathbf{v}}$, $\mathring{\mathbf{i}}$. All branch voltages of the original circuit satisfying KVL (1.21) can be viewed as a picture of the nodal voltages obtained by means of the nodal transformation (1.28). Thus, after the formula (1.28) has been substituted for branch voltages in $\mathring{\mathbf{i}}^t\mathbf{v}$, we obtain an equation of the form $\mathring{\mathbf{i}}^t(\mathbf{A}^t\mathbf{v}_n) = (\mathbf{A}\,\mathring{\mathbf{i}})^t\mathbf{v}_n$, which, due to KCL in the auxiliary circuit (1.18), is equal to zero at any instant. This principle means that the sum of products of the currents and voltages of corresponding branches of two structurally identical circuits is always zero. This rule is only a consequence of KCL and KVL, and hence, holds irrespective of the particular branches involved. This sum has no physical interpretation, except for that case in which these two circuits are the same. Then, $\mathbf{i}^t\mathbf{v}$ is the total power supplied (negative) and dissipated (positive) by the circuit, and stays zero, due to the energy conservation law.

Tellegen's theorem states that two structurally identical circuits satisfy the equation:

$$\mathring{\mathbf{i}}^t\mathbf{v} = \mathring{\mathbf{v}}^t\mathbf{i} = 0. \tag{7.41}$$

Equation (7.41) implies a difference formula $\mathring{\mathbf{i}}^t\mathbf{v} - \mathring{\mathbf{v}}^t\mathbf{i} = 0$. Taking increments in the original branch variables, we can substitute incremented variables $\mathbf{i}+\Delta\mathbf{i}$, $\mathbf{v}+\Delta\mathbf{v}$ for the nominal variables, and side-wise subtract the nominal equation. We obtain $\mathring{\mathbf{i}}^t\Delta\mathbf{v}-\mathring{\mathbf{v}}^t\Delta\mathbf{i}=0$, known as the incremental form of Tellegen's theorem. Substituting differentials for increments produces a differential formula, essential in sensitivity analysis:

$$\mathring{\mathbf{i}}^t d\mathbf{v}-\mathring{\mathbf{v}}^t d\mathbf{i}=0. \tag{7.42}$$

This joins increments in an original circuit response with the nominal response of an auxiliary circuit. This formula is useful in sensitivity analysis, because it provides for the elimination of the differential of one of the variables. Such a manipulation will require a transformation of the variables in the formula (7.42). Hence, the branch variables of (7.42) before this transformation may be called primary variables.

Proceeding to details of this transformation we should notice that increments in primary variables **i**, **v** are in fact an effect of changes in those branch parameters **p**, with respect to which the sensitivities are calculated. To introduce **p** into Tellegen's theorem, we transform the primary variables into so-called secondary variables. They contain n circuit variables **x** (e.g. nodal voltages, and some currents according to MNE), and m independent branch parameters **p** (e.g. R, I_s). Let CD or VD nonlinear branches be described by means of Equation (7.25). Due to the transformation of branch variables (e.g. into modifed nodal variables **x**, as discussed in Section 1.3.4) we can always express branch variables using variables **x** and parameters **p**. For example, taking the conductance $i_a=I_s(v_a/V_T)^3$, and presuming $p=I_s$, we have the primary variables $v_a=v_{nma}-v_{nna}$, $i_a=p[(v_{nna}-v_{nma})/V_T]^3$, where v_{nna}, v_{nma} are nodal voltages (incorporated in **x**) corresponding to terminals m_a, n_a of the conductance involved. We can always write the following general relation between primary and secondary variables:

$$\mathbf{i}=\mathbf{i}(\mathbf{x},\mathbf{p}),\ \ \mathbf{v}=\mathbf{v}(\mathbf{x},\mathbf{p}),$$

which are taken to be continuous and differentiable. Differentiating these relations at the nominal circuit response we obtain

$$\begin{bmatrix} d\mathbf{v} \\ d\mathbf{i} \end{bmatrix}=\begin{bmatrix} \dfrac{\partial\mathbf{v}}{\partial\mathbf{x}} & \dfrac{\partial\mathbf{v}}{\partial\mathbf{p}} \\ \dfrac{\partial\mathbf{i}}{\partial\mathbf{x}} & \dfrac{\partial\mathbf{i}}{\partial\mathbf{p}} \end{bmatrix}\begin{bmatrix} d\mathbf{x} \\ d\mathbf{p} \end{bmatrix}, \tag{7.43}$$

where the four submatrices of partial derivatives stand for matrices of transformations of secondary differentials into primary ones. We substitute (7.43) for the primary variables differentials in the formula (7.42), and then obtain the following formula

$$\begin{aligned}&\mathring{\mathbf{i}}^{t}\left(\frac{\partial \mathbf{v}}{\partial \mathbf{x}}\,d\mathbf{x}+\frac{\partial \mathbf{v}}{\partial \mathbf{p}}\,d\mathbf{p}\right)-\mathring{\mathbf{v}}^{t}\left(\frac{\partial \mathbf{i}}{\partial \mathbf{x}}\,d\mathbf{x}+\frac{\partial \mathbf{i}}{\partial \mathbf{p}}\,d\mathbf{p}\right)=\\ &=\left(\mathring{\mathbf{i}}^{t}\frac{\partial \mathbf{v}}{\partial \mathbf{x}}-\mathring{\mathbf{v}}^{t}\frac{\partial \mathbf{i}}{\partial \mathbf{x}}\right)d\mathbf{x}-\left(-\mathring{\mathbf{i}}^{t}\frac{\partial \mathbf{v}}{\partial \mathbf{p}}+\mathring{\mathbf{v}}^{t}\frac{\partial \mathbf{i}}{\partial p}\right)d\mathbf{p}=0\end{aligned} \tag{7.44}$$

This formula is similar to (7.41), but it incorporates secondary variables **x**, **p**. The formula (7.44) can be expressed as:

$$\mathring{\mathbf{p}}^{t}\,d\mathbf{x}-\mathring{\mathbf{x}}^{t}\,d\mathbf{p}=0. \tag{7.45}$$

The symbol $\mathring{\mathbf{p}}^{t}$, $\mathring{\mathbf{x}}^{t}$ stands for expressions (row-vectors), which in Equation (7.44) have been taken in parentheses. They can be viewed as secondary variables of the auxiliary circuit, and are of the form:

$$\mathring{\mathbf{p}}=\left(\frac{\partial \mathbf{v}}{\partial \mathbf{x}}\right)^{t}\mathring{\mathbf{i}}-\left(\frac{\partial \mathbf{i}}{\partial \mathbf{x}}\right)^{t}\mathring{\mathbf{v}}, \tag{7.46}$$

$$\mathring{\mathbf{x}}=-\left(\frac{\partial \mathbf{v}}{\partial \mathbf{p}}\right)^{t}\mathring{\mathbf{i}}+\left(\frac{\partial \mathbf{i}}{\partial \mathbf{p}}\right)^{t}\mathring{\mathbf{v}}. \tag{7.47}$$

Equation (7.46) is a system of linear branch equations of an auxiliary circuit. These equations are written in terms of primary variables $\mathring{\mathbf{i}}$, $\mathring{\mathbf{v}}$ of the auxiliary circuit. Hence, (7.46) defines branches of the auxiliary circuit, which has so far been considered only as an arbitrary circuit, structurally identical with the original one. The auxiliary circuit branches have been chosen in such a way that the auxiliary $\mathring{\mathbf{x}}$, $\mathring{\mathbf{p}}$ and original **x**, **p** secondary variables obey the formula (7.45). Hence, the auxiliary circuit is not arbitrary anymore, it is now linked with the original circuit, and we call it the adjoint circuit. Its branches can be obtained from original (and linearized at a solution) branches by means of the matrices of derivatives introduced in (7.43). Branches of the adjoint circuit have arbitrary excitations $\mathring{\mathbf{p}}$, dedicated to the particular network function involved. Adjoint counterparts to the original branches can be created in this way for each branch separately. Two adjoint models of standard branches are shown in the next section.

7.2.3.2 Adjoint models of standard branches

To demonstrate how the theory just introduced generates adjoint branch models we discuss CD and VD branches described by means of Equations (7.25).

In the case of the CD branch $i_a = \phi_i(v_a, v_b, i_c, p)$ we consider primary variables $\boldsymbol{i} = [i_a, i_b, i_c]^t$, and $\boldsymbol{v} = [v_a, v_b, v_c]^t$. Selecting $\mathbf{x} = [v_a, v_b, i_c]^t$ (i.e. branch independent variables), and $\mathbf{p} = [p]$, we can calculate:

$$\frac{\partial \mathbf{v}}{\partial \mathbf{x}} = \begin{bmatrix} 1 & 0 & 0 \\ 0 & 1 & 0 \\ 0 & 0 & 0 \end{bmatrix}, \quad \frac{\partial \mathbf{i}}{\partial \mathbf{x}} = \begin{bmatrix} \partial\phi_i/\partial v_a & \partial\phi_i/\partial v_b & \partial\phi_i/\partial i_c \\ 0 & 0 & 0 \\ 0 & 0 & 1 \end{bmatrix},$$

$$\frac{\partial \mathbf{i}}{\partial p} = \begin{bmatrix} \partial\phi_i/\partial p \\ 0 \\ 0 \end{bmatrix}, \quad \frac{\partial \boldsymbol{v}}{\partial p} = \begin{bmatrix} 0 \\ 0 \\ 0 \end{bmatrix}. \tag{7.48}$$

Transposing the above derivatives and substituting into (7.46) we obtain equations of the adjoint branches and formulae for secondary adjoint variables. The latter take the form:

$$x_0 = \frac{\partial \phi}{\partial p} \mathring{v}_a \tag{7.49}$$

while the equations of the adjoint branches

$$\begin{bmatrix} 1 & 0 & 0 \\ 0 & 1 & 0 \\ 0 & 0 & 0 \end{bmatrix} \begin{bmatrix} \mathring{i}_a \\ \mathring{i}_b \\ \mathring{i}_c \end{bmatrix} - \begin{bmatrix} \dfrac{\partial\phi_i}{\partial v_a} & 0 & 0 \\ \dfrac{\partial\phi_i}{\partial v_b} & 0 & 0 \\ \dfrac{\partial\phi_i}{\partial i_c} & 0 & 1 \end{bmatrix} \begin{bmatrix} \mathring{v}_a \\ \mathring{v}_b \\ \mathring{v}_c \end{bmatrix} = \begin{bmatrix} \mathring{p}_a \\ \mathring{p}_b \\ \mathring{p}_c \end{bmatrix} \tag{7.50}$$

are equivalent to three individual formulae, which describe, except for excitations, the adjoint model of the CD branch so far known from Section 7.2.2.2 and drawn in Figure 7.3(a)

$$\mathring{i}_a = \frac{\partial\phi_i}{\partial v_a}\mathring{v}_a + \mathring{p}_a, \quad \mathring{i}_b = \frac{\partial\phi_i}{\partial v_b}\mathring{v}_a + \mathring{p}_b, \quad \mathring{v}_c = \frac{\partial\phi_i}{\partial i_c}\mathring{v}_a - \mathring{p}_c. \tag{7.51}$$

Regarding excitations in the above equations, the problem of assigning their values has not been discussed so far. We will discuss this in the next section, let us meanwhile denote them as $\mathring{\mathbf{p}}$.

For a VD branch $v_a=\phi(i_a, v_b, i_c, p)$ we can write the same primary variables $\mathbf{v}^t=[v_a, v_b, v_c]$, $\mathbf{i}^t=[i_a, i_b, i_c]$. Let $\mathbf{x}=[i_a, u_b, i_c]^t$, and $\mathbf{p}=[p]$ be chosen as secondary variables. Then, the secondary parameter of the adjoint circuit takes the form:

$$x_0 = -\frac{\partial \phi_v}{\partial p}\mathring{i}_a \tag{7.52}$$

while the adjoint branch equations are of the form

$$\frac{\partial \mathbf{v}}{\partial \mathbf{x}} = \begin{bmatrix} \partial\phi/\partial v_a & \partial\phi/\partial v_b & \partial\phi/\partial i_c \\ 0 & 1 & 0 \\ 0 & 0 & 0 \end{bmatrix}, \quad \frac{\partial \mathbf{i}}{\partial \mathbf{x}} = \begin{bmatrix} 1 & 0 & 0 \\ 0 & 0 & 0 \\ 0 & 0 & 1 \end{bmatrix}$$

$$\frac{\partial \mathbf{v}}{\partial p} = \begin{bmatrix} \partial\phi/\partial p \\ 0 \\ 0 \end{bmatrix}, \quad \frac{\partial \mathbf{i}}{\partial p} = \begin{bmatrix} 0 \\ 0 \\ 0 \end{bmatrix}. \tag{7.53}$$

Substituting these derivatives in (7.46) we obtain an adjoint branch, which, except for excitations, is identical with the one demonstrated in Figure 7.3(b),

$$\mathring{v}_a = \frac{\partial \phi_v}{\partial i_a}\mathring{i}_a + \mathring{p}_a, \quad \mathring{i}_b = -\frac{\partial \phi_v}{\partial v_b}\mathring{i}_a + \mathring{p}_b, \quad \mathring{v}_c = \frac{\partial \phi_v}{\partial i_c}\mathring{i}_a + \mathring{p}_c. \tag{7.54}$$

7.2.3.3 Adjoint excitations and sensitivity calculation

Proceeding to the application of the adjoint circuit to sensitivity calculations we recall Equations (7.46), (7.47). Assuming that we look for a row-vector of sensitivites of a certain network function $f(\mathbf{x}, \mathbf{p})$ with respect to all components of $\mathbf{p}$, a differential increment in this function can be written as:

$$\mathrm{d}f = \frac{\partial f}{\partial \mathbf{x}}\mathrm{d}\mathbf{x} + \frac{\partial f}{\partial \mathbf{p}}\mathrm{d}\mathbf{p}. \tag{7.55}$$

To eliminate d**x** from the above equation, we use Equation (7.45) obtained from Tellegen's theorem, and assume that the vector of adjoint excitations $\mathring{\mathbf{p}}$ is equal to $(\partial f/\partial \mathbf{x})^t$. Then, the first component of (7.55) can be substituted for the second component of (7.45), and we obtain sensitivities:

$$\frac{\mathrm{d}f}{\mathrm{d}\mathbf{p}} = \mathring{\mathbf{x}}^t + \frac{\partial f}{\partial \mathbf{p}} \tag{7.56}$$

composed of the direct component $\partial f/\partial \mathbf{p}$, and an indirect component equal to the response $\mathring{\mathbf{x}}$ of the adjoint circuit to the stimulation equal to $\mathring{\mathbf{p}}$. This response can be calculated from (7.47) by means of the adjoint circuit solution.

Recalling the examples given in Section 7.2.3.2, and not defining precisely the form of the network function $f(\mathbf{x}, p)$, we have to take, in both CD and VD branch cases, adjoint excitations according to the equation

$$\mathring{\mathbf{p}}^t = [\mathring{p}_a, \mathring{p}_b, \mathring{p}_c] = (\partial f/\partial \mathbf{x}). \tag{7.57}$$

Therefore, after the adjoint circuit has been solved for $\mathring{\mathbf{x}}$ (that is calculation of $\mathring{x}$ from the circuit of the same topology as the original and incorporating branches (7.46)), the sensitivity of f, according to (7.52), is:

$$\frac{\mathrm{d}f}{\mathrm{d}p} = \frac{\partial f}{\partial p} - \frac{\partial \phi_v}{\partial p}\mathring{i}_a \tag{7.58}$$

in the VD branch case, and according to (7.49)

$$\frac{\mathrm{d}f}{\mathrm{d}p} = \frac{\partial f}{\partial p} + \frac{\partial \phi_i}{\partial p}\mathring{v}_a \tag{7.59}$$

in the CD branch case.

The point of this extension to the adjoint equations method was the selection of such a substitute adjoint excitation, that the sensitivity of a given network function f will be available directly from the adjoint circuit solution. If the network function is the jth component of the response $\mathbf{x}$, then $f = \mathbf{e}_j^t\mathbf{x}$, and hence, the adjoint excitations vector is equal to $\boldsymbol{e}_j$, i.e. the extension presented reduces to the simple approach introduced in Section 7.2.2. More precisely, if we select as the secondary variables those branch voltages and/or currents which are consistent with the circuit variables of the MNE (as in the examples given in Section 7.2.3.2) and all original branches have their adjoint counterparts, then the matrix of the adjoint circuit is a transposed original linearized matrix. Since we can choose secondary variables other than branch voltages and/or currents (e.g. wave variables), the extended approach presented can be viewed as a generalization of the one previously discussed.

However, there is still the problem that adjoint models do not exist for branches in which both primary variables i, v belong to the secondary variables vector $\mathbf{x}$ (e.g. in MNEs when the current of a CD branch is selected as circuit variable). Then, the adjoint model cannot be created. Even so, we can formally transpose the original matrix and hence the approach introduced in Section 7.2.2 from this point of view seems to be more general than described in Section 7.2.3.

7.2.4 Sensitivity analysis of small-signal a.c. network functions

Small-signal a.c. analysis has been discussed in Section 2.2.2. As we have shown, calculation of the small-signal a.c. response consists of modeling nonlinear branches by means of counterparts linearized in the vicinity of the bias. The coefficients of linearized impedance or admittance branches were calculated using operating points. For example, the nonlinear conductance of a diode $i=I_s[\exp(v/V_T)-1]$ introduces a linear conductance $G=(I_s/V_T)\exp(v/V_T)$, dependent on the operating point v, into a substitute immitance circuit. If we want to calculate the sensitivity of an a.c. response with respect to the parameter I_s, then its changes influence both the operating point and the substitute conductance G. Hence, none of the algorithms discussed so far can be used, since our problem is neither analysis of the operating-point sensitivity nor the sensitivity of an a.c. response in a linear circuit. It is a complex problem composed of these two problems together.

Thus, a complex vector of a.c. responses $\mathbf{X}$ (i.e. a solution of the substitute immittance circuit) depends on the coefficients $\mathbf{G}$ of branches of that immittance circuit. As a.c. dedicated linearization of branches is performed in the vicinity of operating points, then the coefficients of immittance branches are functions of variable parameters $\mathbf{p}$ and an operating-point $\mathbf{x}_0$. As for the operating point, it is also a function of the parameters $\mathbf{p}$ through the d.c. circuit equations. Hence, the a.c. response $\mathbf{X}$ can be written as:

$$\mathbf{X}=\mathbf{X}\{\mathbf{G}[\mathbf{p},\mathbf{x}_0(\mathbf{p})]\}. \tag{7.60}$$

We have shown, that there is a double influence of parameters $\mathbf{p}$ on the a.c. response $\mathbf{X}$: direct, due to the appearance of $\mathbf{p}$ in the formulae for the linearized elements G, and indirect, due to the relations between these parameters and the operating point $\mathbf{x}_0$. This is the essential problem in small-signal a.c. sensitivity analysis, since it requires a combination of two sensitivity analyses.

First, we need the d.c. analysis of a nonlinear resistive circuit, obtained after zero is substituted for all derivatives in the original circuit branches. Then, the required sensitivity analysis of the d.c responses $\mathbf{x}_0(\mathbf{p})$ with respect to $\mathbf{p}$ can be performed by means of the incremental or adjoint equations introduced in previous sections.

Now, we proceed to the linearized immittance circuit used in a.c. analysis. Using incremental or adjoint equations, the sensitivities of the complex responses $\mathbf{X}$ to linearized branch parameters $\mathbf{G}$ can also be easily calculated from an immittance circuit which is linear (the immittance circuit is linear in fact). Combination of both sorts of sensitivities is done by means of the derivatives of coefficients $\mathbf{G}$ to d.c. signals $\mathbf{x}_0$, coming from the branch functions. Differentiating (7.60) with respect to $\mathbf{p}$ we obtain the formula

$$\frac{\mathrm{d}\mathbf{X}}{\mathrm{d}\mathbf{p}} = \frac{\partial \mathbf{X}}{\partial \mathbf{G}}\left(\frac{\partial \mathbf{G}}{\partial \mathbf{p}} + \frac{\partial \mathbf{G}}{\partial \mathbf{x}_0}\frac{\partial \mathbf{x}_0}{\partial \mathbf{p}}\right), \tag{7.61}$$

where $\partial\mathbf{X}/\partial\mathbf{G}$ are a.c. sensitivities with respect to the linearized parameters, and $\partial\mathbf{x}_0/\partial\mathbf{p}$ are d.c. sensitivities with respect to $\mathbf{p}$. Both require incremental or adjoint sensitivity analyses. The derivatives of the linearized parameters $\mathbf{G}(\mathbf{p},\mathbf{x}_0)$, i.e. $[\partial\mathbf{G}/\partial\mathbf{p}\,|\,\partial\mathbf{G}/\partial\mathbf{x}_0]$, can be obtained from the analytical formulae for branches, and have to be embedded into the program code.

Recalling the conductance mentioned at the beginning of this section, which, after a.c. linearization, has given the parameter $G=(I_s/V_T)\exp(v/V_T)$, we observe that the small-signal a.c. analysis of sensitivities with respect to I_s requires two auxiliary analytical derivatives:

$$\frac{\partial G}{\partial I_s} = \frac{\exp(v/V_T)}{V_T}, \quad \frac{\partial G}{\partial v} = \frac{I_s \exp(v/V_T)}{V_T^2}.$$

Before using Equation (7.61) we should notice that the above linearized parameter G cannot be treated as the only one in our circuit dependent on I_s, since a change in I_s implies a change in all operating points $\mathbf{x}_0$, which influence all linearized elements $\mathbf{G}$ of the circuit. Hence, we have to take into account the sensitivities $\partial\mathbf{x}_0/\partial I_s$ and all derivatives $\partial\mathbf{G}/\partial\mathbf{x}_0$. Then, we require also all the sensitivities $\partial\mathbf{X}/\partial\mathbf{G}$. If I_s is a parameter of only one branch, then derivatives $\partial\mathbf{G}/\partial I_s$ have only a single nonzero component.

Small-signal a.c. sensitivity analysis requires the following calculations:

(i) D.c. analysis of the original circuit.

(ii) D.c. sensitivity analysis of the original circuit to obtain sensitivities with respect to parameters **p** of those signals influencing nonlinear branches (i.e. currents of nonlinear VD branches, voltages of nonlinear CD branches and signals controlling nonlinear branches).

(iii) Calculation of the derivatives of the substitute linearized immittance coefficients **G** with respect to the operating points of these branches, for nonlinear branches (branches, whose operating points are insensitive to **p** can be omitted).

(iv) Calculation of the derivatives with respect to **p** of their linearized immittance coefficients **G**, for nonlinear branches directly dependent on the parameters **p**.

(v) Linearization of the original circuit and its small-signal a.c. analysis.

(vi) Linear a.c. sensitivity analysis of the linearized immittance circuit with respect to those **p** which are parameters of linear branches, and with respect to all linearized immittance coefficients **G**, possibly except for those omitted in item (iii).

(vii) Calculation of sensitivities from Equation (7.61).

To demonstrate a full process of small-signal sensitivity analysis let us consider the simple circuit shown in Figure 7.4(a). In this case we can perform the calculations analytically.

Two diodes $i_2 = I_{2s}[\exp(v_2/V_T) - 1]$, $i_3 = I_{3s}[\exp(v_3/V_T) - 1]$ driven by a current source, give d.c. voltage responses: $v_{20} = V_T \ln(1 + I_{1DC}/I_{2s})$, $v_{30} = V_T \ln(1 + I_{1DC}/I_{3s})$, and hence we have

$$\frac{\partial v_{20}}{\partial I_{3s}} = 0, \quad \frac{\partial v_{30}}{\partial I_{3s}} = -\frac{V_T}{I_{3s}}.$$

Since v_{20} is insensitive to I_{3s} we can omit the influence of the signal v_{20}.

Now we proceed to the linearized coefficients

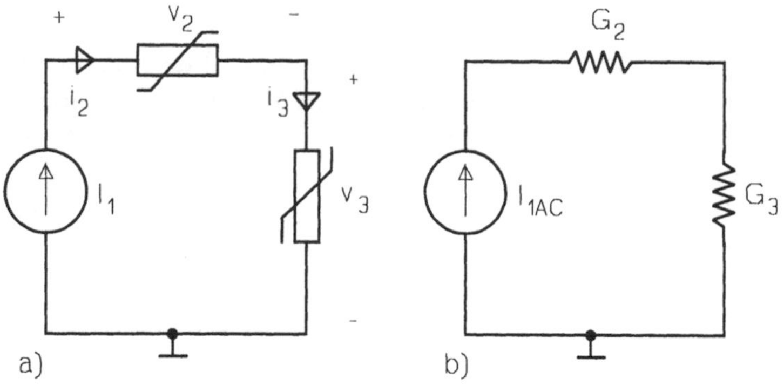

Figure 7.4 (a) A simple two diode circuit and (b) its substitute linearized circuit dedicated to a.c. analysis

$$G_2 = \frac{I_{2s}}{V_T} \exp\left(\frac{v_{20}}{V_T}\right), \quad G_3 = \frac{I_{3s}}{V_T} \exp\left(\frac{v_{30}}{V_T}\right)$$

and calculate their sensitivities with respect to I_{3s} and v_{30} (since v_{20} was insensitive):

$$\frac{\partial G_2}{\partial I_{3s}} = 0, \quad \frac{\partial G_2}{\partial v_{30}} = 0, \quad \frac{\partial G_3}{\partial I_{3s}} = \frac{\exp(v_{30}/V_T)}{V_T}, \quad \frac{\partial G_3}{\partial v_{30}} = \frac{I_{3s}\exp(v_{30}/V_T)}{V_T^2}.$$

Finally, we proceed to the linearized circuit shown in Figure 7.4(b). Considering the a.c. network function and $V_3 = I_{1AC}/G_3$ we find the sensitivities with respect to G_2 and G_3, namely:

$$\frac{\partial V_3}{\partial G_2} = 0, \quad \frac{\partial V_3}{\partial G_3} = -\frac{I_{1AC}}{G_3^2}.$$

Combining the above results according to the formula (7.61) we obtain a surprising result:

$$\frac{dV_3}{dI_{3s}} = \frac{\partial V_3}{\partial G_3}\left(\frac{\partial G_3}{\partial I_{3s}} + \frac{\partial G_3}{\partial v_{30}}\frac{\partial v_{30}}{\partial I_{3s}}\right) =$$

$$-\frac{I_{1AC}}{G_3^2}\left[\frac{\exp(v_{30}/V_T)}{V_T} + \frac{I_{3s}}{V_T^2}\exp(v_{30}/V_T)\,(-V_T/I_{3s})\right] = 0$$

It shows the possibility of compensation between d.c. operating point and a.c. linear sensitivities in the small-signal a.c. sensitivity problem. Now, we can see how far from reality small signal a.c. sensitivity analysis can be, when the influence of the operating point is neglected. Thus, though the algorithm described is troublesome in implementation (requires a lot of memory for storing intermediate derivatives, and a lot of program code to introduce derivatives of branch functions), its importance is considerable. The only analyzer implementing it is SPICE 3.

To check our understanding of sensitivity analysis, let us work through the following problems:

- Explain the concept of the incremental circuit, and using a few examples of linear and nonlinear branches demonstrate how their incremental models are created.
- Explain the concept of adjoint equations and the adjoint circuit. Is the latter defined in all cases of adjoint equations?
- Compare computational the efforts involved in the incremental and adjoint approaches.
- Outline an algorithm for d.c. or a.c. sensitivity analysis of m output signals with respect to n parameters, where switching between incremental and adjoint methods is implemented to minimize computational effort.
- For four types of controlled sources derive adjoint models using the consequences of Tellegen's theorem, as introduced in this section.
- Explain the principle of small-signal a.c. sensitivity analysis and the calculations that are required in this algorithm.

7.3 Sensitivity analysis of time-domain responses

Sensitivity analysis of time-domain responses involves the calculation of the derivatives $\partial \mathbf{x}(t)/\partial \mathbf{p}$ of time domain circuit responses at each instant of numerical integration within a range $[0, T_{max}]$. The sensitivities thus obtained are time-varying quantities. We are looking for methods, by which these quantities arise from a substitute circuit by means of its time-domain analysis (i.e. numerical integration). Such an analysis is quite standard, and has already been discussed in Chapter 5. In this section we rather focus on the generation of substitute equations that can then be integrated using methods referred to in Chapter 5.

7.3.1 The method of variational equations

7.3.1.1 General discussion of a variational circuit

In what follows we are going to discuss a nonlinear dynamic circuit described by means of OADE canonical equations (1.23), which for initial conditions $\mathbf{q}[\mathbf{v}(0)] = \mathbf{q}_0$, $\psi[\mathbf{i}(0)] = \psi_0$ have a continuous solution $\mathbf{i}(t), \mathbf{v}(t)$ within the presumed range $[0, T_{max}]$. Let us select a variable parameter p of a circuit branch. We are looking for the sensitivities of the response $\mathbf{i}(t), \mathbf{v}(t)$ with respect to this parameter. Thus, (1.23) can be rewritten as:

$$\begin{aligned} &\mathbf{A}\,\mathbf{i} = \mathbf{0} \\ &\mathbf{B}\,\mathbf{v} = \mathbf{0} \\ &\mathbf{f}\left[\mathbf{i}, \mathbf{v}, \frac{\mathrm{d}\mathbf{q}(\mathbf{v})}{\mathrm{d}t}, \frac{\mathrm{d}\psi(\mathbf{i})}{\mathrm{d}t}, p, t\right] = \mathbf{0}. \end{aligned} \tag{7.62}$$

Differentiating both sides of these equations with respect to p at a nominal solution $\mathbf{i}(t), \mathbf{v}(t)$ of (7.62), and interchanging the order of differentiation with respect to p and t we obtain equations:

$$\begin{aligned} &\mathbf{A}\frac{\partial \mathbf{i}}{\partial p} = \mathbf{0} \\ &\mathbf{B}\frac{\partial \mathbf{v}}{\partial p} = \mathbf{0} \\ &\frac{\partial \mathbf{f}}{\partial \mathbf{i}}\frac{\partial \mathbf{i}}{\partial p} + \frac{\partial \mathbf{f}}{\partial \mathbf{v}}\frac{\partial \mathbf{v}}{\partial p} + \frac{\partial \mathbf{f}}{\partial \dot{\mathbf{q}}}\frac{\mathrm{d}}{\mathrm{d}t}\left[\frac{\partial \mathbf{q}}{\partial p} + \mathbf{C}(\mathbf{v})\frac{\partial \mathbf{v}}{\partial p}\right] + \\ &\qquad + \frac{\partial \mathbf{f}}{\partial \dot{\psi}}\frac{\mathrm{d}}{\mathrm{d}t}\left[\frac{\partial \psi}{\partial p} + \mathbf{L}(\mathbf{i})\frac{\partial \mathbf{i}}{\partial p}\right] = -\frac{\partial \mathbf{f}}{\partial p} \end{aligned} \tag{7.63}$$

where $\mathbf{C}(\mathbf{v}) = \partial\mathbf{q}/\partial\mathbf{v}$, $\mathbf{L}(\mathbf{i}) = \partial\psi/\partial\mathbf{i}$ are dynamic capacitance and inductance matrices at the solution point. Comparing (7.63) with (2.2) we observe that the former is the canonical equation of a linear substitute circuit structurally identical with the original (7.62). The unknown variables of this substitute circuit are derivatives of the corresponding original variables with respect to p. Substitute branches arise from differentiation with respect to p of original branches along the solution path, which will be discussed later on. Due to this derivation the coefficients of the substitute branches (e.g. $\partial\mathbf{f}/\partial\mathbf{i}$) are functions of a solution $\mathbf{i}$, $\mathbf{v}$, varying in time. Hence, the circuit represented by (7.63) incorporates time-varying elements. Their instantaneous values are defined by means of corresponding instantaneous values of the original solution and they have been calculated before. Thus, this substitute circuit is nonstationary, and we will call it variational or incremental.

7.3.1.2 Variational models of basic elements

The concept of a variational circuit will be explained by means of basic elements. For a linear conductance $i = G\,v$, after differentiation, we obtain $\partial i/\partial p = G\,(\partial v/\partial p) + (\partial G/\partial p)\,v(t)$. Though the new branch variables are their respective derivatives, the branch obtained contains the same conductance, connected in parallel with the current source $(\partial G/\partial p)\,v(t)$. This source varies in time together with the voltage v in the original circuit. This model is similar to the incremental model described in the static sensitivity case (see Section 7.2.2) with one exception: the time variability of its coefficients.

Similarly, for the conductance of a diode $i = p[\exp(v/V_T) - 1]$, after differentiation with respect to the saturation current p, we obtain the equation $\partial i/\partial p = (p/V_T)\exp[v(t)/V_T]\,(\partial v/\partial p) + \{\exp[v(t)/V_T] - 1\}$. This equation describes a linear conductance $(p/V_T)\exp[v(t)/V_T]$ connected in parallel with a current source $\exp[v(t)/V_T] - 1$. Both are time-varying, according to $v(t)$ in the original circuit.

In the same manner, we can built variational models of any linear and nonlinear resistive branches.

Proceeding to reactive branches, we start with a linear capacitor $i = C\,\mathrm{d}v/\mathrm{d}t$. Differentiating its formula with respect to p, and interchanging the order of differentiation, we obtain a variational model

$$\frac{\partial i}{\partial p} = C\frac{\mathrm{d}}{\mathrm{d}t}\frac{\partial v}{\partial p} + \frac{\partial C}{\partial p}\frac{\mathrm{d}v(t)}{\mathrm{d}t} \tag{7.64}$$

which for signals $\partial i/\partial p, \partial v/\partial p$ is equivalent to the original capacitance C connected in parallel with a current source dependent on time by means of the time derivative of v. The latter has to be known *a priori*, and it can be obtained from the original circuit analysis at the same instant. Introducing the differential formula $\dot{x} = \gamma(x - d_x)$ for discretization of the variational circuit, we obtain the companion variational model of linear capacitance

$$\frac{\partial i}{\partial p} = \gamma C\frac{\partial v}{\partial p} - \gamma C d_{\partial v/\partial p} + \gamma\frac{\partial C}{\partial p}(v - d_v). \tag{7.65}$$

It is composed of the conductance γC and an appropriate current source dependent on the past.

In the case a nonlinear capacitor described by means of a charge state variable $i = \mathrm{d}q(v)/\mathrm{d}t$, we differentiate its description with respect to p, and interchange the order of differentiation (first to t, second to p) to obtain a linear nonstationary capacitor

$$\frac{\mathrm{d}i}{\mathrm{d}p} = \frac{\mathrm{d}}{\mathrm{d}t}\left[\frac{\mathrm{d}q[v(t)]}{\mathrm{d}p}\right] = \frac{\mathrm{d}}{\mathrm{d}t}\left[\frac{\partial q[v(t)]}{\partial p} + \frac{\partial q[v(t)]}{\partial v}\frac{\partial v}{\partial p}\right]. \tag{7.66}$$

Its state variable $\mathrm{d}q[v(t)]/\mathrm{d}p$ is the counterpart of charge in the original branch, and it will be referred to as "charge". As can be seen from (7.66), this "charge" is connected with the sensitivity $\partial v(t)/\partial p$ by means of the relation $\mathrm{d}q[v(t)]/\mathrm{d}p = \partial q[v(t)]/\partial p + \{\partial q[v(t)]/\partial v\}\,(\partial v/\partial p)$. This relation stands for a capacitor having a time-varying capacitance $\partial q[v(t)]/\partial v$ and a reference "charge" $\partial q[v(t)]/\partial p$ (corresponding to the case when $\partial v/\partial p = 0$).

Proceeding to discretization with respect to the state variable $\partial q/\partial p$ we obtain a companion variational model described by

$$\frac{\mathrm{d}i}{\mathrm{d}p} = \gamma\frac{\partial q[v(t)]}{\partial v}\frac{\partial v}{\partial p} + \gamma\left[\frac{\partial q[v(t)]}{\partial p} - d_{\mathrm{d}q/\mathrm{d}p}\right]. \tag{7.67}$$

It is composed of the time-varying conductance $\gamma\mathrm{d}q[v(t)]/\mathrm{d}v$ and current source connected in parallel. Notice that a conductance is the same as in time-domain analysis while an excitation is not, since it depends on the variable $\mathrm{d}q/\mathrm{d}p$, storing the past history of the sensitivity of charge.

As an example, we take $i = \mathrm{d}q/\mathrm{d}t$, where $q(v) = q_0\exp(v/V_T)$, and differentiate both sides with respect to q_0. We obtain the formula $\mathrm{d}i/\mathrm{d}q_0 = \mathrm{d}(\mathrm{d}q/\mathrm{d}q_0)/\mathrm{d}t$. The substitute "charge" having the form $\mathrm{d}q/\mathrm{d}q_0 = \exp[v(t)/V_T] + [(q_0/V_T)\exp(v(t)/V_T)](\partial v/\partial q_0)$ incorporates a reference component and a component connected with a linear capacitance $[(q_0/V_T)\exp(v(t)/V_T)]$. Proceeding to discretization we obtain a companion variational model

$$\frac{\mathrm{d}i}{\mathrm{d}q_0} = \gamma\frac{q_0}{V_T}\exp[v(t)/V_T]\frac{\partial v}{\partial p} + \gamma\left[\exp(v(t)/V_T] - d_{\mathrm{d}q/\mathrm{d}p}\right]. \tag{7.68}$$

In this model we can distinguish two terms: one to be introduced into the matrix and another into the RHS vector.

Returning to the general consideration and selecting the capacitor's voltage as a state variable, we write $i = C(v)\mathrm{d}v/\mathrm{d}t$. Therefore, after differentiation, the variational model takes a different form

$$\frac{\mathrm{d}i}{\mathrm{d}p} = C[v(t)]\frac{\mathrm{d}}{\mathrm{d}t}\frac{\partial v}{\partial p} + \left[\frac{\partial C[v(t)]}{\partial v}\frac{\partial v}{\partial p} + \frac{\partial C[v(t)]}{\partial p}\right]\frac{\mathrm{d}v}{\mathrm{d}t} \tag{7.69}$$

which, after discretization of derivatives, gives the model

$$\begin{aligned}\frac{\mathrm{d}i}{\mathrm{d}p} &= \gamma\left[C[v(t)] + \frac{\partial C[v(t)]}{\partial v}(v-d_v)\right]\frac{\partial v}{\partial p} + \\ &+ \gamma\left[\frac{\partial C[v(t)]}{\partial p}(v-d_v) - C[v(t)]\, d_{\partial v/\partial p}\right].\end{aligned} \tag{7.70}$$

This model consists of a time-varying conductance and, connected in parallel, a current source. The former is the same as in time-domain analysis with a voltage state variable, while the latter is specific, depending on variables storing the past of both v and $\partial v/\partial p$.

Recalling the example $i = (q_0/V_T)\exp(v/V_T)\,(\mathrm{d}v/\mathrm{d}t)$, we differentiate this formula with respect to q_0 to obtain the model

$$\begin{aligned}\frac{\mathrm{d}i}{\mathrm{d}q_0} &= \frac{q_0}{V_T^2}\exp[v(t)/V_T]\frac{\mathrm{d}v(t)}{\mathrm{d}t}\frac{\partial v}{\partial q_0} + \\ &+ \frac{q_0}{V_T}\exp[v(t)/V_T]\frac{\mathrm{d}}{\mathrm{d}t}\frac{\partial v}{\partial q_0}\end{aligned} \tag{7.71}$$

which after discretization gives

$$\begin{aligned}\frac{\mathrm{d}i}{\mathrm{d}q_0} &= \gamma\left\{\frac{q_0}{V_T^2}\exp[v(t)/V_T]\,[v(t)-d_v] + \frac{q_0}{V_T}\exp[v(t)/V_T]\right\}\frac{\partial v}{\partial q_0} + \\ &- \gamma\frac{q_0}{V_T}\exp[v(t)]/V_T]\, d_{\partial v/\partial q0}\,.\end{aligned} \tag{7.72}$$

This model consists of a conductance and a past-dependent current source. This model is less useful than (7.68) obtained for a charge-type state variable.

A similar discussion can be performed for a nonlinear flux-defined inductance.

To sum up these examples, we observe that the matrix parts of the companion incremental models are the same as the companion models for normal time-domain analysis at the corresponding instant.

7.3.1.3 Example and implementation notes

The sensitivity analysis introduced in Section 7.3.1.1 is performed simultaneously with time-response analysis. It is drawn up in Algorithm 7.1. This algorithm remains of the same structure as the time-domain analysis Algorithm 5.1.

First, at each time instant the original response is calculated using NR iterations. Then, a companion variational RHS vector is set up using the companion variational models introduced in the preceding section. Companion variational equations are solved by means of a forward and backward substitution using the LU factorized matrix which appears in the last iteration after NR loop convergence. In this way, we obtain the sensitivities of all circuit variables with respect to one parameter. To calculate sensitivities with respect to a number of parameters, that number of companion variational systems of equations have to be subsequently solved.

To calculate companion variational excitations, an additional system of past-storing data structures has to be provided. It should store the past data of sensitivities with respect to p for all state variables. This almost doubles the past-storing memory size.

To start analysis of a variational circuit, we have to know the sensitivities of all state variables with respect to p at instant $t=0$. These initial values of sensitivities can be obtained from an auxiliary d.c. (i.e. static) sensitivity analysis (see Section 7.2), which should precede the time-domain sensitivity analysis. Now, we demonstrate the whole analysis process by means of the example shown in Figure 7.5. The elements introduced are as follows:

$$E_1(t) = A\sin(\omega t),\ \ C_3(v_3) = \frac{q_0}{V_T}\exp(v_3/V_T),\ \ \psi_4(i_4) = \psi_0(i_4/I_0)^3.$$

We want to calculate sensitivities with respect to ψ_0 by creating variational equations. For the voltage source, we obtain the short-circuit $\partial v_1/\partial\psi_0=0$. The capacitance with the state variable v_3 gives the companion variational model (7.72), which can be written as a parallel connection of linear conductance and current source

$$\frac{\partial i_3}{\partial \psi_0} = G_3\frac{\partial v_3}{\partial \psi_0} + I_3,$$

Algorithm 7.1 Analysis of circuit solution and its sensitivity in time-domain

data: data for time-domain analysis
initialization of the past storing tables for time-domain analysis;
d.c. sensitivity analysis for calculation of sensitivities of initial conditions;
initialization of tables storing the past sensitivities by means of sensitivities of initial conditions;

time = 0;
for $n = 1$ **to** *ntimemax* **do** {main loop over time}
begin
calculation of coefficients of DF;
prediction;
b: *time* = *time* + h;
Newton-Raphson loop for calculation of an original solution at *time*
storage of the original solution in the solution buffer or on disk;
setting up a companion variational RHS vector;
solution of the companion variational system of equations;
storage of a sensitivity solution in the sensitivity buffer or on disk;
if (*time* ≥*timemax*) **then stop** {successful termination}
derivation of state variables and their derivatives (using DF) for both original and variational circuits;
updating the past-storing tables for original and variational variables;
selection of a new step h;
updating of the previous steps storing table;
end

For the inductance, after differentiation, we obtain a linear time-varying inductance

$$\frac{\partial v_4}{\partial \psi_0} = \frac{\mathrm{d}}{\mathrm{d}t}\left[\frac{i_4^3}{I_0^3} + \frac{3\,\psi_0[i_4(t)]^2}{I_0^3}\frac{\partial i_4}{\partial \psi_0}\right] = \frac{\mathrm{d}}{\mathrm{d}t}\left[\psi_{40}(t) + L_4(t)\frac{\partial i_4}{\partial \psi_0}\right]$$

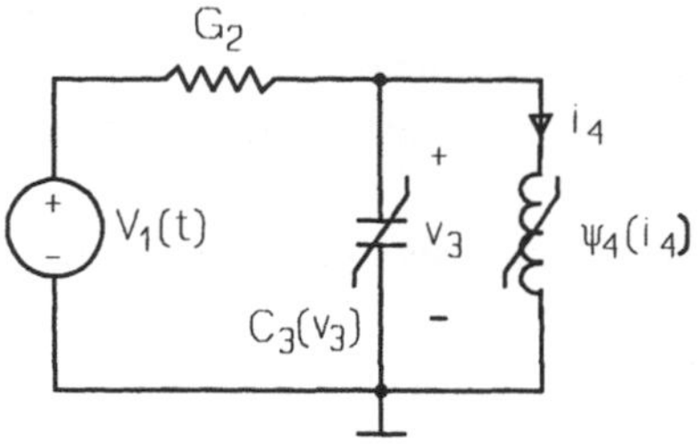

Figure 7.5 A nonlinear resonant circuit

which after discretization gives

$$\frac{\partial v_4}{\partial \psi_0} = \gamma L_4(t)\frac{\partial i_4}{\partial \psi_0} + \gamma\left[\psi_{40}(t) - d_{\partial\psi_4/\partial\psi_0}\right] =$$

$$= R_4\frac{\partial i_4}{\partial \psi_0} + V_4 .$$

The final variational circuit is demonstrated in Figure 7.6. Its MNE are as follows:

$$\begin{bmatrix} G_2 & -G_2 & 1 & 0 \\ -G_2 & G_2+G_3 & 0 & 1 \\ -1 & 0 & 0 & 0 \\ 0 & -1 & 0 & R_4 \end{bmatrix} \begin{bmatrix} \frac{\partial v_{n1}}{\partial \psi_0} \\ \frac{\partial v_{n2}}{\partial \psi_0} \\ \frac{\partial i_1}{\partial \psi_0} \\ \frac{\partial i_4}{\partial \psi_0} \end{bmatrix} = \begin{bmatrix} -I_2 \\ I_2-I_3 \\ 0 \\ -V_4 \end{bmatrix} . \tag{7.73}$$

Now, we describe the algorithm for time-domain sensitivity analysis. The procedure begins from a calculation of initial sensitivities. Short-circuiting the inductor and open-circuiting the capacitor we create linear equations for d.c. analysis of the circuit given in Figure 7.5

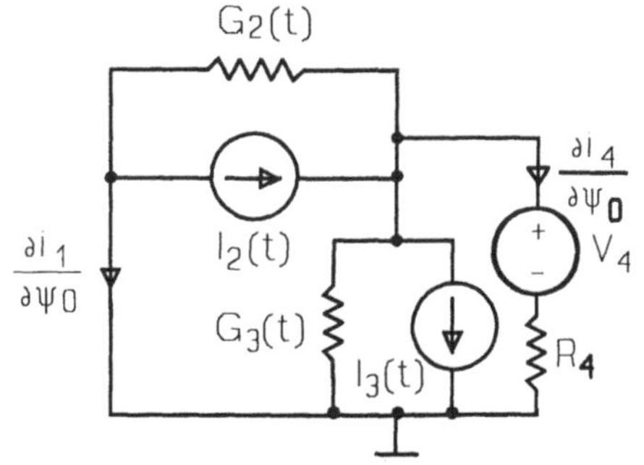

Figure 7.6 A companion variational counterpart of the circuit shown in Figure 7.5

$$\begin{bmatrix} G_2 & -G_2 & 1 & 0 \\ -G_2 & G_2 & 0 & 1 \\ -1 & 0 & 0 & 0 \\ 0 & -1 & 0 & 0 \end{bmatrix} \begin{bmatrix} v_{n1} \\ v_{n2} \\ i_1 \\ i_4 \end{bmatrix} = \begin{bmatrix} 0 \\ 0 \\ -V_{1DC} \\ 0 \end{bmatrix}. \tag{7.74}$$

After the d.c. solution is found, we calculate d.c. sensitivities with respect to ψ_0. Since the d.c. model of the inductor is not dependent on ψ_0, all incremental excitations I_2, I_3, V_4 become zero. Thus, the vector of incremental excitations is equal to zero, and hence, all d.c. sensitivities with respect to ψ_0 are also zero. This result is obvious in this case, since sensitivities with respect to a parameter not influencing d.c. solutions are being calculated.

Then, the d.c. initial conditions v_{n2}, i_4 obtained for the time response, and zero initial conditions for sensitivities $\partial v_{n2}/\partial\psi_0$, $\partial i_4/\partial\psi_0$ are used to start analyses in the time-domain. At each instant t and iteration k, we solve the companion linearized equations

$$\begin{bmatrix} G_2 & -G_2 & 1 & 0 \\ -G_2 & G_2+G_3 & 0 & 1 \\ -1 & 0 & 0 & 0 \\ 0 & -1 & 0 & R_4 \end{bmatrix} \begin{bmatrix} v_{n1} \\ v_{n2} \\ i_1 \\ i_4 \end{bmatrix} = \begin{bmatrix} -I_2 \\ I_2-I_3 \\ -V_1(t) \\ -V_4 \end{bmatrix} \tag{7.75}$$

with elements calculated according to Chapter 5, i.e.

$$
\begin{aligned}
&I_2=0\\
&G_3=\gamma\left[\frac{\partial C(v^{(k)})}{\partial v}(v^{(k)}-d_v)+C(v^{(k)})\right]\\
&I_3=\gamma C(v^{(k)})(v-d_v)-G_3 v^{(k)}\\
&R_4=\gamma\frac{\partial\psi_4(i_4^{(k)})}{\partial i_4},\quad V_4=\gamma[\psi_4(i_4^{(k)})-d_{\psi_4}]-R_4 i_4^{(k)}.
\end{aligned}
\tag{7.76}
$$

After the convergence of NR iterations, we obtain G_3, R_4, and take the excitations I_2, i_3, V_4 from the companion variational models involved to solve the system (7.73). We obtain instantaneous sensitivities, and then proceed to the next instant. As for the past-storing memory, we require the past state of v_{n2}, i_4 and $\partial v_{n2}/\partial\psi_0$, $\partial i_4/\partial\psi_0$. For these quantities, the past-storing variables d_x have been included.

Apart from this variational equations method, time-domain sensitivities can be also calculated by means of adjoint equations. The main theoretical advantage of the latter is the requirement for only one adjoint circuit to obtain sensitivities with respect to all circuit parameters. However, the adjoint circuit usually needs a separate integration. An especially complicated case appears when initial conditions are dependent on variable parameters, as e.g. in d.c. initial state. That is why this method is less popular than the one discussed and we neglect it in this consideration.

We can now work through the following problems to reinforce our understanding of the main topics of this section:

- Explain the concept of the companion variational circuit and show how its equations can be obtained.
- For a few invented elements of different types, calculate companion variational models and compare their stamps with those of companion models for time-domain analysis.
- Explain why the variational method can be implemented using the conventional time-domain analysis.
- Using the examples shown in Section 7.3.1.3 try to show the relation between companion variational models of branches and incremental models for the calculation of static initial values of sensitivities.

7.4 Large-change sensitivity analysis of linear circuits

7.4.1 Introduction to large-change sensitivity problems

By large-change sensitivity analysis we mean the repeated calculation of a circuit response $\mathbf{x}(\mathbf{p})$ for a number of vectors of circuit parameters $\mathbf{p}$: namely $\mathbf{p}_0$, $\mathbf{p}_1$, ..., $\mathbf{p}_m$. If the changes in parameters are small, then this problem can be approximately solved by analysis of the circuit and its small-change sensitivities with respect to all components of $\mathbf{p}$ at the nominal point $\mathbf{p}_0$. Next, using sensitivities and increments $\mathbf{p}_i - \mathbf{p}_0$ $(i=1, 2, ..., m)$, we can evaluate an increment in the network function with respect to $\mathbf{x}(\mathbf{p}_0)$. This small-change sensitivity problem can be solved by means of the adjoint equations method.

Because it is based on the differential approach, small-change sensitivity analysis is often too inaccurate in practice. Thus we need to solve the large-change sensitivity problem. Obviously, it can be solved by $(m+1)$-times repeated analysis of a circuit for each set of parameters p_i. However, this approach is too time-consuming.

Dedicated large-change sensitivity algorithms can calculate increments in the circuit response with a computational effort smaller than full circuit analysis. Such methods have been investigated almost exclusively for linear circuits. They rely on the extraction of variable elements from the circuit and the representation of the fixed part of the circuit by means of a substitute Thevenin-Norton circuit. If the number of variable elements is smaller than the size of the full circuit matrix, than analysis of a circuit composed of variable elements connected to a substitute circuit can be much less time-consuming than full analysis. That is why large-change sensitivity algorithms are suitable only for a small number of variables. When only one parameter is subject to change, large-change sensitivity analysis is especially efficient; we call this the **one-dimensional orthogonal search** (ODOS).

7.4.2 One-variable large-change sensitivity problems

7.4.2.1 Bilinear network functions

Let us consider a linear circuit with such a variable parameter p, that matrix $\mathbf{Y}(p)$ and RHS vector $\mathbf{b}(p)$ are linear functions of p. For example, in the MNE case, p can be a coefficient of the standard branch given in Figure 2.5. Complex a.c. (i.e. immittance) or real d.c. equations of this circuit can be written in the form

$$\mathbf{Y}(p)\,\mathbf{x} = \mathbf{b}(p). \tag{7.77}$$

According to Cramer's rule, the kth solution of the vector $\mathbf{x}$, which is a real d.c. or complex a.c. network function in the p domain, can be expressed as a ratio of two determinants

$$x_k = \frac{\det \mathbf{Y}_k(p)}{\det \mathbf{Y}(p)}, \tag{7.78}$$

where $\mathbf{Y}_k$ is the matrix $\mathbf{Y}$ with the kth column replaced by $\mathbf{b}$. Since the matrix incorporates at most four elements linearly dependent on p and located at vertices of a rectangle, then we can conclude (after Bode) that the circuit function (7.78) takes the bilinear form

$$f(p) = \frac{n(p)}{d(p)} = \frac{n_0 + n_1 p}{d_0 + d_1 p} \tag{7.79}$$

whose coefficients are independent of p.

Bilinearity holds in linear circuits built of branches acceptable to MNE, if each branch coefficient is unique. For instance, if a gyrator appears in a circuit, then it introduces a model composed of two dependent VCCS. Their coefficients are not independent, but described by the same gyration parameter g. In that case, a circuit function is not bilinear with respect to g.

Considering bilinearity with respect to a few parameters, we obtain a multibilinear function, e.g.

$$f(p_1, p_2, p_3) = \frac{n_0 + n_1 p_1 + n_2 p_2 + n_{12} p_1 p_2}{d_0 + d_1 p_1 + d_2 p_2 + d_{12} p_1 p_2}. \tag{7.80}$$

Now, if we introduce synchronous parameters (e.g. $p_1 = p_2$), then the bilinearity of the function vanishes.

Bilinear network functions have very convenient sensitivity properties, which can be easily exploited in large-change sensitivity analysis. To introduce the sensitivity description we have to express the bilinear function (7.79), in the vicinity of a chosen nominal point p_0, by means of the Fidler-Nightingale formula [7.3, 7.4]. It has three sensitivity coefficients calculated at that nominal point: function value $f_0 = f(p_0)$, small-change sensitivity $D_p = \partial f(p_0)/\partial p$, and semi-relative small-change sensitivity of the denominator of (7.79) $Q_p = [\partial d(p_0)/\partial p]/d(p)$. The Fidler-Nightingale formula is as follows:

$$f(p) = f_0 + \frac{D_p (p - p_0)}{1 + Q_p (p - p_0)}. \tag{7.81}$$

It can be represented by means of an incremental Mason's graph, with vertices standing for signals and edges representing transmission coefficients. A variable element can be viewed as a transmitting block whose input signal is y, while its output is the increment $z-z_0=(p-p_0)y$. Thus, the Mason's model demonstrated in Figure 7.7 has a transfer function

$$f(p)=f_0+\frac{ab(p-p_0)}{1+Q_p(p-p_0)}, \tag{7.82}$$

which is identical with (7.81) iff $ab=D_p$.

The transmittance (7.82) can be easily found from a given circuit having any original excitations and one output signal f (a voltage between two nodes or a current through a branch). Setting the original excitations and $p=p_0$ we can analyze the original circuit to calculate:

a as a signal y on a variable element;

f_0 as the output signal f;

Now, we set zero for the original excitations and impose a unitary auxiliary excitation of the same type as signal z, connected in such a way that it adds to the signal z_0. If z is a current, then a 1 A current source is added in parallel with the element p_0. If z is a voltage, then a 1 V voltage source is added in series with the element p_0. Setting $p=p_0$ we can identify:

b as the output signal f;

Q_p as a negative signal y at an input of element p_0.

From the above we obtain three sensitivity coefficients: f_0 and Q_p directly, and D_p as a product of transmittances $D_p=ab$. Thus, three sensitivity coefficients can be calculated by means of two circuit analyses: first with the original excitations, and then, with a unit auxiliary excitation located at the output of the variable element. It is interesting to observe that the denominator sensitivity Q_p is equal to the negative reverse transmittance

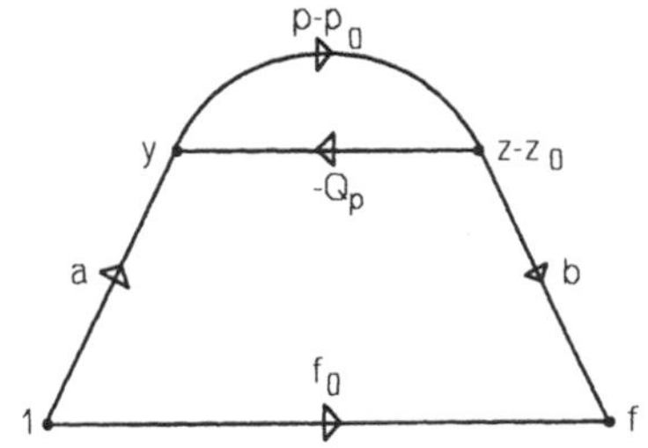

Figure 7.7 Incremental Mason's model of the bilinear transfer function

between the variable element ports. It can be readily shown, that the rule for $D_p = ab$ is equivalent to the incremental equations method discussed in Section 7.2. These rules can be extended to higher order sensitivities.

Calculation of the three sensitivity coefficients of the Fidler-Nightingale formula gives full large-change sensitivity information in a class of circuits having bilinear network functions. These coefficients describe the behavior of the circuit with one parameter varying in the vicinity of the given nominal point. The Fidler-Nightingale formula proves that the nonlinearity of the function $f(p)$ is uniquely described by the denominator sensitivity Q_p. Hence, the nonlinearity index (7.14) is the product $Q_p(p-p_0)$. The maximum increment obeying the criterion of small changes can be evaluated as

$$|p-p_0| \le \frac{\varepsilon}{|Q_p|}. \tag{7.83}$$

This provides that the error of the small-change sensitivity approach remains at a level not higher than ε. This criterion constitutes a linear range around the nominal point, where small-change sensitivities are said to be sufficiently accurate.

7.4.2.2 Example of a three sensitivity coefficients calculation

In this section we demonstrate the use of the algorithm just introduced for calculating the three sensitivity coefficients of the Fidler-Nightingale formula. In Figure 7.8(a) we have a linear circuit. Let $p \equiv g_{42}$, while voltage v_5 is an output signal. To calculate the three sensitivity coefficients for this signal, at a nominal point g_{420}, we solve the original circuit for excitation I_1. The nominal value of f_0 will be equal to the response v_5, i.e.:

$$f_0 \equiv v_5 = \frac{I_1}{G_5 + (G_2 + g_{420})(1 + t_{35})}. \tag{7.84}$$

The variable parameter is of the form $i_4 = g_{42} v_2$. The transmittance a is equal to the response v_2 at the input to this element, i.e.

$$a \equiv v_2 = -\frac{(1 + t_{35})}{G_5 + (G_2 + g_{420})(1 + t_{35})}. \tag{7.85}$$

Now, we create an auxiliary circuit, where $I_1 = 0$ and an additional 1 A

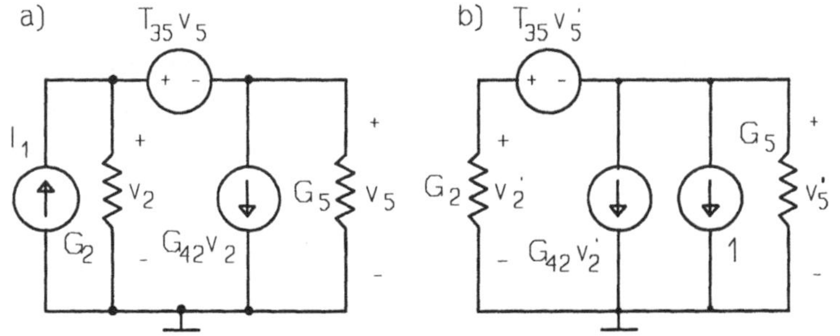

Figure 7.8 (a) A simple linear circuit, and (b) its auxiliary counterpart for calculation of three sensitivity coefficients

current source is connected in series with g_{420}, and directed in accordance with it (hence, it adds up to i_{40}). This circuit is shown in Figure 7.8(b).

Therefore, the transmittance b will be the response at the output

$$b \equiv v_5' = -\frac{1}{G_5+(G_2+g_{420})(1+t_{35})}. \tag{7.86}$$

while Q_p is equal to the negative response at the input to the variable element, i.e. v_2'. Thus, it is equal to

$$Q_{g_{42}} \equiv v_2' = \frac{1+t_{34}}{G_5+(G_2+g_{420})(1+t_{34})}. \tag{7.87}$$

Quantities a and b give the small-change sensitivity

$$D_{g_{42}} = a\,b = -\frac{I_1(1+t_{35})}{[G_5+(G_2+g_{420})(1+t_{35})]^2}. \tag{7.88}$$

Assuming the following data: $G_5=10^{-5}$, $G_2=10^{-4}$, $g_{42}=-9{,}95\cdot10^{-5}$, $t_{35}=9$ and $I_1=10^{-3}$, we obtain $f_0=66.6$, $D_{g_{42}}=4{,}44\ 10^7$, and $Q_{g_{42}}=6{,}66\ 10^5$. From the denominator sensitivity we can obtain the non-linearity index $NI=6.66\ 10^5\,(g_{42}-g_{420})$. Assuming an error level $\varepsilon=1$ per cent, we obtain an acceptable relative perturbation (linear range) of g_{42} equal to ±0.015 per cent. This example shows how large can be the errors in small-change sensitivity calculations for active circuits with a considerable gain.

7.4.2.3 Three sensitivity coefficients calculated by means of the network function samples

The bilinear circuit function is uniquely defined by means of three independent coefficients, for instance the three sensitivity coefficients involved. Determination of these coefficients stands for a solution of the one-parameter large-change sensitivity problem. One method for calculating them has been introduced in the preceding section. Here, we show another approach referred in Reference [7.3]. The three coefficients can be identified using three sample points of the bilinear network function for three distinct values of p: nominal p_0 and additional values p_1, p_2. The corresponding values of the function are: f_0, f_1 and f_2. To obtain these solutions, three circuit analyses are required. The three sensitivity coefficients can be easily expressed in terms of these sample points as follows:

$$D_p^{f(p_0)} = \frac{(f_0-f_1)(f_0-f_2)(p_2-p_1)}{(p_1-p_0)(p_2-p_0)(f_2-f_1)}, \tag{7.89}$$

$$Q_p^{d(p_0)} = \frac{p_0(f_2-f_1)+p_1(f_0-f_2)-p_2(f_0-f_1)}{(p_1-p_0)(p_2-p_0)(f_2-f_1)}. \tag{7.90}$$

Thus, three circuit analyses enable us to find the behavior of the circuit while p is subject to change.

However, the computational effort can be reduced using an algorithm requiring only two circuit analyses. Provided the circuit matrix elements are linearly dependent on p, its determinant is equal to a common denominator of all network functions originating in the vector function $\mathbf{x}(p)$. This determinant can be obtained during the LU solution procedure as a product of pivots. Using the values of the denominator the three sensitivity cefficients can be identified by means of only two sample points. Assuming two values p_0, p_1, and corresponding function and denominator values: f_0, d_0 and f_1, d_1, we can write:

$$D_p^{f(p_0)} = \frac{f_0-f_1}{p_0-p_1}\,\frac{d_1}{d}, \tag{7.91}$$

$$Q_p^{d(p_0)} = \frac{d_0-d_1}{(p_0-p_1)\,d_0}. \tag{7.92}$$

In this case, the computational effort is the same as in the incremental method described in Section 7.4.3.

7.4.2.4 Network functions of amplitude and phase

The solution of the one-variable large-change sensitivity problem by means of complex bilinear network-functions is not sufficient. The performance of linear a.c. circuits is usually described in terms of the amplitude and phase of responses, and hence real network functions of magnitude and phase are of great importance.

Small-change sensitivities of magnitude and phase can be easily determined from the complex sensitivity of a complex bilinear network function. Differentiating a network function of the form $|f|e^{\mathrm{j}\arg f}$, we can prove that its semirelative small-change sensitivity is equal to:

$$Q_p^{f(p_0)} = Q_p^{|f(p_0)|} + \mathrm{j}\, D_p^{\arg f(p_0)}. \tag{7.93}$$

Thus its real part is equal to the semirelative sensitivity of amplitude, while the imaginary part is equal to the sensitivity of phase. This principle holds for any function, not necessarily bilinear, and explains how the sensitivities of amplitude and phase can be obtained from a.c. small-change sensitivities, calculated by means of the methods described in this chapter.

For large-change sensitivities, similar relations between complex sensitivity and sensitivities of amplitude and phase can be written only for complex bilinear functions.

$$SL_p^{|f(p_0)|^2} = \frac{p-p_0}{p_0}\,|SL_p^{f(p_0)}|^2 + 2\,\mathrm{Re}\,SL_p^{f(p_0)}, \tag{7.94}$$

$$DL_p^{\arg f(p_0)} = \arg\left(QL_p^{f(p_0)} + \frac{1}{p-p_0}\right). \tag{7.95}$$

The most convenient way of examining the behavior of an amplitude network function in the variable p domain, is to formulate a biquadratic function of the squared modulus of the complex bilinear function in the form introduced in Reference [7.7, 7.8]

$$|f(p)|^2 = |f_0|^2 + \frac{A_2(p-p_0)^2 + A_1(p-p_0)}{B_2(p-p_0)^2 + B_1(p-p_0) + 1}, \tag{7.96}$$

where

$$A_2 = |D_p|^2 + \mathrm{Re}(D_p f_0^\star Q_p^\star), \quad A_1 = 2\,\mathrm{Re}(D_p f_0^\star),$$
$$B_2 = |Q_p|^2, \qquad B_1 = 2\,\mathrm{Re}\,Q_p, \tag{7.97}$$

and ★ indicates the complex conjugate. The main advantage of this formula is the unified description of the complex function, its modulus (and even phase [7.8]) by means of three sensitivity coefficients easily derivable from two circuit analyses. Expressions (7.81) and (7.96) have become the mathematical foundation for a large-change sensitivity analysis technique called one-dimensional orthogonal search (ODOS). Changing one circuit parameter, while other parameters remain fixed, we obtain a technique for searching the space of parameters along orthogonal straight lines parallel to coordinate axes. If, during this search, we take into account variability of amplitude and phase, then ODOS can be applied to examine the circuit performance. For example, in Reference [7.8] it has been used to calculate regions, where amplitude and phase responses satisfy design constraints.

7.4.3 Multiparameter large-change sensitivity analysis

In this section a method for calculating linear circuit responses at numerous nonnominal points in the variable parameters' space is introduced. It relies on one nominal circuit analysis, and then calculation of nonnominal responses by modification of the nominal response for each set of nonnominal parameters separately. We treat a nominal circuit as a whole, including nominal values for variable parameters. Elements representing increments are extracted outside and connected to a circuit succinctly represented as a multiport. The incremental part of a variable element is a current-defined branch, connected in parallel with a current-defined nominal branch, or a voltage-defined branch connected in series with a voltage-defined nominal branch.

For instance, a nominal conductance $i = Gv$, after incrementation, takes the form $i + \Delta i = (G + \Delta G)v$, which can be viewed as an incremental conductance $\Delta i = \Delta G\, v$ connected in parallel with the original. If we know the value of v after change, and hence the increment Δi, then we can substitute the incremental conductance for the equivalent current source whose value is Δi. We call it a substitute incremental source. This substitution has no influence on the circuit response after change.

Such an approach consists of three stages: (1) representation of a nominal circuit by means of equivalent terminal-wise hybrid equations, (2) calculation of substitute incremental excitations for specified increments in the parameters, (3) solution of the circuit with incremental excitations. This method has been introduced in [7.5] and [7.6].

7.4.3.1 Substitute incremental sources method

Let us consider a linear a.c./d.c. circuit containing a number of independent sources. We collect their values in the vector $\boldsymbol{y}_{in}$ and call them input signals. We assume that the circuit moreover has a number of output signals $\boldsymbol{x}_{out}$, whose values after large parameter changes have to be calculated.

This circuit also incorporates m linear variable elements described by:

$$\mathbf{y}_s = \mathbf{P}_s \mathbf{x}_s = \begin{bmatrix} p_1 & 0 & \dots & 0 \\ 0 & p_2 & \dots & 0 \\ . & . & \dots & . \\ 0 & 0 & \dots & p_m \end{bmatrix} \mathbf{x}_s \tag{7.98}$$

where $\mathbf{x}_s$ and $\mathbf{y}_s$ are the independent and dependent signals of variable elements respectively. If the parameters p_i of these elements are subject to change, then equations (7.98) after incrementation (upper bars indicate signals after incrementation) transform into

$$\bar{\mathbf{y}}_s = \mathbf{y}_s + \Delta\mathbf{y}_s = (\mathbf{P}_s + \Delta\mathbf{P}_s)\,\bar{\mathbf{x}}_s \tag{7.99}$$

where

$$\Delta\mathbf{y}_s = \Delta\mathbf{P}_s \bar{\mathbf{x}}_s = \begin{bmatrix} \Delta p_1 & 0 & \dots & 0 \\ 0 & \Delta p_2 & \dots & 0 \\ . & . & \dots & . \\ 0 & 0 & \dots & \Delta p_m \end{bmatrix} \bar{\mathbf{x}}_s \, . \tag{7.100}$$

The above increments can be treated as substitute signals connected to variable elements. If a variable element is current-defined, then a substitute signal, a current source, is added in parallel with this element. If a variable element is voltage-defined, then the substitute signal is a voltage source connected in series with this element. The added signal can be viewed as an auxiliary input to the circuit. Examples of substitute incremental sources are shown in Figure 7.9(b). We will also require an auxiliary circuit output, whose signal is an independent signal $\mathbf{x}_s$ of the variable element. All signals involved can be voltages or currents. A circuit can be viewed as a transmitting system with a number of inputs and outputs.

Assuming that a circuit after incrementation is described by a system of linear algebraic equations (e.g. MNE)

$$\bar{\mathbf{Y}}\bar{\mathbf{x}} = \bar{\mathbf{b}} \tag{7.101}$$

we can express output signals as combinations of components of $\mathbf{x}$ by means of two 0, ±1 matrices $\mathbf{M}_{si}$, $\mathbf{M}_{out}$:

$$\begin{aligned} \bar{\mathbf{x}}_s &= \mathbf{M}_{si}\bar{\mathbf{x}} \\ \bar{\mathbf{x}}_{out} &= \mathbf{M}_{out}\bar{\mathbf{x}} \end{aligned} \tag{7.102}$$

Similarily, the RHS vector can be joined with circuit excitations via

$$\bar{\mathbf{b}} = \mathbf{M}_{in}\mathbf{y}_{in} + \mathbf{M}_{so}\Delta\mathbf{y}_s \tag{7.103}$$

where two auxiliary 0, ±1 matrices locate the values of excitations, with appropriate signs, at appropriate positions in $\mathbf{b}$. Combining equations (7.101)-(7.103) we obtain two hybrid formulae describing the transmitting properties of a circuit

$$\mathbf{T}_{ss}\Delta\mathbf{y}_s + \mathbf{T}_{sin}\mathbf{y}_{in} = \bar{\mathbf{x}}_s, \tag{7.104}$$

$$\mathbf{T}_{outs}\Delta\mathbf{y}_s + \mathbf{T}_{outin}\mathbf{y}_{in} = \bar{\mathbf{x}}_{out}, \tag{7.105}$$

where hybrid submatrices can be expressed as

$$\begin{aligned} \mathbf{T}_{ss} &= \mathbf{M}_{si}\mathbf{Y}^{-1}\mathbf{M}_{so} \\ \mathbf{T}_{sin} &= \mathbf{M}_{si}\mathbf{Y}^{-1}\mathbf{M}_{in} \\ \mathbf{T}_{outs} &= \mathbf{M}_{out}\mathbf{Y}^{-1}\mathbf{M}_{so} \\ \mathbf{T}_{outin} &= \mathbf{M}_{out}\mathbf{Y}^{-1}\mathbf{M}_{in}. \end{aligned} \tag{7.106}$$

Substituting in (7.104) the incremental excitations $\Delta\mathbf{y}_s$ for (7.100), we obtain the so-called system of incremental equations

$$(\mathbf{1} + \Delta\mathbf{P}_s \mathbf{T}_{ss}) \Delta\mathbf{y}_s = -\Delta\mathbf{P}_s \mathbf{T}_{sin} \mathbf{y}_{in}, \tag{7.107}$$

or alternatively, considering that $\Delta\mathbf{P}_s$ is nonsingular diagonal having an obvious inverse

$$(\Delta\mathbf{P}_s^{-1} + \mathbf{T}_{ss})\Delta\mathbf{y}_s = -\mathbf{T}_{sin}\mathbf{y}_{in}. \tag{7.108}$$

These incremental equations enable us to calculate the values of substitute incremental sources, solving only a system of m equations instead of the larger system (7.101). It can be solved by means of the LU factorization described in Chapter 3. However, for small increments $\Delta\mathbf{P}_s$, an iterative method is also available. For a starting value $\Delta\mathbf{y}_s^{(0)}$, an approximation to the signal $\bar{\mathbf{x}}_s$ can be calculated from (7.104). Then, using (7.100) we obtain the next iterative value of $\Delta\mathbf{y}_s^{(k)}$, namely:

$$\Delta\mathbf{y}_s^{(k+1)} = -\Delta\mathbf{P}_s(\mathbf{T}_{sin}\mathbf{y}_{in} - \mathbf{T}_{ss}\Delta\mathbf{y}_s^{(k)}), \quad k = 0, 1, \ldots \tag{7.109}$$

This procedure is at most linearly convergent. In practice, convergence is provided only for relative increments in parameters not exceeding about 20 per cent.

After the substitute excitations $\Delta\mathbf{y}_s$ have been calculated, we can obtain the final response of the circuit from Equation (7.105), i.e. as if a nominal circuit is being excited by substitute incremental sources.

$$\bar{\mathbf{x}}_{out} = \mathbf{T}_{outin}\mathbf{y}_{in} + \mathbf{T}_{outs}\Delta\mathbf{y}_s. \tag{7.110}$$

In this expression, the first term stands for nominal response calculated only once, while the second is an output increment, due to the increments introduced into parameters $\mathbf{P}_s$.

7.4.3.2 Calculation of the hybrid matrix by circuit excitation

To implement the method introduced above, we have to calculate a set of hybrid submatrices $\mathbf{T}_{outin}$, $\mathbf{T}_{sin}$ and $\mathbf{T}_{outs}$, $\mathbf{T}_{ss}$.

One possible method consists of two circuit analyses. One analysis is performed under the original excitations $\mathbf{y}_{in}$ (i.e. with $\Delta\mathbf{y}_s = \mathbf{0}$), to obtain the output signals $\mathbf{x}_{out}$ and independent signals of variable elements $\mathbf{x}_s$. These will be equal to the signals $\mathbf{T}_{sin}\mathbf{y}_{in}$ and $\mathbf{T}_{outin}\mathbf{y}_{in}$ respectively, which are required on the RHS of (7.107) and (7.110). The second analysis consists of setting $\mathbf{y}_{in} = \mathbf{0}$, and imposing unit substitute excitations $\Delta\mathbf{y}_s = \mathbf{1}$, to obtain signals $\mathbf{x}_{out}$ and $\mathbf{x}_s$, which will be equal to $\mathbf{T}_{outs}$ and $\mathbf{T}_{ss}$ respectively. Hence, for m variable elements, the analysis requires one LU factorization and $m+1$ forward/backward substitutions to calculate the responses to subsequent excitations. In the full matrix technique, the computational effort is proportional to $n^3/3-n/3+(m+1)n^2$ of floating-point multiplications and divisions, where n is the dimension of the circuit matrix.

In the case when the number of outputs is smaller than the number of inputs, it is reasonable to excite an adjoint circuit (i.e. the one described by means of a transposed matrix). It can be readily shown, that the hybrid matrix $\mathbf{T}$ of equations (7.104), together with (7.105), is the same as the transposed matrix of the adjoint circuit $\mathring{\mathbf{T}}^t$. Thus, e.g. for a term T_{ab}, instead of exciting port b of the original circuit, and taking a signal from port a, we can excite port a of the adjoint circuit and take a signal from port b, since the relation $T_{ab}=\mathring{T}_{ba}$ holds. That is why we can excite the outputs of the adjoint circuit to calculate $\mathring{\mathbf{T}}_{sout}=\mathbf{T}_{outs}$, $\mathring{\mathbf{T}}_{inout}=\mathbf{T}_{outin}$, and then excite the terminals of the variable elements to obtain $\mathring{\mathbf{T}}_{ins}=\mathbf{T}_{sin}$, $\mathring{\mathbf{T}}_{ss}=\mathbf{T}_{ss}$. This approach is very convenient when we are interested in a small number of outputs. Then, in estimating the computational effort, we set m equal to the number of outputs.

To demonstrate this method, let us consider the circuit shown in Figure 7.9(a). It is excited by a current source I_1, while v_3 is its output signal. In this circuit we select two variable elements: G_2 and t_{42}. Their description is of the form

$$i_2 = G_2 v_2, \quad v_4 = t_{42} v_2 . \tag{7.111}$$

From the original circuit we calculate

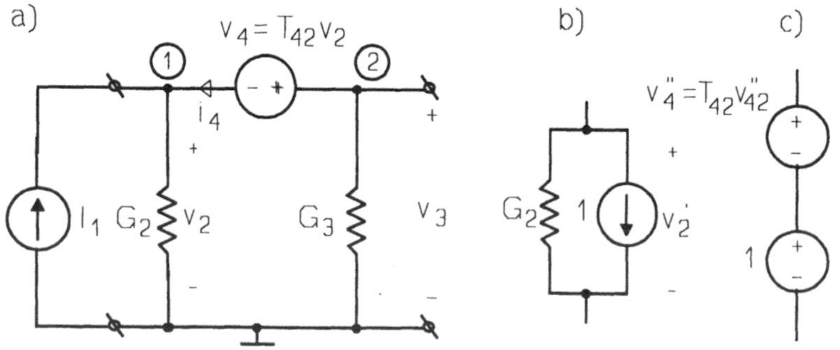

Figure 7.9 (a) A simple linear circuit for demonstration of substitute incremental sources techniques, (b) an incremental model of G_2, and (c) of T_{42}

$$\mathbf{T}_{outin}\,\mathbf{y}_{in} = v_3$$

$$\mathbf{T}_{sin}\,\mathbf{y}_{in} = \begin{bmatrix} v_2 \\ \\ v_2 \end{bmatrix} \tag{7.112}$$

which stand for the response on the output port and the independent signals of the variable elements.

Then, we create two substitute circuits with unit incremental excitations. In both, I_1 is open-circuited. One circuit contains a parallel incremental model, shown in Figure 7.9(b), instead of G_2. Another has a series incremental model, given in Figure 7.9(c), instead of t_{42}. From these two substitute circuits (whose responses we denote by single and double prime superscripts respectively) we calculate

$$\mathbf{T}_{outs} = [\, v_3' \quad v_3'' \,], \tag{7.113}$$

$$\mathbf{T}_{ss} = \begin{bmatrix} v_2' & v_2'' \\ v_2' & v_2'' \end{bmatrix} \tag{7.113}$$

by means of two substitute circuit analyses. Thus, analysis of the hybrid matrix requires a modified nodal matrix formulation, its LU decomposition, and three times RHS vector formulation and forward/backward substitution.

7.4.3.3 Calculation of the hybrid matrix by circuit matrix inversion

Another method of calculating the required transmittances is based on the formulae (7.106), i.e. we invert a circuit matrix **Y**, and then, appropriately combine elements of inversion, according to the numbers of nodes and currents. For the circuit shown in Figure 7.9(a) we have

$$\begin{aligned} &\mathbf{M}_{out} = [0 \;\; 1 \;\; 0], \quad \mathbf{M}_{in} = [1 \;\; 0 \;\; 0]^t, \\ &\mathbf{M}_{si} = [1 \;\; 0 \;\; 0], \quad \mathbf{M}_{so} = [-1 \;\; 0 \;\; 0]^t, \quad \text{for } G_2, \\ &\mathbf{M}_{si} = [1 \;\; 0 \;\; 0], \quad \mathbf{M}_{so} = [0 \;\; 0 \;\; -1]^t, \quad \text{for } t_{42}. \end{aligned} \tag{7.114}$$

Hence, from (7.106), the elements of a hybrid matrix appear to be combinations of the elements of the circuit matrix inverse $\mathbf{Y}^{-1}$. Denoting elements of this matrix by $\tilde{y}_{ij}$, we can write

$$
\begin{aligned}
&T_{out\,in} = [0 \;\; 1 \;\; 0]\; \mathbf{Y}^{-1} [1 \;\; 0 \;\; 0]^{t} = \tilde{y}_{21} \\
&T_{G_2 in} = T_{t_{42} in} = [1 \;\; 0 \;\; 0]\; \mathbf{Y}^{-1} [1 \;\; 0 \;\; 0]^{t} = \tilde{y}_{11} \\
&T_{out G_2} = [0 \;\; 1 \;\; 0]\; \mathbf{Y}^{-1} [-1 \;\; 0 \;\; 0]^{t} = -\tilde{y}_{21}, \\
&T_{out t_{42}} = [0 \;\; 1 \;\; 0]\; \mathbf{Y}^{-1} [0 \;\; 0 \;\; -1]^{t} = -\tilde{y}_{23}, \\
&T_{G_2 G_2} = [1 \;\; 0 \;\; 0]\; \mathbf{Y}^{-1} [-1 \;\; 0 \;\; 0]^{t} = -\tilde{y}_{11}, \\
&T_{t_{42} G_2} = [1 \;\; 0 \;\; 0]\; \mathbf{Y}^{-1} [-1 \;\; 0 \;\; 0]^{t} = -\tilde{y}_{11}, \\
&T^{G_2 t_{42}} = [1 \;\; 0 \;\; 0]\; \mathbf{Y}^{-1} [0 \;\; 0 \;\; -1]^{t} = -\tilde{y}_{13}, \\
&T_{t_{42} t_{42}} = [1 \;\; 0 \;\; 0]\; \mathbf{Y}^{-1} [0 \;\; 0 \;\; -1]^{t} = -\tilde{y}_{13}.
\end{aligned}
\tag{7.115}
$$

In general, hybrid terms can be more complicated, i.e. of the form $\tilde{y}_{ik} - \tilde{y}_{il} - \tilde{y}_{jk} + \tilde{y}_{jl}$. However, it is essential to notice that not all elements of the inverted circuit matrix are required for hybrid analysis calculation. In our case, only four out of nine terms are used. It can be proved that in the inverted matrix, entries of interest lie only in certain rows and columns. Rows of interest are those corresponding to the independent variables $\mathbf{x}_s$ of variable elements, and corresponding to circuit outputs. Columns of interest are those corresponding to signals $\mathbf{y}_s$, and to original excitations of a circuit. Since not all terms of the inverse matrix are required, considerable savings are possible by factorizing $\mathbf{Y}$ into **LU** factors and calculating the inverted factors $\mathbf{U}^{-1}$, $\mathbf{L}^{-1}$ by known methods [7.6]. Having these factors $\mathbf{U}^{-1}\mathbf{L}^{-1}$, we multiply only those rows and columns in these factors, which give the required elements of the inverse matrix. The number of multiplications and divisions is then not greater than about $2n^3/3+4mn$ where n is the circuit dimension. The first term is the cost of LU factorization and inversion, the second is the cost of multiplying the required factors.

Comparing the costs of this method with the one described in Section 7.4.3.2 we observe that its constant part, independent of the number of variable elements, is higher, but the cost per variable parameter is much smaller. That is why the previously introduced method is more advantageous only for a small number of variable elements, i.e. for $m \ll n/3$. Otherwise, the method presented in this section gives better results.

7.4.3.4 Calculation of large-change sensitivities

After a hybrid matrix has been calculated, we can obtain circuit responses for a variety of parameter increments. For each set of increments, first we have to calculate the values of substitute incremental sources $\Delta\mathbf{y}_s$ from the system of m equations (7.108). This requires $m^3/3+m^2-m/3$ multiplications and divisions per set of increments, which is a very small computational effort when the number of variable elements is small in comparison with the circuit dimension n. This computational requirement is the hard limit for large-change sensitivity methods.

In the case of the circuit shown in Figure 7.9(a), we obtain two equations

$$\begin{bmatrix} T_{G_2G_2}+\dfrac{1}{\Delta G_2} & T_{G_2t_{42}} \\ T_{t_{42}G_2} & T_{t_{42}t_{42}}+\dfrac{1}{\Delta t_{42}} \end{bmatrix} \begin{bmatrix} \Delta i_2 \\ \Delta v_4 \end{bmatrix} = \begin{bmatrix} T_{G_2in} \\ T_{t_{42}in} \end{bmatrix} \tag{7.116}$$

giving incremental excitations Δi_2, Δv_4.

To calculate the increment of a circuit response with respect to its nominal value $T_{out\,in}$, we use the formula (7.110) requiring only m multiplications. For this example, we obtain the formula

$$\Delta x_{out} = T_{out\,G_2}\Delta i_2 + T_{out\,t_{42}}\Delta v_4\,. \tag{7.117}$$

The multiparameter method of substitute incremental sources, introduced in Section 7.4.3, is representative of a family of large-change sensitivity methods. Though limited to linear circuits, it is used in simulation and design problems requiring many repeated analyses, e.g. in Monte-Carlo analysis, optimization and the analysis of circuit performance analysis in the space of design parameters.

To check comprehension of the foregoing topics the reader may try to answer the following questions:

- What do we mean by bilinearity of network functions and what assumptions should a linear circuit obey in order to provide bilinearity of its responses?
- How can bilinear functions be determined by three sensitivity coefficients? Explain their definition and two methods of identification: by means of function samples and a circuit excitation.

- Using the biquadratic function of the square of magnitude, examine how to predict the dependence of the amplitude of a complex circuit response on parameter variations.
- For a branch $v = R\,i$, explain the concept of a substitute incremental source, and how it can be used for large-change sensitivity analysis.
- Explain two methods of calculating a hybrid matrix applicable in large-change sensitivity analysis and compare their efficiency.
- Demonstrate how the formulae (7.107) and (7.109), in the case of one variable element, reduce to the Fidler-Nightingale formula. What circuit interpretation of the three sensitivity cefficients arises from this reduction?

References

7.1 Director, S.W., Rohrer, R.A., *Automated Network Design - The Frequency - domain Case,* IEEE Trans. Circ. Theor., vol. 16, no. 3, p. 330-337 (1969).

7.2 El-Turky, F.M., *Efficient Computation of Network Sensitivities,* IEEE Trans. Circ. Syst., vol. 33, no 7, p. 659-664, (1986).

7.3 Fidler, J.K., *Network Sensitivity Calculation*, IEEE Trans. Circ. Syst., vol. 23, no. 9, p. 567-571, (1976).

7.4 Fidler, J.K., Nightingale, C., *Differential-Incremental-Sensitivity Relationships,* Electronics Letters, vol. 8, no. 25, p. 626-627, (14 December 1972).

7.5 Gadenz, R.N., Rezai-Fakhr, M.G., Temes, G.C., *A Method for the Computation of Large-tolerance Effects,* IEEE Trans. Circ. Theory, vol. 20, no. 6, p. 704-708, (1973).

7.6 Krishnan, R., Downs, T., *A Note on Computation of Large-change Multiparameter Sensitivity*, Int. J. Circ. Theory Appl., vol. 4, p. 307-310, (1976).

7.7 Ogrodzki, J., Opalski L., Stybliński M., *Acceptability Regions for a Class of Linear Networks*, Proc. Int. Symp. Circ. Syst., Huston, Tx., USA (1980).

7.8 Ogrodzki, J., *One-dimensional Orthogonal Search - A Method for a Segment Approximation to Acceptability Regions*, Int. J. Circ. Theory Appl., vol. 14, p. 181-194 (1986).

7.9 Tellegen, B.D.H., *A General Network Theorem with Applications,* Philips Res. Rep., no. 7, p. 259-269, (1952).

Chapter 8

Decomposition-based Methods for Large-scale Circuits

8.1 Introductory topics in large-scale circuit simulation

The contemporary technology of Integrated Circuit (IC) manufacturing has caused a spectacular explosion in the complexity of the circuit which can be made on one semiconductor chip. In analog circuits, where bipolar and MOS technologies are both in use, technology has achieved Large-Scale Integration (LSI), where the number of transistors on a single chip can exceed 10^4. In digital circuits, an even larger scale of integration has been reached due to MOS technology, especially dynamic MOS circuits. MOS technology has made possible digital Very Large-Scale Integrated Circuits (VLSI), where one million transistors per chip has been exceeded, while the channel sizes have become lower than one micron. An impressive scale of integration is reached in memories and single-chip processors. This progress is still continuing. The manufacture of VLSI circuits requires a powerful software infrastructure. For decades we observe that it is technology which stimulates the progress in tools for Computer-Aided Design and Manufacturing (CAD/M) of ICs.

A basic software system for assisting the IC designer comprises at least: a layout editor, a circuit extractor, and a circuit analyzer adequate for the circuit involved. Analyzers are an important part of software systems for CAD. Usually, we cannot have one analyzer for all purposes. Large-scale and very large-scale integrated circuits also require such conventional circuit analyses as d.c., a.c. and time-domain transients, however, contemporary ICs are usually too large to be analyzed as a whole.

Since VLSI circuits are mostly digital, we can discard conventional circuit analysis of the whole system and apply a logic simulation, where extremely high efficiency has been obtained by using discretized (digital) signals and trivial models. In more advanced approaches to logic simulation, a variety of levels of logic primitive are introduced. Thanks to such an approach logic simulation is the only one applicable to the whole VLSI circuit. Logic simulation can give accurate static logic states, however the errors in timing evaluation (analysis of propagation delays) can be tens of per cent.

This explains why we cannot give up using circuit simulation and great effort has been devoted to the development of more efficient circuit simulators. This and the next chapter are dedicated to new circuit simulation methods, suitable for concurrent large circuits. Unfortunately, the efficiency of the direct circuit simulation algorithms presented in Chapters 2-7 appears to be insufficient for such large circuits. They can be directly applied only to rather small functional blocks. We need new simulation algorithms. However, these new algorithms for large-scale circuits are not entirely new, but are actually based upon the same foundations discussed in previous chapters. Considerable saving in computer effort is achieved by elimination of redundant calculations, simplification of models, and replacement of time-consuming accurate algorithms by less accurate, though satisfactory relaxation-based ones. Moreover, concurrent algorithms apply more efficient data-structures, advanced programming, and parallelization. The latter is not discussed in this book.

The following groups of circuit simulation methods for large-scale integrated circuits can be distinguished:

1. **Direct simulation**. This method involves the solution of circuit equations by means of such classical methods as: differentiation formula for discretization, modified Newton-Raphson method for solution of nonlinear algebraic equations and sparse Gaussian-elimination-based method for solution of algebraic linear equations. These methods were presented in Chapters 2-8. However, to achieve the efficiency required for large-scale circuits, these methods are combined with decomposition techniques. Redundant calculations are discarded by block macromodeling and latency. The latency exploits the fact that if a block does not change its signals in comparison with the preceding iteration, then recalculation is not required. Direct methods can be about ten times faster than conventional ones, but this gain is insufficient for VLSI circuits. Thus these methods can be applied only to building-blocks of large circuits. They suffer many limitations. First of all their efficiency is dependent on the manner of decomposition. Moreover, they weakly exploit the regular structure of digital circuits. Latency provides savings only if a decomposition is adequate to the latent properties of the real system. It is ubiquitous in digital system blocks, but in analog circuits latency is more rare and difficult to predict. Circuit partitioning should instead be dynamic. Direct decomposition-based methods are covered in the present chapter. Moreover, they can be mixed with relaxation and logic simulation in mixed-mode simulators.

2. **One-point relaxation simulation.** This is a circuit-level simulation method aimed at accurate time-domain analysis and obtained with a drastically smaller computational effort by virtue of exploiting the one-way properties of MOS VLSI circuits. Since in the MOS transistor the

gate is insulated from the channel, the element often works as a unilateral one and the application of node decoupling can give considerable savings in the solution of discretized equations. This leads to so-called relaxation methods of simulation. The basic approach is known as timing simulation based on relaxation methods of integrating circuit equations. If a circuit is first discretized and then nodal decoupling is applied we obtain an algebraic system whose dependency matrix is triangular (or block triangular). Timing simulation algorithms for large-scale circuits give time-profiles of signals about a few hundreds of times faster than direct methods. However, they suffer from limitations introduced by relaxation methods. They basically require unilateral propagation of signals, relatively short stepsizes and grounded capacitors connected to each node. Newer algorithms overcome many of these problems using multi-relaxation and multi-iterative approaches. Accuracy and convergence of relaxation methods can sometimes be unsatisfactory. Thus we have to proceed to block decoupling instead of nodal decoupling. To reduce the computational effort timing methods employ macromodeling of repetitive blocks and so-called look-up tables instead of full MOSFET models.

3. **Waveform-relaxation method.** To improve the stability of integration and convergence of relaxations a waveform relaxation method has been proposed where decoupling of the circuit before discretization is employed. Then conventional integration is applied to decoupled nodes or blocks instead of the relaxation-based integration. A family of one-point and waveform relaxation methods is covered in Chapter 9.

4. **Mixed-mode simulation.** This simulation is the most advanced. It is capable of combining overall logic simulation of a digital system with timing and circuit simulation of those blocks whose importance is such as to require more accurate analysis. The essence of mixed-mode simulation is interfacing (direct and relaxation) circuit simulation methods with logic simulation. Problems of logic and mixed-mode simulation are beyond the scope of this book which is dedicated only to circuit simulation.

In the above classification of concurrent simulation techniques the relaxation techniques are essential. They use decoupling of equations, and the separate solution of each part. This idea originates from such iterative methods of linear algebraic systems solution, as Gauss-Seidel's or Gauss-Jacobi's, introduced in Section 9.2. In these methods a system is solved as a sequence of scalar equations, where each one is solved for one unknown with the others known from previous equations. Thus the Gaussian elimination is discarded and computational effort becomes proportional only to the squared dimension of the system, not to its third power.

Table 8.1. Classification of simulation techniques

	Direct methods		Relaxation methods	
	no decomposition	block decomposition	nodal decoupling	block decoupling
Linearized equation level	*linear algebraic equations* sparse LU methods	*BBD linear algebraic equations* tearing methods	*linear algebraic equations* iterative methods for linear algebraic equations	*BBD linear algebraic equations* block iterative methods for linear algebraic equations
Discretized equations level	*nonlinear algebraic equations* modified Newton–Raphson methods	*nested nonlinear algebraic equations* multi-level Newton-Raphson methods	*nodal nonlinear algebraic equations* relaxation-Newton methods, e.g. GSN	*nested nonlinear algebraic equations* block-relaxation Newton methods, e.g. block GSN
Algebraic differential equation level	*systems of ordinary algebraic differential equations* variable-step integration methods	*system of nested ordinary algebraic differential equations* direct waveform method	*nodal algebraic differential equations* nodal wave-form relaxation method	*nested algebraic differential equations* block waveform-relaxation

Decoupling into individual equations is known as **nodal decoupling**. Moreover, we can also resort to **block decoupling**, where equations are decoupled into groups which are solved with direct methods.
Decoupling can be applied to different levels of equation transformation. We distinguish three such levels:

(i) **The algebraic-differential equations level** is the primary circuit formalism. In direct methods, this formalism has been transformed by discretization using the differentiation formulae (DF) introduced in Chapter 5. Decoupling can be applied to algebraic differential equations prior to their discretization. This leads to Waveform-Relaxation methods.

(ii) **The discretized nonlinear algebraic equations level** arises after the application of DF to algebraic differential equations. Decoupling at this level leads to nonlinear relaxations which are also called relaxation integration techniques.

(iii) **The linearized discretized algebraic equations level** is obtained after Newton-Raphson linearization of discretized equations. Decoupling at this level leads to linear-relaxation methods, where linearized equations are solved by means of the iterative (relaxation) methods introduced in Section 9.2.

In the case of direct methods, a counterpart of decoupling is circuit decomposition.

The introduced classification is shown in Table 8.1. In twelve boxes we have given twelve corresponding types of equations and resulting simulation methods. In the first column entitled "no decomposition" we have the classical direct methods discussed in Chapters 2-7 of this book. The first row "linearized equation level" is dedicated to the solution of linear equations. Its first cell contains classical LU methods covered in Chapter 3. Chapter 8 is dedicated to decomposition-based methods which have been classified in the column of Table 8.1 entitled "block decomposition". The first cell of this column stands for decomposition-based (tearing) methods for linear circuits described in Section 8.2. Nodal and block relaxation methods of the columns entitled "nodal decoupling" and "block decoupling" will be covered in Chapter 9.

8.2 Decomposition based linear circuits analysis

In Section 3.3 the sparsity of the circuit matrices was exploited. Some sparsity features correspond to the circuit structure (topology). In each electrical circuit, especially in a large circuit, some blocks can be selected. By blocks we mean groups of nodes and elements, tightly coupled but connected outside their group with only a rather small number of branches. Such blocks are repesented in a circuit matrix as clusters of nonzero elements, appearing against a background of zero elements. There are algorithms where such a structure is exploited by partitioning the matrix into submatrices representing these blocks and their connections. These matrices are considerably smaller than the original matrix. They can also be sparse, though less sparse than the original matrix. Techniques of partitioning matrices and solving circuits part by part are called circuit analysis by means of decomposition. Decomposed circuits can be solved even when the sparse representation of the whole circuit cannot be stored in memory. Then, only the currently processed block remains in memory, while the rest of the circuit is kept on disk. The efficiency of decomposition methods is strongly dependent on the manner of circuit partitioning. Blocks must not be too small in comparison with the numbers of links between them. The number of such links should be small. Moreover, sparsity within blocks is advantageous.

The efficiency of decomposition methods increases when a circuit is repetitively solved in iterative loops of d.c. and time-domain analyses. In these cases, some elements of matrices are subject to change during the iterative process, while others remain fixed. Fixed blocks we call bypassed, latent or dormant. They do not require numerical processing, since the results of the preceding iteration remain valid.

This section is dedicated to the analysis of linear circuits partitioned into distinct blocks, also called subcircuits. We will introduce algorithms which do not solve the whole circuit at the same time. Each block is considered separately, and then their solutions are combined by specific equations describing connections between subcircuits. Such methods initiated by G. Kron [8.5] were called diacoptics or tearing [8.1, 8.3, 8.9, 8.10, 8.11]. We will call them decomposition methods. They can be thought of as the decomposition of circuits or, equivalently, of systems of algebraic linear equations.

Two main types of circuit decomposition can be distinguished: nodal decomposition, introduced by A. Sangiovanni-Vincentelli in [8.10], and branch decomposition, introduced by F.F. Wu in [8.11]. Both can be viewed as a decomposition of the circuit graph or circuit equations. Decomposition of linear equations can be formally represented as partitioning of graphs (e.g. Coates' graphs) or a system of equations. The partitioned equations have a bordered block diagonal (BBD) structure. This specific structure of equations is the foundation of decomposition-based methods, since a system of equations can be divided into separately solved subsystems.

If all submatrices of a BBD matrix are dense, then the efficiency will be the same as in sparse-matrix techniques. However, blocks are rather sparse, and then the efficiency is usually worse, even if we exploit sparsity in the decomposition-based solution. The decomposition methods have an advantage over common sparse matrices methods if:

(i) circuits are huge, such that storage of their matrices as a whole is impossible,

(ii) circuits incorporate repetitive pieces, whose partial calculations need only be performed once,

(iii) the solution of a linearized circuit is repeated many times, e.g. in an iteration loop, and its matrix has blocks that do not chane during iterations; these fixed blocks can be solved once; then at each iteration we only combine their solution with the rest of the circuit solution.

(iv) the subcircuits are large and sparse enough to solve them efficiently using the sparse matrix technique.

8.2.1 Decomposition of linear circuits

8.2.1.1 The concept of nodal decomposition

Let us consider the topology of a linear circuit C, built of branches from a set B stretched over nodes from a set N including a reference node 0. To divide this circuit into m subcircuits C_i $(i=1, ..., m)$ we arbitrarily select among all N nodes a subset of partition nodes (also called block nodes) N_0, including 0. These nodes are the nodes where selected blocks interact. Then we divide the rest of the nodes $N-N_0$ into m separate subsets N_i. Each N_i is a set of internal nodes of the ith subcircuit. Thus a subcircuit is stretched over its internal nodes and at least one block node. The subcircuit C_i incorporates: (i) branches whose both terminals are connected to nodes within N_i, (ii) branches in which one terminal is connected to a node N_i while the other is connected to a block node N_0, and it can also have (iii) branches connected only to block nodes. A subcircuit cannot have branches joined to internal nodes of other subcircuits. Concerning branches connected only to block nodes, their assignment to a subcircuit is arbitrary since they can be members of any of the subcircuits involved. These branches can also be treated not as members of any subcircuit but as branches belonging directly to the main circuit.

In Figure 8.1 we demonstrate decomposition of a circuit into two subcircuits C_1 and C_2. Four block nodes have been selected: 01, 02, 03 and 00. In subcircuit C_1 we have two internal nodes 11 and 12; in sub-

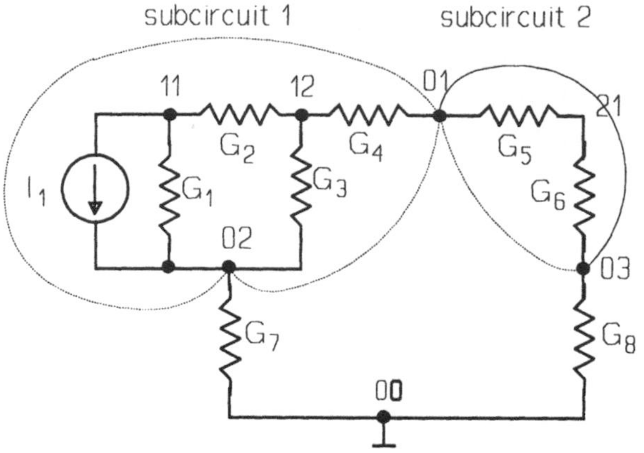

Figure 8.1 The nodal decomposition of a simple circuit into two blocks

circuit C_2 one internal node 21. Hence, C_1 is stretched over nodes 01, 02, 11, 12, while C_2 is over 01, 03, 21. The subcircuit C_1 has branches E_1, G_2, G_3, G_4. The subcircuit C_2 contains G_5, G_6. In this example, we have two other branches joined only to block nodes, i.e. G_7 and G_8. They could be assigned to C_1 or C_2, but we can also treat them as direct members of the main circuit.

In the consideration so far, a circuit has been partitioned, and hence by the node we have meant exactly the circuit node. However, in MNEs we have not only variables corresponding to nodes, but also to some branch currents. If we view such equations as nodal equations of a circuit, then we can perform their nodal decomposition no matter what is the physical sense of the variables in the MNEs. Hence, VD branch currents of MNEs can be treated equivalently as nodal voltages and MNE variables can be viewed as voltages of "generalized nodes". In this way we can apply nodal decomposition also to MNE or any other equations. We notice that in fact this decomposition is a decomposition of a system of linear equations or equivalently of its Coates' graph. In further discussion therefore, a node will be thought of as a generalized node, i.e. possibly also a current variable.

Coming back to the general discussion, let us formulate nodal equations of the whole circuit and reorder the circuit variables, starting from the voltages $\mathbf{v}_i$ of internal nodes of subsequent blocks, and terminating with the voltages $\mathbf{v}_0$ of block nodes. Then the NE matrix takes the Bordered Block Diagonal (BBD) form:

$$
\begin{bmatrix}
\mathbf{Y}_1 & & & & \mathbf{P}_1 \\
& \mathbf{Y}_2 & & & \mathbf{P}_2 \\
& & \ddots & & \vdots \\
& & & \mathbf{Y}_m & \mathbf{P}_m \\
\mathbf{Q}_1^t & \mathbf{Q}_2^t & \cdots & \mathbf{Q}_m^t & \mathbf{R}
\end{bmatrix}
\begin{bmatrix} \mathbf{v}_1 \\ \mathbf{v}_2 \\ \vdots \\ \mathbf{v}_m \\ \mathbf{v}_0 \end{bmatrix}
=
\begin{bmatrix} \mathbf{b}_1 \\ \mathbf{b}_2 \\ \vdots \\ \mathbf{b}_m \\ \mathbf{b}_0 \end{bmatrix}. \tag{8.1}
$$

containing subcircuit submatrices $\mathbf{Y}_i$, interconnection submatrices $\mathbf{P}_i$, $\mathbf{Q}_i$, and a main circuit submatrix $\mathbf{R}$. The above formula becomes obvious if we notice that the KCL has been partitioned into sections describing current balances in internal nodes and in block nodes. The balance in an internal node includes only the relations between the internal voltages involved and block voltages. The balance in block nodes is dependent on the voltages of all circuit nodes.

Below, we demonstrate BBD equations of a circuit presented in Figure 8.1

$$
\left[\begin{array}{cc|c|ccc}
G_2+G_1 & -G_2 & & 0 & -G_1 & 0 \\
-G_2 & G_2+G_3+G_4 & & -G_4 & -G_3 & 0 \\
\hline
 & & G_5+G_6 & -G_5 & 0 & -G_6 \\
\hline
0 & -G_4 & -G_5 & G_4+G_5 & 0 & 0 \\
-G_1 & -G_3 & 0 & 0 & G_3+G_7 & G_1 \\
0 & 0 & -G_6 & 0 & 0 & G_6+G_8
\end{array}\right]
\left[\begin{array}{c} v_{11} \\ v_{12} \\ \hline v_{21} \\ \hline v_{01} \\ v_{02} \\ v_{03} \end{array}\right]
=
\left[\begin{array}{c} I_1 \\ 0 \\ \hline 0 \\ \hline 0 \\ -I_1 \\ 0 \end{array}\right].
$$

where its characteristic BBD structure can be easily seen.

The bordered block diagonal structure is not dependent on the physical meaning of the equations and the manner of their formulation. For any equations, i.e. any Coates' graph, a nodal decomposition is possible iff there is such an order of rows and columns that the matrix takes the BBD form.

To demonstrate how we can select a current variable as a block node, let us turn to Figure 8.2, where the circuit comprises a voltage-defined branch and a current-controlled branch. Hence two currents have been

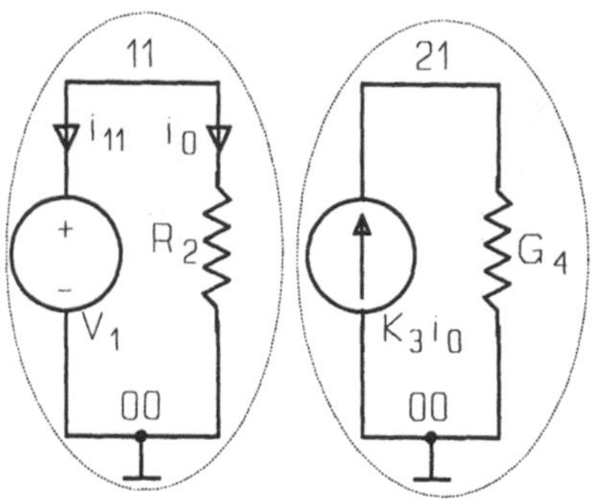

Figure 8.2 Example of a circuit including currents selected as block variables

selected as circuit variables of MNEs. We can treat them as generalized nodes. If the current i_0 is chosen as a block variable, then we partition the rest of the variables into two groups $\{v_{11}, i_{11}\}$ and $\{v_{21}\}$, corresponding to two subcircuits. The first subcircuit is constituted by the KCL at node 11 and the equation of the branch V_1. The second one is generated by the KCL at node 21. The equation of interconnections will be the equation of the branch conducting the block current i_0. We obtain:

$$\left[\begin{array}{cc|c|c} -1 & 0 & & 0 \\ 0 & 1 & & 1 \\ \hline & & G_4 & -K_3 \\ \hline -1 & 0 & 0 & R_2 \end{array}\right] \left[\begin{array}{c} v_{11} \\ i_{11} \\ \hline v_{21} \\ \hline i_0 \end{array}\right] = \left[\begin{array}{c} -V_1 \\ 0 \\ \hline 0 \\ \hline 0 \end{array}\right]. \tag{8.3}$$

8.2.1.2 A concept of branch decomposition

The main difference between branch and nodal decomposition lies in the choice of block variables. In branch decomposition, we perform the partition across a set of branches instead of cutting the circuit across the block nodes. Thus we introduce the concept of block branches, which become interconnections between subcircuits. The currents of these branches are added to the circuit variables and enumerated as the block variables.

Branch decomposition starts from the choice of a set of such branches

B_0, that after their deletion from the circuit C, it splits into m distinct subcircuits C_i having no common nodes, possibly except for the reference 0. For each subcircuit C_i, we select its local reference node, if it has no reference node 0. The voltages of these local references, with respect to the global reference 0, we denote as $\mathbf{v}_0$. The rest of its nodes are internal nodes of the block. Their voltages are denoted as $\mathbf{v}_i$. The block branches B_0, irrespective of their type of definition, we treat as branches that introduce currents to a vector of circuit variables and denote them $\mathbf{i}_0$. We enumerate the circuit variables in such a way that the internal variables of successive blocks are considered first, and then the block variables $\mathbf{i}_0$ and $\mathbf{v}_0$. Thus, we obtain MNE equations of the BBD form:

$$\begin{bmatrix} \mathbf{Y}_1 & & & & \mathbf{Y}_{10} & \mathbf{A}_1 \\ & \mathbf{Y}_2 & & & \mathbf{Y}_{20} & \mathbf{A}_2 \\ & & \ddots & & \vdots & \vdots \\ & & & \mathbf{Y}_m & \mathbf{Y}_{m0} & \mathbf{A}_m \\ \mathbf{Y}_{10}^t & \mathbf{Y}_{20}^t & \cdots & \mathbf{Y}_{m0}^t & \mathbf{R} & \mathbf{A}_0 \\ \mathbf{A}_1^t & \mathbf{A}_2^t & \cdots & \mathbf{A}_m^t & \mathbf{A}_0^t & \mathbf{Z} \end{bmatrix} \begin{bmatrix} \mathbf{v}_1 \\ \mathbf{v}_2 \\ \vdots \\ \mathbf{v}_m \\ \mathbf{v}_0 \\ \mathbf{i}_0 \end{bmatrix} = \begin{bmatrix} \mathbf{b}_1 \\ \mathbf{b}_2 \\ \vdots \\ \mathbf{b}_m \\ \mathbf{b}_0 \\ \mathbf{s}_0 \end{bmatrix}. \quad (8.4)$$

This differs from a nodal decomposition in the interconections, which combine the block variables involved. To clarify the above decomposition let us consider the circuit shown in Figure 8.1. To divide it into two subcircuits we choose branches G_4 and G_7. The decomposed circuit is shown in Figure 8.3. The first block is not grounded. We select its local reference voltage v_0. The second block is grounded. Currents i_4, i_7 and

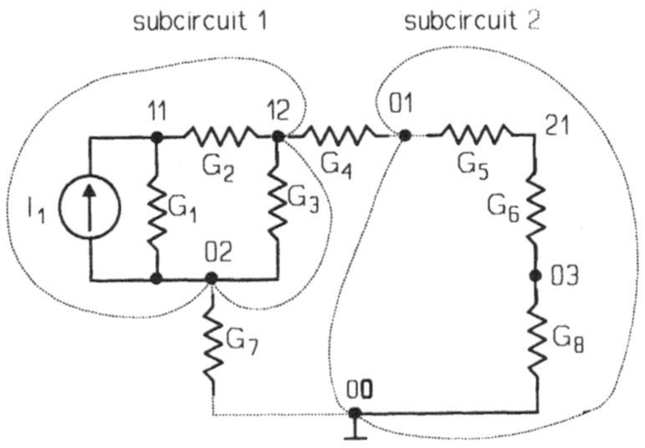

Figure 8.3 Example of branch decomposition of the circuit whose nodal decomposition was given in Figure 8.1

the voltage v_0 are the block variables. We obtain the following MNE of the BBD form:

$$
\left[\begin{array}{cc|ccc|ccc}
G_2+G_1 & -G_2 & & & & -G_1 & 0 & 0 \\
-G_2 & G_2+G_3 & & & & -G_3 & 1 & 0 \\
\hline
 & & G_5+G_6 & -G_5 & -G_6 & 0 & 0 & 0 \\
 & & -G_5 & G_5 & 0 & 0 & -1 & 0 \\
 & & -G_6 & 0 & G_8+G_6 & 0 & 0 & 0 \\
\hline
-G_1 & -G_3 & 0 & 0 & 0 & G_1+G_3 & 0 & 1 \\
0 & G_4 & 0 & -G_4 & 0 & 0 & -1 & 0 \\
0 & 0 & 0 & 0 & 0 & G_7 & 0 & -1
\end{array}\right]
\left[\begin{array}{c}
v_{11} \\ v_{12} \\ \hline v_{21} \\ v_{22} \\ v_{23} \\ \hline v_0 \\ i_4 \\ i_7
\end{array}\right] =
$$

$$
= [\, I_1 \;\; 0 \mid 0 \;\; 0 \;\; 0 \mid -I_1 \;\; 0 \;\; 0 \,]^t .
$$

Branch decomposition is more inconvenient than nodal decomposition since it introduces additional currents into the set of circuit variables. So the difference between the equations of a nondecomposed and decomposed circuit lies not only in the order of equations. That is why in circuit analysis we prefere to use the nodal approach rather than the branch approach. In the next section we will discuss some general properties of the nodal decomposition.

8.2.1.3 Nodal decomposition - its properties and automation

In circuit analysis we prefer to use nodal decomposition, viewed as the decomposition of circuit equations, or equivalently their Coates' graph. Such decomposition consists of transforming equations into the BBD form by means of row and column reordering. However, this reordering is not unique. A different selection of block variables gives different subcircuits, interconnected by a different number of common nodes. Efficient solution of the decomposed circuit requires that each block have a number of block variables small in comparison with the number of internal variables. From the reordering point of view, decomposition is the task of selecting appropriate diagonal blocks with a possibly narrow border responsible for interconnections. In Section 8.3.2 we will estimate the computational effort and decide which decomposition is optimal.

Unfortunately, in decomposition-based circuit simulation the computational effort is strongly dependent on the manner of decomposition, and hence it is difficult to implement this method as a general-purpose approach. In programs where decomposition is done manually, e.g. by defining subcircuits in a way similar to SPICE, the user can be surprised to find a considerable difference in efficiency from a slight change in circuit partitioning. Of course neither SPICE nor OPTIMA use decomposition-based methods. They simply develop any subcircuits introduced into a one-level circuit.

Decomposition-based analyzers require an efficient automatic decomposition algorithm, which should reorganize the partitioning introduced by a user, in order to minimize numerical complexity. However, the efficiency of decomposition, and hence the expected efficiency of analysis in such programs is strongly dependent on circuit topology features. In analog circuits most elements are tightly coupled, and hence it is often difficult to obtain an efficient decomposition. However, in logic circuits blocks can be easily decoupled. In this domain decomposition methods are universally used.

Automatic circuit decomposition algorithms are usually developed using results from graph theory, because a graph is a spectacular representation of linear algebraic equations. As we know, nodal decomposition is equivalent to matrix reordering into the BBD form. A weak point of such a procedure is that we cannot check global optimality criteria. A similar problem in pivoting algorithms was discussed in Chapter 3. In decomposition the problem is much more serious because the iterative decomposition algorithm works only locally. It is started from a given initial set of internal block nodes, and tries to collect neighboring nodes into a cluster, to minimize the number of its nodes interacting with the exterior. This is carried out without any knowledge of the structure of the remaining part of the circuit. Hence, optimality of decomposition is considerably dependent on the choice of starting nodes. The best known algorithm, of A. Sangiovanni-Vincentelli *et al* [8.3, 8.9], will be explained briefly below.

Let N be the set of circuit nodes. If we select from N a node $\{n\}$ and initialize the set of internal nodes of a block $N_i = \{n\}$, then within $N - N_i$ we can determine a set S_i of nodes neighboring with N_i (connected to its nodes by a branch). Its cardinality is l_i. Then for each node n of S_i, we can test how many neighboring nodes would appear if we added n to the set N_i. That node n, which minimizes this number, is added to N_i, i is increased by one and the process is repeated until a maximum size of block *imax* is reached. Now we select within an assumed range of acceptable block sizes [*imin*, *imax*] a block N_r where r is an index of a minimal $l_{imin}, ..., l_{imax}$. The block N_r obtained this way locally minimizes the number of block nodes and limits the sizes of blocks to between *imin* and *imax*. After the first block has been found we can proceed to extract the next one from the rest of the nodes. This method is presented in Algorithm 8.1.

Algorithm 8.1. Nodal decomposition of a circuit

N=all nodes; $S^{(0)}=\varnothing$; $N^{(0)}=\varnothing$; $k=0$
while (N is nonempty)
begin
  $k=k+1$; $i=0$;
  select N_i from N;
  while ($i \le imax$ **and** $i \le \|N-N_i\|+1$)
  begin
    find in $N-N_i+S^{(k-1)}$ a subset S_i of nodes neighboring with N_i;
    $l_i=\|S_i\|$;
    find m minimizing over $n\in S_i$ the number of nodes of $N+S^{(k-1)}-N_i-\{n\}$ neighboring with $N_i+\{n\}$;
    $i=i+1$; $N_i=N_{i-1}+\{m\}$;
  end
  determine r such that $l_r=\min\{l_{imin}, ..., l_{imax}\}$;
  as internal nodes of the kth block $N^{(k)}$ take N_r, as its block nodes take S_r;
  $N=N-N^{(k)}-S^{(k)}$; $k=k+1$;
end

8.2.2 Finite methods for the solution of decomposed circuits

8.2.2.1 Reduced block equations and their circuit interpretation

In the preceding section we have introduced the idea of circuit partitioning by utilizing the concept of block variables. Since in the BBD matrix internal variables are the first to be enumerated, they will be also be the first to be eliminated during the LU factorization. If we isolate the few initial steps from the LU factorization, we can treat it as a partial factorization, as mentioned in Section 3.3.6. Let the LU factorization be stopped after elimination of all internal variables. Such a partial factorization gives circuit equations stretched over block variables only. These equations have a spectacular circuit interpretation. We will discuss it first, to clarify further consideration.

Let a circuit have a number of variables, which have been partitioned into a vector of m block variables $\mathbf{x}_0$, and a vector $\mathbf{x}_i$ including the rest, i.e. $n-m$ internal variables. Elimination of $\mathbf{x}_i$ from the circuit equations gives a reduced circuit which is equivalent to the original circuit from the block variables point of view. This means that the solution at the block nodes is the same as in the original circuit, while internal variables become unavailable. This reduced circuit can be viewed as a substitute Thevenin-Norton circuit stretched over the block nodes.

To formalize it, let us write circuit nodal equations in the form

$$\begin{bmatrix} \mathbf{Y}_i & \mathbf{P}_i \\ \mathbf{Q}_i^t & \mathbf{R}_0 \end{bmatrix} \begin{bmatrix} \mathbf{x}_i \\ \mathbf{x}_0 \end{bmatrix} = \begin{bmatrix} \mathbf{b}_i \\ \mathbf{b}_0 \end{bmatrix} \tag{8.6}$$

where the matrix and the RHS have been partitioned into two subsets: KCL corresponding to internal (index i) and block (index 0) nodes respectively. In particular, excitations $\mathbf{b}_i$ are connected to internal nodes, while $\mathbf{b}_0$ are connected to block nodes. Now we perform elimination of $\mathbf{x}_i$, solving the first equation for $\mathbf{x}_i$, and substituting it into the second one. This is equivalent to $n-m$ steps of the partial Gaussian elimination and gives substitute block equations:

$$\left(\mathbf{R}_0 - \mathbf{Q}_i^t \mathbf{Y}_i^{-1} \mathbf{P}_i\right) \mathbf{x}_0 + \mathbf{Q}_i^t \mathbf{Y}_i^{-1} \mathbf{b}_i = \mathbf{b}_0, \tag{8.7}$$

which can be written briefly as:

$$\hat{\mathbf{Y}}\, \mathbf{x}_0 + \hat{\mathbf{b}}_0 = \mathbf{b}_0. \tag{8.8}$$

Their dimension is equal to the number of block variables m. The left-hand side of (8.8) is a substitute Norton circuit, seen from the block variables. It has substitute excitations $\hat{\mathbf{b}}_0$, arising from transformation of the circuit excitations $\mathbf{b}_i$ onto block nodes, and substitute admittances $\hat{\mathbf{Y}}_0$, which derive from the internal circuit admittances also transformed onto block terminals. The RHS $\mathbf{b}_0$ is the vector of original excitations at block nodes.

To clarify this discussion, let us consider a circuit shown in Figure 8.4(a), where as a block variable the voltage v_0 has been selected, while v_1 remains the internal variable. The nodal equations of this circuit can be written as

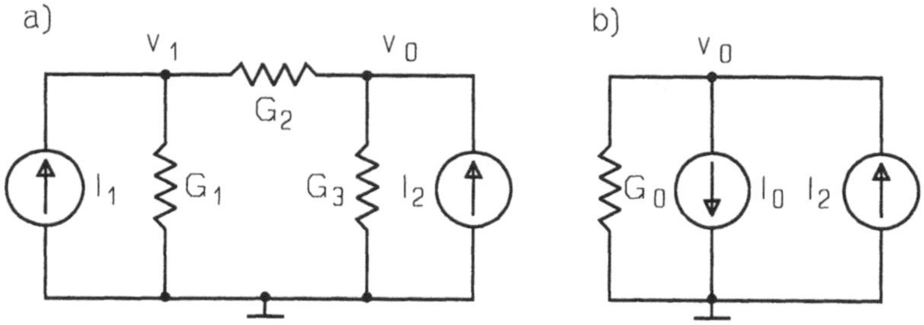

Figure 8.4 (a) A simple circuit and (b) its substitute Norton equivalent, visible from the node with voltage v_0

$$\begin{bmatrix} G_1+G_2 & -G_2 \\ -G_2 & G_2+G_3 \end{bmatrix} \begin{bmatrix} v_1 \\ v_0 \end{bmatrix} = \begin{bmatrix} I_1 \\ I_2 \end{bmatrix}. \tag{8.9}$$

We perform partial Gaussian elimination, i.e. elimination of v_1. This is equivalent to solving the first equation for v_1 and substituting the result into the second one. We obtain a triangular system:

$$\begin{bmatrix} G_1+G_2 & -G_2 \\ & G_2+G_3-\dfrac{G_2^2}{G_1+G_2} \end{bmatrix} \begin{bmatrix} v_1 \\ v_0 \end{bmatrix} = \begin{bmatrix} I_1 \\ I_2+I_1\dfrac{G_2}{G_1+G_2} \end{bmatrix} \tag{8.10}$$

where the second equation is the substitute equation for the circuit drawn in Figure 8.4(b), which is the original circuit reduced to the block variable v_0. We can rewrite the equation of this circuit as

$$G_0 v_0 + I_0 = I_2 \tag{8.11}$$

which is the case of (8.8), with the following substitute block conductance and current source

$$\begin{aligned} G_0 &= G_3 + \frac{G_1 G_2}{G_1+G_2} \\ I_0 &= -I_1 \frac{G_2}{G_1+G_2}. \end{aligned} \tag{8.12}$$

The same results can easily be obtained when we calculate Norton equivalents. The equivalent conductance G_0 is visible from the block terminal, and hence it is conductance G_3, connected in parallel with conductances G_2 and G_1 connected in series. An equivalent current I_0 can be taken as the one flowing to the block node through a short-circuit grounding the node. Therefore, it arises from the current divider G_1, G_2. We obtain the circuit shown in Figure 8.4(b).

This example demonstrates how partial Gaussian elimination leads to block equivalents of circuits. This partial elimination is not limited to nodal equations and Norton equivalents. We can also use hybrid equations, e.g. MNEs, and their reduction to admittance/impedance Thevenin-Norton equivalents. In what follows, we will assume such a general approach.

The Gaussian elimination can be expressed in terms of a partial LU factorization mentioned in Section 3.3.6. Coming back to (8.6), we perform $n-m$ steps of partial LU factorization, equivalent to premultiplying both sides by

$$\bar{\mathbf{L}}^{-1} = \begin{bmatrix} \mathbf{L}_i^{-1} & \mathbf{0} \\ -\mathbf{Q}_i^t \mathbf{Y}_i^{-1} & \mathbf{1} \end{bmatrix} \tag{8.13}$$

where $\mathbf{Y}_i = \mathbf{L}_i \mathbf{U}_i$. This gives factorization of the matrix of (8.6) into the partially triangular factors $\bar{\mathbf{L}}\,\bar{\mathbf{U}}$, namely

$$\bar{\mathbf{L}} = \begin{bmatrix} \mathbf{L}_i & \mathbf{0} \\ \mathbf{Q}_i^t \mathbf{U}_i^{-1} & \mathbf{1} \end{bmatrix}, \quad \bar{\mathbf{U}} = \begin{bmatrix} \mathbf{U}_i & \mathbf{L}_i^{-1}\mathbf{P}_i \\ \mathbf{0} & \mathbf{R}_0 - \mathbf{Q}_i^t \mathbf{Y}_i^{-1} \mathbf{P}_i \end{bmatrix}. \tag{8.14}$$

After this factorization we can perform the forward substitution $\bar{\mathbf{L}}\,\tilde{\mathbf{b}} = \mathbf{b}$, which can be decomposed into equations

$$\begin{aligned} &\mathbf{L}\,\tilde{\mathbf{b}}_i = \mathbf{b}_i \\ &\tilde{\mathbf{b}}_0 \equiv \mathbf{b}_0 - \hat{\mathbf{b}}_0 = \mathbf{b}_0 - \mathbf{Q}^t\mathbf{U}^{-1}\tilde{\mathbf{b}}_i \end{aligned} \tag{8.15}$$

giving the total excitation $\tilde{\mathbf{b}}_0$ which has as a component the substitute Thevenin-Norton component $\hat{\mathbf{b}}_0$. Then we proceed to a backward substitution $\bar{\mathbf{U}}\,\mathbf{x} = \tilde{\mathbf{b}}$, which takes the form

$$\left(\mathbf{R}_0 - \mathbf{Q}_i^t \mathbf{Y}_i^{-1} \mathbf{P}_i\right) \mathbf{x}_0 = \tilde{\mathbf{b}}_0$$
$$\mathbf{U}_i \mathbf{x}_i = \tilde{\mathbf{b}}_i - \mathbf{L}_i^{-1} \mathbf{P}_i \mathbf{x}_0 . \tag{8.16}$$

The first equation above is exactly the block substitute formula (8.7), while the second enables us to calculate the internal variables.

8.2.2.2 Gaussian algorithm for solution of equations having the BBD matrix structure

The solution of a circuit described by equations of BBD structure, consists of the reduction of the circuit with respect to block nodes, by means of partial LU factorization limited to the area of diagonal blocks. In this area the partial LU factorization can be performed for each diagonal block separately. To explain this fact, let us extend the partial factorization (8.14) to the multi-blocks case, and hence, express factors in the form:

$$\bar{L} = \begin{bmatrix} \mathbf{L}_1 & & & \\ & \cdots & & \\ & & \mathbf{L}_m & \\ \mathbf{Q}_1^t \mathbf{U}_1^{-1} & \cdots & \mathbf{Q}_m^t \mathbf{U}_m^{-1} & 1 \end{bmatrix},$$
$$\bar{\mathbf{U}} = \begin{bmatrix} \mathbf{U}_1 & & \mathbf{L}_1^{-1} \mathbf{P}_1 \\ & \cdots & \cdots \\ & \mathbf{U}_m & \mathbf{L}_m^{-1} \mathbf{P}_m \\ & & \mathbf{R} - \sum_i \mathbf{Q}_i^t \mathbf{Y}_i^{-1} \mathbf{P}_i \end{bmatrix}. \tag{8.17}$$

We observe that the above factors are superpositions of components, which can be created separately. This can be done by creation of a simple auxiliary matrix for each block

$$\begin{bmatrix} \mathbf{Y}_i & \mathbf{P}_i \\ \mathbf{Q}_i^t & \mathbf{R}_i \end{bmatrix} \tag{8.18}$$

where $\mathbf{R}_i$ is a component of $\mathbf{R} = \sum_i \mathbf{R}_i$ corresponding to the *i*th block. Then, by a partial factorization, we eliminate the internal variables of the block to obtain

$$\bar{\mathbf{L}}_i = \begin{bmatrix} \mathbf{L}_i & \mathbf{0} \\ \mathbf{Q}_i^t \mathbf{U}_i^{-1} & \mathbf{1} \end{bmatrix}, \quad \bar{\mathbf{U}}_i = \begin{bmatrix} \mathbf{U}_i & \mathbf{L}_i^{-1}\mathbf{P}_i \\ \mathbf{0} & \mathbf{R}_i - \mathbf{Q}_i^t \mathbf{Y}_i^{-1} \mathbf{P}_i \end{bmatrix}. \tag{8.19}$$

Such an approach means a decomposition of the LU factorization process, since using components of (8.19) we can formulate the factorization (8.17) of the whole circuit. Namely, the bottom-left and the top-right submatrices give the borders of (8.17), while the bottom-right submatrix of $\bar{U}_i$ is a component of the sum appearing in the bottom-right corner of $\bar{\mathbf{U}}$ in (8.17).

Therefore, to perform a partial factorization of (8.1) we have to factorize small matrices (8.18), and then add up bottom-right submatrices. We obtain a substitute matrix of the circuit, reduced to block variables:

$$\hat{\mathbf{R}} = \mathbf{R} - \sum_{i=1}^{m} (\mathbf{Q}_i^t \mathbf{U}_i^{-1})(\mathbf{L}_i^{-1}\mathbf{P}_i). \tag{8.20}$$

As the next stage, we perform the forward substitution, i.e. solution of the system $\bar{\mathbf{L}}\tilde{\mathbf{b}} = \mathbf{b}$, whose matrix is given in (8.17); the RHS stands for excitations, while the unknowns are auxiliary quantities. If we partition $\mathbf{b}$ into internal excitations $\mathbf{b}_i$ and block excitations $\mathbf{b}_0$, then forward substitutions split into auxiliary equations for each block:

$$\mathbf{L}_i \tilde{\mathbf{b}}_i = \mathbf{b}_i. \tag{8.21}$$

Their solutions form reduced excitations

$$\tilde{\mathbf{b}}_0 = \mathbf{b}_0 - \hat{\mathbf{b}}_0 = \boldsymbol{b}_0 - \sum_{i=1}^{m} \mathbf{Q}_i^t \mathbf{U}_{ii}^{-1} \tilde{\mathbf{b}}_i. \tag{8.22}$$

comprising original block excitations $\boldsymbol{b}_0$ and substitute Thevenin-Norton excitations $\hat{\mathbf{b}}_0$.

Now, proceeding to backward substitutions $\bar{\mathbf{U}}\mathbf{x} = \tilde{\mathbf{b}}$, we observe that they split into $m+1$ separate equations. The first of them

$$\hat{\mathbf{R}}\mathbf{x}_0 = \tilde{\mathbf{b}}_0 \tag{8.23}$$

is the substitute equation of the circuit reduced to block nodes as in (8.16). Now, to obtain a circuit solution we have to factorize the matrix $\hat{\mathbf{R}} = \hat{\mathbf{L}}\hat{\mathbf{U}}$. This factorization can be thought of as completion of the partial factorization performed so far. Now, we proceed to the usual forward and backward substitutions $\hat{\mathbf{L}}\mathbf{z}_0 = \tilde{\mathbf{b}}_0$, $\hat{\mathbf{U}}\mathbf{x}_0 = \mathbf{z}_0$ to obtain the final block variables $\mathbf{x}_0$.

Algorithm 8.2. Gaussian solution of BBD equations

{LU factorization of BBD matrix}
for i=1 **to** m **do** {loop over blocks}
  partial factorization of the auxiliary matrix (8.34) giving the submatrices $\mathbf{L}_i^{-1}\mathbf{P}_i$, $\mathbf{Q}_i^t\mathbf{U}_i^{-1}$ and $-\mathbf{Q}_i^t\mathbf{U}_i^{-1}\mathbf{L}_i^{-1}\mathbf{P}_i$;
creation of $\hat{\mathbf{R}} = \mathbf{R} - \sum_{i=1}^{m}\mathbf{Q}_i^t\mathbf{U}_i^{-1}\mathbf{L}_i^{-1}\mathbf{P}_i$;
LU factorization of $\hat{\mathbf{R}} = \hat{\mathbf{L}}\hat{\mathbf{U}}$;
{forward and backward substitution}
for i=1 **to** m **do** {loop over blocks}
begin
  {calculation of substitute excitations of blocks}
  solution of $\mathbf{L}_i\tilde{\mathbf{b}}_i = \mathbf{b}_i$; {forward substitutions}
  multiplication by the matrix $\hat{\mathbf{b}}_i = \mathbf{Q}_i^t\mathbf{U}_i^{-1}\tilde{\mathbf{b}}_i$;
end
{solution of the substitute block equations}
solution of $\hat{\mathbf{L}}\mathbf{z}_0 = \mathbf{b}_0 - \sum_{i=1}^{m}\hat{\mathbf{b}}_i$; {forward substitutions}
solution of $\hat{\mathbf{U}}\mathbf{x}_0 = \mathbf{z}_0$; {backward substitutions}
for i=1 **to** m **do** {loop over blocks}
begin
  multiplication of matrices $\mathbf{b}'_i = \mathbf{L}_i^{-1}\mathbf{P}_i\mathbf{x}_0$;
  solution of $\mathbf{U}_i\mathbf{x}_i = \tilde{\mathbf{b}}_i - \mathbf{b}'_i$ {backward substitutions};
end

The remaining equations of the backward substitution discussed here enable us to calculate the internal variables:

$$\mathbf{U}_i \mathbf{x}_i = \tilde{\mathbf{b}}_i - \mathbf{L}_i^{-1} \mathbf{P}_i \mathbf{x}_0. \tag{8.24}$$

This method referred to in the work [8.4], has been briefly written as Algorithm 8.2. It is not the only one possible, since partial triangular factorization of matrices is not unique. However, in this presentation the Gaussian version of factorization has been chosen.

8.2.2.3 Efficiency of decomposition-based solution methods

In this section we will estimate the numerical complexity of the algorithm introduced. Though presented for one particular case, the conclusions are representative of a class of decomposition-based methods. We can distinguish three basic problems of decomposition optimality: the interconnection problem, the fill-ins problem and the latency problem.

Interconnection problem

Let us estimate the numerical complexity of Algorithm 8.2. We will count the number of operations: "a/b" and "a-bc" performed during the process of the solution of a circuit having N nodes. First, let the matrix sparsity be not exploited. For simplicity we assume a regular decomposition, i.e. such decomposition that after the selection of m block nodes, we divide the remaining $N-m$ nodes into p subcircuits having the same number of internal nodes.

It can be proved that n_i steps of a partial LU factorization of an $n \times n$ matrix require $n_i^3/3 + n_i n(n-n_i) - n_i/3$ operations, where $n_i \leq n-1$, while $n_i = n-1$ occurs in the full LU factorization case. Moreover, taking into account all forward-backward substitutions, factorization of a reduced block matrix, and multiplications of a matrix by a vector we obtain the numerical complexity

$$K_{\text{tot}} = \frac{1}{3}\frac{(N-m)^3}{p^2} + \frac{(m+1)(N-m)^2}{p} + (N-m)\left(m^2 + m - \frac{1}{3}\right) + \frac{m(N-m)}{p} + \frac{m^3}{3} - \frac{m}{3} + m^2 \tag{8.25}$$

We observe that it decreases quadratically with the number of blocks, p, and cubically increases with the number of block nodes. Therefore, efficient decomposition should have a maximum number of blocks with a minimum number of block variables. Hence, the number of block variables to each block should be small, i.e. blocks should be connected with the smallest number of block nodes. Of course, the minimum is one block node per block.

In large circuits we can assume that $m \ll N$. Then, (8.25) can be approximated by the formula

$$K_{tot} = \frac{1}{3} N^3 \left[\frac{1}{p^2} + \frac{3(m+1)}{N} \frac{1}{p} + \frac{3m(m+1)}{N^2} \right] \tag{8.26}$$

which approaches the minimum $Nm(m+1)$ as p increases. Notice that this minimum is about $3N^2$ smaller than rough complexity, when a whole circuit is solved with the full-matrix LU method. In practice, we reach this optimum already for $p \gg N/m$. This condition implies that we have very many blocks connected with a very small number of block nodes. Thus the quantity $N/(mp)$ is a good measure of decomposition quality. It should be small in comparison with unity.

In practice, subcircuits are large and we solve them using the sparse matrix technique presented in Section 3.3. Though submatrices of interconnections are also sparse, we will handle them as if they were full. Moreover, if the number of block variables is small in comparison with N, then a reduced block matrix $\hat{\mathbf{R}}$ is dense. Approximating the linear complexity of sparse partial LU factorization by $K = \alpha n_i$, where α is dependent on matrix sparsity, instead of (8.25) we can write the formula

$$K_{tot} = \frac{m(N-m)}{p} + (m+\alpha)(N-m) + \frac{m^3}{3} - \frac{m}{3} + m^2 \tag{8.27}$$

and observe qualitative properties similar to those in the dense matrix case. The complexity varies inversely with p and linearly increases with N. For $m \ll N$ it tends to the minimum $(m+\alpha)N + m^3/3 + m^2$ for increasing p. However, if the sparse matrix technique applied to the matrix of a whole circuit gives a complexity equal only to αN, then the decomposition based method may appear worse then the sparse matrix method, even for a decomposition with satisfying efficiency criteria.

Fill-ins problem

Another problem arising in decomposition-based methods is caused by structural pivoting. As we know from Section 3.3, this pivoting takes into account the Markowitz measure to minimize the number of fill-ins. If we take a whole circuit matrix, then a diagonal selection of pivots can efficiently minimize their number n_f. However, in decomposition methods we perform p partial factorizations of matrices (8.18), where pivots can be searched for only within an area corresponding to the internal nodes. Since we select pivots from a considerably smaller set of candidates, the pivoting can be much worse, and the total number of fill-ins in all blocks can exceed n_f considerably. Thus, decomposition should take into acount not only the number of interconnections, but also fill-ins. Matrix positions corresponding to internal nodes (i.e. diagonal blocks) should have small values of the Markowitz measure. Matrix entries having large Markowitz measures are better candidates for block nodes than for internal nodes.

Latency problem

Decomposition becomes more attractive when blocks are latent, and this latency can be exploited. In practice, we usually solve linear equations arising from a nonlinear circuit discretization and linearization, and therefore we have three categories of matrix elements: constants, elements dependent on signals and changing within Newton-Raphson (NR) loops, elements dependent on time and changing only within time-loops. Moreover, during the course of NR iterations some elements or blocks can converge faster than others, and then remain constant in NR iterations. There are also cases where some signals remain unchanged in time due to some circuit properties, and hence some matrix elements, usually variant, can temporarily become fixed in time. If all elements of a block are fixed during the course of iterations, then we call this block latent. Decomposition of a circuit leading to many frequently-latent blocks is extremely attractive, since we need perform calculations for latent blocks only once and then combine these results with repetitive parts of the calculations.

The main problem is to identify unchanged elements efficiently, and hence find latent blocks. The quickest method is to check the relative change of block variables between two iterations. If this change is small, e.g. satisfies the d.c. test for convergence, then we can schedule the block as being latent.

When a block is latent, we can easily exploit latency in Algorithm 8.2. For a latent block we do not have to repeat the partial LU factorization of (8.18) but can use the factors (8.19) known from the last iteration. We can also omit forward substitutions (8.21) and the calculation of a component of the substitute block excitations $\hat{\mathbf{b}}_0$ (see Equation (8.22)). Thus we build block equations (8.23) combining the results of solutions of nonlatent blocks, and stored solutions from latent blocks. Other calculations remain the same. The latency provides considerable savings in numerical complexity. If many blocks are latent, then time consumption becomes almost proportional to the number of blocks *m*.

To summarize our discussion, we observe that the efficiency of the decomposition-based methods is very sensitive to the manner of decomposition, which is a great weak point of this approach. On the other hand, in analog circuits it is often difficult to find partitions which simultaneously obey interconnection, fill-ins and latency criteria. That is why actually we use these methods mostly when blocks are so large that a multi-block circuit cannot be stored, as a whole, in memory. However, this is not true for up-to-date relaxation-based methods for MOSFET circuits, as shown in Chapter 9. In these methods partitioning is very advantageous since it is free of fill-ins problems, blocks are often one-way and latency is very common due to logical signals.

In the literature we can come across hierarchical decomposition methods. However, as we know from recent studies and experiments, they are not competitive in comparison with the sparse matrix technique.

8.3 Direct methods with macromodularization and latency

8.3.1 Introduction

In this section we address a direct method of nonlinear decomposed circuit solution represented in the second row of the "block decomposition" column of Table 8.1. This method consists of the application of accurate algorithms and accurate models of elements. Due to the complicated models, elements are usually bilateral and hence all nodes of the circuit are tightly coupled. Therefore, neither the nodal nor the block relaxation method is available. We can use only the circuit decomposition and direct analysis whose foundations were introduced in Chapters 2-7.

SPICE [4.9] and OPTIMA are examples of simulators where direct methods without decomposition have been implemented. Though users can define subcircuits, these subcircuits are all expanded onto a single level, and the sparse matrix technique of solution is used. A similar approach has been employed in CAzM [4.4], though this simulator allows full models or more efficient macromodels. CAzM builds linearized circuit equations by assembling components corresponding to particular elements. If an element is macromodeled, then its characteristics are approximated by a multivariate spline stretched over sample-points obtained from an exact model of this element during an analysis.

In the simulator SLATE [8.12], decomposition of linear equations and latency, referred to in Section 8.2, have been applied to discretized and linearized circuit equations. Appropriate decomposition is a weak point of these methods, since it should provide an optimal compromise between a small number of interconnections, of fill-ins in blocks, and the latent behavior of the circuit blocks in time and in Newton-Raphson iterations.

Another experimental simulator MEDUSA [8.2] has used a latent macromodular approach, where decomposition is applied at the level of nonlinear algebraic equations. This gives a multilevel Newton-Raphson algorithm, which treats blocks on the lower level as macromodels, and exploits their latency. We will discuss this approach in this section.

In concurrent simulators, direct methods are combined with relaxation-based methods dedicated to specific classes of circuits, especially to unilateral circuits.

8.3.2 Nodal decomposition of algebraic nonlinear equations

Let a dynamic nonlinear circuit be described by a set of equations $\mathbf{F}[\mathbf{x}, \dot{\mathbf{z}}(\mathbf{x}), t] = \mathbf{0}$, where $\mathbf{z}$ are state variables, while $\mathbf{x}$ are circuit variables. We will analyze this circuit by means of a decomposition-based method. Hence, provided a common step-size is adopted for all subcircuits, we can discretize the whole circuit by means of a Differentiation Formula (DF) discussed in Chapter 5, and obtain nonlinear algebraic equations $\mathbf{F}\{\mathbf{x}, \gamma_0[\mathbf{z}(\mathbf{x}) - \mathbf{d}_z]\} = \mathbf{0}$ at instant t_n (we will omit the index n in the following discussion). These discretized circuit equations can be briefly written as $\mathbf{F}(\mathbf{x}) = \mathbf{0}$. We can apply to these algebraic nonlinear equations decomposition techniques referred to in Section 8.2. Linearization for Newton-Raphson (NR) iterations will be appplied after decomposition. This idea leads to a multilevel NR method.

Nodal and branch decomposition methods were introduced in Section 8.2. Both of them can be applied to algebraic nonlinear circuit equations. Due to the advantages of nodal decomposition, we will focus on it only in further discussion.

Given a set of block nodes having voltages $\mathbf{v}_0$, we partition the rest into m subsets of nodes with voltages $\mathbf{v}_i$ which become internal variables of subcircuits C_i. The nonlinear Modified Nodal Equations (MNE) (see Section 1.3.4) of a discretized circuit, can be grouped in m sections $\mathbf{B}_k$ corresponding to the internal variables of each block and a section $\mathbf{N}$ corresponding to block variables:

$$
\begin{aligned}
&\mathbf{B}_1(\mathbf{v}_1, \mathbf{v}_0) = \mathbf{0} \\
&\mathbf{B}_2(\mathbf{v}_2, \mathbf{v}_0) = \mathbf{0} \\
&\cdots\cdots \\
&\mathbf{B}_m(\mathbf{v}_m, \mathbf{v}_0) = \mathbf{0} \\
&\mathbf{N}(\mathbf{v}_1, \mathbf{v}_2, \ldots, \mathbf{v}_0) = \mathbf{0}
\end{aligned}
\qquad (8.28)
$$

If we assume, for simplicity, that there are only current-defined voltage-controlled branches, then discussion can be limited to Nodal Equations (NE), which are KCL balances at nodes. Then sections $\mathbf{B}_k$ are composed of KCL equations at internal nodes of blocks, while the section $\mathbf{N}$ incorporates KCL equations at block nodes.

However, if we accept also current variables, then MNEs can be partitioned taking currents as block or internal variables. If a current is a block variable, then the equation of a branch including this current falls into the section $\mathbf{N}$. Otherwise, this branch equation belongs to one of sections $\mathbf{B}_k$. In this case, it is essential to have a nonsingular successfully pivotable Jacobian of section $\mathbf{B}_k$, and hence initial interchanges of rows, described in Section 3.3.4, have to be performed within these sections. This considerably limits decomposition, since we notice that it is better to have voltage-defined branches not connected to block nodes.

In a decomposed circuit (8.28) we have m sets of equations of blocks, each one of a dimension $\dim \mathbf{v}_k$

$$\mathbf{B}_k(\mathbf{v}_k, \mathbf{v}_0) = \mathbf{0}\,. \qquad (8.29)$$

It is essential to notice that the equations of the kth block depend only on block variables and internal variables of this block. In fact, they are an implicit description of how the internal variables depend on the block variables, i.e. $\mathbf{v}_k = \mathbf{v}_k(\mathbf{v}_0)$. Thus, if block variables were known we could calculate internal variables by applying a NR iterative process to (8.29). Thus a Jacobian of these equations may not be singular.

Decomposed equations (8.28) yet have $\dim \mathbf{v}_0$ interconnection equations, not mentioned so far

$$\mathbf{N}(\mathbf{v}_1, \mathbf{v}_2, \ldots, \mathbf{v}_0) = \mathbf{0}\,. \qquad (8.30)$$

They join all internal and block variables. However, as we have shown, block equations (8.29) describe internal variables in terms of block variables, and therefore we can discard internal variables obtaining a system reduced to block variables only:

$$\mathbf{N}\,[\mathbf{v}_1(\mathbf{v}_0),\,...,\,\mathbf{v}_m(\mathbf{v}_0),\,\mathbf{v}_0\,]=\mathbf{0}\,. \tag{8.31}$$

Unfortunately, this reduction is mainly theoretical since we have implicit relations (8.29) instead of functions $\mathbf{v}_k(\mathbf{v}_0)$. However, it is essential that if we performed such a reduction, we could solve (8.31) for $\mathbf{v}_0$, and then obtain from (8.29) the internal variables. Obviously, the Jacobi matrix of (8.31) must be nonsingular.

Essential to the above discussion is a structure of dependences of equations on variables. This can be formalized using the dependency matrix **A**, where a_{ij} is zero if the *i*th equation is not dependent on the *j*th variable, while otherwise it is unity. This matrix will be helpful to explain the structure of the decomposition of nonlinear circuits.

To make the discussion more clear, let us consider the circuit shown in Figure 8.5. It contains companion models of diodes: $i_3(v_3)$, $i_4(v_4)$, where $v_3 = v_{01} - v_{11}$, $v_4 = v_{01} - v_{21}$, while

$$\begin{aligned} i_3(v_3) &= I_{s3}\exp[(v_3/V_T)-1]+\gamma_0[Q_{03}\exp(v_3/V_T)-d_{q3}] \\ i_4(v_4) &= I_{s4}\exp[(v_4/V_T)-1]+\gamma_0[Q_{04}\exp(v_4/V_T)-d_{q4}] \end{aligned} \tag{8.32}$$

We select two block nodes: 00 (the ground), and 01 with the voltage v_{01}. Then, we divide the circuit into two blocks with internal voltages v_{11} and v_{21} respectively. The source I_1 between block nodes can be viewed as belonging to the main circuit or to any subcircuit. We obtain the following system of equations

$$\begin{aligned} &(1)\quad G_2 v_{11} - i_3(v_{01}-v_{11}) = 0 \\ &(2)\quad G_5 v_{21} - i_4(v_{01}-v_{21}) = 0 \\ &(3)\quad i_3(v_{01}-v_{11}) + i_4(v_{01}-v_{21}) = I_1 \end{aligned} \tag{8.33}$$

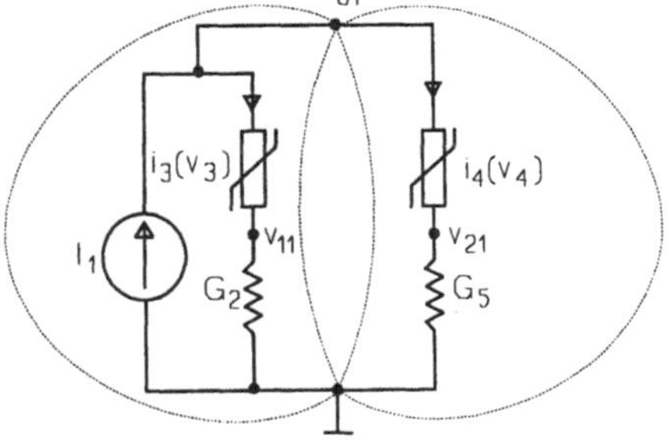

Figure 8.5 A nonlinear circuit decomposed into two subcircuits

where the KCL equation (1) written for the node "11" is the equation of block 1, the KCL equation (2) written for the node "21" is the equation of block 2, and finally the KCL equation at the node "01" is the interconnection equation. A dependency matrix for these equations has the BBD form suitable for nodal decomposition:

$$\mathbf{A} = \begin{bmatrix} 1 & 0 & 1 \\ 0 & 1 & 1 \\ 1 & 1 & 1 \end{bmatrix}. \tag{8.34}$$

The decomposition presented can be easily generalized to a hierarchical approach. Let the tearing be termed level 1. Hierarchical decomposition consists of further partitioning blocks of level 1. We select, e.g. among variables $\mathbf{v}_k$, a subset of level 2 block variables $\mathbf{v}_{k0}$, and partition the rest of them into a number of level 2 internal subsets. In this way we obtain a partition of a level 1 subcircuit into a number of level 2 subcircuits. This corresponds to partitioning the equation (8.29) according to (8.28). This process can be continued recursively.

8.3.3 A multilevel Newton-Raphson algorithm with macromodeling and latency

Equations (8.28) could be solved using the NR method introduced in Chapter 4. However, due to their decomposition, we can also decompose the iterative process into levels. Such an approach has been devised by N.B. Rabbat, H.Y. Hsieh and A. Sangiovanni-Vincentelli [8.7], [8.8]. If a decomposition is hierarchical, then we can have multilevel iterations. However, to be more clear, we will concentrate further discussion on a one-level decomposition, and therefore two-level iterations. The advantages expected from splitting the NR iterations into levels are: bypassing and latency of lower level blocks, possible storing and solution of each block separately, introduction of macromodels, and the possibility of parallel solution of blocks on vector computers. The original approach by G. Rabbat utilized node tearing with the terminal currents of blocks as additional variables and with a hybrid circuit description. In our treatment, a pure form of nodal decomposition is employed.

8.3.3.1 A two-level algorithm

To explain a two-level method of solution of the circuit (8.28), let us discuss how the solution can exploit the separation of equations (8.29) and (8.30). To solve (8.29) for each block, first we have to know the block variables. Thus, theoretically we can start by solving the whole circuit for block variables. We can do it only by means of (8.31), where a relation between internal and block variables is presumed. This is an upper level of calculations. After the block variables have been calculated, we proceed subsequently to each block and solve its equations (8.29) for internal variables. This is a lower level of calculations. Unfortunately, this procedure is only a theoretical concept, since we have no finite method for solving each of the equations involved. Both upper and lower level equations have to be solved by means of the iterative NR procedure.

We solve the upper level equations (8.31) by iterations over $\mathbf{v}_0$. However, at each iteration we require updated local Jacobi matrices of blocks. For each iterative value $\mathbf{v}_0^{(i)}$, this requires that for each block a nested loop of lower level iterations over $\mathbf{v}_k$ has to be arranged. This loop, distinct for each block, gives the internal solutions of this block and its Jacobian at this solution. On the other hand, lower level iterations for each block are performed with the assumption that the block variables are already known. Until the upper level iterations have been terminated, these block variables are as yet inaccurate, so we require an external loop of upper level iterations for their adjustment. Both arguments prove that two nested iterative processes are necessary. In an l-level decomposition we have l nested iterative processes.

To explain lower level iterations let us assume a certain value of block variables $\mathbf{v}_0^{(i)}$ and concentrate on the kth block described by Equation (8.29). We solve this equation for internal variables $\mathbf{v}_k^{(i)}$ by means of the NR method. So, assuming a value $\mathbf{v}_k^{(i)(j)}$ for these variables, we linearize (8.29) with respect to $\mathbf{v}_k$ in the vicinity of this value. We obtain an iterative equation

$$\frac{\partial \mathbf{B}_k(\mathbf{v}_k^{(i)(j)}, \mathbf{v}_0^{(i)})}{\partial \mathbf{v}_k}(\mathbf{v}_k^{(i)(j+1)} - \mathbf{v}_k^{(i)(j)}) = -\mathbf{B}_k(\mathbf{v}_k^{(i)(j)}, \mathbf{v}_0^{(i)}),\ j=0, 1, 2, \dots \quad (8.35)$$

to be solved for $\mathbf{v}_k^{(i)(j+1)}$, which is the $j+1$ iterative value of internal variables. This equation defines lower level iterations. After their convergence internal variables $\mathbf{v}_k^{(i)(j)}$ reach certain values $\mathbf{v}_k^{(i)}$, while in general all terms with superscripts $(i)(j)$ reach their counterparts with superscripts (i).

Let us recall the two-block circuit shown in Figure 8.5. By linearization of equations (1) and (2) of (8.32) we obtain two lower-level iterative equations, the first one

$$(G_2 + G_3^{(i)(j)})\,(v_{11}^{(i)(j+1)} - v_{11}^{(i)(j)}) = I_3^{(i)(j)} \tag{8.36}$$

and the second

$$(G_5 + G_4^{(i)(j)})(v_{21}^{(i)(j+1)} - v_{21}^{(i)(j)}) = I_4^{(i)(j)} \tag{8.37}$$

where

$$G_3^{(i)(j)} \equiv \frac{\partial i_3(v_{01}^{(i)} - v_{11}^{(i)(j)})}{\partial v_3} = \frac{I_s + \gamma_0 q_0}{V_T} \exp\left(\frac{v_{01}^{(i)} - v_{11}^{(i)(j)}}{V_T}\right)$$

$$I_3^{(i)(j)} \equiv i_3(v_{01}^{(i)} - v_{11}^{(i)(j)}) - I_1 = (I_s + \gamma_0 q_0)\exp\left(\frac{v_{01}^{(i)} - v_{11}^{(i)(j)}}{V_T}\right) - \gamma_0 d_{q3} - I_1$$

$$G_4^{(i)(j)} \equiv \frac{\partial i_4(v_{01}^{(i)} - v_{21}^{(i)(j)})}{\partial v_4} = \frac{I_s + \gamma_0 q_0}{V_T} \exp\left(\frac{v_{01}^{(i)} - v_{21}^{(i)(j)}}{V_T}\right) \tag{8.38}$$

$$I_4^{(i)(j)} \equiv i_4(v_{01}^{(i)} - v_{21}^{(i)(j)}) = (I_s + \gamma_0 q_0)\exp\left(\frac{v_{01}^{(i)} - v_{21}^{(i)(j)}}{V_T}\right) - \gamma_0 d_{q4}$$

while $v_{01}^{(i)}$ is fixed, obtained from the upper level.

Coming back to the general discussion, we notice that it is necessary to calculate and transmit onto the upper level not only the values $\mathbf{v}_k^{(i)}$ but also the Jacobians $\partial \mathbf{v}_k^{(i)} / \partial \mathbf{v}_0$ after convergence of Equations (8.35).

To find these Jacobians, we can differentiate (8.29) with respect to $\mathbf{v}_0$ at the current lower-level solution. This gives

$$\frac{\partial \mathbf{B}_k^{(i)(j)}}{\partial \mathbf{v}_k} \frac{\partial \mathbf{v}_k^{(i)(j)}}{\partial \mathbf{v}_0} + \frac{\partial \mathbf{B}_k^{(i)(j)}}{\partial \mathbf{v}_0} = \mathbf{0}\,. \tag{8.39}$$

Now, the Jacobian of the block takes the form

$$\frac{\partial \mathbf{v}_k^{(i)(j)}}{\partial \mathbf{v}_0} = -\left(\frac{\partial \mathbf{B}_k^{(i)(j)}}{\partial \mathbf{v}_k}\right)^{-1} \frac{\partial \mathbf{B}_k^{(i)(j)}}{\partial \mathbf{v}_0}\,. \tag{8.40}$$

However, to find this Jacobian numerically, we have to decompose the

Jacobian of equations (8.28) in the same manner as in Section 8.2. The interconnection equation (8.30) can be assembled from m components corresponding to m subcircuits: $\mathbf{N}=\sum_1^m \mathbf{N}_k$. Since each subcircuit is dependent only on the block variables and its own internal variables, therefore $\partial\mathbf{N}/\partial\mathbf{v}_k=\partial\mathbf{N}_k/\partial\mathbf{v}_k$. For each block we create an auxiliary matrix, including a diagonal component corresponding to the interior of the block and two off-diagonal components standing for interconnections

$$\mathbf{H}_k^{(i)(j)} = \begin{bmatrix} \dfrac{\partial\mathbf{B}_k^{(i)(j)}}{\partial\mathbf{v}_k} & \dfrac{\partial\mathbf{B}_k^{(i)(j)}}{\partial\mathbf{v}_0} \\ \dfrac{\partial\mathbf{N}_k^{(i)(j)}}{\partial\mathbf{v}_k} & \dfrac{\partial\mathbf{N}_k^{(i)(j)}}{\partial\mathbf{v}_0} \end{bmatrix}. \tag{8.41}$$

Elimination of internal variables by partial LU factorization gives

$$\begin{bmatrix} \mathbf{L}_k\backslash\mathbf{U}_k & \mathbf{L}_k^{-1}\dfrac{\partial\mathbf{B}_k^{(i)(j)}}{\partial\mathbf{v}_0} \\ \dfrac{\partial\mathbf{N}_k^{(i)(j)}}{\partial\mathbf{v}_k}\mathbf{U}_k^{-1} & \mathbf{R}_k^{(i)(j)} \end{bmatrix} \tag{8.42}$$

where $\mathbf{L}_k\mathbf{U}_k=\partial\mathbf{B}_k^{(i)(j)}/\partial\mathbf{v}_k$.

During partial LU factorization we obtain in the top-left corner of (8.42) the matrix $\mathbf{L}_k\backslash\mathbf{U}_k=\mathbf{L}_k+\mathbf{U}_k-\mathbf{1}$, which is a product of LU factorization of the matrix $\partial\mathbf{B}_k^{(i)(j)}/\partial\mathbf{v}_k$ essential for lower-level iterations (8.35). Now to solve (8.35), having the LU factors, we still require forward and backward substitutions only.

Regarding the bottom-right corner of $\mathbf{H}_k^{(i)(j)}$, the LU factorization gives $\mathbf{R}_k^{(i)(j)}=(\partial\mathbf{N}_k^{(i)(j)}/\partial\mathbf{v}_0)-(\partial\mathbf{N}_k^{(i)(j)}/\partial\mathbf{v}_k)\mathbf{U}_k^{-1}\mathbf{L}_k^{-1}(\partial\mathbf{B}_k^{(i)(j)}/\partial\mathbf{v}_0)$ or equivalently $(\partial\mathbf{N}_k^{(i)(j)}/\partial\mathbf{v}_0)-(\partial\mathbf{N}_k^{(i)(j)}/\partial\mathbf{v}_k)(\partial\mathbf{v}_k^{(i)(j)}/\partial\mathbf{v}_0)$. After lower-level iterations, this matrix converges to $\mathbf{R}_k^{(i)}$, while $\mathbf{N}_k^{(i)(j)}$ converges to $\mathbf{N}_k^{(i)}$. These two terms stand for a block representation useful on the upper level.

Once more recalling our example, we create matrices (8.41) of the two blocks involved:

$$\begin{bmatrix} G_2+G_3^{(i)(j)} & -G_3^{(i)(j)} \\ -G_3^{(i)(j)} & G_3^{(i)(j)} \end{bmatrix}, \quad \begin{bmatrix} G_5+G_4^{(i)(j)} & -G_4^{(i)(j)} \\ -G_4^{(i)(j)} & G_4^{(i)(j)} \end{bmatrix} \tag{8.43}$$

After their partial LU factorization, equivalent to the elimination of variables v_{11}, v_{21}, in the bottom-right corners, we obtain

$$R_1^{(i)(j)} = \frac{G_2 G_3^{(i)(j)}}{G_2 + G_3^{(i)(j)}}, \quad R_2^{(i)(j)} = \frac{G_5 G_4^{(i)(j)}}{G_4^{(i)(j)} + G_5} \tag{8.44}$$

which are Norton equivalent conductances of the iterative subcircuits viewed from the block node "01".

Now we proceed to a general consideration of upper-level iterations. The interconnections equation (8.30) can be linearized in the vicinity of the point $\mathbf{v}_1^{(i)}$, $\mathbf{v}_2^{(i)}$, ..., $\mathbf{v}_m^{(i)}$, $\mathbf{v}_0^{(i)}$, giving

$$\mathbf{N}^{(i)} + \sum_{k=1}^{m} \frac{\partial \mathbf{N}^{(i)}}{\partial \mathbf{v}_k}(\mathbf{v}_k^{(i+1)} - \mathbf{v}_k^{(i)}) + \frac{\partial \mathbf{N}^{(i)}}{\partial \mathbf{v}_0}(\mathbf{v}_0^{(i+1)} - \mathbf{v}_0^{(i)}) = \mathbf{0}\,. \tag{8.45}$$

In this equation we substitute increments of internal variables for increments of $\mathbf{v}_0$ multiplied by the derivative (8.40). This gives an iterative upper level equation:

$$\mathbf{R}^{(i)}(\mathbf{v}_0^{(i+1)} - \mathbf{v}_0^{(i)}) = -\mathbf{N}^{(i)} \quad (i=0, 1, 2, ...) \tag{8.46}$$

with the Jacobi matrix

$$\mathbf{R}^{(i)} = \frac{\partial \mathbf{N}^{(i)}}{\partial \mathbf{v}_0} - \sum_{k=1}^{m} \frac{\partial \mathbf{N}^{(i)}}{\partial \mathbf{v}_k}\frac{\partial \mathbf{v}_k^{(i)}}{\partial \mathbf{v}_0} = \sum_{k=1}^{m} \hat{\mathbf{R}}_k^{(i)} \tag{8.47}$$

and the RHS

$$-\mathbf{N}^{(i)} = \sum_{k=1}^{m} (-\mathbf{N}_k^{(i)}) \tag{8.48}$$

both reduced to block variables $\mathbf{v}_0$. This proves, that the upper-level equations (8.45) are a superposition of m stamps

$$\left[\mathbf{R}_k^{(i)} \mid -\mathbf{N}_k^{(i)}\right] \tag{8.49}$$

including a component of the matrix $\mathbf{R}^{(i)}$ and the RHS $-\mathbf{N}^{(i)}$. Hence, they require the summation in a loop of the stamps (8.49) of all blocks involved. Calculation of each stamp requires the inner NR procedure (8.35).

For the two blocks of Figure 8.5, due to (8.44) and (8.33), we have the following stamps

$$\left[\frac{G_2 G_3^{(i)}}{G_2+G_3^{(i)}} \;\middle|\; -i_3^{(i)}+I_1\right], \quad \left[\frac{G_5 G_4^{(i)}}{G_5+G_4^{(i)}} \;\middle|\; -i_4^{(i)}\right] \tag{8.50}$$

available after convergence of two lower-level NR processes. Their LHSs and RHSs are the Norton equivalent conductances and currents of the blocks, respectively, viewed from the block node "01". Assembling these stamps we obtain the upper-level equation of the type (8.46)

$$R^{(i)}(v_{01}^{(i+1)}-v_{01}^{(i)}) = -I^{(i)} \tag{8.51}$$

where

$$\begin{aligned} R^{(i)} &= R_1^{(i)}+R_2^{(i)} = \frac{G_2 G_3^{(i)}}{G_2+G_3^{(i)}} + \frac{G_4^{(i)} G_5}{G_4^{(i)}+G_5}\,. \\ I^{(i)} &= -i_3^{(i)} - i_4^{(i)} + I_1 \end{aligned} \tag{8.52}$$

8.3.3.2 Convergence properties and tests for convergence

It is easy to notice, that if the lower level solution were accurate, then the upper level would be provided with accurate Jacobi matrices and solutions, and hence the upper level iterations would be NR iterations having quadratic convergence. This was explained in Chapter 4. A problem, with the character of convergence of the multilevel algorithm, arises since the lower-level iterative process is not continued indefinitely but is stopped after satisfying a test for convergence. Hence at the lower level we obtain an unavoidable error in solutions $\mathbf{v}_k^{(i)}$ and in the stamp (8.49). Due to this error, the Jacobi matrix (8.47) of the upper-level iterations is inadequate. As we know from Chapter 4, the NR method is weakly sensitive to perturbations in the Jacobi matrix; however, such perturbations always worsen the convergence of (8.46). Hence, to maintain quadratic convergence, we have to strictly control the error of the lower-level iterations to ensure that it is not too large. The problem is, how inaccurate can the lower-level iterations be and yet ensure quadratic convergence at the upper level.

Algorithm 8.3 Two-level NR solution of a discretized nonlinear circuit

```
data: upper-level criterion of convergence δ; τ=δ;
initialization of starting values of block variables v_0^0;
for i=1 to imax do {upper-level NR loop}
begin
   for k=1 to m do {the loop over blocks}
   begin
      prediction of starting values of internal variables v_k^0;
      -N=0;
      for j=1 to jmax do {inner NR loop}
      begin
         formulation of submatrices ∂B_k/∂v_k, ∂B_k/∂v_0, ∂N_k/∂v_k,
         ∂N_k/∂v_0 and the RHS's -B_k for signals v_k^0, v_0^0;
         partial LU factorization of (9.13) giving (9.14);
         forward/backward substitutions to solve
         L_k U_k (v_k^1 - v_k^0) = -B_k for v_k^1;
         {lower-lever test for convergence}
         if (||v_k^1 - v_k^0|| ≤ τ) then break
      end
      calculation of -N_k for v_k^1, v_0^0 and addition to -N;
   end
   {now in place of ∂N/∂v_0 we have the jacobian R^(i) (9.19)}
   LU factorization of R;
   solution of (9.18) LU(v_0^1 - v_0^0) = -N for v_0^1 by means of for-
   ward/backward substitutions;
   τ=min{τ_max, ||v_k^1-v_k^0||^2};
   {upper-level test for convergence}
   ||v_0^1 - v_0^0|| ≤ δ;
end
```

At the lower level we apply a test for convergence

$$\| \mathbf{v}_k^{(i)(j+1)} - \mathbf{v}_k^{(i)(j)} \| \le \tau^{(i)} \tag{8.53}$$

where the index $\tau^{(i)}$ comes from the upper level. The problem is, how to determine it during the course of upper-level iterations. An intuitive concept is to guarantee that the inaccuracy of data calculated at the lower level and used on the upper level is not worse than the current inaccuracy of the upper level $\| \mathbf{v}_0^{(i+1)} - \mathbf{v}_0^{(i)} \|$. Since lower-level iterations are stopped before we get the solution $\mathbf{v}_0^{(i+1)}$, $\tau^{(i)}$ should be expressed in terms of the error $\| \mathbf{v}_0^{(i)} - \mathbf{v}_0^{(i-1)} \|$ of the preceding upper-level iteration. To obtain quadratic convergence we impose $\tau^{(i)}$ equal to the square of the preceding upper-level error. In [8.8] it has been proved that this intuitive consideration in fact provides quadratic convergence. Thus in the upper-level loop, after we have calculated $\mathbf{v}_0^{(i)}$, an updated test for lower-level convergence can be determined

$$\tau^{(i)} = \min\,[\tau_{\max}, \|\mathbf{v}_0^{(i)} - \mathbf{v}_0^{(i-1)}\|^{\gamma}] \tag{5.54}$$

and transferred to the lower level iterations. Theoretically, we should set $\gamma=2$. In practice, for $1<\gamma<2$ we can also obtain good convergence. This form of management of tests for convergence is easily extended to multilevel iterations.

A basic implementation of this two-level algorithm is summarized in Algorithm 8.3.

The two-level algorithm causes considerable problems with the upper-level convergence if there are tight couplings between different blocks. Therefore, we find another decomposition criterion: not to cut the circuit through feedback paths.

8.3.3.2 Lower-level calculations as macromodeling processes

In the inner NR loop of the two-level algorithm presented we have a procedure based on a partial LU factorization. It is used for the reduction of the companion iterative model of a subcircuit. This reduction gives its Thevenin/Norton equivalent visible from its terminals. When this inner loop reaches convergence, we obtain a linear equivalent of the subcircuit, appropriate to a solution in response to block signals $\mathbf{v}_k^{(i)}$ coming from the upper level and hence treated as excitations. This equivalent we can represent as the stamp (8.49), and call the macromodel of the subcircuit. On the upper-level, instead of using the full subcircuit we use only its macromodel, free of internal nodes.

The macromodel is given in a form linearized with respect to block variables. Its stamp (8.49) is a hybrid description of the block; in particular, if all $\mathbf{v}_k$ are voltages, then (8.49) stands for an admittance description of the block $\mathbf{Y}\mathbf{v}_0 = \mathbf{i}_0$. The linearized macromodel (8.49) has a resistive nonlinear origin, which after linearization gives (8.49), but we are not interested in these nonlinear macromodels and will assign the term macromodel to the linearized equivalent circuit whose MNE stamp is (8.49). As is seen in Equation (8.50), macromodels of the blocks of Figure 8.5 are a parallel connection of conductances and a current source.

By the way, notice that to calculate them we have to know in advance the voltages at internal nodes.

Several methods of generating macromodels are available. Some of them provide savings in computational effort.

(i) In Algorithm 8.3 we have the basic method, where macromodels are generated by inner NR loops including at each iteration a partial LU factorization and forward/backward substitutions to solve (8.35).

(ii) Since the Jacobi matrix in (8.29) is sparse and the solution of (8.35) is repeated, we can employ an efficient sparse matrix procedure or even generate a symbolic code for solving it. We have presented the problems of code generation in Section 3.3. However, as we know from decomposition methods, a reordering for fill-ins minimization is difficult if the number of block nodes is not much less than the number of internal nodes.

(iii) For further savings, macromodels can be approximated by means of piecewise linear or spline interpolation similarly to CAzM [4.4]. We construct auxiliary circuits where subcircuits are separately excited. Sweeping the excitations, while sampling and storing the terminal responses enables us to create forms of look-up table. Then we use these samples for a piecewise linear or spline multivariate interpolation of the characteristics of the block. The stamp (8.49) can be obtained from this interpolation with a slight computational effort in comparison with inner NR iterations. This approach is very memory-consuming and directly applicable to d.c. analysis only.

(iv) If the signals at the terminals of the block do not change either during NR iterations or between time steps, then the stamp or macromodel also remains unchanged and need not be recalculated. This property, called bypassing (in iterations) or latency (in time) provides considerable savings of computational effort. We will discuss these problems later on.

8.3.3.3 Bypassing the recalculation of macromodels in NR iterations

Bypassing the recalculation of iterative models when their terminal signals reach convergence is a powerful technique for saving computational effort. In the single-level NR methods described in Chapter 4 it was applied to a particular nonlinear element when its terminal signals converged faster than other signals. If the terminal signals of an element change little between two successive NR iterations, then we can bypass calculation of a new iterative model, but use the model obtained at the preceding iteration. Since the calculation of models consumes about 70 per cent of total simulation time, bypassing is essential for efficiency.

This idea can be easily extended to macromodels of subcircuits. If during the course of upper-level iterations, the terminal variables of a block remain unchanged in iterations $(i-1)$ and (i), then the internal variables of this block will also remain the same, and hence we can bypass updating the macromodel (8.49) of this block but at the ith iteration use the macromodel recalled from the $i-1$ iteration. This requires storage of (8.49) from the preceding iteration.

To implement bypassing of blocks in Algorithm 8.3, we have to introduce the following procedure at the begining of the loop over the blocks

(I) If the kth block is not bypassed, then it will be bypassed while calculating $\mathbf{v}_0^{(i+1)}$ if the following two conditions are simultaneously satisfied:

(i) variables at its terminals are little changed in comparison with the $i-1$ solution, i.e.

$$|v_{0j}^{(i)} - v_{0j}^{(i-1)}| \le \delta + \varepsilon \max\{ |v_{0j}^{(i)}|, |v_{0j}^{(i-1)}|\} \tag{8.55}$$

where j are indexes of terminals of the block,

(ii) at the most recent lower-level iteration all nonlinear branches of this block have been bypassed according to the bypassing principles presented in Chapter 4;

(II) If the block was bypassed so far, then after its solution for $\mathbf{v}_0^{(i+1)}$ we decide to maintain this status if the terminal signals are little changed from those at the lth upper-level iteration, when the block obtained the status of being bypassed, i.e.

$$|v_{0j}^{(i+1)} - v_{0j}^{(l)}| \le \delta + \varepsilon \max\{ |v_{0j}^{(i+1)}|, |v_{0j}^{(l)}|\} \tag{8.56}$$

Requirement (i) means that the external status of the block is stable, while checking of the criterion (ii) ensures that also the internal signals of the block have been stabilized in NR iterations. The macromodel of the bypassed block is not calculated, but recalled from memory.

8.3.3.4 Latent subcircuits

In time-domain analysis, multilevel NR algorithms are performed within the time-loop. If at instants t_{n-2} and t_{n-1} the signals of a block are the same, then we may hope that these signals will remain unchanged also at instant t_n. Then, we can expect that at this instant under some conditions the block will have the same macromodel as at previous instants. Thus we can we can omit updating of its macromodel but apply in all upper-level iterations of instant t_n the macromodel already known from instant t_{n-1}. Blocks which can be treated in this way we call latent. The following criteria of latency, introduced in SLATE [8.12], are representative.

(I) If the block has not been latent so far, then we make it latent from instant t_n if:

(i) all terminal signals of the block have been little changed at t_{n-1} in comparison with instant t_{n-2}, i.e.

$$|v_{0j}(t_n) - v_{0j}(t_{n-1})| \le \delta + \varepsilon \max\{|v_{0j}(t_n)|, |v_{0j}(t_{n-1})|\} \tag{8.57}$$

for all indexes j of the block terminals

(ii) currents of all capacitors and voltages of all inductors of the block have been little changed at t_{n-1} in comparison with instant t_{n-2}, i.e.

$$\begin{aligned} |i_{Cj}(t_{n-1}) - i_{Cj}(t_{n-2})| &\le \delta + \varepsilon \max\{|i_{Cj}(t_{n-1})|, |i_{Cj}(t_{n-2})|\} \\ |v_{Lj}(t_{n-1}) - v_{Lj}(t_{n-2})| &\le \delta + \varepsilon \max\{|v_{Lj}(t_{n-1})|, |v_{Lj}(t_{n-2})|\} \end{aligned} \tag{8.58}$$

for all indexes j of capacitors and inductors of the block.

(iii) step-size is large in comparison with the time constant introduced by each capacitor and inductor, i.e.

$$\frac{t_n - t_{n-1}}{\tau} \le 1 \tag{8.59}$$

where, since in the linear RC circuit we can relate a time constant with the charge and current of capacitor by $\tau = -q/i$, therefore in any nonlinear case we can perform a rough estimation by

$$\begin{aligned} 1/\tau_C &= |i_{Cj}(t_n) - i_{Cj}(t_{n-1})| / |q_{Cj}(t_n) - q_{Cj}(t_{n-1})| \\ 1/\tau_L &= |v_{Lj}(t_n) - v_{Lj}(t_{n-1})| / |\psi_{Lj}(t_n) - \psi_{Lj}(t_{n-1})| \end{aligned} \tag{8.60}$$

where to be sure that estimation of the time constant is accurate we require that the increment of current i_C and voltage v_L be greater that a minimum level $|\Delta i_C| \ge \delta = 1\text{pA}$, $|\Delta v_L| \ge \varepsilon = 1\text{nV}$.

(II) If a block was latent since t_{n-l} and we have to decide on its latency at instant t_n, then we decide to maintain its latency at this instant if

$$|v_{0j}(t_n) - v_{0j}(t_{n-l})| \leq \delta + \varepsilon \max\{|v_{0j}(t_n)|, |v_{0j}(t_{n-l})|\} \tag{8.61}$$

If this condition does not hold we cancel the latent status of the block.

Latency is a powerful tool in circuits where most of the variables remain fixed, while only a small portion change. For instance in digital circuits signals are usually at their stable logical levels, and vary only during a relatively short phase of switching. Morever, switching may occur only along some paths in the circuit which incorporate a small fraction of the total number of logical elements. Therefore, almost the whole circuit is usually latent and we do not require to calculate it. In analyzers like SPICE we always calculate the whole circuit, while latency exploiting simulators can be up to ten times faster. Unfortunately, latency is not so efficient in analog circuits, where the latent fraction of the circuit is small.

In this section we have presented direct methods exploiting decomposition at the companion nonlinear equations level and multilevel NR iterations. In Section 8.1 methods based on decomposition at the linearized companion equations level have been demonstrated. Both groups of methods exploit macromodels having a generated sparse symbolic code, bypassing and latency, and can be about ten times faster than classical direct methods. Unfortunately, in analog circuits with tightly coupled nodes, the effects of latency are not so spectacular. The true explosion of efficiency appears in circuits with one-way blocks, where whole strings of cascaded blocks can become latent.

A tenfold improvement in simulation efficiency is not good enough for concurrent VLSI ICs. In Chapter 9 we will present new-generation methods whose acceleration is obtained by using relaxation-based solutions instead of the Newton-Raphson procedure with the LU solution.

References

8.1 Chua, L.O. Chen Li-Kuan, *Diacoptic and Generalized Hybrid Analysis.* IEEE Trans. CAS, vol. CAS-23, no. 12, pp. 694-705 (1976)

8.2 Engl, W.I., Laur, R., Dirks, H., *MEDUSA A Simulator for Modular Circuits,* IEEE Trans. CAD, vol. 1, no. 4, p. 85-93 (1982).

8.3 Guardabassi, G., Sangiovanni-Vincentelli A. *A Two Levels Algorithm for Tearing.* IEEE Trans. CAS, vol. CAS-23, no. 12, pp. 783-791 (1976).

8.4 Hajj, I.N. *Sparsity Consideration in Network Solution by Tearing.* IEEE Trans. Circ. Syst., vol. 27, no. 5, p. 357-366 (1980).

8.5 Kron G. *A Set of Principles to Interconnect the Solution of Physical Systems.* Journal of Applied Physics, vol. 24, no 8, pp. 965-980 (1953).

8.6 De Micheli, G., Hsieh, H.Y., Hajj, I.N., *Decomposition Techniques for Large Scale Circuit Analysis and simulation.* In: *Circuit Analysis Simulation and Design, VLSI Circuit Analysis and Simulation,* Ed.: A.E. Ruehli. Amsterdam: North-Holland (1987).

8.7 Rabbat, N., Hsieh, H.Y., *A Latent Macromodular Approach to Large-Scale Sparse Networks,* IEEE Trans. Circ. Syst., vol. 22, no. 12, p. 745-752 (1976).

8.8 Rabbat, N.B.G., Sangiovanni–Vincentelli, A., Hsieh, H.Y., *A Multilevel Newton Algorithm with Macromodelling and Latency for the Analysis of Large Scale Nonlinear Circuits in the Time Domain,* IEEE Trans. Circ. Syst., vol. 26, no. 9, p. 733-740 (1979).

8.9 Sangiovanni – Vincentelli A., Bickart T.A. *On the Reduction of a Matrix into an Optimal Bordered Trangular Form.* Proc. IEEE Int. Symp. Circ. Syst., p. 76-80 (1978).

8.10 Sangiovanni – Vincentelli A., Chen L.K., Chua L.O. *A New Tearing Approach–Node Tearing Nodal Analysis.* Proc. IEEE Int. Symp. Circ. Syst., p. 143-147 (1976).

8.11 Wu F.F. *Solution of Large-Scale Networks by Tearing.* IEEE Trans. Circ. Syst., vol. 23, no 12, p. 706-713 (1976).

8.12 Yang, P., Hajj, I.N., Trick, T.N., *SLATE: A Circuit Simulation Program with Latency Exploitation and Node Tearing,* Proc. IEEE Int. Conf. Circuits and Computers, Port Chester, N.Y., no 10, p. 353-355 (1980).

Chapter 9

Relaxation-based Simulation Methods

9.1 Introductory topics in relaxation methods

Since VLSI circuits so far have been mostly digital, the problem of their simulation has been resolved first at the level of digital signals. This has given us logic simulators. However, logic simulators do not accurately model either analog signals or their time profiles (so-called timing relationships). Logic simulation is capable of handling the VLSI digital circuit as a whole. However, the accuracy of this simulation is usually not sufficient to verify the performance of some critical circuit fragments. By critical fragments we mean those where feedback and bidirectional elements play an important role or design specifications are so tight that simulation must be more accurate than logic simulation provides. Therefore, independently of logic simulation, analog circuit simulation of large-scale circuits has been continuously developed.

Considerable progress has been achieved thanks to relaxation-based methods, also called third-generation simulation methods. The concept of relaxation methods arises from the iterative (also called relaxation) methods of linear algebraic equations solution which will be discussed in Section 9.1 below.

The main idea used in relaxation methods is decoupling of circuit equations. Let a circuit be described by a system of p OADEs with p unknowns $\mathbf{v}$: $\mathbf{F}[\mathbf{v}, \dot{\mathbf{z}}(\mathbf{v})] = \mathbf{s}(t)$, where $\mathbf{z}$ are state variables (voltages and currents or charges and fluxes). These equations can be written in a form stressing their composition as p scalar equations, namely:

$$
\begin{aligned}
&F^1(v^1, v^2, ..., v^p, \dot{z}^1, ..., \dot{z}^m) = s^1(t) \\
&F^2(v^1, v^2, ..., v^p, \dot{z}^1, ..., \dot{z}^m) = s^2(t) \\
&\dots\dots\dots\dots\dots\dots\dots\dots\dots\dots \\
&F^p(v^1, v^2, ..., v^p, \dot{z}^1, ..., \dot{z}^m) = s^p(t)
\end{aligned}
\qquad (9.1)
$$

where superscripts indicate the numbers of unknowns and equations. If we treat each scalar equation F^i separately and solve it only for the unknown v^i, with remaining unknowns taken from the solutions of other equations, then such an approach to the system we call **decoupling**. In the case of equations formulated according to a nodal transformation of variables (see Section 1.3.1), each equation of (9.1) stands for the KCL at a node, and hence decoupling can be said to be **nodal**.

If Equations (9.1) are partitioned into q groups $\mathbf{F}^i$, each group including a number of scalar equations, the unknowns are partitioned respectively into groups $\mathbf{v}^i$, and similarly the state variables $\mathbf{z}^i = \mathbf{z}^i(\mathbf{v}^i)$, where $i = 1, ..., q$, then we can write (9.1) in block form

$$
\begin{aligned}
&\mathbf{F}^1(\mathbf{v}^1, \mathbf{v}^2, ..., \mathbf{v}^q, \dot{\mathbf{z}}^1, ..., \dot{\mathbf{z}}^q) = \mathbf{s}^1(t) \\
&\mathbf{F}^2(\mathbf{v}^1, \mathbf{v}^2, ..., \mathbf{v}^q, \dot{\mathbf{z}}^1, ..., \dot{\mathbf{z}}^q) = \mathbf{s}^2(t) \\
&\cdots\cdots\cdots\cdots\cdots\cdots\cdots\cdots \\
&\mathbf{F}^q(\mathbf{v}^1, \mathbf{v}^2, ..., \mathbf{v}^q, \dot{\mathbf{z}}^1, ..., \dot{\mathbf{z}}^q) = \mathbf{s}^q(t)
\end{aligned}
\tag{9.2}
$$

Provided the ith group of equations is solved for the ith group of unknowns with other unknowns given from the solution of previous groups of equations, we obtain **block-decoupling** of equations. Each group can be solved by the direct, i.e. LU method.

The differences between relaxation methods begin with the level at which decoupling takes place. Decoupling of OADEs before discretization, and discretization (using differentiation formulae, see Chapter 5) applied to each decoupled component separately lead to **waveform-relaxation** (WR) analysis. If we first discretize the whole circuit and then apply decoupling, we obtain **one-point relaxation** methods. Among these we have methods where decoupling is applied to the algebraic nonlinear equations of a substitute companion circuit (see Chapter 5), known as **nonlinear relaxation** analysis. However, if we linearize the companion equations for d.c. and time-domain analysis, and apply decoupling to linearized companion equations, then we have **linear-relaxation** analysis.

Except for this choice of levels, differences between relaxation methods stem from different strategies for contributing variables other than v_i to the ith decoupled equation. In Section 9.2 we will present **Gauss-Seidel** (GS) and **SOR** methods. These are known as the most efficient and popular relaxation-methods. Another common method, that is usually (but not always) slightly less efficient than GS and SOR, is the **Gauss-Jacobi** (GJ) method.

Relaxation methods have two main advantages. (i) The costly formulation and direct solution of a system of equations is replaced by a considerably less costly solution of decoupled equations. (ii) Due to decoupling of nodes or blocks, their independence and latency can be easily exploited. Only a few of the possible combinations of levels and strategies give a real computer-time savings and satisfactory accuracy.

The iterative linear relaxation method, implemented within the common Newton-Raphson (NR) analysis as an inner loop instead of the LU algorithm, gives only a small time profit since we have the overhead of repetitive formulation of the whole Jacobi matrix. This overhead vanishes in nonlinear relaxation methods. Moreover, due to the inaccuracy of the inner relaxation loop (terminated by a test for convergence), convergence of the NR loop is no longer quadratic. However it is at least linear and hence its rate ranges from 1 to 2. Linear relaxation methods are of little practical importance in circuit simulation.

It is interesting to note that if we reduce the computational effort of a linear relaxation method, by carrying out only one linear relaxation, then convergence becomes only linear. This so-called **one-sweep relaxation** can be exploited also in nonlinear relaxation methods giving a timing analysis. In newer simulators one-sweep has been replaced by **iterated relaxation** methods, which appear to be more successful for real circuits with different types of strong coupling.

We distinguish four groups of practically important methods:

(i) timing analysis, also termed the relaxation-based integration method,

(ii) iterated timing analysis,

(ii) one-step relaxation analysis,

(iii) waveform relaxation analysis,

These methods will be discussed in the following sections.

In general it can be said that relaxation methods can be at most about one hundred times faster than the conventional direct methods of simulators such as SPICE. However, they suffer from certain limitations and have weak points which seem to be at last overcome (possibly not fully yet) after more than ten years of investigation:

(i) In sparse implementations of relaxation methods the computational cost per iteration is proportional to the number of equations p, while in sparse direct methods the total cost is proportional to p^{γ} ($\gamma \in \{1.1 \div 1.5\}$). Hence the superiority of relaxation methods is limited to cases where the number of iterations increases no faster than $p^{\gamma-1}$. It can be proved that relaxations are convergent iff the eigenvalues of their characteristic matrix are limited to the open unit ball. Their asymptotic rate of convergence is only linear. Hence, we may be afraid that for circuits whose matrices are far from having diagonal dominance, convergence will be slow and their superiority over direct methods doubtful.

(ii) Relaxation methods are suitable for MOS circuits due to the gate separation from drain and source (one-way property [9.15]) and the saturating character of drain-characteristics means there is only a small error acccumulation. Adequate one-way macromodels of many subcircuits can be created. However, efficiency decrease arises when there are clusters of nodes strongly bidirectionally coupled. In these cases the methods operate very slowly (sometimes more slowly then SPICE) and give inaccurate results.

(iii) Relaxation methods require protection against strong couplings, which appear due to transmission gates, pass transistors, feedback loops and, especially, floating capacitors. In older simulators the occurrence of floating elements was restricted. In newer programs we apply more robust algorithms and often strongly coupled nodes are handled by block-relaxation methods in which a direct solution of the block is mixed with relaxations at the block level.

(v) Due to the increasing size of VLSI circuits, the best relaxation methods of yesterday appear too slow for the practical needs of today. Different accelerators that resort to such tricks as clever model simplification, are being invented. An example of one such a simplifying trick is included in the simulator ELDO-XL, which is capable of handling circuits with as many as 120 000 MOS transistors. Acceleration can also be obtained by parallelization on general-purpose computers and especially on dedicated hardware that runs only one simulation algorithm. This is termed the **simulation engine**.

9.2 Linear relaxation methods

The considerable numerical complexity of finite methods leads to a need for other methods of solving linear simultaneous systems, the number of operations in finite methods solution being the third power of the system size order ($O(n^3)$) (certainly for the full matrix technique). Among alternative methods the most common are direct iteration methods whose numerical complexity is about n^2 operations per iteration. If convergence can be reached in a number of iterations much smaller than the system size, then this method will be much more efficient then the LU method. Hence, the iterative approach is convenient for huge systems once convergence is good. It is not obvious that good convergence is attainable since the direct iteration methods are only of linear convergence. Good, albeit linear, convergence occurs in electrical systems of unilateral (or nearly unilateral) signal propagation, as e.g. in logical systems. The literature describes other gradient methods with better convergence [9.7, 9.22] but they require a greater number of numerical operations per iteration.

9.2.1 Simple iteration algorithms and their properties

To introduce a class of simple iteration methods for solving the system of simultaneous linear equations $\mathbf{A}\,\mathbf{v}=\mathbf{b}$ we assume a nonsingular matrix $\mathbf{B}$. The system can be written equivalently in the form $\mathbf{B}\,\mathbf{v}=\mathbf{b}-(\mathbf{A}-\mathbf{B})\mathbf{v}$, which easily yields a simple recursive iteration rule

$$\mathbf{B}\,\mathbf{v}^{k+1}=\mathbf{b}-(\mathbf{A}-\mathbf{B})\mathbf{v}^{k}. \tag{9.3}$$

It enables $\mathbf{v}^{k+1}$ to be obtained from $\mathbf{v}^{k}$, if (9.3) is solvable with reasonable computational effort and the generated sequence $\{\mathbf{v}^{k}\}_{k=0}^{\infty}$ is convergent. It can be readily shown that if it is convergent, then its limit is equal to the only solution of our system. A solution at the k+1th iteration may be expressed in a form derivable from (9.3)

$$\mathbf{v}^{k+1}=(\mathbf{1}-\mathbf{B}^{-1}\mathbf{A})\,\mathbf{v}^{k}+\mathbf{B}^{-1}\mathbf{b}. \tag{9.4}$$

It implies the rate of convergence depending on properties of the characteristic matrix $\mathbf{C}=\mathbf{1}-\mathbf{B}^{-1}\mathbf{A}$. This can be formalized as follows.

The direct iteration method (9.3) *is convergent for any initial vector* $\mathbf{v}^0$ *if and only if the spectral radius of* $\mathbf{C}$ (*known as the maximum modulus of its eigenvalues and denoted as* $\rho(\mathbf{C})$) *is less then one. Moreover it can be proved that if for any induced norm the inequality* $\|\mathbf{C}\|<1$ *is obeyed, then convergence of the simple iteration process is guaranteed.*

Calculating the error at the kth iteration $\Delta\mathbf{v}^{k}=\mathbf{v}^{k}-\mathbf{A}^{-1}\mathbf{b}$ from (9.4) we obtain the relation between errors in subseqent iterations $\Delta\mathbf{v}^{k}=\mathbf{C}^{k}\Delta\mathbf{v}^{0}$. Hence, it can be proved that a sequence of these errors is convergent to zero for any initial solution vector if and only if all eigenvalues of $\mathbf{C}$ are of moduli smaller than one. This property occurs, if the induced norm of $\mathbf{C}$ is smaller than 1 because this norm is always an upper specification of the eigenvalue moduli.

The matrix $\mathbf{C}$ also controls the rate of convergence, i.e. the maximum (over initial values $\Delta\mathbf{v}^0$) limit of the sequence

$$\sqrt[k]{|\Delta\mathbf{v}^{k}|\,/\,|\Delta\mathbf{v}^{0}|} \tag{9.5}$$

is equal to $\rho(\mathbf{C})$ as k approaches infinity. This rule shows that the spectral radius of **C** is also the rate of convergence and in the vicinity of solution it realistically overestimates the ratio of an error decreasing in two subsequent iterations. The smaller the radius $\rho(\mathbf{C})$, the faster is the convergence of the method. Hence, we have to make **C** have eigenvalues of small modulus. Defining a difference matrix $\mathbf{E} = \mathbf{B} - \mathbf{A}$ we can see that $\mathbf{C} = \mathbf{B}^{-1}\mathbf{E}$, and hence small eigenvalues of **C** arise from **B** being close to **A**. To have simultaneously the system (9.3) that might be solved with relative ease (e.g. in about n^2 operations) it is convenient to make **B** triangular. This yields the so-called, Gauss-Seidel method. We present this method first as it is the most important among iterative methods.

9.2.2 The Gauss-Seidel method

9.2.2.1 Formalization of the Gauss-Seidel method

The Gauss-Seidel method consists of taking as the matrix **B** the lower triangular part of **A** together with its diagonal. Let

D be a matrix containing only the diagonal compound of **A**,

L be the subdiagonal matrix portion,

U be the above diagonal part of **A**.

Then $\mathbf{B} = \mathbf{D} + \mathbf{L}$. Hence the Gauss-Seidel (GS) iteration takes the form

$$(\mathbf{D} + \mathbf{L})\,\mathbf{v}^{k+1} = \mathbf{b} - \mathbf{U}\,\mathbf{v}^{k}. \qquad (9.6)$$

The system (9.6) is to be solved by means of forward substitution. Hence diagonal components have to be nonzero and possibly of large modulus as is known from the previous discussion of finite methods. To yield this numerical advantage, the equations have to be reordered prior to the solution process. The best results will be achieved if the components with the largest moduli have been located in the diagonal by interchanging rows and columns, before the matrix is divided into the triangular part. Let us notice that interchanges are carried out once at the very beginning, instead of the dynamic strategy of pivoting during the solution process which is incorporated in the finite methods.

To demonstrate the GS method let us consider the system

$$\begin{bmatrix} 1 & 50 & 2 \\ 2 & 5 & 40 \\ 100 & 4 & 3 \end{bmatrix} \begin{bmatrix} v_1 \\ v_2 \\ v_3 \end{bmatrix} = \begin{bmatrix} 107 \\ 132 \\ 117 \end{bmatrix}. \qquad (9.7)$$

To maximize the diagonal elements we interchange the 1st and the 3rd row and simultaneously the 2nd and the 3rd column yielding

$$\begin{bmatrix} 100 & 3 & 4 \\ 2 & 40 & 5 \\ 1 & 2 & 50 \end{bmatrix} \begin{bmatrix} v_1 \\ v_3 \\ v_2 \end{bmatrix} = \begin{bmatrix} 117 \\ 132 \\ 107 \end{bmatrix}. \tag{9.8}$$

Now, to produce the recursive GS formula we extract the above diagonal part of the matrix and move the corresponding component to the RHS

$$\begin{bmatrix} 100 & & \\ 2 & 40 & \\ 1 & 2 & 50 \end{bmatrix} \begin{bmatrix} v_1^{k+1} \\ v_3^{k+1} \\ v_2^{k+1} \end{bmatrix} = \begin{bmatrix} 117 - 3v_3^k - 4v_2^k \\ 132 - 5v_2^k \\ 107 \end{bmatrix}. \tag{9.9}$$

For k=0 and the initial solution $v_1^0 = 1, v_2^0 = 1, v_3^0 = 1$ we get the next approximation of the solution from the system

$$\begin{bmatrix} 100 & & \\ 2 & 40 & \\ 1 & 2 & 50 \end{bmatrix} \begin{bmatrix} v_1^1 \\ v_3^1 \\ v_2^1 \end{bmatrix} = \begin{bmatrix} 110 \\ 127 \\ 107 \end{bmatrix}. \tag{9.10}$$

After forward substitution we achieve the result

$$\begin{aligned} v_1^1 &= 1.1 \\ v_3^1 &= (127 - 2v_1^1)/40 = 3.12 \\ v_2^1 &= (107 - v_1^1 - 2v_3^1)/50 = 1.99 \end{aligned} \tag{9.11}$$

which nearly approaches the exact solution $v_1=1, v_3=3, v_2=2$. The reader can perform the next step him/herself to observe the convergence.

Turning to general formalization let us neglect the reordering (so that the enumeration is taken to correspond to the post-reordering system). Thus, the ith equation (9.6) may be written in the form

$$\sum_{j=1}^{i-1} a_{ij} v_j^{k+1} + a_{ii} v_i^{k+1} + \sum_{j=i+1}^{n} v_j^{k} = b_i . \tag{9.12}$$

Hence, forward substitution yields the following expression for components of the vector $\mathbf{v}^{k+1}$

$$v_i^{k+1} = \frac{1}{a_{ii}}\left(b_i - \sum_{j=1}^{i-1} a_{ij} v_j^{k+1} - \sum_{j=i+1}^{n} a_{ij} v_j^{k}\right), \quad \text{for } i=1, \ldots, n \,. \tag{9.13}$$

With relative ease we notice that after v_i^{k+1} has been calculated, the value v_i^{k} is not necessary any more. Hence, the GS algorithm may manipulate elements of an n-element table, which after the ith ieration contains new values (derived from the k+1th iteration) at the i beginning positions, while at other positions there are old values derived from the kth iteration.

GS iterations may be carried out not only in the forward form (9.6) but also in a dual backward form by taking an upper-diagonal part of the matrix as the auxiliary **B** and using backward substitution instead of (9.13)

$$v_i^{k+1} = \frac{1}{a_{ii}}\left(b_i - \sum_{j=1}^{i-1} a_{ij} v_j^{k} - \sum_{j=i+1}^{n} a_{ij} v_j^{k+1}\right), \quad \text{for } i=n, \ldots, 1 \,. \tag{9.14}$$

The equations (9.13) and (9.14) are called forward and backward GS iterations respectively. They may also be used alternately, to reduce the numerical error of our algorithm.

9.2.2.2 Convergence of the Gauss-Seidel method

Convergence of the simple iteration method has been shown in the preceding subsection to be dependent on the eigenvalues of the matrix **C**. In the case of the GS method, maintaining the former notation, the matrix **C** is of the form $\mathbf{C} = -(\mathbf{L}+\mathbf{D})^{-1}\mathbf{U}$. As the necessary and sufficient condition of convergence $\rho(\mathbf{C}) < 1$ cannot be easily checked, a convenient sufficient condition is usually applied instead. It derives from the inequality stating an upper estimation of the maximum norm (that is l^{∞}) of the matrix **C**

$$\|\mathbf{C}\|_{\infty} \le \max_i \frac{1}{|a_{ii}|} \sum_{j \ne i} |a_{ij}| \tag{9.15}$$

and also from the sufficient condition of convergence $\|\mathbf{C}\|_{\infty} < 1$, introduced in Section 9.2.1. Thus, we get the following sufficient condition of convergence of the GS algorithm

If matrix **A** *is of strong diagonal dominance, that is for its all rows the inequality* $|a_{ii}| > \sum_{j \neq i} |a_{ij}|$ *holds, then the GS method is convergent.*

After some mathematical development, it can be proved that, if there is such a symmetric reordering (i.e. the same for rows and columns) that all zero components of the matrix **A** fall into a rectangular zero submatrix remaining in the lower left corner of **A**, then weak diagonal dominance is sufficient to guarantee convergence of the GS method. Weak diagonal dominance is derived from the definition of strong diagonal dominance by replacing the sign $<$ by $\leq$ in at most all rows except that one where the strong inequality is left.

Diagonal dominance is a common feature of matrices describing circuits composed of bidirectional. However, this sufficient condition is not necessary. Another sufficient condition might require triangular dominance which is common in circuits built of unidirectional elements (e.g. gates, MOS transistors). That is why the reader studying circuit simulation needs some familiarity also with iterative methods. These problems will be continued in the following sections of this chapter that are dedicated to relaxation techniques. Notice that the instructive example (9.8) was of strong diagonal dominance, and hence even after the first iteration a quite good approximation to a solution has been obtained.

9.2.2.3 How to implement the Gauss-Seidel method

Algorithm 9.1 demonstrates a simple implementation of the GS method. It assumes a maximum number of iterations *kmax*, a minimum diagonal element level *tol*, a relative *eps* and absolute *del* accuracy index.

The GS method is not of superior numerical quality, i.e. its error cannot be viewed as the exact result of a certain substitution data perturbation because its exact solution is not reached until after an infinite number of iterations. To reduce arithmetic errors by an appropriate initial reordering, diagonal elements of the matrix should be made as large as possible. There is a considerable numerical error at each iteration due to the unsymmetric influence of matrix elements on the solution. To minimize the accumulation of error a strategy of alternate forward and backward iterations is convenient; this is known as symmetric displacement.

Similar to the GS method is the Jacobi method, where the auxiliary matrix $\mathbf{B} = \mathbf{A} - \mathbf{D}$ (the whole matrix **A** without diagonal elements) is assumed. However, the Jacobi method has not been described here since its numerical properties are somewhat inferior to those of the GS method.

Algorithm 9.1 Solution of linear equations by means of the Gauss-Seidel method

```
for k=1 to kmax do {main iterations loop}
begin
  ifl = 0; {convergence flag}
  for i=1 to n do {rows loop}
  begin
    p = 0;
    for j=1 to i-1 do p = p - a_ij v_j;
    for j=i+1 to n do p = p - a_ij v_j;
    p = p + b_i;
    if |a_ii|<tol then stop {too small diagonal element};
    p = p / a_ii;
    if |v_i - p|>eps max(|v_i|,|p|)+del then ifl = 1;
                                        {the ith stop test}
    v_i = p;
  end
  if ifl=0 then stop {convergence reached}
end
stop {no convergence}
```

9.2.3 Simple over-relaxation method

The simple over-relaxation (SOR) method involves a slight modification of the Gauss-Seidel algorithm aimed at reducing the spectral radius of the matrix **C**. If the forward Gauss-Seidel iteration (9.13) is written in a form which stresses an additive correction $\Delta\mathbf{x}_{k+1}$, then $v_i^{k+1} = v_i^k + \Delta v_i^{k+1}$, ($i$=1,...,$n$) is obtained, where

$$\Delta v_i^{k+1} = \frac{1}{a_{ii}}\left(b_i - \sum_{j=1}^{i-1} a_{ij} v_j^{k+1} - \sum_{j=i}^{n} a_{ij} v_j^k\right). \tag{9.16}$$

The SOR method introduces a relaxation index ω, that is a scaling factor for $\Delta\mathbf{v}^{k+1}$ transforming it into $\omega\Delta\mathbf{v}^{k+1}$. Hence, after some algebra, the following expression for the ith component of the solution at the k+1th iteration is gained

$$v_i^{k+1} = (1-\omega)\, v_i^k + \frac{\omega}{a_{ii}}\left(b_i - \sum_{j=1}^{i-1} a_{ij} v_j^{k+1} - \sum_{j=i}^{n} a_{ij} v_j^k\right). \tag{9.17}$$

It corresponds to the simple iteration (9.3) but includes the slightly modified matrix $\mathbf{B}(\omega)=\mathbf{D}/\omega+\mathbf{L}$. Convergence is dependent on the characteristic matrix $\mathbf{C}=\mathbf{B}(\omega)^{-1}[\mathbf{B}(\omega)-\mathbf{A}]$, which in our case takes the form $\mathbf{C}=(\mathbf{D}+\omega\mathbf{L})^{-1}[(1-\omega)\mathbf{D}-\omega\mathbf{U}]$. It can be shown that for any matrix $\mathbf{A}$ the inequality $\rho(\mathbf{C})\geq|\omega-1|$ holds. It proves that convergence occurs only when $\rho\in(0,2)$. For $\omega=1$, the method reduces to the Gauss-Seidel approach. Once SOR is convergent, then there is the question whether a slight adjustment of ω in the vicinity of 1 may reduce $\rho(\mathbf{C})$, and hence improve convergence. It has been proved that the character of the function $\rho(\omega)$ in the range (0, 2) depends on the type of the matrix $\mathbf{A}$. A plot of this function usually has a minimum in a "valley" whose left slope descends rapidly, while the right is more gentle. Thus, when we do not know the optimal value of ω, it is better to take a slightly too large value than a value that is too small.

By making stronger assumptions about the matrix **A**, we are able to determine ω_{opt}. Let that matrix be transformable by symmetric reordering into the form

$$\begin{bmatrix} D_1 & M_1 \\ M_2 & D_2 \end{bmatrix}$$

where $\mathbf{D}_1$, $\mathbf{D}_2$ are diagonal. Moreover, let the eigenvalues of $\mathbf{D}^{-1}(\mathbf{L}+\mathbf{U})$ be real, less then 1. Then,

$$\omega_{opt}=\frac{2}{1+\sqrt{1-\lambda}}, \tag{9.18}$$

where λ is the greatest eigenvalue of the above given matrix. Hence, the rate of convergence is

$$\rho[\mathbf{C}(\omega_{opt})]=\omega-1=\left[(\sqrt{\text{cond}(\mathbf{A})}-1)/(\sqrt{\text{cond}(\mathbf{A})}+1)\right]^2.$$

In practice, it is quite impossible to find an optimal ω for all matrices. Thus we have to assume a certain value slightly exceeding 1 and observe the convergence. In fact, the greater the condition number the slower will be the convergence. Experiments can be carried out to count the number of iterations necessary for decreasing by ten times the solution error. For tasks whose condition number is of the order of some tens, both methods need some tens of iterations. Hence, their efficiency is almost the same. In cases when the condition number is several thousands the GS methods performs several thousands of iterations while the SOR does only several hundreds. Thus, SOR is something like ten times more efficient. It can be concluded that SOR is a method for ill-conditioned tasks.

We may modify Algorithm 9.1 with relative ease to implement the SOR method. The relaxation index should be chosen arbitrarily from the range (1,2), for example 1.05 or 1.1.

Before we end this section let us repeat the main topics and answer the following questions

- Transforming the equation $av - c = 0$ into the form $bv = c + (b - a)v$ a simple iteration scheme for solving it may be generated. Write a corresponding recursion.
- For the above equation try to work out the relation between errors in two subsequent iterations. For a given a find such b that the algorithm be convergent. Compare with the general principle given in Section 9.2.1.
- Write the forward and backward Gauss-Seidel and the forward and backward SOR algorithms in a programming language that you are familiar with or using the Pascal-like language used in this book.
- Explain the concept of strong and weak diagonal dominance and its influence on the convergence of all the methods considered.

9.3 One-point nonlinear relaxation methods

9.3.1 General formulation of nonlinear relaxation methods

In the above discussion linear-relaxation methods have been found not to be efficient enough. We proceed to nonlinear-relaxation methods, which rely on discretization of circuit equations (9.1) by means of a differentiation formula (DF) (see Chapter 5). Then, the derivatives of state variables $\mathbf{z}$ are substituted for the DF and we obtain a system of p nonlinear algebraic equations representing a companion circuit at a given instant t_n. We can write it in the form

$$\begin{aligned} &f^1(v_n^1, v_n^2, \ldots, v_n^p) = 0 \\ &f^2(v_n^1, v_n^2, \ldots, v_n^p) = 0 \\ &\cdots\cdots\cdots\cdots\cdots\cdots \\ &f^p(v_n^1, v_n^2, \ldots, v_n^p) = 0 \end{aligned} \tag{9.19}$$

where the subscript stands for time, while the superscript enumerates variables and equations.

In relaxation methods (9.19) is solved iteratively. In one iteration (called relaxation) we enter subsequent equations, namely $i = 1, ..., p$, and solve the ith one for v^i. The most common relaxation strategy is Gauss-Seidel (GS), where, during solution of the ith equation, values of variables $v_n^1, ..., v_n^{i-1}$ stem from the solutions of previous equations in this relaxation, while $v_n^{i+1}, ..., v_n^p$ not yet calculated, stem from the preceding relaxation. Introducing the relaxation count superscript (k) we can outline the GS solution process of Equation (9.19) in time in the form of Algorithm 9.2. In practice some method of solution of the scalar equation and a test for relaxations convergence has to be added.

Algorithm 9.2 A general nonlinear GS relaxation method

for $n = 1$ **to** *nmax* **do** {time-loop}
begin
  prediction of $v_n^{1(0)}, ..., v_n^{p(0)}$; {starting point for relaxations}
  for $k = 1$ **to** *kmax* **do** {relaxations loop}
  begin
    for $i = 1$ **to** p **do** {decoupled nodes loop}
    solve $f^i(v_n^{1(k)}, v_n^{2(k)}, ..., v_n^{i-1(k)}, v_n^{i(k)}, v_n^{i+1(k-1)}, ..., v_n^{p(k-1)}) = 0$
      for $v_n^{i(k)}$;
    if $(\| \mathbf{v}_n^{(k)} - \mathbf{v}_n^{(k-1)} \| \leq \varepsilon \max\{ \| \mathbf{v}_n^{(k)} \|, \| \mathbf{v}_n^{(k-1)} \| \} + \delta)$ **then break**
  end
end

The scalar equation can be solved by means of NR iterations over $v_n^{i(k)}$ starting from a certain predicted value (e.g. the solution from the last relaxation).

To provide good definition of the steps of Algorithm 9.2 we require that the scalar equation can be uniquely solved. If it were linear (linear relaxation case) a nonzero value of its ith coefficient a_{ii} would be required. Then, convergence is guaranteed iff eigenvalues of the characteristic matrix $\mathbf{L}^{-1}\mathbf{U}$ are inside the unit circle. In the nonlinear case, the conditions for convergence take the following form [9.14]

Let

(i) *there exist a solution* $\mathbf{v}^*$ *of* (9.19),

(ii) *the Jacobi matrix* $\partial \mathbf{f}(\mathbf{v})/\partial \mathbf{v}$ *of* $\mathbf{f}$ *be continuously differentiable in an open neighborhood* S_0 *of this solution,*

(iii) $\partial \mathbf{f}(\mathbf{v}^*)/\partial \mathbf{v}$ *be split into* $\mathbf{L}+\mathbf{U}$, *where* $\mathbf{L}$ *is a lower triangular matrix with diagonal;* $\mathbf{U}$ *is an upper triangular matrix without diagonal, diagonal elements of* $\mathbf{L}$ *are nonzero,*

(iv) *eigenvalues of* $\mathbf{M} = \mathbf{L}^{-1}\mathbf{U}$ *are inside the unit circle,*
then

there exists an open ball $S \subset S_0$ *such that the nonlinear GS starting from any point of this ball are well defined and converge to* v^*.

A similar theorem holds for the Gauss-Jacobi method. We split the Jacobi matrix into $\mathbf{L} + \mathbf{D} + \mathbf{U}$ ($\mathbf{D}$ is diagonal), and define $\mathbf{M} = \mathbf{D}^{-1}(\mathbf{L} + \mathbf{U})$.

Convergence of the nonlinear Gauss-Seidel (and Gauss-Jacobi) method is linear, i.e. the results of two subsequent relaxations satisfy the inequality

$$\| \mathbf{v}^{(k)} - \mathbf{v}^* \| \leq \gamma \| \mathbf{v}^{(k-1)} - \mathbf{v}^* \| . \tag{9.20}$$

Therefore, inner NR iterations (which are quadratically convergent) are purposeless, since, if we carry out only one NR iteration, then convergence is linear as well. Hence, though the one-NR iteration process takes slightly more relaxations it is much more efficient. This so-called Gauss-Newton (Gauss-Seidel-Newton or Gauss-Jacobi-Newton) algorithm is ubiquitous in relaxation methods.

Though Gauss-Newton methods are linearly convergent, the number of relaxations required for a given level of accuracy is highly dependent on the location of eigenvalues of the characteristic matrix $\mathbf{M}$ within the unit circle. As we know from Section 9.2 a sufficient condition for location of them inside this circle is strict diagonal dominance of the Jacobi matrix of (9.19). However, this is not a necessary condition. For instance, if the Jacobi matrix is lower-triangular (LT) then certainly one Gauss-Seidel relaxation is sufficient to obtain a solution, although its diagonal entries may not be dominant.

In general, the structure of dependence of the equations affects the rate of convergence. This structure can be represented by the **dependency matrix A**, composed of zeros and units. An entry (i, j) is equal to unit iff the equation f^i of (9.19) is dependent on the unknown v^j. Hence, if the dependency matrix is LT, then one relaxation gives an accurate solution.

If in the upper-triangular area we have a number of entries then one relaxation does not provide a solution, however if the diagonal entries of the Jacobi matrix are dominant, then convergence will be reached. The stronger the diagonal dominance, the smaller will be number of relaxations required, and simultanously the smaller will be the error, if we stop the relaxation process before convergence. This discussion shows that two things are essential for good convergence and accuracy:

(i) a lower-triangular (or almost lower-triangular) dependence matrix which can be provided by an appropriate ordering of the nodes in accordance with the direction of signal flow, and

(ii) diagonal dominance of the Jacobi matrix at a solution.

First we must try to reorder nodes as optimally as possible, and then to improve diagonal dominance by shortening the integration step. This problem will be discussed later on.

Now we focus on item (i). Let us consider the circuit shown in Figure 9.1, composed of cascaded stages, where the first stage is driven by the primary input I_1, the second stage by the solution of the first one, the third stage by the solution of the second and first stage, *etc.* The nodal equations

$$\begin{aligned} &f^1(v^1) = I_1 \\ &g^2(v^1) + f^2(v^2) = 0 \\ &g^3(v^1, v^2) + f^3(v^3) = 0 \\ &\dots\dots\dots\dots\dots\dots \\ &g^p(v^1, \dots, v^{p-1}) + f^p(v^p) = 0 \end{aligned} \tag{9.21}$$

have a LT dependency matrix. In the example circuit we have only one direction of signal flow, since the nodal voltage v^i depends only on the voltages at preceding nodes. The solution of this unilateral circuit can be carried out by solving for v^1 the stage connected to the first node. Then the second stage is solved for v^2 using the calculated v^1, *etc.* After p steps we have the whole circuit solved. This is exactly the one-relaxation GS process. To exploit this property, nodes have to be enumerated according to a structure of dependency, i.e. the signal flow. Certainly in

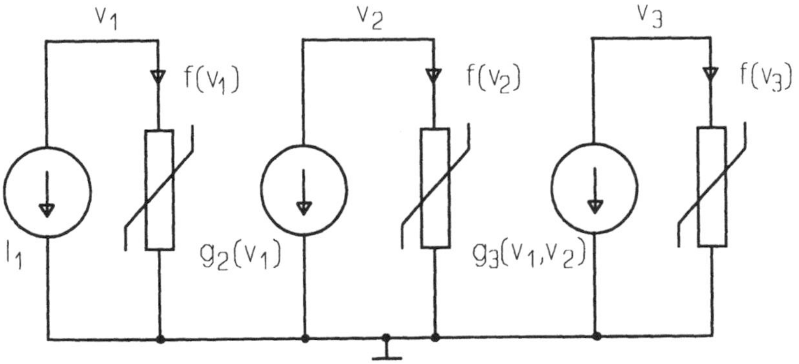

Figure 9.1 Example of a cascaded circuit

practical circuits dependency is not so universally forward. Existence of feedback in this circuit would introduce a dependence of v^i on one of next nodal voltages. A stage dependent on a feedback signal could not be solved before calculation of this signal and hence one relaxation is not sufficient.

As we can see from this consideration, the ordering of nodes (or blocks) is critical to the operation and efficiency of relaxation methods. These problems are common to different relaxation methods and we will discuss them separately later on.

Relaxation methods basically use only nodal voltages and a nodal circuit description. Hence they accept current-defined and voltage controlled branches. However, we can easily adapt a voltage-defined branch by collapsing its terminals and writing a common KCL balance. Then, the first of the terminal voltages is calculated from relaxation, while the second one is enumerated after it and obtained by means of the branch function. The controlling voltage has to be numbered as the previous one in the numbering sequence.

9.3.2 Timing analysis as relaxation-based integration

The development of timing analysis was started in 1975 by the creators of a simulator MOTIS [9.3], which was revolutionary for those days and whose purpose was to analyze MOS circuits in the time-domain more accurately than by logic simulation but more efficiently than by direct methods. At the very beginning, timing analysis was a very inaccurate simulation, but since then researchers have been trying to improve it and have produced a variety of relaxation methods.

Timing analysis was a first attempt at analog simulation modeled on the simplicity of logic analysis of gate circuits. In such a circuit if we know the signals on a gate's inputs, then we can find in one step the signal on its output, since the gate is unilateral. Similar one-sweep propagation of signals through MOS transistors has been exploited in MOTIS. It is satisfactory for transistors operating in gate-driven configuration. For pass-transistors the decoupling of drain and source (MOTIS-C) is rather artificial and rough. A one-sweep approach can be identified as a single relaxation of a nonlinear relaxation algorithm. In the first timing simulator MOTIS it was a *regula falsi* relaxation, but in MOTIS-C, SPLICE1 and many further simulators better convergent GSN iterations were used. In such rough analysis, complicated differentiation formulae are pointless. The simple and A-stable first-order implicit Euler formula has become the most popular in timing analysis. In early simulators only one relaxation was used in an attempt to match the speed of gate-level logic simulators; as a result they suffered many limitations, which we will discuss later on.

Timing analysis methods are adequate for circuits described in terms of nodal equations, where at the *i*th node we can distinguish a current charging capacitors connected to this node $\mathbf{C}^i(\mathbf{v})\,\dot{\mathbf{v}}$ and a current flowing out of the node through resistive elements connected to it $f^i(\mathbf{v}, t)$.For the whole circuit we have the following NE

$$\mathbf{C}(\mathbf{v})\dot{\mathbf{v}} + \mathbf{f}(\mathbf{v}, t) = \mathbf{0}, \quad \mathbf{v}(0) = \mathbf{v}_0 \tag{9.22}$$

where **f** stands for a resistive part of the circuit, while **C** is the capacity matrix representing those capacitors visible from circuit nodes. If a circuit incorporates only nonzero grounded capacitors connected to all nodes, then the capacity matrix is diagonal nonsingular. Such a case is the only one acceptable to basic timing analysis methods. Usually, circuits also contain floating capacitors, and then the capacity matrix is nondiagonal and this causes many serious accuracy and stability problems.

Let the circuit have only capacitors at each node, so that the capacity matrix is diagonal. The *i*th equation can be written in the form $C^i(v^i)\,\dot{v}^i + f^i(v^1, ..., v^p, t) = 0$, where $i = 1, ..., p$. Writing this equation at instant t_n and applying the implicit Euler DF we obtain $C^i(v_n^i)\,(v_n^i - v_{n-1}^i) + hf^i(v_n^1, ..., v_n^p, t_n) = 0$, $i = 1, ..., p$. Now this system is solved in time using at each instant only one relaxation of the GS Algorithm 9.2, namely

for $n = 1$ **to** *nmax* **do** {time-loop}
begin
  prediction of $v_n^{1(0)}, ..., v_n^{p(0)}$; {starting point for relaxations}
  for k=1 **to** 1 **do**
  for $i = 1$ **to** p **do** {decoupled nodes loop}
    solve
    $C^i(v_n^{i(k)})\,(v_n^{i(k)} - v_{n-1}^i) + hf^i(v_n^{1(k)}, ..., v_n^{i-1(k)}, v_n^{i(k)}, v_n^{i+1(k-1)}, ..., v_n^{p(k-1)}, t_n) = 0$
    for $v_n^{i(k)}$;
end

Given the zero-order prediction $v_n^{i(0)} = v_{n-1}^{i(1)}$ where the starting value for the relaxation stems from the preceding time point t_{n-1} we can rewrite the procedure, omitting the relaxations superscripts

for $n=1$ **to** *nmax* **do** {time-loop}
  for $i=1$ **to** p **do** {decoupled nodes loop}
    solve
    $C^i(v_n^i)(v_n^i - v_{n-1}^i) + hf^i(v_n^1, \ldots, v_n^{i-1}, v_n^i, v_{n-1}^{i+1}, \ldots, v_{n-1}^p, t_n) = 0$
    for v_n^i;

and solve the inner equation by one NR iteration. After linearization with respect to v_n^i at the point v_{n-1}^i, known from the preceding step, we obtain Algorithm 9.3 with a very simple modification step.

Algorithm 9.3 Forward Gauss-Seidel-Newton timing analysis

for $n=1$ **to** *nmax* **do** {time-loop}
  for $i=1$ **to** p **do** {decoupled nodes loop}

$$v_n^i = v_{n-1}^i - \frac{f^i(v_n^1, \ldots, v_n^{i-1}, v_{n-1}^i, v_{n-1}^{i+1}, \ldots, v_{n-1}^p, t_n)}{\frac{1}{h}C^i(v_{n-1}^i) + \frac{\partial f^i(v_n^1, \ldots, v_n^{i-1}, v_{n-1}^i, v_{n-1}^{i+1}, \ldots, v_{n-1}^p, t_n)}{\partial v^i}};$$

To obtain satisfactory accuracy for each node voltage by means of only one modification in Algorithm 9.3, strong dominance of the diagonal term of the Jacobi matrix is required, since this term is a divider in the denominator of the modification formula. Derivatives of the resistive part **f** may not provide this diagonal dominance, but an additional diagonal capacitive term can easily enforce diagonal dominance by choice of an appropriately small time step h. However, this strong diagonal dominance is possible only by virtue of the fact that all nodes have their grounded capacitors and we do not accept any floating capacitors at all.

In Algorithm 9.3 the inner loop repeated at each time point can be viewed as a single integration step of the circuit equations. Although it is based upon the implicit Euler DF, it is not the exact Euler integration, since the circuit solution is not completed at each instant. This method approximates a solution by means of a single relaxation and one NR iteration. Such an approach has to be thought of as a specific integration method. Its numerical properties are quite different from the implicit Euler DF. For instance, though the Euler DF is A-stable (stable for any h), the one-sweep GSN can easily give oscillating responses for steps which are too large [9.13]. As we know from Chapter 5, numerical integration algorithms generate several numerical components, where only one is correct while the others are spurious and should be strongly damped. Timing analysis easily generates spurious weakly-damped oscillations for excessively large steps.

The numerical properties of relaxation-based integration methods have been explored by G. de Micheli, A.R. Newton and A. Sangiovanni-Vincentelli [9.11], [9.12]. They use a slightly generalized (compared with (5.98)) linear test circuit $\dot{\mathbf{v}} = \mathbf{Y}\mathbf{v}$, $\mathbf{v}(0) = \mathbf{v}_0$, whose matrix $\mathbf{Y}$ has eigenvalues in the open left-half complex plane. We will demonstrate some numerical properties exhibited by the one-sweep GSN method for this test circuit.

In this test, Newton linearization is not required due to its linearity. If we partition $\mathbf{Y}$ into a lower-diagonal component $\mathbf{L}$ and upper component $\mathbf{U}$, then one step of the one-sweep GSN method can be expressed as

$$\mathbf{v}_n = \frac{\mathbf{1} + h\mathbf{U}}{\mathbf{1} - h\mathbf{L}} \mathbf{v}_{n-1} \equiv \mathbf{M}\, \mathbf{v}_{n-1} \tag{9.23}$$

where $\mathbf{M}$ is the companion or characteristic matrix.

The integration method is known as consistent if $\mathbf{M}(h)$ can be expanded in power series $\mathbf{1} + h\mathbf{Y}$ + the 2nd order residuum, and hence if $h \to 0$, then $\mathbf{v}_n \to \mathbf{v}_{n-1}$, $\dot{\mathbf{v}}_n \to \mathbf{Y}\mathbf{v}_{n-1}$.

The integration method is known as stable if there exists $\delta > 0$ and $r > 0$ such that for all starting points $\mathbf{v}_0$ there exists such a $\bar{n} > 0$ that $\|\mathbf{v}_n\| < r$ for all $n \geq \bar{n}$, and $h \in [0, \delta]$.

It can be proved [9.11], [9.12], that the GSN method is consistent and stable for the test circuit having only grounded capacitors at each node. As is seen, stability provides a limited response for a limited step-size. Integration methods stable for any steps are called $\tilde{A}$-stable. The GSN method is not $\tilde{A}$-stable. If we admit floating capacitors, then the GSN method is in general neither consistent nor stable.

If we assume that the preceding solution is accurate $\mathbf{v}_{n-1} = \mathbf{v}(t_{n-1})$ then the norm of difference $\|\mathbf{v}_n - \mathbf{v}(t_n)\|$ is called the local truncation error. If it can be estimated by h^{l+1} then l is the an order of integration method. It can be shown that the GSN method is of the first order, as is the implicit Euler DF.

In timing analysis some additional improvement in stability, which was defined for linear circuits, stems from the saturating shape of MOS transistor characteristics.

To improve the stability and accuracy of timing analysis, a symmetric-displacement timing method has been introduced [9.12]. It is also slightly more resistant to floating elements. The step of symmetric-displacement integration between instants t_{n-1} and t_n consists of a forward GSN iteration over the halved time-step from t_{n-1} to $t_{n-1/2}$, and then of a similar backward GSN iteration from $t_{n-1/2}$ to t_n. The GSN procedure introduced in Algorithm 9.3 was forward. The backward one is based on handling equations and variables in the opposite direction. Superposition of these two procedures improves accuracy considerably and provides the 2nd order and even $\tilde{A}$-stability for circuits with only grounded capacitors.

In the first half-step the derivative at point $t_{n-1/2}=t_{n-1}+h/2$, takes the form $\dot{v}_{n-1/2}=2(v_{n-1/2}-v_{n-1})/h$. Then, we linearize the ith equation at the point v^i_{n-1}, and solve it for $v^i_{n+1/2}$, with $i=1,...,p$. In the second half-step a derivative at point t_n takes the form $\dot{v}_n=2(v_n-v_{n-1/2})/h$, and we can linearize the ith equation at the point $v^i_{n-1/2}$ and solve for v^i_n, with $i=p, ..., 1..$

Algorithm 9.4 Symmetric GSN timing analysis

for $n=1$ **to** *nmax* **do** {time-loop}
begin
  for $i=1$ **to** p **do** {forward loop}

$$v^i_{n-1/2}=v^i_{n-1}-\frac{f^i(v^1_{n-1/2},...,v^{i-1}_{n-1/2},v^i_{n-1},v^{i+1}_{n-1},...,v^p_{n-1},t_{n-1/2})}{\frac{2}{h}C^i(v^i_{n-1})+\frac{\partial f^i(v^1_{n-1/2},...,v^{i-1}_{n-1/2},v^i_{n-1},v^{i+1}_{n-1},...,v^p_{n-1},t_{n-1/2})}{\partial v^i}};$$

  for $i=p$ **to** 1 **do** {backward loop}

$$v^i_n=v^i_{n-1/2}-\frac{f^i(v^1_{n-1/2},...,v^{i-1}_{n-1/2},v^i_{n-1/2},v^{i+1}_n,...,v^p_n,t_n)}{\frac{2}{h}C^i(v^i_{n-1/2})+\frac{\partial f^i(v^1_{n-1/2},...,v^{i-1}_{n-1/2},v^i_{n-1/2},v^{i+1}_n,...,v^p_n,t_n)}{\partial v^i}};$$

end

There are two main weak points in all, even the symmetrical, timing analysis methods.

(i) They fail to analyze circuits with floating capacitors. Considering Algorithms 9.3 and 9.4 we observe that in this case the capacity matrix becomes nondiagonal and in the numerator of the modification formula, except for f^i, we find also a component corresponding to off-diagonal capacitive entries, proportional to $1/h$ and hence increasing as h decreases. This means, that neither consistency nor stability can be proved if relaxation is broken after the first step. Moreover, the step predicted from accuracy criteria is often too large from a stability point of view, especially when floating capacitors appear. As we know from Chapter 5, the usefulness of any integration methods which are not stable for $h\rightarrow\infty$ is doubtful.

(ii) It is very difficult to assign the time step efficiently, since the local truncation error stems from both the LTE of Euler's DF and the one-sweep solution of discretized equations. Both errors are important and the latter difficult to estimate. Methods used were heuristic and based on the rate of voltage change and estimation of the time constants.

All these problems made timing analysis programs unreliable and so of little practical value. So though this seemed to be inefficient, researchers decided to employ more than one relaxation at each time point.

9.3.3 Iterated timing analysis

The main source of stability and accuracy problems, especially in the floating-capacitors case, is the fact that we give up continuing relaxations until convergence. Hence, J. Kleckner, R. Saleh and A.R. Newton [9.17], [9.18] decided to carry out calculations according to Algorithm 9.2 with relaxations continued until convergence. This method is called Iterated Timing Analysis (ITA). If the matrix is strongly diagonally dominant then the number of relaxations in ITA will be small, otherwise it increases and sometimes ITA, e.g. implemented in SPLICE2, operate even more slowly than SPICE.

Let the circuit be described by means of NE (9.22). However we no longer assume diagonality of the capacitance matrix **C**. Now this matrix has all nonzero diagonal entries, and possibly some off-diagonal ones. It is symmetric and positive-definite. In circuit terms, all nodes have grounded capacitors and any floating capacitors are accepted. Each capacitor is positive, nonlinear dependent on its own voltage only. Thus **C** is strictly diagonally dominant, because the diagonal entry is equal to the sum of absolute values of off-diagonal entries plus the value of grounded capacitance. Letting the equations be written in the form

$$\dot{\mathbf{q}}(\mathbf{v})+\mathbf{f}(\mathbf{v})=\mathbf{0} \tag{9.24}$$

and discretized using a sufficiently stable DF $\dot{\mathbf{v}}_n=\gamma_0(\mathbf{v}_n-\mathbf{d}_v)$, as is known from Chapter 5 (e.g. trapezoidal or BDF), the ith row of a discretized equation can be written in the form $\gamma_0[q_n^i(\mathbf{v}_n)-d_{qi}]+f^i(\mathbf{v}_n)=0$. Now we substitute this equation into Algorithm 9.2 obtaining

for $n = 1$ **to** *nmax* **do** {time-loop}
begin
prediction of $v_n^{1(0)}, \dots, v_n^{p(0)}$; {starting point for relaxations}
for $k = 1$ **to** *kmax* **do** {relaxations loop}
begin
for $i = 1$ **to** p **do** {decoupled nodes loop}
solve

$$\gamma_0 [q_n^i(v_n^{1(k)}, v_n^{2(k)}, \dots, v_n^{i-1(k)}, v_n^{i(k)}, v_n^{i+1(k-1)}, \dots, v_n^{p(k-1)}) - d_q^i] + \\ + f^i(v_n^{1(k)}, v_n^{2(k)}, \dots, v_n^{i-1(k)}, v_n^{i(k)}, v_n^{i+1(k-1)}, \dots, v_n^{p(k-1)}) = 0$$

for $v_n^{i(k)}$;
if $(0\| \mathbf{v}_n^{(k)} - \mathbf{v}_n^{(k-1)} \| \le \varepsilon \max\{\|\mathbf{v}_n^{(k)}\|, \|\mathbf{v}_n^{(k-1)}\|\} + \delta)$ **then break**
end
end

To solve the inner scalar equation we use a single NR iteration, hence linearize with respect to $v_n^{i(k)}$ in the vicinity of $v_n^{i(k-1)}$ (zero order prediction). A higher order prediction is also available. Finally the GSN ITA takes the form given in Algorithm 9.5.

Algorithm 9.5 The GSN ITA method

for $n = 1$ **to** *nmax* **do** {time-loop}
begin
prediction of $x_n^{1(0)}, \dots, x_n^{p(0)}$; {starting point for relaxations}
for $k = 1$ **to** *kmax* **do** {relaxations loop}
begin
for $i = 1$ **to** p **do** {decoupled nodes loop}
begin

$$\tilde{\mathbf{v}}_n^{(k-1)} = [v_n^{1(k)}, v_n^{2(k)}, \dots, v_n^{i-1(k)}, v_n^{i(k-1)}, v_n^{i+1(k-1)}, \dots, v_n^{p(k-1)}];$$

$$v_n^{i(k)} = v_n^{i(k-1)} - \frac{\gamma_0 [q_n^i(\tilde{\mathbf{v}}_n^{(k-1)}) - d_q^i] + f^i(\tilde{\mathbf{v}}_n^{(k-1)})}{\gamma_0 C^i(\tilde{\mathbf{v}}_n^{(k-1)}) + \dfrac{\partial f^i(\tilde{\mathbf{v}}_n^{(k-1)})}{\partial v^i}};$$

end
if $(\| \mathbf{v}_n^{(k)} - \mathbf{v}_n^{(k-1)} \| \le \varepsilon \max\{\|\mathbf{v}_n^{(k)}\|, \|\mathbf{v}_n^{(k-1)}\|\} + \delta)$ **then break**
end
end

Due to relaxations being carried out until convergence and a stable DF applied, the ITA is consistent, stable, and accurate. Its order stems from the order of DF. It can be proved that if the capacity matrix obeys the assumptions already introduced (where the most important was a nonzero grounded capacitor to each node), then the ITA is convergent for sufficiently small step-sizes.

To be more clear let us demonstrate the GSN iterations for the circuit shown in Figure 9.2. Its ODEs

$$
\begin{aligned}
&(C_3+C_{12})\dot{v}^1 - C_{12}\dot{v}^2 + G_2 + I_s\{\exp[(v^1-v^2)/V_T]-1\}/-I_1 = 0 \\
&-C_{12}\dot{v}^1 + (C_6+C_{12})\dot{v}^2 + G_5 - I_s\{\exp[(v^1-v^2)/V_T]-1\} = 0
\end{aligned}
\tag{9.25}
$$

after discretization $\dot{v}_n^i = \gamma_0(v_n^i - d_v^i)$ give Equation (9.26)

$$
\begin{aligned}
&\gamma_0(C_3+C_{12})(v_n^1-d_{vn}^1) - \gamma_0 C_{12}(v_n^2-d_{vn}^2) + G_2 + I_s\{\exp[(v_n^1-v_n^2)/V_T]-1\}/-I_1 = 0 \\
&-\gamma_0 C_{12}(v_n^1-d_{vn}^1) + \gamma_0(C_6+C_{12})(v_n^2-d_{vn}^2) + G_5 - I_s\{\exp[(v_n^1-v_n^2)/V_T]-1\} = 0 .
\end{aligned}
$$

Hence the GSN ITA relaxation process will consist of calculating (9.27)

$$
v_n^{1(k)} = v_n^{1(k-1)} - \frac{\gamma_0(C_{12}+C_3)(v_n^{1(k-1)}-d_v^1) + C_{12}\gamma_0(v_n^{2(k-1)}-d_v^2) + G_2 + I_s[\exp\dfrac{v_n^{1(k-1)}-v_n^{2(k-1)}}{V_T}-1] - I_1(t_n)}{\gamma_0(C_{12}+C_3) + \dfrac{I_s}{V_T}\exp\dfrac{v_n^{1(k-1)}-v_n^{2(k-1)}}{V_T}}
$$

$$
v_n^{2(k)} = v_n^{2(k-1)} + - \frac{\gamma_0(C_{12}+C_6)(v_n^{2(k-1)}-d_{vn}^2) - C_{12}\gamma_0(v_n^{1(k)}-d_{vn}^1) + G_2 - I_s[\exp\dfrac{v_n^{1(k)}-v_n^{2(k-1)}}{V_T}-1] - I_1(t_n)}{\gamma_0(C_{12}+C_6) + \dfrac{I_s}{V_T}\exp\dfrac{v_n^{1(k)}-v_n^{2(k-1)}}{V_T}}
$$

repeated for $k = 1, 2, \ldots$ until the test for convergence is fulfilled.

To increase ITA efficiency the test for convergence has to be checked for each node by separately comparing $v_n^{i(k)}$, $v_n^{i(k-1)}$. If it is fulfilled at the ith node, then modification of this node can be bypassed. Relaxations are continued until convergence is not reached at least for one node.

Efficient implementation of ITA and other relaxation methods requires node-driven circuit representation lists and relaxation-orientated stamps.

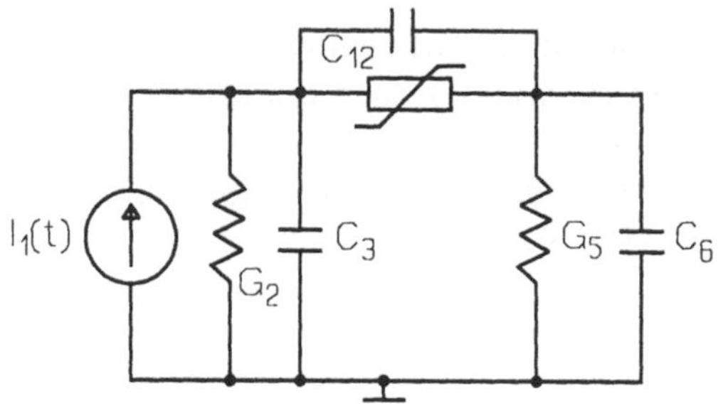

Figure 9.2 A simple rectifier - instructive example of GSN iterations

After a circuit net-list is read in, its representation is stored in lists of structures. In direct simulation they are ordered according to types of elements. In relaxation methods we require lists of elements connected to each node. There are two types of elements connected to each node, the nonlinear capacitor shown in Figure 9.3(a)

$$i_a = \dot{q}(v_a) \tag{9.28}$$

and the nonlinear CD voltage-controlled branch of Figure 9.3(b)

$$i_a = \varphi_a(v_a, v_b, t) \tag{9.29}$$

where i_a flows from the node m_a to n_a. The modification formula of relaxation methods takes the form

$$v_n^{i(k)} = v_n^{i(k-1)} - \frac{F}{Y} \tag{9.30}$$

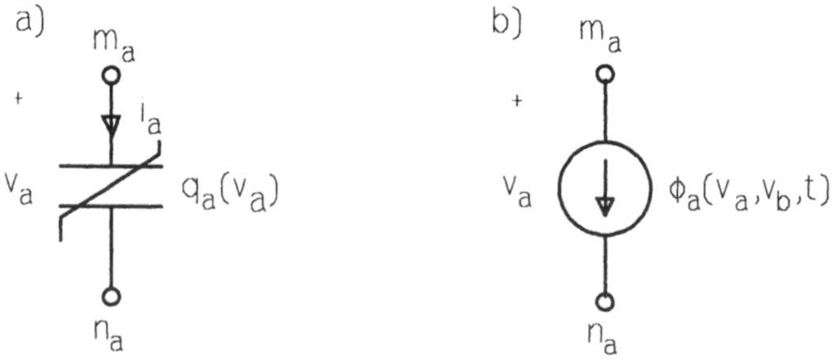

Figure 9.3 (a) Capacitive and (b) resistive branches accepted by relaxation analyses

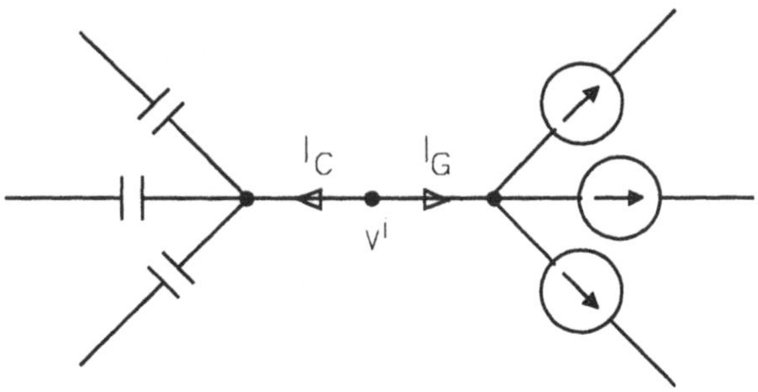

Figure 9.4 Capacitive and conductive components of KCL in a node

where $F = I_C + I_G$ is the sum of resistive and capacitive components of nodal currents in the KCL, as is seen in Figure 9.4. Similarly $Y = (\partial I_C / \partial v^i) + (\partial I_R / \partial v^i)$. Both F and Y stem from the sum of elementary stamps $F(j)$, $Y(j)$ of branches $j = 1, 2, \ldots$ of type (9.28), (9.29) connected to the ith node. These stamps take the form

branch type	$F(j)$	$Y(j)$
capacitor	$s_i \gamma_0 [q_a(v_a) - d_{qa}]$	$\gamma_0 C_a(v_a)$
CD resistive branch	$s_i \varphi_a(v_a, v_b, t_n)$	$\partial \varphi_a(v_a, v_b, t_n) / \partial v_a$

where $s_i = 1$ for $i = m_a$, and $s_i = -1$ for $i = n_a$. In these stamps voltages are calculated from nodal voltages $v_a = v_{ma} - v_{na}$, $v_b = v_{mb} - v_{nb}$, where nodal voltages are from the vector $\tilde{\mathbf{v}}_n^{(k-1)} = [v_n^{1(k)}, \ldots, v_n^{i-1(k)}, v_n^{i(k-1)}, \ldots, v_n^{p(k-1)}]^t$ for forward GSN and $\tilde{\mathbf{v}}_n^{(k-1)} = [v_n^{1(k-1)}, \ldots, v_n^{i(k-1)}, v_n^{i+1(k)}, \ldots, v_n^{p(k)}]^t$ for backward GSN.

9.3.4 One-step relaxation timing analysis (OSR)

A method slightly improved in comparison with ITA was introduced in 1985 by B. Hennion and P.Senn. It is known as One-Step Relaxation (OSR) timing analysis and is implemented in the simulator ELDO [9.8], [9.9]. It differs from ITA in that solution of the decoupled equation in Algorithm 9.2 is not stopped after one NR iteration but continued until convergence. This increases computational effort but provides convergence even in the case of strongly coupled nodes. To reduce the computational overhead some auxiliary tricks have been adopted: (i) efficient prediction for each node of the starting value $v_n^{i(0)}$ for relaxations, (ii) elimination of derivatives by secant solution of the decoupled equation, (iii) mixture with direct methods by means of block decoupling. The OSR analysis can be outlined in Algorithm 9.6.

Regarding item (i), the prediction of $v_n^{i(0)}$ can be carried out assuming weak coupling of the *i*th node with neighboring nodes. Given a capacitor connected to this node and neglecting floating capacitors we can write the KCL equation $C^i(v^i)\dot{v}^i = -f^i(\mathbf{v}, t)$ and after discretization using a DF obtain the predictor

$$v_n^{i(0)} = d_v^i - \frac{f^i(\mathbf{v}_n)}{\gamma_0 C^i(v_n^i)} \tag{9.31}$$

If a coupling is weak, then the prediction is accurate and only one relaxation is needed to check it. If coupling is strong a few relaxations are required.

Algorithm 9.6 The GS OSR analysis

```
for n = 1 to nmax do {time-loop}
begin
   prediction of v_n^{1(0)}, ..., v_n^{p(0)}; {starting point for relaxations}
   for k = 1 to kmax do {relaxations loop}
   begin
      for i = 1 to p do {decoupled nodes loop}
      begin
         ṽ_n^{i(k-1)} = [v_n^{1(k)}, v_n^{2(k)}, ..., v_n^{i-1(k)}, v_n^{i(k)}, v_n^{i+1(k-1)}, ..., v_n^{p(k-1)}];
         solve γ_0 [q_n^i(ṽ_n^{(k-1)}) - d_q^i] + f^i(ṽ_n^{(k-1)}) = 0 for v_n^{i(k)};
      end
      if (‖v_n^{p(k)} - v_n^{p(k-1)}‖ ≤ ε max{‖v_n^{p(k)}‖, ‖v_n^{p(k-1)}‖} + δ) break
   end
end
```

As for item (ii), we can solve the decoupled equation using any iterative method continued until convergence. One possibility is to use NR iterations according to (9.30), as in Algorithm 9.5. Then to formulate the decoupled iterative equations, the stamps given in Section 9.3.2.3 can be used. However, to save time in calculating derivatives and simplify the models an alternative secant (*regula falsi*) iteration is also available. Letting the *i*th decoupled equation (9.24) be of the form $\dot{q}^i(v^i) + f^i(v^i) = 0$ and carrying out the secant linearization of charge and current of the resistive part at the $j-1$ secant solution we obtain the iterative equation

$$\gamma_0[q^i(0) + C^i v_{n(j)}^{i(k)} - d_q^i] + f^i(0) + G^i v_{n(j)}^{i(k)} = 0 \qquad (9.32)$$

which can be easily solved for the next $v_{n(j)}^{i(k)}$ secant solution. Substitute secant capacitance and conductance is evaluated from the secant at the $j-1$ solution. They can be created as sums of secant stamps of branches connected to the ith node, similarly to the NR stamps used in Equation (9.30).

In practice NR (Algorithm 9.5) and secant (Equation (9.32)) methods should be mixed. For this purpose we designate several categories of nodes.

(i) Nodes having only one linear element connected require only one NR step of Algorithm 9.5 or of (9.32) equivalently.

(ii) Nodes connected with MOS transistors can be efficiently solved with secant iterations. To improve efficiency of analysis a simplified MOSFET model with a piecewise-linear drain-source conductance G has been proposed in ELDO-XL [9.2]. Below threshold G is zero, above V_{on} it is constant, between these two values it is a linear function of V_{GS}. Then in the constant conductance region of transistor operation a single iteration (9.32) is sufficient. In a linearly variable conductance region only a few secant iterations are necessary. This is possible due to the weak nonlinearity of the MOSFET.

(iii) Nodes having strongly nonlinear elements require NR iterations (9.30) carried out until convergence.

(iv) Nodes strongly coupled require block decoupling and direct solution within the block as described in the next section.

Though the ELDO simulator does not exploit event-driven methods and the time-step is the same for the whole circuit, an event-driven implementation of OSR analysis similar to ITA is fully available.

9.3.5 Nonlinear-relaxation methods on the block level

As we have mentioned in the above consideration dedicated to ITA and OSR analyses, there is a problem with strongly coupled blocks. It is that circuit equations cannot have a LT dependency matrix but have above-diagonal terms, even if there are no "far" feedbacks. Such a structure leads to convergence of the relaxation methods in several iterations, instead of one as in the LT case. However, if we group bidirectionally or backward coupled variables in one block, then the dependency matrix can take the block LT (BLT) form, where the *i*th block is affected only by blocks from 1 to *i*-1. If we treat a block as one node (i.e. solve it using direct methods) and perform relaxations at the block level, then these will be convergent in one relaxation only.

In Figure 9.5(b) we observe a partitioning such that blocks are strongly coupled through the pass-transistor, and there is feedback from node 3 to node 2. This leads to the non-BLT matrix shown. In Figure 9.5(a) partitioning of the same circuit is performed otherwise and this block-decoupling provides a one-way flow of signal through the blocks. The BLT matrix now appears. Solution of the former 9.5(b) requires many relaxations. Just as many would be required in the case of nodal decoupling. However, in the case of block decoupling given in Figure 9.5(a), convergence is reached in one relaxation only. In order to obtain simpler matrices, we have neglected the coupling through capacitors introduced by MOSFET models in a real circuit, as shown in Figure 9.7.

OSR (or ITA) relaxation Algorithm 9.6 can be easily extended to the block level. Assuming that circuit equations (9.24) have been partitioned into *m* blocks similarly to (9.2) we can write the decoupled nodal equation of the *i*th block

$$\dot{\mathbf{q}}^i(\mathbf{v}^1, ..., \mathbf{v}^m) + \mathbf{f}^i(\mathbf{v}^i, ..., \mathbf{v}^m) = \mathbf{0} \quad (i = 1, ..., m) \tag{9.33}$$

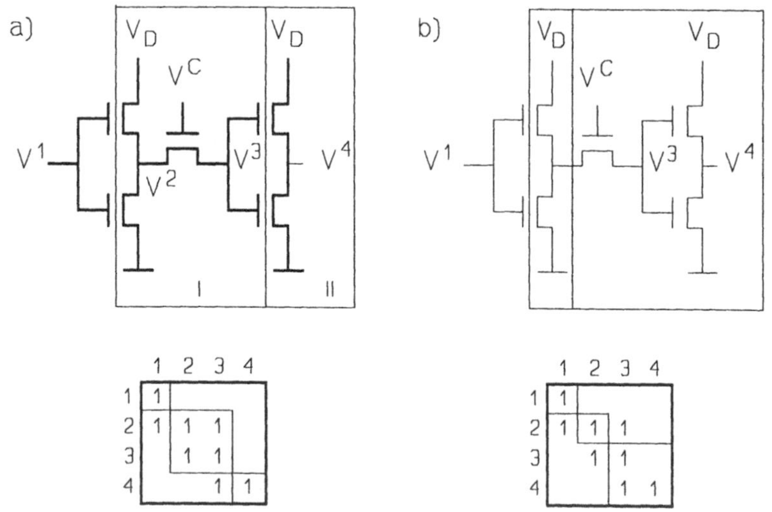

Figure 9.5 Two ways of partitioning into blocks I and II: (a) blocks are strongly coupled, (b) blocks are one-way coupled

where $\mathbf{q}^i$ is the vector of the charges of capacitors connected to nodes of the ith block, while $\mathbf{f}^i$ stands for the currents of resistive/conductive branches connected to nodes of the ith block. Discretization $\dot{\mathbf{v}}_n = \gamma_0(\mathbf{v}_n - \mathbf{d}_v)$ leads to the companion nonlinear equations of the ith block. Assuming the GS decoupling process, where in the kth relaxation $(k=1,2,\ldots)$ the ith block is solved for $\mathbf{v}_n^{i(k)}$ with solutions $\mathbf{v}_n^{1(k)}, \ldots, \mathbf{v}_n^{i-1(k)}$ previously calculated and $\mathbf{v}_n^{i+1(k-1)}, \ldots, \mathbf{v}_n^{m(k-1)}$ known from the preceding relaxation. To calculate variables of the ith block we have to solve a system of n_i equations with unknowns $\mathbf{v}_n^{i(k)}$ using such a direct approach as NR iterations and the LU method of solution of linear equations. We obtain the block GSN Algorithm 9.7 where we do not explicitly describe the direct solution of the block.

Algorithm 9.7 The block GSN analysis

for $n=1$ **to** *nmax* **do** {time-loop}
begin
  prediction of $x_n^{1(0)}, \ldots, x_n^{p(0)}$; {starting point for relaxations}
  for $k=1$ **to** *kmax* **do** {relaxations loop}
  begin
    for $i=1$ **to** p **do** {decoupled blocks loop}
    begin
      $\tilde{\mathbf{v}}_n^{i(k-1)} = [\mathbf{v}_n^{1(k)}, \mathbf{v}_n^{2(k)}, \ldots, \mathbf{v}_n^{i-1(k)}, \mathbf{v}_n^{i(k)}, \mathbf{v}_n^{i+1(k-1)}, \ldots, \mathbf{v}_n^{m(k-1)}]$;
      solve $\gamma_0[\mathbf{q}_n^i(\tilde{\mathbf{v}}_n^{(k-1)}) - \mathbf{d}_q^i] + \mathbf{f}^i(\tilde{\mathbf{v}}_n^{(k-1)}) = \mathbf{0}$ for $\mathbf{v}_n^{i(k)}$;
    end
    if $(\|\mathbf{v}_n^{m(k)} - \mathbf{v}_n^{m(k-1)}\| \le \varepsilon \max\{\|\mathbf{v}_n^{m(k)}\|, \|\mathbf{v}_n^{m(k-1)}\|\} + \delta)$ **break**
  end
end

Recalling the circuit partition shown in Figure 9.5(a) we notice that in the kth relaxation a solution of the first block (I) can be viewed as a separate analysis of its companion subcircuit, with the states at nodes connected with surrounding subcircuits fixed at values drawn from previous solutions. As we observe in Figure 9.6 which represent the block I of the circuit in Figure 9.5(a), with the first and the 4th node voltages substituted for voltage sources, a signal of the earlier numbered block 1 is taken from the kth relaxation, while the later one 4 is taken from the $(k-1)$th relaxation. We have a full analogy with the nodal decoupling shown in Figure 9.4, where a solution for one nodal voltage is calculated with surrounding nodes fixed.

Of course the block GSN Algorithm 9.7 is convergent if the circuit satisfies sufficient conditions for ITA convergence, i.e. sufficiently small

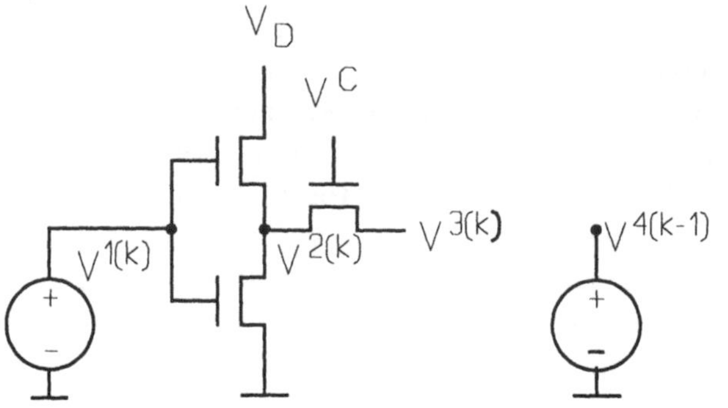

Figure 9.6 Circuit interpretation of one block-relaxation

step-size and capacitors connected to each node. However these conditions are too exacting. Much weaker sufficient conditions are also possible. To introduce them we realize that for a given discretization of the circuit equations the GSN algorithm is convergent if the eigenvalues of the discretized matrix of (9.1) are inside the unit circle. A similar condition holds also for the Jacobi matrix of the block partitioned system (9.2) after its discretization. This takes a general form $\mathbf{F}(\mathbf{v}) = \mathbf{0}$ or a partitioned form $\mathbf{F}^i(\mathbf{v}^1, ..., \mathbf{v}^m) = \mathbf{0}$, $i=1, ..., m$. If we denote

$$\mathbf{H}_L(\mathbf{v}) = \begin{bmatrix} \mathbf{F}'_{11} & \mathbf{0} & \dots & \mathbf{0} \\ \mathbf{F}'_{21} & \mathbf{F}'_{22} & \dots & \mathbf{0} \\ \dots & \dots & \dots & \dots \\ \mathbf{F}'_{m1} & \mathbf{F}'_{m2} & \dots & \mathbf{F}'_{mm} \end{bmatrix} \tag{9.34}$$

$$\mathbf{H}_U(\mathbf{v}) = \begin{bmatrix} \mathbf{0} & \mathbf{F}'_{12} & \dots & \mathbf{F}'_{1m} \\ \mathbf{0} & \mathbf{0} & \dots & \mathbf{F}'_{2m} \\ \dots & \dots & \dots & \dots \\ \mathbf{0} & \mathbf{0} & \mathbf{0} & \mathbf{0} \end{bmatrix} \tag{9.35}$$

where $\mathbf{F}'_{ij} = \partial \mathbf{F}^i(\mathbf{v}) / \partial \mathbf{v}_j$, then the following GSN convergence theorem holds [9.14]

If

(i) *the function* $\mathbf{F}$ *of a domain* $D \subset R^p$ *and values in* R^p *has a solution* $\mathbf{v}^* \in D$ *such that* $\mathbf{F}(\mathbf{v}^*) = \mathbf{0}$, *and is continuously differentiable on an open neighborhood* $S_0 \subset D$ *of* $\mathbf{v}^*$,

(ii) *this function has a partition* (9.2) *and corresponding partition of the Jacobi matrix,*

(iii) $\mathbf{F}'_{ii}$ $(i = 1, ..., m)$ *are nonsingular,*

(iv) *the spectral radius of* $\mathbf{H}_L^{-1}(\mathbf{v}^*)\mathbf{H}_U(\mathbf{v}^*)$ *is smaller than* 1

then

Block-GSN iterations: solve for $\mathbf{v}^{i(k)}$ *the equation*

$$\mathbf{F}^i(\mathbf{v}^{1(k)}, ..., \mathbf{v}^{i(k)}, \mathbf{v}^{i+1(k-1)}, ..., \mathbf{v}^{m(k-1)}) = \mathbf{0} \quad (k = 1, 2, ...), \ (i = 1, ..., m)$$

are well defined and convergent to $\mathbf{v}^*$ *from any point of an open ball* $S \subset S_0$.

This theorem is not convenient for relaxation circuit analysis since it has no direct circuit interpretation. Sufficient conditions of convergence for block-relaxation methods can also be formulated using the concept of block diagonal dominance. However, the most useful conditions in circuit simulation are the very realistic sufficient conditions of M.P. Desai and I.N. Hajj [9.5]. They have the following topological form.

If

(i) *the circuit equation* (9.24) *are decoupled as stated by* (9.2) *and have solution* $\mathbf{v}^*$,

(ii) $\mathbf{F}'_{ii}$ $(i = 1, ..., m)$ *introduced in* (9.34), (9.35) *are nonsingular and have the inverse of bounded norms,*

(iii) $\partial \mathbf{f}(\mathbf{v}^*)/\partial \mathbf{v}$ *is independent of the time step and has bounded entries,*

(iv) *there is a path consisting entirely of capacitors from each feedback node to the reference node,*

then

the GSN converges to the solution $\mathbf{v}^*$ *for a small enough time step and an initial guess close enough to this solution.*

In this theorem the concept of a feedback node is used. A node of the ith block is called the feedback node, if the KCL in this node is affected by some nodal voltage in a block j numbered after the ith block, i.e. $i < j \leq m$. According to the above theorem, all feedback nodes require a path, consisting entirely of capacitors, joining these nodes with the ground.

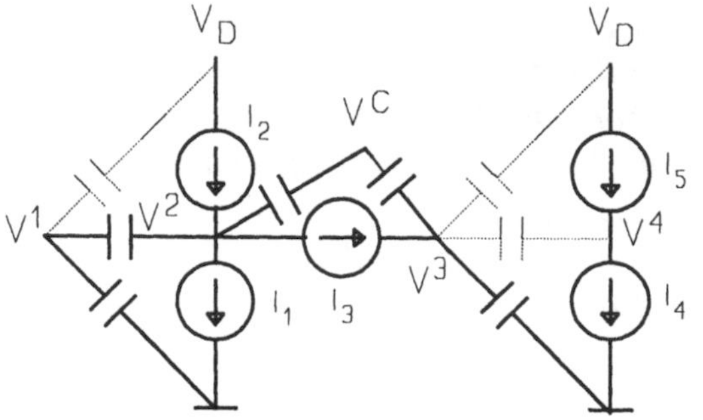

Figure 9.7 Circuit representing the system of Figure 9.6(a) after introduction of MOSFET models

Recalling the partition shown in Figure 9.5(b), we observe that block I is bidirectionally coupled with block II due to the pass transistor. So node 2 is the feedback node. After expansion of the MOSFET models to include a drain current source and capacitors between gate and drain, as well as between gate and source, we obtain the circuit of Figure 9.7. Node 2 even has two grounded paths $\{C_{12}, C_{1gnd}\}$ and $\{C_{2c}, C_{c3}, C_{3gnd}\}$, drawn with continuous lines. This circuit will be convergent for a sufficiently short time-step. Convergence is maintained even if we delete most of these capacitors, such that only one of these paths remains. However, without any capacitors convergence would not occur.

If we perform the better partition shown in Figure 9.5(a) then we have the feedback node 3 due to the floating capacitor C_{34} only. No resistive elements affect the backward signal flow. In this case convergence occurs thanks to the path $\{C_{3gnd}\}$ and probably with slightly longer steps than in the preceding partition where the step must be shortened to overcome the feedback through the pass transistor. Even if we removed all capacitors except for C_{34}, C_{3gnd}, convergence would be preserved. If we also removed C_{34}, then convergence would take one relaxation. However, if we retain C_{34} but remove C_{3gnd}, the relaxation process will not converge due to there being no capacitor path to ground.

If we partition the circuit as shown in Figure 9.5(a), then, given an input $v^1(t)$, formulae for GSN relaxations take the form:

for $k = 1, 2, \dots$
begin
solve a system

$$-\gamma_0 C_{12}[v^1(t_n) - d_v^1 - v_n^{2(k)} + d_v^2] +$$
$$+\gamma_0 C_{2c}[v_n^{2(k)} - d_v^2 - v^c(t_n) + d_v^c] + I_1[v^1(t_n), v_n^{2(k)}] +$$
$$-I_2[v^1(t_n) - v_n^{2(k)}, V_D - v_n^{2(k)}) + I_3[v^c(t_n) - v_n^{3(k-1)}, v_n^{2(k)} - v_n^{3(k-1)}] = 0$$
$$-I_3[v^c(t_n) - v_n^{3(k)}, v_n^{2(k)} - v_n^{3(k)}] + \tag{9.37}$$
$$-\gamma_0 C_{c3}(v_n^{3(k)} - d_v^3 - v^c(t_n) - d_v^c) + \gamma_0 C_{3D}(v_n^{3(k)} - d_v^3) +$$
$$+\gamma_0 C_{34}(v_n^{4(k-1)} - d_v^4 - v_n^{3(k)} + d_v^3) + \gamma_0 C_{3\text{gnd}}(v_n^{3(k)} - d_v^3) = 0$$

for $v_n^{2(k)}$, $v_n^{3(k)}$;

solve

$$I_4(v_n^{3(k)}, v_n^{4(k)}) - I_5(v_n^{3(k)} - v_n^{4(k)}, v_D - v_n^{4(k)}) - \gamma_0 C_{34}(v_n^{3(k)} - d_v^3 - v_n^{4(k)} - d_v^4) = 0 \tag{9.38}$$

for $v_n^{4(k)}$;
end

Due to the occurrence of feedback, the convergence of this algorithm will take several relaxations. It requires a sufficiently small time step that reflects the large values of the coefficient γ_0.

If all blocks of the circuit are strongly bidirectionally coupled, then partitioning leads to nodal decomposition, referred to in Chapter 8, where the dependency matrix will be BBD. All block nodes are now feedback nodes, and require grounded capacitive paths, which leads to block diagonal dominance. In this case a relaxation solution of the circuit can be achieved in a small number of relaxations.

These one-point block-relaxation methods have overcome problems of strong bidirectional couples by taking couples into blocks. However, the larger the blocks, the closer will be the efficiency of block-relaxation methods to that of direct methods. Thus, in practical programs it is essential to take into blocks the minimum number of critical nodes which really cause serious trouble with convergence. Blocks have to be as small as possible.

In concurrent third generation analyzers block relaxations are ubiquitous. We call such an analysis hybrid or mixed-mode, with circuit and timing analyses mixed (in mixed-mode analyzers it is also integrated with a logic simulation). In hybrid analyzers decoupling should be adequate to the circuit specifics. Usually nodal decoupling is applied to all nodes except for those extremely strongly coupled, which are treated as a single block. It is necessary to lock up in single blocks the nodes of bjts, diodes, small resistors, inductors, and voltage sources. Such an approach has been applied in ELDO [9.8]. Regarding the drain and source of a MOSFET, it is not reasonable to join them in a block, since in digital circuits many channels are cut off and their block solution would be an unnecessary overhead considerably reducing efficiency.

9.4 Waveform-relaxation analysis

9.4.1 Basic waveform-relaxation analysis

The principle of waveform relaxation analysis is to perform decoupling directly on the level of circuit OADEs [9.4]. Let the circuit be described by a system of NEs (9.24) consisting of KCL balances at nodes

$$\begin{aligned} &\dot{q}^1(v^1, ..., v^p) + f^1(v^1, ..., v^p) = 0 \\ &\dot{q}^2(v^1, ..., v^p) + f^2(v^1, ..., v^p) = 0 \\ &\cdots\cdots\cdots\cdots\cdots\cdots \\ &\dot{q}^p(v^1, ..., v^p) + f^p(v^1, ..., v^p) = 0 \end{aligned} \tag{9.39}$$

and a set of initial conditions $v^1(0), ..., v^p(0)$. If we know some approximation to solutions $v^{1(k-1)}(t), ..., v^{p(k-1)}(t)$ on the interval $[0, T]$ satisfying these initial conditions, then it can be gradually adjusted solving the first OADE (9.39) for the function $v^{1(k)}(t)$ with the initial condition $v^1(0)$ and the rest of the waveforms known from the $k-1$ approximation. Then we solve the socond equation (9.39) for $v^{2(k)}(t)$ using the previously calculated $v^{1(k)}(t)$ and the rest of the waveforms from $k-1$ relaxation. This is the GS procedure applied directly to OADEs. After the loop over equations is completed we proceed to the next relaxation. Such relaxations can converge to nodal waveforms which are solutions of the circuit. This procedure is called **waveform-relaxation analysis (WRA)**.

Each ith decoupled equation is a scalar OADE with one unknown function of time $v^i(t)$. This equation can be solved using common discretization methods discussed in Chapter 5. If all nodal equations are solved at the same set of instants, then we obtain **discretized WRA**, which takes the form of Algorithm 9.8. It is based on the nodal circuit description (9.39).

Algorithm 9.8 The GS WR analysis

> assume initial conditions $v^1(0), ..., v^p(0)$;
> **for** $k=1$ **to** *kmax* **do** {relaxations loop}
> **begin**
>   **for** $i=1$ **to** p **do** {decoupled nodes loop}
>   **for** $n=1$ **to** *nmax* **do** {time-loop}
>   **begin**
>   predict a starting value $\mathbf{v}_{n(0)}^{i(k)}$ for NR iterations;
>   let $\tilde{\mathbf{v}}_n^{i(k-1)} = [v_n^{1(k)}, ..., v_n^{i-1(k)}, v_n^{i(k)}, v_n^{i+1(k-1)}, ..., v_n^{p(k-1)}]$;
>   solve $\gamma_0 [q_n^i(\tilde{\mathbf{v}}_n^{i(k-1)}) - d_q^i] + f^i(\tilde{\mathbf{v}}_n^{i(k-1)}) = 0$ for $v_n^{i(k)}$ using linearization and NR iterations;
>   **end**
>   **end**
>   **if** (for $n=1, ..., nmax$ inequality
> $\|\mathbf{v}_n^{p(k)} - \mathbf{v}_n^{p(k-1)}\| \le \varepsilon \max\{\|\mathbf{v}_n^{p(k)}\|, \|\mathbf{v}_n^{p(k-1)}\|\} + \delta$ holds) **break**
> **end**

The discretized WRA differs in the order of loops over time, relaxations, and equations in comparison with OSR (Algorithm 9.6). In WRA we first perform relaxations; at each relaxation we iterate over nodes, and finally for each node we solve an OADE using discretization and NR iterations. Hence the numerical properties and efficiency of WRA are very similar to OSR. Recalling the Figure 9.5 we can interpret a node solution in one relaxation of WRA as the calculation over [0, T] of a voltage $v^{i(k)}(t)$, with neighboring nodal waveforms $v^{1(k)}(t), ..., v^{i-1(k)}(t)$, $v^{i+1(k-1)}(t), v^{i+2(k-1)}(t), ...$ treated as time-domain voltage supplies. The scalar OADE is solved using time discretization. Given uniform time discretization at all nodes E. Lelarasmee, A. Ruehli and A. Sangiovanni-Vincentelli [9.10] have proved sufficient conditions for discretized WRA.

If

(i) *the charge characteristics of capacitors* $\mathbf{Q}(\mathbf{v})$ *are Lipschitz continuous with respect to* $\mathbf{v}$,

(ii) *the controlled currents* $\mathbf{f}(\mathbf{v},t)$ *of conductors and drains of MOSFETS are Lipschitz continuous and uniformly bounded throughout the relaxation process,*

(iii) *all nodes have grounded capacitors which for any permissible value of voltage are positive,*

(iv) *all floating capacitors are finite for any permissible voltage across them,*

then

for any given set of initial conditions $v^i(0)$ $(i=1, ..., p)$ *and piecewise continuous stimuli, WRA is convergent to a circuit solution satisfying given initial conditions.*

Differences between fixed-step OSR and fixed-step discretized WRA are small, and efficiency, after using all the speeding-up tricks introduced in Section 9.3.3.2, is similar. Only the requirement for capacitors at all nodes, due to nodal decoupling, is an inconvenient limitation. However, this problem can be avoided by the introduction of block-decoupling and block WRA, easily extendable from Algorithm 9.8 by introducing block partitioning, replacing a loop over nodes by a loop over blocks and solving at the most inner loop a system of OADEs of a block instead of a single equation. Hence, in the *i*th stage of a relaxation, as in Figure 9.5(b), we have to solve a block for a vector of waveforms $v^{2(k)}(t)$, $v^{3(k)}(t)$ with neighboring nodal waveforms $v^{1(k)}$, $v^{4(k-1)}$ known. M.P. Desai and I.N. Hajj [9.5] have proved sufficient conditions for convergence of the discretized block WRA for a sufficiently small time-step.

If

(i) *the capacitance matrix* $d\mathbf{Q}/d\mathbf{v}$ *of* (9.24) *is Lipschitz continuous with respect to* $\mathbf{v}$,

(ii) $\mathbf{f}$ *is differentiable,*

(iii) *there is a capacitive path between every feedback node and ground,*

then

for any set of initial conditions $v^i(0)$ $(i=1, ..., p)$ *the block WRA is convergent to a circuit solution satisfying given initial conditions.*

However, a uniform time-step is not acceptable, while the most efficient approach is a variable step technique, with separate step values for each node or block. Such time-domain analysis we call multirate. In direct analysis a multirate approach is not available. It becomes available as a consequence of decoupling nodes or blocks. General multirate algorithms consist of repeating in the relaxation and equations (or blocks) loop a variable step solution of OADEs with given voltage waveforms from the neighboring nodes. In a variable step solution of the *i*th OADE for $v^{i(k)}(t)$ we use synchronous samples of the stored $v^{1(k)}(t)$, ..., $v^{i-1(k)}(t)$,

$v^{i+1(k-1)}(t), ..., v^{p(k-1)}(t)$, which are discretized in quite another way; we usually apply quadratic interpolation. Sufficient conditions for convergence of the multirate algorithm are similar to those introduced above. The great advantages of multirate block WRA are mild assumptions for convergence and a rather quick convergence. In strictly one-way block circuits WRA is convergent in one relaxation, in common digital MOS circuits with weak backward coupling between blocks we need a few relaxations. For strongly coupled blocks the number of relaxations exceeds ten.

9.4.2 Speed-up techniques for waveform-relaxation analysis

Techniques for improving the efficiency of WRA are basically the same as those associated with timing-analysis and direct methods: **bypassing** (omitting recalculation over relaxations) and **latency** (omitting recalculations over time). Specific to WRA is a decomposition in time known as **windowing**. Moreover, for efficient solution it is essential to have a proper **starting waveform**, while also critical are the manner of our circuit **partitioning**, **ordering**, i.e. numbering of variables (or their groups), and the order of submitting nodes (blocks) to analysis, known as **scheduling**.

Bypassing in WRA consists of exploiting the partial convergence of some of the nodal waveforms during relaxations. Recalling the circuit in Figure 9.5, we notice for instance that the KCL equation at node 2 is affected by voltages v^1, v^2 and v^c. If in some time range (t_a, t_b) differences $|v^{1(k-1)}(t)-v^{1(k-2)}(t)|$, $|v^{2(k-1)}-v^{2(k-2)}|$, $|v^{c(k-1)}-v^{c(k-2)}|$ from the two preceding relaxations are sufficiently small, then the voltage $v^{2(k)}(t)$ in this range will certainly be close to $v^{2(k-1)}(t)$ and we can bypass analysis of this node within this range in the kth relaxation.

In general, given the ith block

$$\dot{\mathbf{q}}^i(\mathbf{v}) + \mathbf{f}^i(\mathbf{v}) = 0 \tag{9.40}$$

if all its fanins $v^1, ..., v^i$ after relaxations $k-2, k-1$, and in range $[t_a, t_b]$ satisfy bypassing conditions

$$\|\mathbf{v}^{j(k-1)} - \mathbf{v}^{j(k-2)}\| \le \varepsilon \max\{\|\mathbf{v}^{j(k-2)}\|, \|\mathbf{v}^{j(k-1)}\|\} + \delta \quad (j=1, ..., i) \tag{9.41}$$

then the solution $\mathbf{v}^{i(k)}$ in the range $[t_a, t_b]$ can be dropped in the kth relaxation and taken to be the same as in the $(k-1)$th.

Latency is a similar effect though in one relaxation. Returning to the example, we notice that if during integration over a range $[t_a, t_b]$, the voltage v^2 appears to be almost constant in a few initial points of this range, and in the whole of this range fanins v^1, v^c are also almost constant, then certainly v^2 will be almost constant in the whole of this range and its analysis can be dropped until t_b.

In general, given the ith block in the kth relaxation

$$\dot{\mathbf{q}}^i(\tilde{\mathbf{v}}^{i(k)}) + \mathbf{f}^i(\tilde{\mathbf{v}}^{i(k)}) = 0 \tag{9.42}$$

if all its fanins $\mathbf{v}^{1(k)}(t), ..., \mathbf{v}^{i-1(k)}(t), \mathbf{v}^{i+1(k-1)}(t), \mathbf{v}^{p(k-1)}(t)$ in the range $[t_a, t_b]$ and $\mathbf{v}^{i(k)}(t)$ in a few points at the beginning of this range are almost constant, then solution $\mathbf{v}^{i(k)}(t)$ is also almost constant in the range $[t_a, t_b]$ and can be dropped in the kth relaxation and taken to be the same as in the $(k-1)$th one.

Bypassing and latency enable us to drop time-domain analysis of a node (block) in certain ranges, i.e. to select an appropriately long step and jump directly to the breakpoint t_b. In practical WRA, ranges of bypassing and latency cover a quite large fraction of the analysis range [0, T] and hence time-domain analysis during one relaxation becomes very quick. In Figure 9.9 we have waveforms corresponding to the circuit of Figure 9.5, convergent in 3 relaxations. In the first relaxation $v^2(t)$ can be decided to be latent in intervals [0, 1], [1, 5], [5, 8], [10, 12] where $v^{1(1)}(t)$, $v^{c(1)}(t)$ are constant. In the 2nd relaxation, latency intervals are [0, 1], [2, 5], [5, 8], [10, 12]. In the 3rd relaxation the solution $v^2(t)$ is the same as in the first. Moreover $v^1(t)$, $v^c(t)$, used in the 3rd relaxation, are the same as in the 2nd, and hence in the 3rd relaxation analysis in ranges [0, 1], [2, 5], [5, 8] and [10, 12] can be bypassed. The 3rd relaxation requires integration only in intervals [1, 2] and [8, 10]. In the first relaxation we have integrated only over 2 units, in the 2nd and 3rd over 3 time units. The saving, measured as the ratio of time where integration is carried to the total analysis time (T multipled by the number of relaxations), is 2/9 (i.e. about 80 per cent of profit).

To speed up analysis we can perform at the first relaxation a simplified analysis, where backward influences are not taken into account. For instance solution $v^{2(1)}(t)$ of the block in Figure 9.6 for the first relaxation is performed with a given fanin waveform $v^1(t)$ but with constant fanouts v^3, v^c. The waveform thus obtained is a good starting waveform for next relaxation.

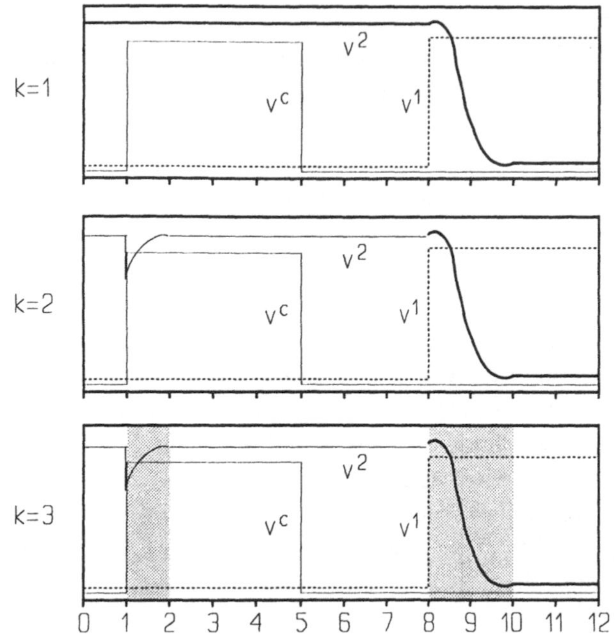

Figure 9.8 Waveforms obtained in three subsequent relaxations. Illustration of bypassing and latency application

In circuits with logic feedback loops the number of relaxations required increases considerably. It can be shown that it is proportional to the length of the simulation interval $[0, T]$. To reduce this number we can apply the concept of dividing this interval into segments $[0, t_1]$, $[t_1, t_2]$, ..., $[t_l, T]$. This is known as **windowing**. We analyze a circuit in a time interval (window) $[t_{i-1}, t_i]$ using the WRA, and, after its convergence, the value of solution obtained at point t_i is taken as an initial condition for WRA in the window $[t_i, t_{i+1}]$, and so on. We notice that the bypassing and latency techniques were in fact also methods of automatic determination of windows in which it is reasonable to perform time-domain analysis of circuit components based on time profiles of signals. Hence, if we employ bypassing and latency the application of fixed windows is of little advantage.

The most advantageous technique is a dynamic windowing. First of all it requires detection of latency and bypassing regions. Then, during the WRA in regions of signal transition, as is seen in Figure 9.9, we introduce windows. Their length should cover a few or a few tens of integration-steps. As a window covers a smaller number of time steps, so the

WRA becomes closer to the one-point relaxation method. For the one-step window the WRA transforms into the one-point relaxation method. In one-point relaxation it is straightforward to use one step-size for the whole circuit. If not, we have to synhronize the time points of different components by interpolation. In windowed WRA we can use a multirate integration to calculate waveforms over a given window and then synchronization within the window is straightforward. The shorter the windows, the smaller is the number of relaxations.

Windows are especially advantageous in circuits with global feedback, if we select a window length shorter than the forward propagation time through the feedback loop. Then we obtain a quick WRA convergence limited to the forward circuit part. The influence of the feedback signal, which considerably increases the number of relaxations, appears delayed by this propagation time. Hence, it can be handled by a WRA in another window. The best known WRA simulator with block decoupling and windowing is RELAX 2 [9.20], [9.21].

The next techniques, such as partitioning, ordering and scheduling are not specific to WRA, but also apply to timing, ITA, and OSR analyses. We will discuss them in a separate section.

9.4.3 Waveform-relaxation Newton analysis (WRNA) and Newton waveform-relaxation analysis (NWRA)

A slight improvement of WRA introduced by R.A. Saleh and J.K. White [9.19] is known as waveform-relaxation Newton analysis (WRNA). It is based on the concept of waveform Newton (WN) iterations which can be used for time-domain analysis instead of the one-point NR iterations presented in Chapter 5. The WN iterations are not attractive in direct analysis. However, in relaxation methods they are more useful.

The concept of WN iterations can be explained using the previously introduced nodal description (9.24) with given initial conditions $\mathbf{q}(0)=\mathbf{q}_0$. In direct analysis this equation has been first discretized by means of a DF and then linearized for NR iterations. We obtain two iterative loops. In the outer one, time is subject to change. In the inner NR loop, voltages converge quadratically to a solution. However, the order of these two iterative processes can be inverted. First, a NR linearization of circuit OADEs can be carried out within the space of waveforms in the vicinity of a given waveform $\mathbf{v}^k(t)$. These linearized equations can be viewed as a description of a substitute circuit - dynamic, linear, and time-variant, consisting of elements dependent on time through the function $\mathbf{v}^k(t)$, namely

$$\frac{d}{dt}\left[\mathbf{q}[\mathbf{v}^k(t),\mathbf{u}] + \frac{\partial \mathbf{q}[\mathbf{v}^k(t),\mathbf{u}]}{\partial \mathbf{v}}[\mathbf{v}^{k+1}(t) - \mathbf{v}^k(t)]\right] + \\ + \frac{\partial \mathbf{f}[\mathbf{v}^k(t),\mathbf{u},t]}{\partial \mathbf{v}}[\mathbf{v}^{k+1}(t) - \mathbf{v}^k(t)] + \dot{\mathbf{q}}(\mathbf{v}^k,\mathbf{u}) + \mathbf{f}[\mathbf{v}^k(t),\mathbf{u},t] = \mathbf{0} \tag{9.43}$$

This circuit can be analyzed in the time-domain, i.e. its linear time-variant OADEs can be integrated for $\mathbf{v}^{k+1}(t)$, using the discretization methods presented in Chapter 5. For such linear circuits several other solution methods are also available.

The WN approach is not useful in direct circuit analysis because of two main disadvantages: large memory consumption for storage of waveforms $\mathbf{v}^{k+1}(t)$, $\mathbf{v}^k(t)$, and difficulty in predicting a starting waveform $\mathbf{v}^0(t)$, for WN iterations. The latter means that WN is convergent only in a large number of iterations. However, this method can also be very attractive. It is quadratically convergent in the vicinity of the solution, and moreover it is globally convergent, given some mild assumptions about the functions $\mathbf{q}(\mathbf{v},\mathbf{u})$, $\mathbf{f}(\mathbf{v},\mathbf{u})$, and provided that the starting waveforms obey given initial conditions $\mathbf{q}_0$. [9.19]

The WN iterations can be applied in WRA for the solution of each decoupled circuit component because these components are relatively small and storage of waveforms is normal in WRA. Since relaxations are linearly convergent, we can, as in ITA [9.18], carry out only one NR iteration for each decoupled component. After such linearization as in Equation (9.43), though applied to a decoupled component, this component becomes a linear subcircuit, whose integration enables a longer step size to be used, and hence is more efficient, than integration of a nonlinear subcircuit as is the case of WRA. Moreover, in WRNA integration accuracy can easily be controlled. In beginning relaxations integration can be more coarse and then gradually refined.

This WRNA can be extended to the Newton waveform-relaxation approach (NWRA) of D.J. Erdman and D.J. Rose [9.6]. In this approach a NR linearization in function space is applied to the equations of the whole circuit, as given in (9.43). Then we obtain a huge dynamic linear time-variant circuit. This circuit can be decoupled and repetitively solved by WRA, i.e. all components are discretized and subsequently solved. After convergence of relaxations we update the circuit linearization and repeat relaxations for the next NR iteration. To efficiently exploit windowing in NWRA we can introduce a window which dynamically slides forward in time during the course of WN iterations.

The main difference between WRNA and NWRA is that in NWRA, when components are iterated in a relaxation loop they are still inaccurate due to NR iterations, since both types of iterations operate on the whole circuit and are responsible for convergence. In WRNA, Newton iterations (one or even more) are local within one circuit component and give its solution only in a slightly more efficient direct manner. NWRA storage requirements are considerably greater than WRA and WRNA, since a matrix of waveforms of partial derivatives of circuit components has to be stored.

From publications so far known [9.6], [9.19] WRNA in different examples can be from 10 to 70 per cent quicker than NWRA. However, NWRA is only slightly more efficient in a few cases. In general, WRNA is only a few times more efficient than direct methods. There are examples where the saving is better than for ITA and OSR but this is not the rule. The efficiency of WRA is strongly dependent on the latent and unidirectional properties of a circuit.

9.5 Partitioning, ordering and scheduling for relaxation circuit analysis

In Section 9.3 we have so far introduced the idea of decoupling circuit equations. Assuming an initial solution we subsequently solve each decoupled part as a separate subcircuit. After all parts have been visited, the relaxation is completed, giving an adjusted solution. Such relaxations can be repeated until convergence. The separate solution of each decoupled part is the main advantage of relaxation methods. The decoupled component obtained from a dedicated partitioning process can be a node with its neighboring elements (in the nodal decoupling case) or a subcircuit (in the block decoupling case). Since solution at a node requires only a scalar equation, while a subcircuit introduces a system of simultaneous equations for direct solution, efficient simulation requires the partition of a circuit into pieces as small as possible. For efficient relaxation analysis, decoupled components should be unidirectional or weakly bidirectional. Then relaxation analysis can converge in one or at most a few relaxations. Thus, relaxation analysis begins from an appropriate circuit partitioning into unidirectional or almost unidirectional components.

When partitioning is completed we can formulate a graph of interactions among subcircuits and try to prepare a schedule for visiting them during analysis, which is known as ordering and scheduling. This can be done statically or dynamically.

9.5.1 Circuit partitioning

A circuit is usually partitioned into groups of nodes strongly coupled by bidirectional elements. Bidirectional coupling between terminal nodes is introduced by such elements as bjts, inductors, resistors, diodes, and MOSFET channels. This coupling can be weak, if the conductance between nodes is small. However, since coupling between a gate and channel (drain and source) is unidirectional (for d.c. signals, i.e. when capacitors are neglected), gates can be separated from channels and this allows a partition into d.c. components.

This universal partitioning method consists of finding d.c. components, which are disjoined parts into which a circuit is broken down by insulating gates from channels and splitting the supply voltage and the ground node. Such a partitioning will be satisfactory as being almost unidirectional even if gate capacitors are taken into account. Partitioning into d.c. components with supply and ground splitting is shown in Figure 9.9.

Efficient circuit partitioning based on the union-find method [9.1] visits successive bidirectional circuit two-poles and classifies their nodes and elements into strongly-coupled groups as presented in Algorithm 9.9. In this algorithm, circuit elements are read from *ElementList*, and nodes of each element are checked and classified according to the group to which they belong. Nodes already visited are collected on *NodeList*.

Circuit partitioning into d.c. components can be too coarse, so we then try to find a partition of a d.c. connected component by checking all floating resistors and classifying them into strong couples and weak couples through which a cut-set can be established. A resistor will be a strong connection if it is sufficiently small in comparison with the Thevenin equivalent resistances visible from its nodes. Then it is assigned to a group. Otherwise, we remove this element from the circuit and assign

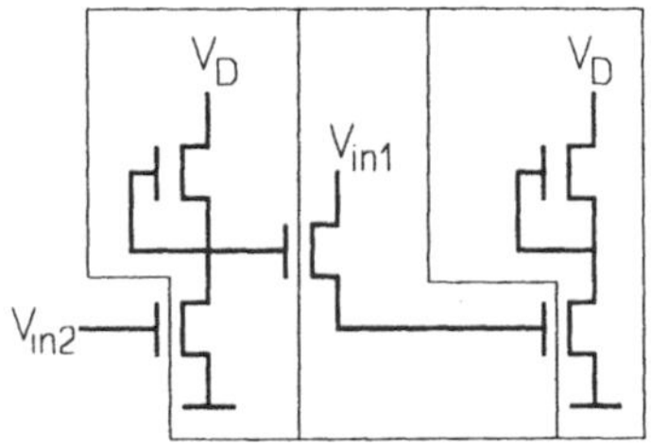

Figure 9.9 Circuit partitioning into three d.c. components. Supplies and reference nodes have been split

its two nodes to two different groups. After a number of such manipulations a circuit can split into several components. In this algorithm we have not taken into account supply and ground nodes. If M or N is the supply or ground, split it and assign it to all groups where other elements belong which are connected with these nodes.

Algorithm 9.9 Circuit partitioning by the union-find method

```
initialize NodeList by two nodes of an element classified as belong-
ing to group G(1); i=1;
for (all elements from ElementList)
begin
   read element from ElementList, let its nodes be M and N;
   if (M and N belong to NodeList)then
      if(both M and N belong to an already known group G(i))
      then
         add this element to group G(i);
      elseif (M belong to G(i) and N belong to G(j)) then
         join group G(i) with G(j);
         all node assignments to group G(j) change to G(i);
      endif
   elseif (one of M, N (let say M) belongs to NodeList and a
   known group G(i)) then
      add the other node N to NodeList and assign it to group G(i);
      add this element to group G(i);
   else
      create a new group G();
      add nodes M, N to NodeList and assign them to G();
      add this element to G();
   endif
end
```

A criterion for treating the resistor as small is explained in Figure 9.10. After the R is selected, we substitute for all nonlinear resistances (MOSFET channels) resistors equal to their maximum resistance possible in varying conditions. This is to prevent cutting through resistors which can be large in comparison with the resistances of channels of other MOSFETs for some D-S voltages, but not necessarily so for other D-S voltages. Then, we calculate two substitute Norton conductances visible from nodes M and N, R_M, R_N. In the relaxation process $v^{N(k-1)}$ affects $v^{M(k)}$ through a voltage division ratio $\alpha = R_M/(R_M + R)$, and similarly

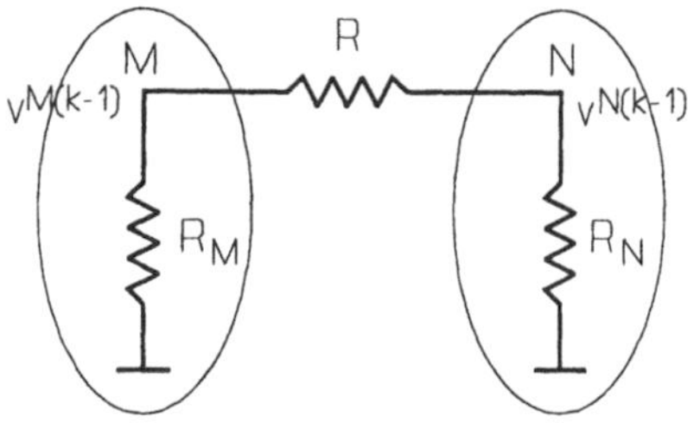

Figure 9.10 The concept of decoupling a d.c. component into parts by deleting sufficiently large resistors R

$v^{M(k-1)}$ affects $v^{N(k)}$ through a voltage division ratio $\beta = R_N/(R_N + R)$. If R is large enough and hence $k = \alpha\beta$ small enough (e.g. ≤ 0.3), the connection between M and N is weak and alternate relaxations in these nodes will converge quickly. Hence, we can decouple these nodes, assigning them to different groups. Otherwise, we must treat them as belonging to the same group. Such a partition is called diagonal dominance partitioning.

After this partitioning we can define the terminals of subcircuits. Some of them are only inputs, some others are only outputs and we may also have terminals which are simultaneously inputs and outputs. The latter introduce a strong connection between subcircuits, as for instance in diagonal dominance partitioning. Similar bidirectional coupling is introduced by floating gate to channel capacitors. To handle a partitioned circuit we proceed to ordering its components.

9.5.2 Ordering of circuit components

Ordering is applied to a circuit, which is partitioned into components, to introduce their unique order. In practice, components can have the form of one node with neighboring elements or of a subcircuit; however, we will treat both types the same, and call them simply components. The purpose of ordering is to obtain the best efficiency in analysis. As has been shown in previous sections the order of components is critical to the efficiency of relaxation methods. If the circuit is one-way, as for instance in Figure 9.1, then the GS strategy gives a result in one relaxation if components are ordered in accordance with the signal flow. Hence, one-way circuits play the most important role in ordering.

The most convenient way of representing a structure of dependency among circuit components (or equivalently decoupled subsets of circuit equations) is a dependency matrix and a corresponding dependency

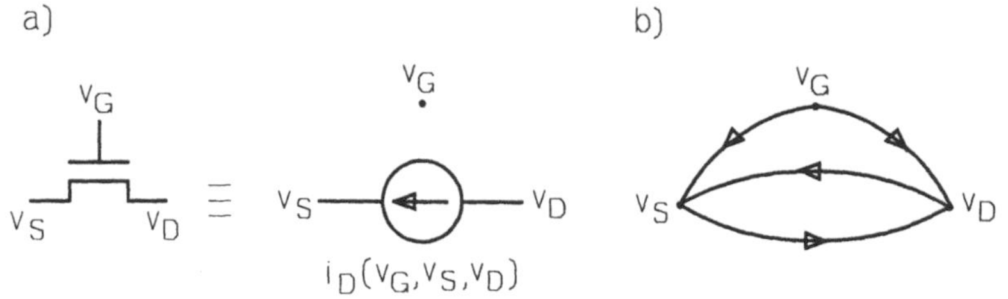

Figure 9.11 (a) MOS field effect transistor and (b) its dependency graph

graph. If the *i*th component is dependent on solution of the *j*th component, then the dependency matrix **A** has the element $a_{ij} = 1$. Equivalently, we can draw the signal-flow graph, with as many vertices as there are components in the circuit, and set in this graph a directed edge from *j* to *i* for each pair of dependent components. This edge shows that the solution $\mathbf{v}_j$ of the fanin component affects the *i*th component, and hence its solution $\mathbf{v}_i$.

Let a circuit be partitioned into nodal components. The ubiqitous MOS transistor of the Shichman-Hodges model (see (2.57)) is drawn in Figure 9.11(a). It is characterized by a drain current $i_D(v_D - v_S, v_G - v_S)$ dependent on voltages v_D, v_S, v_G. KCL equations at drain (D) and source (S) are dependent on voltages v_G, v_S, and v_G, v_D, respectively. This dependency is represented by arrows in the graph of Figure 9.11(b). The initial vertex of the edge is termed the fanin of the ending vertex. The ending vertex is termed the fanout of the beginning vertex. For the CMOS gate of Figure 9.12(a) the dependency graph is demonstrated in Figure 9.12(b). From this we can notice that circuit partitioning into d.c.

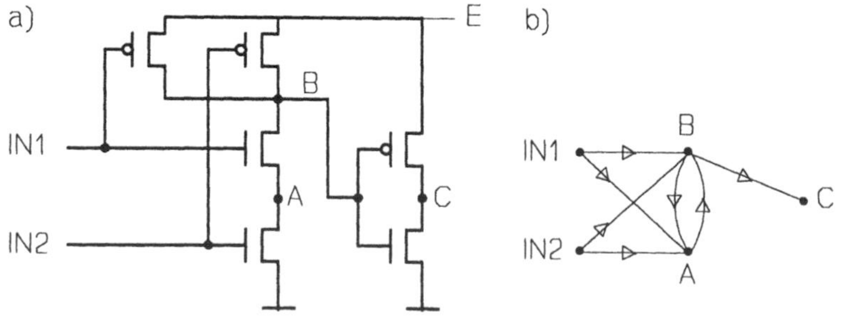

Figure 9.12 (a)CMOS NAND gate and (b) its dependency graph

components can be easily interpreted as the extraction of clusters of nodes having bidirectional connections such as A and B.

The dependency graph representation is also adequate for partitioning into subcircuit components. A component may have a number of input vertices, output vertices and both input-output vertices. It is represented by a set of edges joining each input with each dependent output and directed from an input to an output. Between vertices *i*, *j* being both an input and an output we have two edges, one from *i* to *j* and the other from *j* to *i*.

An acyclic dependency graph stands for a one-way circuit, where no feedbacks occur. In Figure 9.5(a) we have shown components coupled in the one-way manner. This case can be characterized by the dependency graph of Figure 9.13(a). If vertices are ordered according to the signal flow, then the dependency matrix has lower-triangular (LT) form and GS relaxations can be convergent in one step.

A confusing situation occurs if the dependency graph incorporates pairs of bidirectionally connected components. This means that components are strongly connected and affect each other. Recalling the example given in Figure 9.5(b), we notice that dependencies among three blocks take the form shown in Figure 9.13(b) which has a strong connection.

A second, more general confusing situation arises when the dependency graph includes directed loops incorporating any number of components, as we demonstrate in Figure 9.14. This figure can be used to explain the concept of the strongly connected component (SCC).

The SCC is a maximal subgraph of the dependency graph which includes for any pair of nodes $[a,b]$ a directed path from a to b and another from b to a. In the figure mentioned we have two $SCC_1 = \{3,4\}$ and $SCC_2 = \{5,6,7\}$. If SCCs are condensed (collapsed to one component) then all loops are removed and we obtain an acyclic dependency graph, as if a one-way circuit were involved.

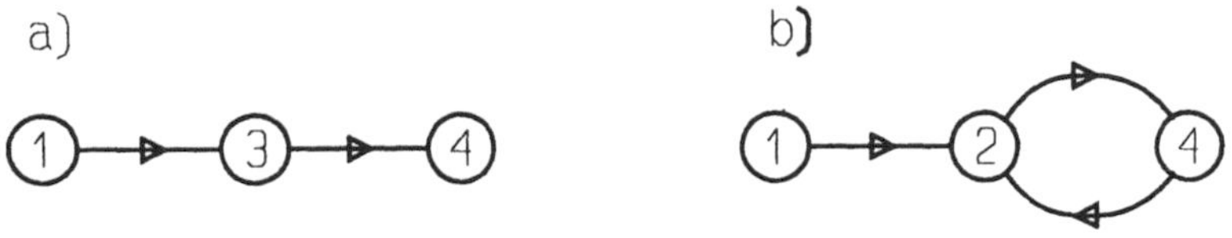

Figure 9.13 Dependency graph for partitioning given in (a) Figure 9.5(a) and (b) Figure 9.5(b)

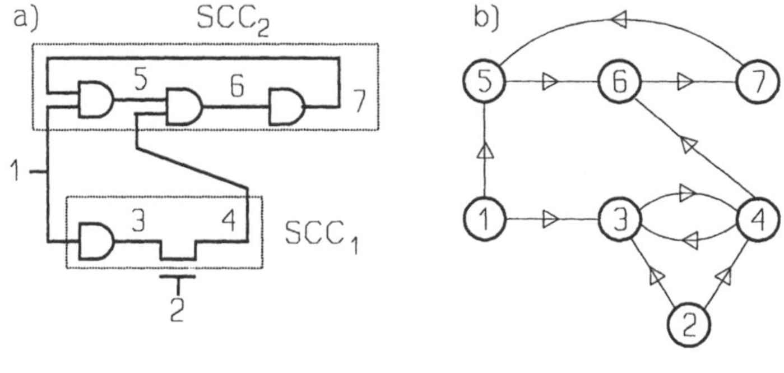

Figure 9.14 (a) A digital circuit and (b) its dependency graph with SCCs drawn with thickned lines

Regarding the ordering problem, we can first enumerate vertices (nodes or blocks) of an acyclic graph. To transform a circuit containing feedback loops to the acyclic form we can:

(i) find an SCC and join its components into one component (this increases the size of the components and decreases the efficiency of simulation),

(ii) condense components of SCC, i.e. not join them but keep them together during ordering and solve by means of a dedicated scheduling scheme,

(iii) select feedback edges and cut them to obtain an acyclic graph.

After the acyclic graph is obtained we enumerate its vertices according to the signal flow, i.e. a fanin is always numbered before its fanout. For the circuit in Figure 9.14(a) we can find two SCCs shown. Ordering starts from the primary input 1. We realize a depth-first idea, proceeding from a fanin to fanout as soon as all fanins of a given fanout have been already visited. After 1 we cannot visit either SCC_2 or SCC_1 before 2. Then, since the output 4 has not been visited yet, we must first order SCC_1 before SCC_2. Algorithm 9.10 implements this idea using the stack. We obtain the order: 1, 2, SCC_1, SCC_2.

Since each SCC comprises a number of condensed components, we have to introduce their local order inside the SCC. This is usually performed by automatic selection and virtual cutting of feedback edges. Then an SCC becomes acyclic, and we can number its vertices according to Algorithm 9.10 for one-way circuits. However, for efficient analysis it is convenient to number the input vertex of each feedback edge after the output vertex. For the circuit of Figure 9.14(a), if we select $4 \rightarrow 3$ and $7 \rightarrow 5$ as feedback branches, then local orders {3,4} in SCC_1 and {5,6,7} in SCC_2 are obtained. Finally, the order of vertices is 1, 2, {3, 4}, {5, 6, 7}.

Algorithm 9.10 Ordering of an acyclic graph

```
for (all vertices v in a graph) do
begin
   NumFanIn(v) initialize with a number of fanins of v;
   if (NumFanIn(v) = 0) push v onto Stack;
end
k=1;
repeat
begin
   pop out a vertex w from Stack;
   order w as the kth one;
   for (for all fanins v of the vertex w) do
   begin
      NumFanIn(v) = NumFanIn(v)-1;
      if(NumFanIn(v) = 0)push v onto Stack;
   end
end
until (all vertices of the graph are ordered);
```

9.5.3 Scheduling of circuit components for relaxation analysis

The ordering introduced here enables us to determine the order in which circuit components are analyzed. This order is known as the schedule. Since in relaxation algorithms components are solved separately and the order of their visiting is critical to efficiency, a relaxation program can be viewed as being driven by the schedule which determines which component has to be visited after each one.

All the relaxation algorithms introduced in Sections 9.3 and 9.4 comprise a loop over all decoupled components (nodes or blocks). This strategy is called the **basic scheduling**. We perform ordering of all circuit components and visit them always in the same order corresponding to their numbers. Each component is visited once at each relaxation. After each relaxation, a test for convergence is checked. In one-point relaxations reaching convergence means that a solution at the current point has been obtained and we can proceed to the next instant. In WRA, we check the proximity of waveforms in two relaxations over the whole analysis range [0, *T*]. If this his test is fulfilled a final solution has been reached.

Problems with scheduling appear when a circuit incorporates feedback loops. The simplest resolution consists of performing ordering with feedback edges cut off, as if no feedback occurred in the circuit. Then we apply the basic scheduling. All components are scheduled and analyzed according to the order involved and all feedback signals are calculated. Then these signals are used for updating values at points where feedback signals are applied.

An alternative scheduling for circuits with SCC is epsilon-scheduling, where in one relaxation not all components but a subset of a fixed number is involved. The choice of these components varies dynamically during the course of relaxations.

Another possibility is overlapping scheduling useful in strongly connected components. To explain this, let us take three components A, B, and C, which are strongly connected. During a relaxation process we solve A and B for their unknowns, treating them together as one decoupled component. Then B and C are joined and together solved as the next decoupled component, using the known solution of A. Such an overlapping provides a very efficient solution in the case of strong connections. If we have only two subcircuits strongly connected, then it may be useful to introduce an auxiliary interface B, not changing the circuit solution but enabling overlapping.

9.5.4 Dynamic scheduling using an event-driven technique

In the relaxation methods so far, the order of visiting circuit components was fixed, and corresponded to the predicted direction of signal flow. However, in practice the signal flow and distribution of latent components in the circuit is highly dependent on the signals established. This leads to a requirement for dynamic selection of components for analysis in an appropriate order. This we call dynamic scheduling. This scheduling is also useful in WRA with dynamic windows. The main technique for dynamic scheduling of phenomena happening in time is the event-driven technique. We will show how to implement an event-driven one-point relaxation algorithm first with nodal decoupling. Then we explain some problems of event-driven block relaxation analysis. The event-driven technique can be easily extended to WRA with windowing, since analysis over a window can be viewed as an event.

9.5.4.1 Event-driven ITA or other time-point relaxation-methods

Latency of circuit components in practical relaxation analysis is dependent on signal propagation which stems from particular stimuli at primary inputs. To exploit this dynamic latency in circuits with decoupled nodes, a very efficient event-driven method can be introduced. The event is a particular circuit component (here, a node with its neighboring elements) calculation at a particular time, and is described by the two [*time*, *node*]. By virtue of decoupling, each event can be separately processed, i.e. modification (9.30) for i corresponding to *node* is calculated for the current state of the remaining voltages. This current state stems from the fact that previously scheduled nodes are already known from the k relaxation, while the unprocessed nodes, are taken from the $k-1$ relaxation.

We will explain the event-driven organization of the ITA, since it is one of the most efficient timing analysis methods so far. The whole algorithm is driven by the next event list (NEL), where events wait for processing, and the current event list (CEL) where we have events presently being processed. Moreover, the external event list (EEL) is used, where a sequence of input events (excitation turning points at primary inputs) is stored. The NEL contains chronologically ordered structures containing a pointer to a list of nodes to be processed at this instant and to a structure corresponding to the next instant. Events newly scheduled have to comply with the chronological order.

The event-driven algorithm starts from empty NEL and CEL, while EEL is filled by external events at input nodes. As a current time t_n we take the earliest event of the external or next events lists. We extract from these lists events occurring at the current time, and put them onto the current event list (CEL). Events on this list have to be scheduled in appropriate fanin before fanout order, since it contains information as to which node and in which order they have to be processed at the current time.

Now, successively, for each node i from CEL we formulate and run Equation (9.30), and check the test for convergence. If it is satisfied, then we estimate the LTE for this node (see Chapter 5), and evaluate the time step h. From its value we find that the ith node has to be scheduled on NEL for its next calculation at time $t+h$. However, if the test for convergence is not satisfied (this is always the case in the first relaxation), then we realize that the next relaxation for this node is necessary. Then we put this ith node onto the auxiliary list (AL) of nodes to be processed in the next relaxation and schedule all its fanouts on CEL in fanout after fanin sequence. If i is a circuit output, then no fanouts are added. When processing of the ith node is completed, we proceed to the next node of CEL and the procedure is continued. We notice that this next node will be in accordance with fanout after fanin ordering.

Algorithm 9.11 Event-driven implementation of ITA

> initialize NEL and AL to empty;
> schedule all nodes that are connected to independent sources to EEL;
> $t_n = t_0$;
> **while** (*true*)
> **begin**
>   select current time t_n from NEL and EEL;
>   **if** ($t_n > tstop$) **break**
>   take current events out of NEL and EEL, and schedule them on CEL;
> $k = 1$;
>   **while** (CEL is not empty)
>   **begin**
>   **for subsequent** i **in CEL**
>   **begin**
>   calculate $v_n^{i(k)}$ from modification (9.??);
>   **if** ($|v_n^{i(k)} - v_n^{i(k-1)}| \le \varepsilon \max\{ |v_n^{i(k)}|, |v_n^{i(k)}| \} + \delta$) **then**
>   predict time step h^i and $t^i = t_n + h^i$;
>   schedule the node i on NEL for time t^i;
>   **else**
>   schedule the node i on AL;
>   schedule fanout nodes of i on CEL if they are not already on;
>   **endif**
>   **end**
>   clear CEL; copy AL to CEL; clear AL;
> $k = k + 1$;
>   **end**
> **end**

When all events of a CEL have been carried out, it means that signal propagation in a relaxation has reached all primary outputs. Then we clear CEL, copy onto it nodes from AL, and clear AL. Now, if CEL is not empty it means that AL has contained nodes where the voltage has not still reached convergence, and hence further relaxation is required. Therefore, we proceed to the next relaxation. However, if CEL is empty it means that AL was empty, and hence a solution at t_n has been obtained. Then, we find the next current instant from the external or next event lists, take current events out of these lists, put onto CEL and start a new relaxation process. The algorithm is terminated when the current time reaches a final value *tstop*. This is briefly presented in Algorithm 9.11.

In this algorithm, time-step prediction can be almost the same as in direct methods, however it is calculated for each node separately, using the past and prediction for this node only. This prediction can be certainly used as a starting point of $v_n^{i(0)}$ for relaxations.

Now we want to stress the main difference between a direct method of time-domain analysis and the event-driven method. In the direct solution the time-step is common to the whole circuit. At a uniquely determined instant we solve companion equations of the whole circuit and obtain its solution.

In the event-driven approach a particular ith node, dependent on its fanin i-1, is extracted, as demonstrated in Figure 9.15(a). Now, at the ith step only a part of the circuit concentrated around the ith node is separately solved for v_n^i at an instant t_n^i that has been reached from a step previously selected for this node. The value of fanin v^{i-1} required for calculation of v^i is already known, but not necessarily at the same instant. It is very probable that it has been calculated at the preceding instant t_{n-1}^{i-1} but is not used before t_n^i, as demonstrated by Figure 9.15(b). To improve accuracy at t_n^i it would be reasonable to extrapolate the solution $v(t_{n-1}^{i-1})$ forward to instant t_n^i. This so-called dormant model will be discussed in the next section.

The main weak point of event-driven simulation is efficient implementation of NEL. It can be implemented as the circular list of structures shown in Figure 9.16, whose structure stands for chronologically ordered events. Each event EV includes a time value, a pointer to the next event, and a pointer to a list of nodes to be processed at this time. This list contains structures corresponding to nodes (the number of the compo-

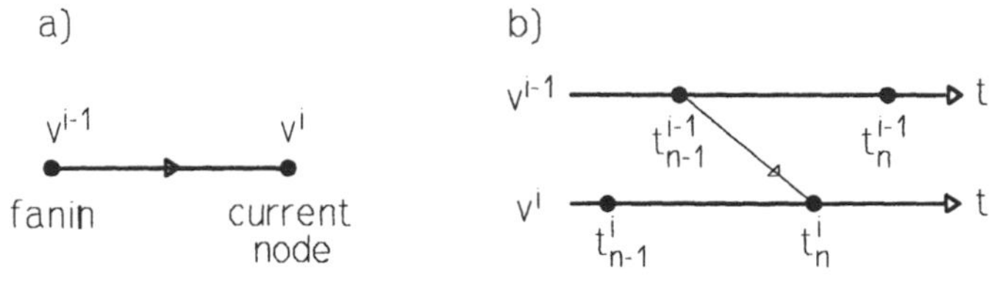

Figure 9.15 Event-driven analysis: (a) order of components (b) asynchronism of instants

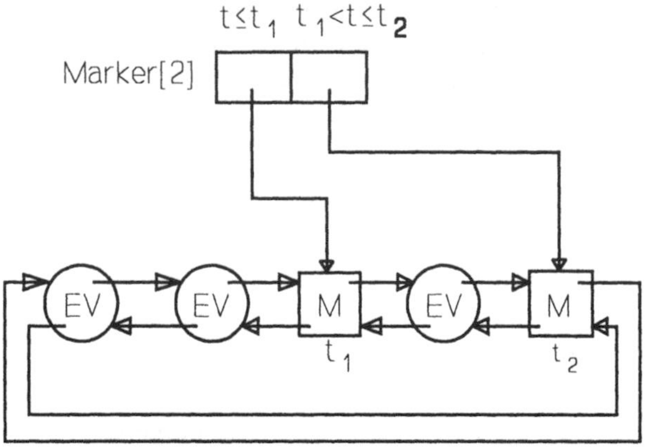

Figure 9.16 Example of implementation of next-event list (NEL)

nent, a place for its voltage, a pointer to the next node). Nodes are ordered in fanout after fanin order. Since each node has a separate step estimation, simultaneous events on NEL are not frequent. (Simultaneous nodal events will not be considered before current events are taken out of NEL and EEL, put on CEL and generate fanouts for simultaneous calculation.) NEL has to have as few events as possible, since the most time-consuming action is the proper scheduling of new events in chronological order. This requires a search of the NEL. To speed up this process we can divide NEL into intervals by means of markers M shown in Figure 9.16. The marker denotes the location of a fixed instant within the list. Markers are accessible by pointers. To schedule a new event at point t we have to enter the nearest marker corresponding to a time greater or smaller then t and linearly search the section between markers.

Finally let us observe how this event driven algorithm operates if we have two nodes i, j strongly coupled, i.e. each is the fanout to the other. Let i be ordered before j. Then, at a particular time point we first enter i and calculate v^i using v^j predicted by means of a solution at the last instant. Since the v^i obtained has not converged, we add i to AL and the fanout j to CEL, proceed to j and calculate v^j using the already obtained v^i. Since it also has not converged we add it to AL. In the next relaxation we enter once more both nodes i and j registered on AL. First, we recalculate v^i using v^j from the last relaxation and then v^j using the most recent v^i. This may have to be repeated many times due to strong coupling and hence can prolong simulation. This explains why clusters of strongly coupled nodes cause inefficiency even of ITA and hence require more efficient block relaxation methods, which will be discussed later on.

9.5.4.2 Event-driven analysis of a circuit partitioned into decoupled subcircuits

The event-driven technique can be applied to relaxation algorithms with circuit partitionıng into subcircuits. This block decoupling is even more attractive than nodal decoupling due to its milder convergence conditions. As we know from Section 9.3.5 these conditions are the existence of grounded capacitors connected to all feedback nodes. In relaxation analysis each component (subcircuit) is scheduled for analysis separately, as we have discussed in Section 9.4. A subcircuit can be viewed as in Figure 9.17, surrounded by neighboring elements through which its block nodes are connected to the neighboring nodes of the subcircuit. In relaxation analysis the voltages at neighboring nodes are known from the solution of other subcircuits, in the same or preceding relaxation. We solve the subcircuit for its internal and block node voltages.

In many circuits, especially digital, most subcircuits are latent in time (see Section 8.3). Latency is usually strongly dependent on the current signals. A subcircuit can be latent for some input signals, but a small change in these signals can alter the situation considerably, activate this latent subcircuit but make another, so far active subcircuit, latent. For a latent subcircuit its block and internal voltages are maintained from the previous solution. This block has not to be scheduled for analysis and hence the number of blocks so scheduled is considerably reduced. Event-driven analysis consists of deciding which subcircuits have to be scheduled for analysis during the course of analysis.

Further simplification comes from each component requiring different time steps. This so-called multirate analysis, difficult to employ in the direct methods of Chapter 8, can be easily implemented using relaxation techniques by virtue of decoupling components. For each subcircuit S_i we have an independent sequence of instants $\Xi_i = [t_{i1}, t_{i2}, ...]$, at which this subcircuit has to be analyzed, as seen in Figure 9.15. Summing up these invividual sets of instants we obtain an overall set of instants $\Xi = \Xi_1 \cup \Xi_2 \cup ... \cup \Xi_m$ at which the whole circuit has to be analyzed. However not all subcircuits are analyzed at all instants Ξ. At any one particular instant we have a set of components to be analyzed, and a set of components which are in the intermediate state between two instants of analysis and await analysis, as we see in Figure 9.15(b). The former subcircuits we call active, while the latter are dormant. In true multirate analysis the number of active components is small. In Figure 9.15(b) at instant t^i_{n-1} the v^i component (or node) is active while the v^{i-1} component (or node) is dormant.

In event-driven analysis an event is the fact of analysis of a particular decoupled subcircuit at a particular instant. After a subcircuit S_i has been analyzed at instant t and its local truncation error has been estimated, we can predict its appropriate step length h_i and schedule this subcircuit on the next event list (NEL) for next analysis at time $t+h_i$. This mechanism of scheduling next events continuously fills the NEL with events.

Analysis consists of taking all simultaneous current events out of the NEL and scheduling them for analysis. All other blocks are latent or dormant. Let us focus on the subcircuit shown in Figure 9.17 which has been scheduled for analysis. Neighboring nodes are members of neighboring subcircuits connected with this subcircuit. These subcircuits can be latent, dormant or active at any given instant. Hence, neighboring nodes can also be latent, dormant or active.

The voltage of a latent neighbouring node is known from a previous solution and maintained for the current instant. However, if after solution of all current active subcircuits the terminal voltages of a latent block appear not to satisfy latency maintaining conditions, then we have to cancel its latency, and start to treat this subcircuit as active. Then all currently active subcircuits adjacent to this latent subcircuit have to be recalculated.

The voltages of dormant neighboring nodes are described by a dormancy model, which is a polynomial extrapolation streatched over a number of preceding solutions of the dormant block.

$$\mathbf{w}_k(t_n) = \sum_{k=1}^{\text{order}+1} \alpha_k \mathbf{w}_k(t_{n-k}) \tag{9.44}$$

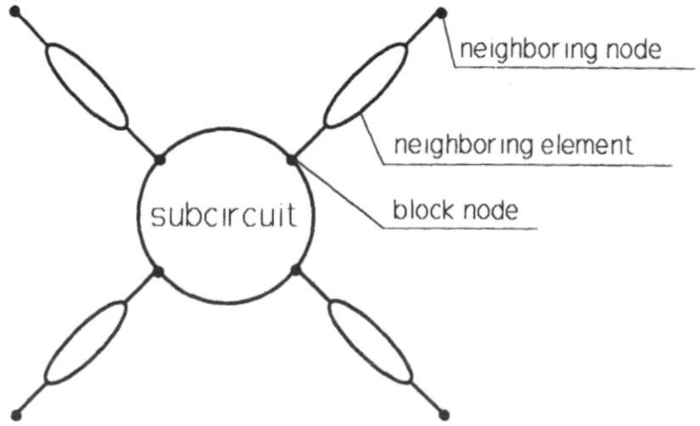

Figure 9.17 Interfacing a subcircuit with surrounding blocks in event-driven simulation

This extrapolation is performed for all internal nodes of a dormant block. Dormancy introduces an error. To control this error, after analysis of all current active blocks we can identically extrapolate the voltages of the block nodes of a dormant subcircuit and compare them with the current solution obtained. If the error is too large we cancel dormancy and make this block active. Then we recalculate all current active subcircuits.

The voltages of active neighboring nodes are known from the solution of other active subcircuits according to their order of scheduling, as it is known from the principles of decoupling equations.

Each active subcircuit is separately solved from the known voltages of neighboring nodes. Hence, it is essential to decide the order of scheduling current events, i.e. subcircuits which are simultanously active. In event-driven simulation we can schedule current events according to principles similar to those described in Section 9.5.4.1. However, since they are strongly separated by virtue of latency and dormancy, they usually do not form strongly connected components with feedbacks. Hence, we can schedule fanins before fanouts and obtain a solution after scheduling each subcircuit once. This case corresponds to one GS relaxation. Such an approach has been implemented in the simulator SAMSON [9.16]. If there is a strong connection between simultaneously solved subcircuits, then we have to schedule them several times, i.e. perform several GS relaxations.

Simulation methods using block decoupling and hence interfacing between direct and relaxation circuit simulations are known as mixed-mode simulation. Practical mixed-mode simulators also incorporate simulation on the logic signals level. The results of logic simulation are interfaced with circuit simulation by means of dedicated A/D and D/A software converters. Since logic simulation is beyond the scope of our book, we shall not discuss further aspects of mixed-mode simulators.

To check the absorption of topics covered in this section we can explore the following problems.

- Explain the concept of unidirectional connection of two nodes; give an example using a controlled source and another example consisting of a MOS transistor. Then explain and demonstrate bidirectional coupling of nodes. Explain the similar properties of a subcircuit stretched over a number of block nodes. How can we represent uni- and bidirectional coupling by means of signal-flow graphs?

- Explain the concept of decoupling circuit equations on three levels: those of OADEs, of discretized equations, and level of discretized and linearized equations. In relaxation methods the solution of one decoupled equation is used in the solution of another. There are different strategies for this interaction between equations. Explain the Gauss-Seidel strategy.

- In general relaxation analysis we have four loops, over: time, relaxations, decoupled components, and NR iterations. How are they ordered and how many iterations of each type are performed in the following relaxation analyses: TA, ITA, OSR and WR?

- Explain the concept of block decoupling and how to extend any relaxation algorithm, e.g. ITA, to the case of block relaxation. What are the advantages of block decoupling? Compare the sufficient conditions of convergence.

- Provided a circuit contains such elements as are available in the SPICE input language, explain how to partition a circuit into elementary d.c. components. If a d.c. component is too large how can we divide it into unidirectional or weakly coupled parts?

- Why do we order the decoupled components of a circuit? Give an example of a circuit where a proper ordering provides convergence of the GS procedure in one relaxation, while other orderings do not.

- Explain the concept of condensing elementary circuit components into strongly connected components. How can we order SCCs? How can we order and schedule elementary components in an SCC to minimize the influence of feedbacks? Explain cutting feedback edges, epsilon and overlapping scheduling.

- Explain the concept of event-driven analysis as a dynamic ordering and scheduling. What do we mean by the event? How are events managed by means of the Next Event List and a mechanism of scheduling of new events and taking current events out of the list?

- What is happening in event-driven analysis with components which are not active at the current instant? Explain the concept of the latent and dormant circuit components. How are they involved in the analysis of an active component?

References

9.1 Aho, A.V., Hopcroft, J.E., Ullman, J.D., *The Design and Analysis of Computer Algorithms*, Reading, MA: Addison-Wesley (1974).

9.2 Bensard, J. Benkoski, J., Hennion, B., *ELDO-XL: A Software Accelerator for the Analysis of Digital MOS Circuits by an Analog Simulator*, IEEE Int. Conf. CAD, p. 136-141 (1991).

9.3 Chawla, B.R., Gummel, H.K., Kozak, P., *MOTIS-An MOS Timing Simulator,* IEEE Trans. Circ. Syst. vol. 22, no. 12, p. 901-909 (1975).

9.4 Debefve, P., Odeh, F., Ruehli, A.E., *Waveform Techniques*. In: *Circuit Analysis Simulation and Design, VLSI Circuit Analysis and Simulation.* Ed.: A.E. Ruehli, Amsterdam: North-Holland (1987).

9.5 Desai, M.P., Hajj, I.N., *On the Convergence of Block-Relaxation Methods for Circuit Simulation*, IEEE Trans. Circ. Syst., vol. 36, no. 7, p. 948-958 (1989).

9.6 Erdman, D.J., Rose, D.J., *Newton Waveform Relaxation Techniques for Tightly Coupled Systems*, IEEE Trans. CAD, vol. 11, no. 5, p. 598-606 (1992).

9.7 Forsythe, G.E., Moler, C.E., *Computer solution of linear algebraic systems*, Englewood Cliffs, NJ: Prentice-Hall. (1967).

9.8 Hennion, B., Senn, P., *ELDO: A New Third Generation Circuit Simulator Using the One-STEP Relaxation Method*, Int. Symp. Circ. Syst., p. 1065-1068 (1985).

9.9 Hennion, B., Senn, P., Coquelle, D., *A New Algorithm for Third Generation Circuit Simulators: the One-Step Relaxation Method*, IEEE Design Automat. Conf. (1985).

9.10 Lelarasmee, E., Ruehili, A.E., Sangiovanni-Vincentelli, A., *The Waveform Relaxation Method for Time-Domain Analysis of Large-Scale Integrated Circuits*, IEEE Trans. CAD of Integ. Circ. & Syst., vol. 1, no. 8, p. 131-145, (1982).

9.11 De Micheli, G., Sangiovanni-Vincentelli, A., *Characterization of Integration Algorithms for Timing analysis of MOS VLSI circuits*, Int. J. Circ. Theory Appl., no. 10, p. 299-309 (1982).

9.12 De Micheli, G., Newton, A.R., Sangiovanni–Vincentelli, A. *Symmetric Displacement Algorithms for the Timing Analysis of MOS VLSI Circuits,* IEEE Trans. Circ. Syst., no. 7, p. 167-179 (1983).

9.13 Newton, A.R., Sangiovanni-Vincentelli, A.L., *Relaxation-Based Electrical Simulation*, IEEE Trans. Electron Dev., vol. 30, no. 9, p. 1184-1207 (1983).

9.14 Ortega, J.M., Rheinboldt, W.C., *Iterative Solution of Nonlinear Equations in Several Variables*, New York: Academic Press (1970).

9.15 Ruehli, A., Sangiovanni-Vincentelli, A.L., Rabbat, G., *Time Analysis of Large-Scale Circuits Containing One-Way Macromodels*, IEEE Trans. Circ. Syst., vol. 29, no. 3, p. 185-191 (1982).

9.16 Sakallah, K.A., Director, S.W., *SAMSON: A Mixed Circuit-Logic-Level Simulator.* In: *Advances in Computer-Aided Engineering Design*, Ed. A. Sangiovanni–Vincentelli, Greenwich, Connecticut, USA: Jai Press Inc. (1985).

9.17 Saleh, R., *Iterative Timing Analysis and SPLICE 1.* M.S. Thesis, Univ. of California, Berkeley (1983).

9.18 Saleh, R.A., Kleckner, J.E., Newton, A.R., *Iterated Timing Analysis in SPLICE1*, IEEE Int. Conf. CAD (November 1983).

9.19 Saleh, R.A., White, J.K., *Accelerating Relaxation Algorithms for Circuit Simulation Using Waveform-Newton and Step-Size Refinment,* IEEE Trans. CAD, vol. 9, no. 9, p. 951-958 (1990).

9.20 White, J.K., Sangiovanni-Vincentelli, A., *RELAX-2 A Modified Waveform Relaxation Approach to the Simulation of MOS Digital Circuits*, Int. Symp. Circ. Syst., Newport Beach, p. 756-759 (May 1983).

9.21 White, J.K., Sangiovanni–Vincentelli, A., *Relaxation Techniques for the Simulation of VLSI Circuits*, Boston: Kluwer (1986).

9.22 Young D. M., *Iterative solution of large linear systems*, New York: Academic Press (1971).

Index

Milton Keynes UK
Ingram Content Group UK Ltd.
UKHW031125141024
449569UK00006B/444

9 780849 378942